Engineering Plasticity and Its Applications

Engineering Plasticity and Its Applications

Proceedings of the 10th Asia-Pacific Conference

Wuhan, China 15 – 17 November 2010

editors

Jianjun Li

Huazhong University of Science and Technology, China

Zhenhuan Li

Huazhong University of Science and Technology, China

Xia-Ting Feng

Institute of Rock and Soil Mechanics, Chinese Academy of Sciences, China

W.B. Lee

The Hong Kong Polytechnic University, China

Huamin Zhou

Huazhong University of Science and Technology, China

World Scientific

NEW JERSEY • LONDON • SINGAPORE • BEIJING • SHANGHAI • HONG KONG • TAIPEI • CHENNAI

Published by

World Scientific Publishing Co. Pte. Ltd.

5 Toh Tuck Link, Singapore 596224

USA office: 27 Warren Street, Suite 401-402, Hackensack, NJ 07601

UK office: 57 Shelton Street, Covent Garden, London WC2H 9HE

British Library Cataloguing-in-Publication Data
A catalogue record for this book is available from the British Library.

ENGINEERING PLASTICITY AND ITS APPLICATIONS
Proceedings of the 10th Asia-Pacific Conference

ISBN-13 978-981-4324-04-5
ISBN-10 981-4324-04-3

Printed in Singapore by B & Jo Enterprise Pte Ltd

PREFACE

This is the proceedings of The 10th Asia-Pacific Conference on Engineering Plasticity and Its Applications (AEPA2010) held in Wuhan, China, November 15-17, 2010, which is built upon the previous successful conferences in Hong Kong (1992), Beijing (1994), Hiroshima (1996), Seoul (1998), Hong Kong (2000), Sydney (2002), Shanghai (2004), Nagoya (2006) and Daejeon (2008).

Since the convocation of the first conference in Hong Kong (1992), AEPA has gained and enjoyed a burst of popularity among researchers and scientists not only in the Asia-Pacific region but across the world. The primary objective of AEPA is to provide a high-level international forum for researchers and scientists working in the broad disciplines of plasticity including solid mechanics, material science, metal forming, structural analysis, as well as geo-mechanics.

This collection contains 91 of 240 accepted papers, selected from over 500 papers submitted after peer review by two experts on the scientific committees and the organizing committee one by one, and presented orally or exhibited at poster sessions during AEPA2010. The papers cover a broad range of engineering plasticity such as constitutive modeling, damage/fracture/fatigue and failure, dynamic loading and crash dynamics, structural plasticity, multi-scale plasticity and modeling, experimental and numerical techniques, crystal plasticity, phase transformations, super plasticity, plasticity in advanced materials, plasticity in fracturing processes of rocks and rock engineering, plasticity in nano/meso/macro-machining and materials processing, plasticity in porous/cellular and composite materials, novel materials processing technology, time-dependent plasticity, plastic instability and strain localization, engineering applications and case studies.

We are grateful to all the authors, committee members, conference sponsors, and all volunteers for their great contributions to AEPA2010. Special thanks are due to the reviewers for their timely and remarkable efforts in reviewing a large number of papers. We are especially grateful to Huazhong University of Science and Technology(HUST), Institute of Rock Soil Mechanics(CAS), The Hong Kong Polytechnic University, as well as National Natural Science Foundation of China(NSFC). Without their supports, the success of the conference would not be possible.

Jianjun Li
State Key Laboratory of Materials Processing and Mold Technology,
Huazhong University of Science and Technology,
Wuhan, P.R. China
E-mail address: ljj@263.net.cn
Chair of AEPA2010

Zhenhuan Li
Department of Mechanics,
Huazhong University of Science and Technology,
Wuhan, P.R. China
E-mail address: zhli68@263.net
Co-chair of AEPA2010

Xia-Ting Feng
State Key Laboratory of Geomechanics and Geotechnical Engineering,
Institute of Rock and Soil Mechanics, Chinese Academy of Sciences,
Wuhan, P.R. China
E-mail address: xtfeng@whrsm.ac.cn
Co-chair of AEPA2010

W. B. Lee
Department of Industrial and Systems Engineering,
The Hong Kong Polytechnic University,
Hong Kong, P.R. China
E-mail address: wb.lee@inet.polyu.edu.hk
Co-chair of AEPA2010

Huamin Zhou
State Key Laboratory of Materials Processing and Mold Technology,
Huazhong University of Science and Technology,
Wuhan, P.R. China
E-mail address: hmzhou@hust.edu.cn
Chair of AEPA2010 Program Committee

CONTENTS

Preface v

Part I Elastic-Plastic Mechanics

Pushover analysis of structures considering strain rate effects........................ 3
Wenming Wang, Hongnan Li

Dynamical response of recycled concrete under cyclic loading...................... 8
Xingwen Luo, Hailin Yao

Theoretical study on the dynamic mechanical properties of the NBR composites under static pressure 13
Bo Qiao, Guorong Zhang, Xiuying Zhao, Sizhu Wu

Investigation on computational method for crack of reinforced concrete pipeline under internal pressure retrofitted with FRP 19
Wang Suyan, Liu Fei, Yu Wenhua

Study on K_R-curves of the concrete beam reinforced with CFRP 25
Dong Wei, He Huanan, Wu Zhimin

Effect of loading rate on reinforced concrete member 30
Min Li, Hongnan Li

Seismic response of concrete structure reinforced with superelastic shape memory alloy 35
Di Cui, Hong-Nan Li, Ping Guan

Analysis of three-dimensional microscopic stress distributions at a free edge of a unidirectional CFRP laminate... 40
Keita Goto, Tetsuya Matsuda

The dynamic property and liquefaction analysis of earth dam........................ 45
Deyong Wang, Xianqi Luo, Xiurun Ge

Identification the interface fracture parameter by inverse analysis: Application of Kalman-filter technology 50
Bin Liu, Zhou Li He, Min Ge Duan, Fei Xu

Modeling of thermal stress during casting solidification process 56
Dunming Liao, Liliang Chen, Jianxin Zhou, Ruixiang Liu

Prediction of forming limit diagrams of DP590 steel based on the M-K model with experimental verification.. 61
Yanshan Lou, Seokbong Kim, Chansoo Lee, Hoon Huh

Numerical simulation of radome with fluid-structure interaction 66
Xu Kun-Mei, Hu Yong, Tang Ye

In-plane stability of derrick of inclined drilling rig.. 71
Z. Liang, G. H. Zhao, F. G. Jiang, J. L. Tian, J. J. Deng

Torsional stiffness of laminated glass elements in structural applications — influence of a elasto-viscoplastic ionomer interlayer on the pre-breakage behaviour 76
Dieter Callewaert, Jan Belis, Didier Delincé, Delphine Sonck, Rudy Van Impe

Bending stiffness of laminated glass elements in structural applications — influence of a elasto-viscoplastic ionomer interlayer on the pre-breakage behaviour 81
Dieter Callewaert, Jan Belis, Didier Delincé, Delphine Sonck, Rudy Van Impe

On the inelastic constitutive equations of plates and shells made of foams.. 86
Holm Altenbach, Victor A. Eremeyev

Experimental investigation of the influence of temperature on local bridging behaviour in laminated glass elements in post-breakage state.......... 91
Dieter Callewaert, Didier Delincé, Delphine Sonck, Jan Belis, Rudy Van Impe

Prediction of the plastic deformation of circular cylindrical shells................ 96
A. Darvizeh, H. Rajabi

Plastic collapse of steel water towers 103
Wesley Vanlaere, Delphine Sonck, Rudy Van Impe, Arne Jansseune, Wouter De Corte

A discussion on elastoplastic model of double yield surfaces........................ 108
Xianqi Luo, Deyong Wang

The numerical simulation on the cracks of the concrete cover under hoop reinforcement corrosion and expansion 114
Qiu Zhaoguo, Zhang Fengpeng, Liang Li

Nonlinear analysis of FRP reinforced RC slab carrying capacity 119
Liu Jun, Zhang Fengpeng, Li Qing

The time dependence research of buckling for delamination with elastic bridge in laminar composite material 124
Zhang Xiaochun, Liu Shushu

Effect of loading rate on dynamic characteristics of reinforced concrete frame column 132
Debin Wang, Hongnan Li

Energy-absorbing performance of nested tube systems under compression load 137
Chen Zong, Xue Pu

Part II Materials Processing Engineering

Causes and preventive measures for cracks in anchored ends of prestressed concrete hollow slab beams 145
Yang Zhong, Liu Aiqun, Yin Yihui

New improvements in the multi-step inverse FEM for the fast formability prediction of multi-step autobody staming process 151
Bingtao Tang, Xiaoyang Lu

An study on the flexural performance of reninforced concrete beams considering loading history with pre-stressed CFRP sheets 156
Zhe Li, Gang Li

Modeling of shear fracture in the sheet metal forming processes using FEM approach 161
Yu Song, Zhao Jun

Effects of mean stress and stress concentration on high strength bolt 166
Zhou Congling, Li Jianyu

Implementation and application of a real-time control system using Matlab/xPC 171
Xiaofeng He, Zhengyi Jiang, Christopher Cook, Yinshui Liu

Study on stretching efficiency of heavy forged shaft based on orthogonal experimental design 177
Qinxiang Xia, Guangming Wei, Yulin Huo, Su Huang, Ming Li

Storage method of FE equations and its application in numerical simulation of sheet metal forming 182
Bao Yidong, Du Guo Kang, Chen Wen Liang, Wang Zhi Guo

Establishment of springback prediction model for the rotary-draw bending of thin-walled rectangular tube considering variation of Young's modulus 188
W. Q. Dong, Y. L. Liu, G. Y. Zhao, H. Yang

Optimization of gas-assisted injection molding process for a handle product based on orthogonal experiment 195
Jiang Shaofei, Li Jiquan, Lu Congda

The effect of ion assisting deposited TiN/Ti coating on the fretting fatigue behavior of Ti-811 alloy at elevated temperature 201
Xiaohua Zhang, Liudao Xin

Wear characteristics of traditional and new generation brake discs 206
Dariusz Rudnik, Andrzej Wojciechowski, Jerzy Sobczak

Microstructure variations of cast iron affecting its mechanical behavior........ 211
Krystyna Pietrzak

Researches on flow behavior and constitutive equation of A390 aluminum alloy 216
Haiting Guo, Xinyun Wang, Lei Deng, Juchen Xia

Research on model of on-line measurement for vertical ring rolling mill........ 221
Xiaokai Wang, Lin Hua

Research on analytical algorithm of contact problem using finite element analysis for bulk forming 226
Zhong-Jin Wang, Bin-Xian Yuan

Amendment and reconstruction of empirical formulas calculating bursting pressure of special spherical vessels(I) 232
Yin Yihui, Liu Xinfu, Sun Yin, Tan Yun, Mei Jun, Liu Yuandong, Yu Shaorong

Amendment and reconstruction of empirical formulas calculating bursting pressure of special spherical vessels(II) 239
Yin Yihui, Liu Xinfu, Sun Yin, Tan Yun, Mei Jun, Liu Yuandong, Yu Shaorong

Deduction of stress and elastic strain distribution for torsional circular shaft with minimum energy principle 245
Li Zhihui, Shi Junping, Tang Anmin

Plasticity-driven damage detection using lamb waves for stiffened plate 250
Shen Yanfeng, Qu Wenzhong, Xiao Li

Stamping-forging process of an aluminum alloy pan 255
Luo Jian-Cheng, Wang Xin-Yun, Guo Mei-Ling, Xia Ju-Chen, Luo Yun-Hua

3D simulation of electromagnetic forming for tube bulging based on FEM 260
Xiaohui Cui, Jianhua Mo

Analysis of light-weight structure for vehicle using metal and carbon fiber 265
Jung Ho Kim, Jung Min Bang, Ji Hoon Kim

Study on springback of high strength steel MPV front side rail based on BP neural network model 270
Wenping Li, Yunwang Ma, Miao Jin, Baofeng Guo Shaomin Nie, Zhong Wang, Jiyan Liu

Study on safety evaluation methods and failure mechanism of high roller compacted concrete dam 275
Su Pei-Fang, Wang Wei Ming, Lu Xing-Li

Analytical study on the mechanical behavior of truss core structures under compressive and shear loads 281
Sang-Woo Kim, Kwang-Seok Lee, Beom-Soo Kang

Instrumentation of infrared thermography during tensile test of glass fiber reinforced polymer 287
Jin Woo Kim, Dong Gi Lee, Jae Ki Sim, Jae Yeol Kim, Seung Hyun Choi

Deformation characteristics in cold rotary forging of a 20CrMnTi spur bevel gear 292
Xinghui Han, Lin Hua

Numerical simulations and optimization of B-pillar forming based on different binders 301
Jun Liu, Hua-Jie Mao, Rong-Chuang Chen, Wei Cao

High temperature compression deformation behavior of 7050 aluminum alloy........ 306
Huie Hu, Xiaodong Kong, Debin Chen

CFRP/foam circular members for energy absorption characteristics under axial compression........ 311
Ju-Ho Choi, Jae-Ki Sim, Gil-Young Han, Hyun Kook, In-Young Yang

The comparative study of intensity and rigidity of the hybrid member and carbon fiber........ 316
Ji Hoon Kim, Jung Min Bang, Jung Ho Kim, Hyun Goo Kang

Effect of initial state on grain refinement strengthening of AZ31 magnesium alloy 321
Jianmin Yu, Zhimin Zhang

Superplastic forming of AZ80A magnesium alloy automobile wheel........ 326
Yang Yongshun, Yang Xi

The effect of die curvature and material anisotropy in U-bending of AZ31 magnesium alloy sheet on springback........ 331
H. H. Kim, C. G. Kang

Influence of preformed billet on the multi-directional forging of the blade with a damper platform........ 336
Y. L. Liu, Y. Wang, H. Yang, Q. Guo

Embedded delamination growth in interlaminar short-fiber reinforced composite panels under compressive load 342
Zhang Fengpeng, Qiu Zhaoguo, Liu Ddcheng

Analysis of the forming defects and its influencing factors of the poly-wedge pulley spinning 347
Xiuquan Cheng, Jiazi Wang, Qinxiang Xia

The buckling of the stretchable electronic structure........ 352
Zhang Xiaoqing, Ou Zhicheng, Yao Xiaohu, Wang Weitong

Closed form stress distribution in 2D elasticity for all boundary conditions 357
Youzhen Yang

The compound materials defect test that used ultrasound thermography 362
Yong Ha Yang, Seung Hyum Choi, Sung Hyun Kim, Jae Yeol Kim

The 20MN energy savings fast forging press hydraulic system design 368
Dai Min-Qiang, Zhao Sheng-Dun, Xu Fan, Yuan Xiao-Mei

Experimental research on shear fatigue performance of CFRP-strengthened high-strength concrete structure 373
Hong Hai, Yinghua Zhao, Guangshan Zhu

Influence of process conditions on processing surface in shaving 378
Mitsuhiro Asao, Hidenobu Mano, Yoshinori Yoshida, Zhigang Wang, Yuki Sato, Hironobu Suzuki

A model of plastic deformation and surface roughening of polycrystalline metal based on R-value of grains 386
Takeji Abe

Study on the casting-forging process for low plasticity aluminum alloy connecting rods 391
Wang Yihua, Hu Chengliang, Zhang Haiying, Lv Jinglin

A reshaping roll forming technique for square/rectangular steel tube manufacturing: Numerical study and process design 398
Dayong Li, Liming Li, Ding Tang, Yinghong Peng, Jilong Yin, Feizhou Wang

The SSV constraint for 0/1 solutions in density-based topological design 403
Xiangyang Zhou, Liangzhi Fan

Galling behavior in square cup drawing of high tensile strength steel sheet 409
Keisuke Tanabe, Mingxu Yang, Yasuharu Yoshikawa, Zhigang Wang

Development of bottom compression drawing process for solid boss forming 414
Zhigang Wang, Toru Ando

Development of bottom compression drawing process for solid boss forming 419
Zhigang Wang, Masanao Karasawa

A newly devised forming method for triple cup 424
Zhigang Wang, Yasuharu Yoshikawa, Isamu Kato

The deformation simulation of plastic injection mold based on boundary element method 429
Yun Zhang, Li Dai, Huamin Zhou, Weiyi Li

Elasto-plastic deformation behavior of cemented carbide die in cold forging 434
Zhigang Wang, Haojie Li, Shinya Nakano

Part III Rock and Soil Mechanics

Model test and numerical simulation for dynamic characteristics of rock slope under blasting loads 441
Zhong Dong-Wang, Wu Liang

Deformation of soft soil roadbed on-site monitoring and FLAC simulation 446
Li Qinyuan, Ren Jianxi, Liuhui, Wangyang, Sunyang

Natural crack-face boundary conditions in the fracture mechanics of piezoelectric solids 451
Xian-Ci Zhong, Ke-Shi Zhang

Analysis of cumulative plastic deformation of road on soft subsoil induced by vehicle traffic loadings 456
Lu Zheng, Yao Hailin, Liu Jie, Hu Mengling

The theoretic analysis and experimental research on the time-dependency of TC piles 461
Y. H. Chen, L. Chen, X. D. Zhou, X. Q. Wang

The impact of polypropylene fibre content on mechanical properties of reinforcement soil 467
Pan Hongke, Zhu Yanzhi

On effect of non-persistent joints on hydro-mechanical behavior of jointed rock mass in UDEC 472
Wu Yuexiu, Liu Quansheng, Liu Xiaoyan

The creep behaviour of fracture granite under hydro-mechanical condition.......... 477
Lin-Fang Shen, Xia-Ting Feng, Peng-Zhi Pan, Hui Zhou

Numerical analysis of a transmission tower with compound foundation under horizontal ground motion.......... 482
Ju-Ke Bai, Long-Zhu Chen

Findings in stress evolution of surrounding rocks in gob-side entry retaining.......... 487
Zhang Nong, Kan Jia-Guang, Li Gui-Chen, Zheng Xi-Gui

Effects of temperature and wheel loading on rutting deformation of asphalt mixture.......... 492
Xinhua Yang, Jin Shang, Anyi Yin, Shengfeng Yang, Yong Ye

Part I

Elastic-Plastic Mechanics

PUSHOVER ANALYSIS OF STRUCTURES CONSIDERING STRAIN RATE EFFECTS

WENMING WANG †

Faculty of Infrastructure Engineering, Dalian University of Technology, Dalian, China

HONGNAN LI

Faculty of Infrastructure Engineering, Dalian University of Technology, Dalian, China

Most materials used in civil engineering are sensitive to strain rate. Therefore, strain rate effects should be considered in seismic analysis. Pushover analysis is a static method that deals with dynamic problems. How to consider strain rate effects in the method rationally needs research. The theory and approach of pushover analysis considering strain rate effects are presented in the paper. The process of pushover analysis can be regarded as a 1/4 cycle, and the duration of a 1/4 cycle can be estimated approximately. Using average strain rate, performing pushover analysis for the structure twice, strain rate effects can be considered rationally. According to the procedure presented in the peper, a four-story reinforced concrete frame structure is analysed. The result shows that the capacity of the structure increases after considering strain rate effects.

Keywords: Pushover analysis; strain rate effects; 1/4 cycle; average strain rate.

1. Introduction

Pushover analysis is a static method that deals with dynamic problems. In the past deacdes, many experts studyed the method. The focuses of their studies are the distribution of lateral force and its application scope[1-4]. The constitutive relations, moment-curvature properties and failure criteria used in the method are mostly based on quasi-static theory. Most materials used in civil engineering are sensitive to strain rate. At different strain rate, the mechanical properties of materials including strength, stiffness and ductility are different[5-6]. To improve the precision of the method, strain rate effects should be considered. Bracci and co-workers proposed that strain rate effects can be accounted for in an equivalent manner by adjusting the moment-curvature properties of the members[7]. But how to adjust the moment-curvature properties needs research. The theory and approach of pushover analysis considering strain rate effects are presented in the paper.

---

† Corresponding author, email: wangwenming87110@163.com

2. The Theory of Pushover Analysis considering Strain Rate Effects

2.1. *The duration of a 1/4 cycle*

The structures reciprocate in earthquake. The time-history curve of top displacement(or storey drift) has many zero-crossings. The process between two zero-crossings is named semicycle. The duration of a semicycle correspond with relatively large or maximum top displacement(or storey dtift) is approximately equal to half of the fundamental period(T_1) of the structure whose responses are dominated by the fundamental mode on condition that there is no big difference between the predominant period of the ground motion(T_g) and T_1. A semicycle can be divided into two 1/4 cycle, the duration of a 1/4 cycle is approximately equal to 1/4 of T_1.

2.2. *The variation law of strain in a 1/4 cycle*

The variation law of strain in a semicycle is studyed by nonlinear time-history analysis of a reinforced concrete structure by ABAQUS(no elaboration here). It is discovered that the shape of strain curve is most like sine curve in a semicycle.

2.3. *How to consider strain rate effects*

By pushover analysis, we can obtain the roof displacement-base shear curve or storey drift-base shear curve. Before applying lateral load, the structure is subjected to gravity load. As the applied lateral load becomes larger, the roof displacement or storey drift becomes larger. When the target displacement is reached, the process is over. The process can be regarded as a 1/4 cycle. In a 1/4 cycle, the time-history curve of strain is most like sine curve. Because the duration of a 1/4 cycle and variation law of strain are approximate, it is impossible to consider strain rate effects precisely in pushover analysis. Using average strain rate is the most convenient and rational method. It is convenient to adjust the moment-curvature properties of the members according to average strain rate.

2.4. *Rationality of using average strain rate*

Before applying lateral load, the structure is subjected to gravity load. Every point of the structure has initial strain. Before the target displacement is reached, the strain changes all the time. When the target displacement is reached, the variation of strain reaches the maximum. In most cases, the initial strain is micro. Therefore, it is neglected in the following.

We will illustrate the rationality of using average strain rate accoring to the constitutive relations of rebar. The strain of rebar during pushover analysis can be expressed as follows:

$$\varepsilon(t) = \varepsilon_{\max} \sin \frac{2\pi}{T_1} t \qquad (1)$$

Where $\varepsilon_{\max}$ is the peak value of strain in a semicycle; T_1 is the fundamental period, $0 \le t \le \frac{T_1}{4}$.

The strain rate of rebar during pushover analysis can be expressed as follows:

$$\dot{\varepsilon}(t) = \varepsilon_{\max} \frac{2\pi}{T_1} \cos \frac{2\pi}{T_1} t \qquad (2)$$

The average strain rate can be expressed as follows:

$$\overline{\dot{\varepsilon}} = \frac{4\varepsilon_{\max}}{T_1} \qquad (3)$$

Based on the equations above, we can obtain some conclusions: when the strain is zero, the strain rate arrive at the peak value; when the strain arrive at the peak value, the strain rate is zero; the maximun strain rate is 1.57 times as large as average strain rate.

As the strain rate increases, the variation of rebar's elastic modulus is micro; the yield strength increases and can be expressed as follows[6]:

$$f_{yd} = (1 + c_f \lg \frac{\dot{\varepsilon}}{\dot{\varepsilon}_0}) f_{ys} \qquad (4)$$

$$c_f = 0.1709 - 3.289 \times 10^{-4} f_{ys} \qquad (5)$$

Where $\dot{\varepsilon}$ is the current strain rate; $\dot{\varepsilon}_0$ is the quasi-static strain rate, $\dot{\varepsilon}_0 = 2.5 \times 10^{-4} / s$; f_{ys} is the yield strength at quasi-static strain rate; f_{yd} is the dynamic yield strength at the current strain rate.

Elastic-perfectly plastic model for the rebar is shown in Fig.4. f_{yd1} is the dynamic yield strength at the average strain rate in a 1/4 cycle, and f_{yd2} is the dynamic yield strength at the maximum strain rate in a 1/4 cycle; ε_{ys} is the yield strain at the quasi-static strain rate; $\varepsilon_{\max}$ is the maximun strain in the 1/4 cycle; ε_{yd1} is the yield strain at the average strain rate; ε_{hs} is the ultimate strain.

If $\varepsilon_{\max}$ is known, we can obtain the relations of stress and strain in the 1/4 cycle using (1) to (5). Before the strain exceed ε_{ys}, the Constitutive relations at different strain rate is identical(line AB in Fig.1); when the strain is equal to $\varepsilon_{\max}$, a point F can be obtained according to the constitutive relations at the quasi-static strain rate; there is another point E that can be determined using (1) to (5) in line BD; E and F can be connected with a straight line approximatly. Curve ABEF is the relations of stress and strain in the 1/4 cycle. When $\varepsilon_{\max} = 1.297\varepsilon_{yd1}$ (there is no big difference between ε_{yd1} and ε_{ys} in earthquake), E coincides with C; when $\varepsilon_{\max} > 1.297\varepsilon_{yd1}$, E can be determined between C and D. Generally, it is rational to use average strain rate.

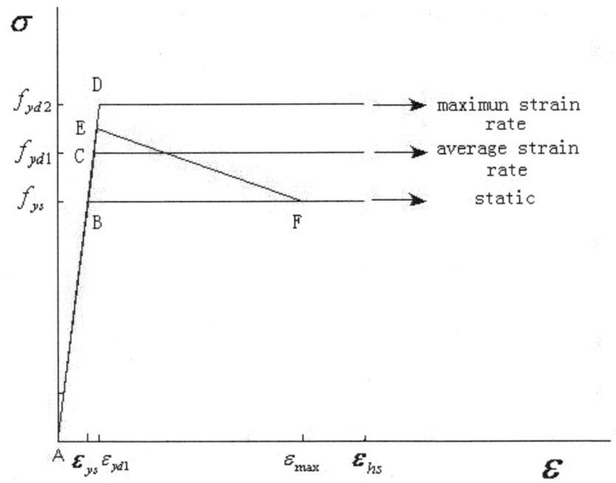

Fig.1. Constitutive relations of rebar.

3. The Approach of Pushover Analysis considering Strain Rate Effects

According to the theory above, strain rate effects can be considered in pushover analysis. The initial state of structure and duration of a 1/4 cycle is known, but the end-state is unknown. However, the end-state can be abtained approximately by pushover analysis without considering strain rate effects, which is called pseudo end-state. The step-by-step approach of this method is as follows:

1. Determine the initial state of the structure and model.
2. Obtain the pseudo end-state by pushover analysis without considering strain rate effects.
3. The average strain rate of every point in the 1/4 cycle can be obtained according to the initial state, the pseudo end-state and the duration of the process, then the model can be modified according to the average strain rate of every point. In the end, perform pushover analysis for the second time. Plastic-hinge modle is used widely in pushover analysis. Therefore, it is necessary to modify the settings of plastic-hinge only.

4. Numerical Simulation

According to the procedure above, a four-story reinforced concrete frame structure is analysed by SAP2000(no elaboration here). We considered strain rate effects by adjusting the moment-curvature properties of the members. The pushover curve is shown in Fig.2.

By comparing the two curves, some conclusions can be obtained. First, the base shear and story drift increase when the structure shows nonlinear characteristics after considering strain rate effects. Second, the structure can endure larger base shear after considering strain rate effects.

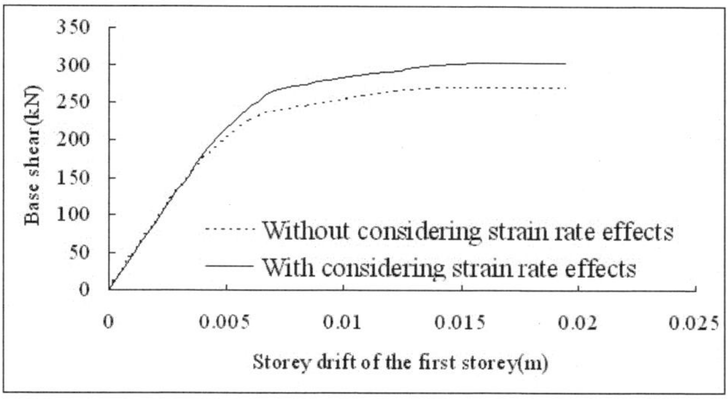

Fig.2. Pushover Curve of the Structure.

References

1. P. Fajfar, P. Gaspersic: *Earthquake Engineering and Structural Dynamics*, 1996, **25**, (1), 31 - 46.
2. J. M. Bracci, S. K. Kunnath, A. M. Reinhorn: *Journal of Structural Engineering*, 1997, **123**, (1), 3 - 10.
3. B. Gupta, S. K. Kunnath: *Earthquake Spectra*, 2000, **16**, (2), 367 - 391.
4. A. K. Chopra, R. K. Goal: *Earthquake Engineering and Structural Dynamics*, 2002, **31**, (3), 561–582.
5. P. H. Bischoff, S. H. Perry: *Materials and Structures*, 1991, **24**, (6), 425-440.
6. M. Li, H. N. Li: *China Civil Engineering Journal*, 2008, **43**, (4), 70-75.
7. P. Pankaj, E. Lin: *Engineering Structures*, 2005, **27**, (7), 1014-1023.

DYNAMICAL RESPONSE OF RECYCLED CONCRETE UNDER CYCLIC LOADING

XINGWEN LUO†, HAILIN YAO

State Key Laboratory of Geomechanics and Geotechnical Engineering, Institute of Rock and Soil Mechanics, Chinese Academy of Sciences, Wuhan, Hubei 430071, China

In order to research the dynamical response and fatigue damage mechanism of recycled concrete under cyclic loading, we have made cyclic loading tests on four types of recycled concrete specimens respectively. Test results show the evolution of fatigue total strain of recycled concrete undergoes three stages. The cycle times in the three stages accounted for 10%, 80% and 10% of the fatigue life respectively. These stages can be simulated by power function equations, first-order linear equations and exponential functions accordingly. The internal cracks of recycled concrete developed constantly during the tests. The development of the external deformation of recycled concrete is in line with the growing of its internal cracks, and they are unanimous essentially.

Keywords: Recycled concrete; Dynamical response; Damage mechanism.

1. Introduction

Recycled concrete is made from waste concrete after a series of treatments, such as crushing, washing, screening and grading. Recycled concrete aggregate is used to substitute for part or all of the natural aggregate in concrete. The development of the technology of recycled concrete adapts to the demands of the development of our society. Recycled concrete is a kind of sustainable green concrete.[1-2] Recycled concrete has been tried to use in highway and construction engineering currently, but its dynamic mechanical characteristic has not been researched before. Compared with natural aggregate, recycled aggregate is not only multi-edges and corners, but also its surface is often wrapped with a layer of mortar. The interfaces between aggregate and paste structure of recycled concrete are more complex. The number of the interfaces in recycled concrete is bigger than that in nature concrete. If concrete is looked as aggregate scattered with mortar to form a continuous phase of composite materials, there are some new internal interfaces between recycled concrete aggregate and mortar that

† Corresponding author: XINGWEN LUO, email: xwluo@whrsm.ac.cn

recycled concrete differentiate to nature concrete.[3-4] There are a lot of hydration products in the interfaces between recycled concrete aggregate and mortar because the water-cement ratio in the regions of recycled concrete is significantly higher than that in nature concrete. Therefore, initiation cracks are most likely to generate and grow from these region. The transition zone is the weakest link, and this is an important factor that affect the mechanical properties and durability of recycled concrete directly.[5-7]

In order to determine the amplitude of loading stress, we have made complete stress-strain static loading tests on the four types of the recycled concrete specimens at first. Then we made cyclic loading on the four types of recycled concrete specimens respectively. The development of the evolution of fatigue total strain can be get after the tests. At last the fatigue damage mechanism of recycled concrete has been researched.

2. Material and Method

The materials that recycled concrete made from were bellow.

1. Recycled concrete coarse aggregate derived from the concrete whose grade was *C30* and the particle scattering was smaller than 4 centimeter.
2. The particle scattering of natural coarse aggregate was smaller than 4 centimeter.
3. Fine aggregate were made form natural continuous gradation river sand which the fineness modulus was 2.7.
4. Slag cement's grade was 325#.
5. Water was form urban tap water.

The mix proportion of recycled concrete were designed as follow. Cement: water: fine aggregate sand: coarse aggregate was equal to 1:0.5:1.5:3.0.

We have made four types of recycled concrete of which the substitute ratio were different. There were at least 12 blocks samples in every type of recycled concrete specimens. The loading frequency on recycled concrete specimens in cyclic loading tests was 1.0 *HZ*. A single-stage and constant-amplitude cyclic loading mode was adopted in the tests.

3. Result and Discussion

3.1. *Evolution characteristics of fatigue total strain of recycled concrete*

The loading stress and its amplitude on the four types of recycled concrete were the same in the tests. The minimal stress was 2.7*MPa*, and the maximal stress was 24.5*MPa*.

Test results show that the development of the fatigue total strain and fatigue residual strain of recycled aggregate concrete all undergo three stages under cyclic loading. The evolution of the fatigue total strain of the four types of recycled concrete are shown in Figure 1.

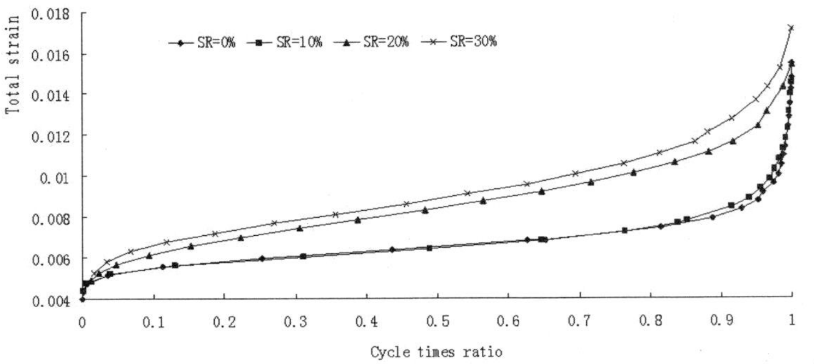

Fig. 1. Evolution of the fatigue total strain of recycled concrete.

As can be seen from Figure 1, the evolution of the fatigue total strain of the four types of recycled concrete all undergo a three-stage development model under cyclic loading. The proportion of cycle times on every type of recycled concrete account for the fatigue life of every stage are almost the same. The proportion is about 10% in the first stage under cyclic loading, and the proportion in the second stage is 80%. The proportion in the last stage is about 10%. The deformation of the four types of recycled concrete all undergo three stage under cyclic loading. How the substitute ratio of recycled concrete affect their deformation is still not known. Table 1 gives the evolution equation of the fatigue total strain of every stage of the four types of recycled concrete.

Table 1. Evolution function of the fatigue total strain of recycled concrete.

substitute ratio	the first stage	the second stage	the third stage
0%	$y=0.0061\lambda^{0.0474}$	$y=0.0028\lambda+0.0050$	$y=0.00004e^{5.8011}\lambda$
10%	$y=0.0063\lambda^{0.0587}$	$y=0.0032\lambda+0.0052$	$y=0.00003e^{6.0872}\lambda$
20%	$y=0.0078\lambda^{0.1034}$	$y=0.0059\lambda+0.0056$	$y=0.001e^{2.6863}\lambda$
30%	$y=0.0088\lambda^{0.1256}$	$y=0.0065\lambda+0.0058$	$y=0.0011e^{2.7103}\lambda$

3.2. *Damage mechanism of recycled concrete*

The development of fatigue damage of recycled concrete also undergo three stage. By optical microscopy and electron microscopy, the damage process of recycled concrete under cyclic loading are mainly as bellow.

In the first stage, the initiation cracks exist in the interfaces between aggregate and cement mortar. The medium term are the stage of the stable extension of cracks. Cracks link up in the last stage, and eventually leading to the complete destruction of recycles concrete.

Concrete is non-homogeneous and multi-phase complex. Recycled concrete aggregate is made partly or all from recycled aggregate. This lead to its structure is very complex. If the change of the environment such as the temperatures and humidity due to the difference of elastic module of aggregate and mortar, the inconsistent expansion deformation between aggregate and mortar will happen. The generation of internal initial stress lead to the subtle cracks generate in the interfaces between aggregate and mortar. Especially in the interface between recycled aggregate and new mortar cracks is more susceptible to grow. A lot of cracks already exist in the interface between aggregate and mortar before cyclic loading tests, especially in the interfaces between recycle aggregate and mortar. The higher the substitute ratio of recycled concrete, the more original cracks in recycled concrete. Water absorbent of recycle aggregate is stronger than that of natural aggregate. Water is likely move to the hydrophilic porous surfaces of recycled aggregate. Water films will be formed between the surface of recycled aggregate and cement, and this lead to the micro-cracks generate in recycled concrete.[8-9]

The second stage is a phase of the stead expansion of cracks. Cracks in the interfaces between aggregate and mortar extend or expand further under cyclic loading in this stage. Sometimes cracks stretch into the mortar house. Longer cracks will formed by some short cracks bayonet connecting. At the same time some new cracks generate. The strengthen of the interfaces between recycled aggregate and mortar are relatively poor. The higher the substitution ratio of recycled concrete, the more quickly cracks in the interfaces between aggregate and mortar grow. The cycle times in this stage accounts for about 80% of the fatigue life.

Cracks expanse rapidly in the last stage. In this stage cracks in the body of mortar increase dramatically. They develop very fast, and this lead to the complete destruction of recycled concrete. The cycle times in this stage accounts for about 10% of the fatigue life of recycled concrete.

4. Conclusion

Following conclusions can be drawn by the tests.

1. The evolution of the fatigue total strain of recycled aggregate concrete undergoes a three-stage development model under cyclic loading. They can

be simulated by power function equations, first-order linear equations and exponential functions. The greater the substitute ratio of recycled concrete, more quickly the fatigue total strain develops.

2. The internal cracks of recycle concrete grow during the whole cyclic loading tests. The density of recycled concrete decrease with the increase of cycles times.
3. With the development of the external deformation of recycle concrete, the growing of its internal cracks also undergo three stages under cyclic loading correspondingly. The development of the external deformation of recycled concrete is in line with the growing of its internal cracks, and they are unanimous essentially.

References

1. Wan Huiwen, Shui Zhonghe, Lin Zhongshou and Xiao Kaitao: Environmental assessment for regenerated concrete [J], *Journal of Wuhan University of technology*, 2003, **25**,(4),17-21.
2. Ryu J S: An experimental study on the effect of recycled aggregate concrete properties[J], *Magazine of Concrete Research*, 2002, **54**,(1), 7-12.
3. Chen Yungang, Sun Zhenping and Xiao Jianzhuang: Characteristics and strengthening methods of interfacial zone between aggregate and cement paste in recycled-aggregate concrete[J], *Concrete*, 2004,2,10-13.
4. Dhir R K, Limbachiya M C and Leelawat: Suitability of recycled aggregate for use in BS 5328 designated mixes[J], *Proceedings of the Institution of Civil Engineers: Structures and Buildings*, 1999, **134**,(3),257-274.
5. LI Chaoyang, Song Yupu and Zhao Guofan: Study of residua strain of concrete under fatigue loading[J], *Journal of Dalian University of Technology*, 2001,**41**,(3), 355-358.
6. Jo Byung-Wan, Park Seung-Kook and Park Jong-Chil: Mechanical properties of polymer concrete made with recycled PET and recycled concrete aggregates[J], *Construction and Building Materials*, 2008,**22**,(12), 2281- 2291.
7. Otsuki Nobuaki, Miyazato Shin-Ichi and Yodsudjai Wanchai: Influence of recycled aggregate on interfacial transition zone, strength, chloride penetration and carbonation[J], *Journal of Materials in Civil Engineering*, 2003, **15**, (5), 443-451.
8. Wan Huiwen, Xu Jinlong, Shui Zhonghe and Jiang Jian: Study on the Structure and Properties of Interfacial Transition Zone(ITZ)of the Regenerated Concrete[J], *Journal of Wu Han University of Technology*, 2004, **26**,(11), 29-32.

THEORETICAL STUDY ON THE DYNAMIC MECHANICAL PROPERTIES OF THE NBR COMPOSITES UNDER STATIC PRESSURE

BO QIAO, GUORONG ZHANG, XIUYING ZHAO, SIZHU WU†

Key Laboratory of Beijing on

"Preparation and Processing of Novel Polymer Materials",

Beijing University of Chemical Technology,

Beijing 100029, P.R. China

Time-temperature superposition is shown that at certain frequency the higher temperature, the lower modulus, which indicates that enhanced temperature and reduced frequency has the same effects. Based on this, the mechanical properties of rubber materials at wide range of frequency can be obtained in different temperature by time-temperature superposition. To study the basic theory of the damping properties of rubber under static pressure, a modified Williams-Lendel-Ferry (WLF) equation has been proposed. According to the loss factor tgδ from DMA experiments, the shift factors under different pressure could be calculated. Then, the mast curve of loss factor versus frequency can be plotted. The result showed that the peak of tgδ has been appeared after the reference temperature. With the increasing of the pressure, the glass transition temperature moved towards higher frequency of 1.5 orders of magnitude. This method might offer a theoretical method to quantitatively characterise the damping properties of rubber materials under pressure.

Keywords: Pressure; Williams-Lendel-Ferry equation; Grass transition temperature; DMA; NBR; Shift factor.

1. Introduction

The damping rubber has been widely used in submarines as muffler materials and the study on its properties under static pressure is an important basic

† Corresponding author, email: wusz@mail.buct.edu.cn

scientific research.[1] However, the quantitative explanation is far from enough.[2-3] Consequently, the precise prediction of the change of the performances under pressure has been a significant challenge, which is the objective of this study.

According to the deduced modified WLF equation, this paper attempts to characterise the dynamic mechanical properties of nitrile-butadiene rubber (NBR) at different pressures by the dynamic mechanical analyser (DMA) tests at atmospheric pressure. This method provides a theoretical background for characterization of the dynamical properties of damping materials.

2. Experimental Method

2.1. *Materials*

The polymers were selected: NBR N230S containing 41% acrylonitrile supplied by Japan JSR Company and the antioxidants AO-80 as a functional additive in this study supplied by Japan Solectron sprinkling of Afghanistan (Shanghai) Co., Ltd. Carbon black N330 was purchased from Longxing Co., Ltd. Hebei and other agents were all regular in rubber industry.

2.2. *Apparatus and test methods*

DMA experiments were performed at the VISCOANALY SEUR VA3000 instrument owed by French 01dB-Metravib Company. The temperature and frequency scanning of the NBR specimens were carried out under the tensile mode. The specific test conditions are presented in Table 1.

Table 1. Test conditions of dynamic mechanical analysis.

Test conditions	Temperature scanning	Frequency scanning
Delta force	0.1%	0.1%
Frequency	10Hz	1～70Hz
Temperature	-70～150°C	-50～60°C
Scanning rate	3°C/min	3°C /min

3. The modified WLF Equation under Pressure

The free volume of visco-elastic polymer is highly affected under static pressure. As the pressure can squeeze out some free volume, the glass transition temperature will increase correspondingly, which shows in Eq. (1) for rubber materials. (∂Tg/∂P) (K/MPa) of the rubber materials usually is 0.25. [4]

$$T_g^P = T_g^0 + \Delta P \kappa_f / \alpha_f = T_g^0 + 0.25\Delta P \tag{1}$$

Eq. (2) is the common WLF equation.

$$\begin{aligned} \log a_T &\approx \left[-B(T - T_g)/2.303 f_g\right] / \left[f_g/\alpha_f + (T - T_g)\right] \\ &= C_1 (T - T_g) / \left[C_2 + (T - T_g)\right] \end{aligned} \tag{2}$$

Based on equations (1) and (2), it can be deduced the modified WLF equation under pressure as Eq. (3).

$$\begin{aligned} \log a_T &= -C_1 (T - T_g^P) / \left[C_2 + (T - T_g^P)\right] \\ &= \left[-C_1 (T - T_g^0 - 0.25\Delta P)\right] / \left[C_2 + (T - T_g^0 - 0.25\Delta P)\right] \end{aligned} \tag{3}$$

If the reference temperature is selected as room temperature, the shift factor $a_{T\text{-}R}$ from test temperature to room temperature can be calculated by the following Eq. (4):

$$\lg a_{T-R} = -C_1 * C_2 (T - 298) / \left[C_2 + (T - T_g^0 - 0.25\Delta P)\right] * \left[C_2 + (298 - T_g^0 - 0.25\Delta P)\right] \tag{4}$$

4. Results and Discussion

The frequency of DMA measurement was 1-70Hz. By scanning, the frequency spectrums at 21 different temperatures are shown in Figure 1. Based on time-temperature superposition, the shift factor at room temperature 25℃ is obtained from the WLF equation. Thus, the data in Figure 1 could be plotted to have the master curve.

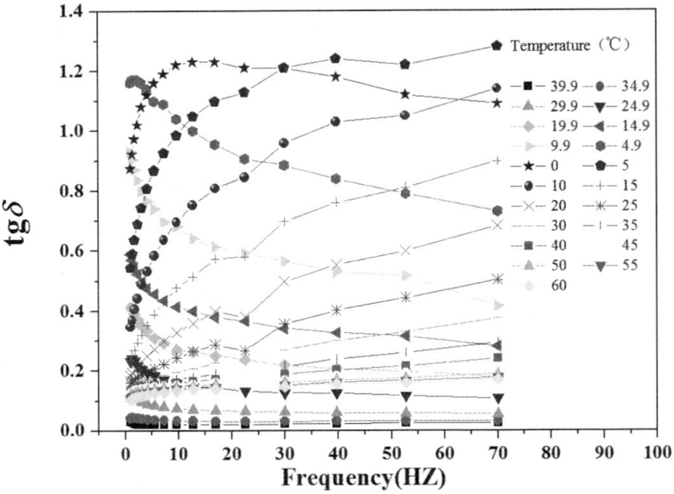

Fig. 1. Loss factor tgδ - frequency logω spectra of NBR under different temperatures.

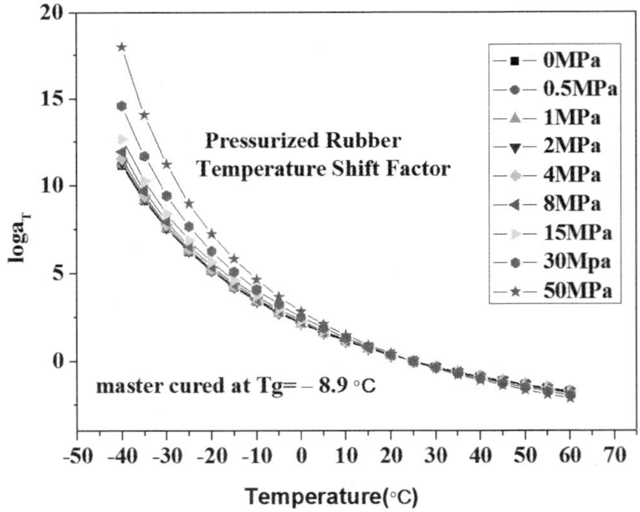

Fig. 2. Shift factor-temperature spectra of NBR under different pressures.

Considering the effect of pressure to the visco-elastic properties, the pressure conditions of 0.5MPa, 1MPa, 2MPa, 4MPa, 8MPa, 15MPa, 30MPa, and 50MPa was selected. Then all data is substituted into Eq. (18) and the frequency shift factors $loga_T$ in different pressures are shown in Fig.4. According to time - temperature superposition, each temperature curve will shift along the main curve. Due to higher temperature and lower frequency equivalence, $tg\delta$ curve still keeps a peak for showing visco-elasticity. At low temperature, the shift factor is positive. On the contrary, the shift factor is found to be negative at high temperature. It should be noted that the shift factor is larger at high-pressure and low-temperature. From above discussion we can conclude that loss factor $tg\delta$ on the frequency axis do not move in the master fitting curve at the reference temperature. Afterwards, rise in temperature leads to the increase of loss factor $tg\delta$. However, the loss factor moved towards low frequency when temperature is below the reference temperature.

At room temperature, the parameters of NBR are C_1 = 5.2, C_2= 87.2 in Eq.(2). By which the curve of loss factor versus frequency under different pressures can be gained in Figure 3. The results in Figure 3 show that when pressure rise up to 50MPa, the glass transition temperature shifts to high frequency on the axis with about 1.5 orders of magnitude, which means pressure makes the tgδpeak extended.

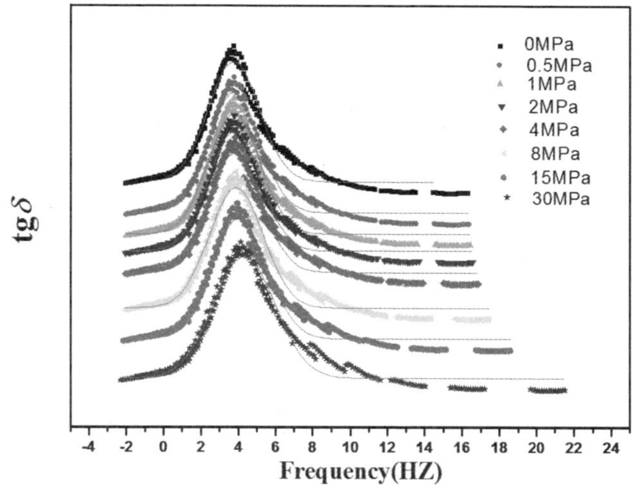

Fig. 3. Spectra of NBR fitting curve of loss factor plectra of NBR under different pressures.

5. Conclusion

This study provides a modified WLF equation including pressure parameters, by which the shift factors and the curve of loss factor versus frequency under different pressures can be plotted. According to the results from DMA experiments, the glass transition temperature of NBR can be calculated, which is -8.9℃ around. The analysis results show that if increase pressure to 30MPa, the glass transition temperature will increase and the peak will move to the high frequency. This paper offered a theoretical quantitative method to characterise the damping properties of rubber materials. It is helpful to develop the structural materials with high damping capacity and to solve the problem caused by vibration and noise under static pressure.

References

1. X. Y. Zhao, P. Xing, M. Tian, F. Hao, R. Q. Jin, L.Q. Zhang: *Polymer*, 2007, 48, (20), 6056-6063
2. A. V. Amirkhizi, J. Isaacs, J. McGee, S. Nemat-Nasser: *Philos. Mag.*, 2006, 86, (36), 5847-5866
3. S. Kasapis, S. S. Sablani: *Carbohydrate Polymers*, 2008, 72, (3), 537-544
4. A. Eisenberg: *J. Phys. Chem.*, 1963, 67, 1333-1336

INVESTIGATION ON COMPUTATIONAL METHOD FOR CRACK OF REINFORCED CONCRETE PIPELINE UNDER INTERNAL PRESSURE RETROFITTED WITH FRP

WANG SUYAN

Faculty of Infrastructure Engineering, Dalian University of Technology, Dalian 116023, China

LIU FEI[†]

Jiangsu Transportation Research Institute CO., LTD, Nanjing 210017, China

YU WENHUA

Faculty of Infrastructure Engineering, Dalian University of Technology, Dalian 116023, China

In order to investigate the strengthening problem of reinforced concrete pipeline sufferring from the seismic damage, the computational formula for the crack of reinforced concrete pipeline retrofitted with the Fiber Reinforced Polymer (FRP) was deduced. For simulating the damage of the medium or small earthquake, four pre-damage reinforced concrete pipelines under the internal pressure were loaded to damage, and then strengthened with the different FRP. Compared with the experiments of three undamaged reinforced concrete pipelines and one strengthened reinforced concrete pipeline without being damaged, the results show that the strengthening method with FRP can improve mechanical properties of pipeline, and delay the developing speed of crack. The serviceability ultimate load of the seismic damage pipelines not only could fully satisfy the application requirements before the damage, but also was improved. The theoretical and experimental data agree well. And it indicates that the computational model has the good applicability and may supply the reference for engineering applications.

Keywords: Pipeline under internal pressure; Damage; Fiber reinforced plastic; Strengthening; Crack.

1. Introduction

In the factors affecting the reinforced concrete pipeline security, the earthquake is one of the main causes which bring about its structural failure. Wang & Kennedy pointed out that the earthquake caused sudden increase in dynamic pressure of pipes, and the longitudinal cracking along reinforced concrete drain pipe section is one of the most common modes of damage[1].

2. Deduction of Crack Formula

When the pipeline bears internal pressure, the FRP pasted on its inner wall and hoop reinforcement in it is all considered taken as steel loops. The following formula is presented in Reference 2.

1.Calculation of the structural radial displacement after cracking

$$\left.\begin{aligned} U_{out} &= \frac{P_1 - P_1^{'}}{E_s t_1} r_1^2 \\ P_1^{'} &= K_1 P_1 \\ K_1 &= \frac{1}{(1 + \frac{t_1}{t_2} \bullet \frac{r_2}{r_1} + \frac{E_s}{E_c} \frac{t_1}{r_1} \ln \frac{r_2}{r_1})} \end{aligned}\right\} \quad (1)$$

2.Width sum of all the longitudinal cracks $\sum W$

$$\sum W = 2\pi (U_{out} - R_{out} \varepsilon_{t\,\max}) \quad (2)$$

According to experimental value, $\varepsilon_{t\,\max} = 7.5 \times 10^{-5}$ mm.

3.The average spacing between cracks l_{cr}

$$l_{cr} = 2(C + \frac{S}{10}) + K_1 K_2 \frac{d}{\rho_{te}} \quad (3)$$

4.The number of the longitudinal cracks N

$$N = \frac{2\pi \overline{R}}{l_{cr}} \quad (4)$$

5.Crack average width $\overline{W}$

$$\overline{W} = \frac{\sum W}{N} = \frac{(U_{out} - R_{out}\, \varepsilon_{t\,\max}) l_{cr}}{\overline{R}} \quad (5)$$

6.crack maximum width $W_{\max}$

$$W_{max} = 1.3\overline{W} \tag{6}$$

3. Test Survey

3.1. *Specimen design*

The design dimensions of tested pipeline specimens are as follows: the outer radius 200mm, the inner radius 160mm, the wall thickness t=40mm and the pipe length L=1000mm. In the pipeline, there are cool drawn low carbon steel wires with B ϕ^{b}, longitudinal bars of 6 ϕ^{b}5, spiral reinforcement of ϕ^{b}4 with spacing s=74mm. And the center of spiral reinforcement is at a distance of two-fifths wall thickness from the inner surface. Specimen concrete is C30 of gravel concrete.

3.2. *Test program*

The pipelines were loaded by uniform internal pressure. For simulating the longitudinal cracking in the section of the pipeline due to the medium or small earthquake,firstly four pre-damage reinforced concrete pipelines were loaded to damage, and then strengthened with one layer CFRP or GFRP pasted on the inside of the pipelines[3]. The specimen numbering and description are seen in Table 1. In the experiment, the cracking load was taken as the pre-damage load.

Table 1. Specimen numbering and description.

	Specimen number	Reinforcing material	Description	Loading program
The contrastive pipelines	TR	—	Without being reinforced	Being directly loaded to the serviceability limit state
	GTR	E450	Reinforced without damage	
	CTR	HITEX-C300	Reinforced without damage	
reinforced pipelines with damage	GRa1	E450	Reinforced after pre-damage at level a	Being firstly loaded to the pre-damage load a and then being loaded to the serviceability limit state after being reinforced
	GRa2	E450		
	CRa1	HITEX-C300		
	CRa2	HITEX-C300		

4. Experimental Results and Analysis

4.1. *Load results*

The serviceability ultimate load is the corresponding load when the first crack with the width more than 0.2mm formed[4]. The cracking load and the serviceability ultimate load can be seen in Table 2.

Table 2. Comparison of load.

Specimens number	The cracking load (MPa)	The serviceability ultimate load (MPa)
TR	0.413	0.442
GTR	0.452	0.539
CTR	0.468	0.553
GRa1	—	0.507
GRa2	—	0.511
CRa1	—	0.517
CRa2	—	0.527

4.2. *Crack computation*

Metal pipelines under internal pressure strengthened by FRP winding were studied by Yong-xin Yang[5]. The author assumed that damage be caused by the uniform reduction of metal pipelines wall thickness. And this method obtained good results. For the reinforced concrete pipelines damage caused by the earthquake, we made the same assumption. For the crack analysis of strengthened pipes with damage, the model of the uniform reduction of pipeline wall thickness was used. In the experiment, damage load was defined as cracking load. At the moment, the depth was assumed as 2mm.That was the uniform reduction amount of pipeline wall thickness. Model and experiment crack maximum width of CTR can be seen in Figure 1, The model can reflect the ultimate state of crack.

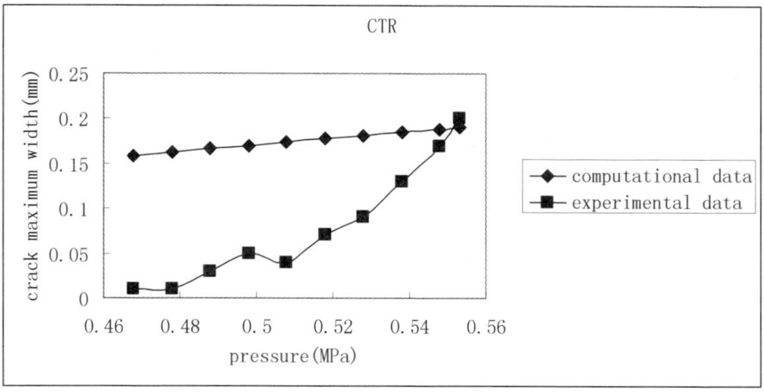

Fig. 1. Model and experiment curves of CTR.

The calculation results are seen in Table 3.

Table 3. Comparison analysis of maximum crack in the outer wall.

Specimen number	Experimental Value (mm)	Theoretical value (mm)
GTR	0.20	0.3237
CTR	0.20	0.1903
GRa1，GRa2	0.22	0.3098
CRa1，CRa2	0.21	0.1623

5. Conclusion

The strengthening method pasting the FRP on the inner wall of the pipeline is very effective. And it can improve mechanical properties of pipeline and obviously inrease the serviceability ultimate load. At the same time, the strengthening effect on delaying the development speed of crack and changing the crack patterns are very obvious. The crack computational model derived in the paper is suitable for the circular pipelines under the internal pressure strengthened by the FRP.

References

1. Wang Ru-liang. Damage, seismic checking calculation, seismic design and measures of underground[G].Translated by Wang Luo-jia, Wang Qian-xin, Wang dianjiechun, Li Hong-nan and so on. Beijing:Earthquake Press, 2007:207-217.

2. Wang Liyang. The Experimental Study on Pressure Conduits of Faber Reinforced Concrete[D].Master Thesis, Wuhan University,2004.
3. Zhang Wei-song, Cheng Shang-jian, Liu Yi-min. The Experimental Study on Pressure Conduits of Faber Reinforced Concrete[J].China Rural Water and Hydropower,2005(6):54-57.
4. GB/T 16572-2006 Concrete and reinforced concrete drain test methods[S].
5. Yang Yong-xin, Peng Fu-ming,Yue Qing-rui. Stress Analysis of Metal Pipelines Strengthened by FRP Winding[J].Port Engineering Technology, 2005(4):17-19.

STUDY ON K_R-CURVES OF THE CONCRETE BEAM REINFORCED WITH CFRP

DONG WEI†, HE HUANAN, WU ZHIMIN

State Key Laboratory of Coastal and Offshore Engineering, Dalian University of Technology, Dalian 116024, PR China

A new method based on the fictitious crack model to calculate K_R-curves of the concrete beam reinforced with CFRP is proposed in this paper. In this method, the resistance varied with crack propagation can be divided into three parts: the crack initial toughness that is the inherent toughness of the concrete material, the contribution from the cohesive force along the fictitious crack zone during the complete processes of fracture, and the crack resistance providing by CFRP. Considering the contribution of CFRP as a closing force, the analytic solution of the K_R-curves of the reinforced concrete was computed according to the *P-CMOD* curves measured from tests. Afterwards, the method was used to calculate the K_R-curves of the concrete beams reinforced with CFRP with heights of 200mm and the initial relative crack lengths of 0.2、0.3、0.4 respectively. The results indicated that the fracture process zone length also increased from zero to the maximum, and then decreases, which showed the same behaviors as that of the plain concrete.

Keywords: CFRP composites; K_R-curves; cracked concrete beam; fracture characteristics.

1. Introduction

Carbon fiber reinforced polymer (CFRP) has been widely used in rehabilitation and strengthening of concrete structures. It is crucially important to investigate the fracture characteristics and ductility of CFRP-strengthened cracked concrete beams. The analytic solution of the K_R curves considering the cohesive force that acts on the fracture process zone (FPZ) was approached by Reinhardt and Xu et al.[1] Hu and Wittmann proposed a new approach to calculate the K_R-curve by introducing a parameter to consider the residual deformation.[2] Wu studied the fracture resistance of a cracked concrete beam post-strengthened with FRP sheets.[3]

This paper presents a theoretical method to calculate K_R curves of the concrete beam reinforced with CFRP. With different initial crack lengths, a series of three-point bending tests were carried out to examine the fracture characteristics

† Corresponding author, email: dlutdw@yahoo.com.cn

of concrete beams externally bonded with CFRP sheets. According to the *P-CMOD* curves measured from tests, the K_R-curves were calculated and the contribution from variety of FPZ length during crack propagation was discussed.

2. Evaluation of Crack Extension Resistance Curves (K_R-curves)

For a CFRP post-strengthened concrete beam shown in Figure 1, the tensile force due to extension of the CFRP sheets is considered as a closing force to resist the crack propagation.

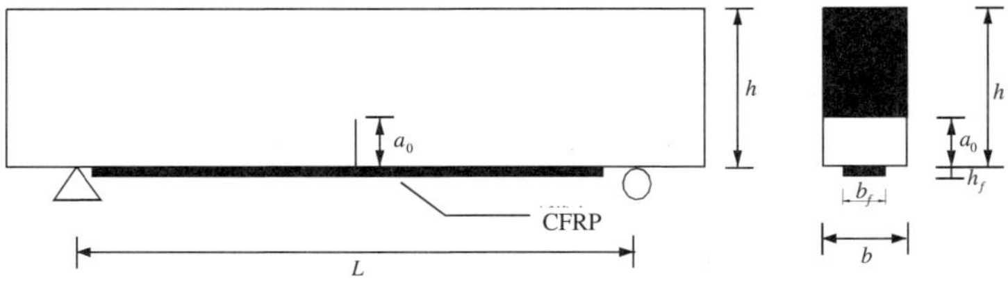

Fig. 1. CFRP strengthened three-point bending notched concrete beam.

Considering that the crack extension resistance is the sum of three parts: the crack initial toughness K_{IC}^{ini}, the contribution from the cohesive force along the fictitious crack zone $K_c(a)$, and the crack resistance providing by CFRP, $K_f(a)$. The expression that is

$$K_R(a) = K_{IC}^{ini} + K_f(a) + K_c(a) \tag{1}$$

According to the double-*K* theory,[4] K_{IC}^{ini} can be obtained by

$$K_{IC}^{ini} = K_{IC}^{un} - K_{IC}^{c} \tag{2}$$

where K_{IC}^{un} is the unstable fracture toughness, K_{IC}^{c} is the stress intensity factor caused by a cohesive force σ(x) in the critical state(a=a_c). Therefore, Equation (1) can be written as following

$$K_{IC}^{ini} = K_{IC}^{un} - K_{IC}^{c} - K_{IC}^{f} \tag{3}$$

in which

$$K_{IC}^{un} = \frac{3P_{1\max} L\sqrt{a_c}}{2h^2 b} F_1\left(\frac{a_c}{h}\right) \tag{4}$$

For the three-point bending notched beam with L/h=4,

$$F_1\left(\frac{a_c}{h}\right) = \frac{1.99 - \frac{a_c}{h}\left(1-\frac{a_c}{h}\right)\left[2.15 - 3.93\frac{a_c}{h} + 2.7\left(\frac{a_c}{h}\right)^2\right]}{\left(1+\frac{2a_c}{h}\right)\left(1-\frac{a_c}{h}\right)^{3/2}} \tag{5}$$

where, $P_{1\max}$ is the first peak loading in *P-CMOD* curve, a_c is the critical crack length and can be evaluated by inserting the measured $P_{1\max}$ and the critical crack mouth opening displacement $CMOD_c$ into the following equation:

$$a_c = \frac{2}{\pi}(h+H_0)arctg\sqrt{\frac{E_c b}{32.6P_{1\max}}CMOD_c - 0.1135} - H_0 \tag{6}$$

where H_0 is the thickness of the knife edge of a clip gauge, E_c is the modulus of elasticity in concrete. If the tensile force provided by CFRP is regarded as a closing force, the contribution of the tensile CFRP sheet is given by

$$K_f(a) = \frac{2\sigma_f h_f b_f}{\sqrt{\pi a} \cdot b} F\left(0, \frac{a}{h}\right) \tag{7}$$

$$F\left(0, \frac{a}{h}\right) = \frac{3.52}{(1-a/h)^{3/2}} - \frac{4.35}{(1-a/h)^{1/2}} + 2.13(1-a/h) \tag{8}$$

The effect crack length can be calculated with the measured *CMOD* and corresponding load *P* during the crack propagation.

$$a = \frac{2}{\pi}(h+H_0)arctg\sqrt{\frac{E_c b}{32.6P}CMOD - 0.1135} - H_0 \tag{9}$$

Assuming that plane sections remain plane after deforming, a linear distribution of strain over the beam depth is obtained.

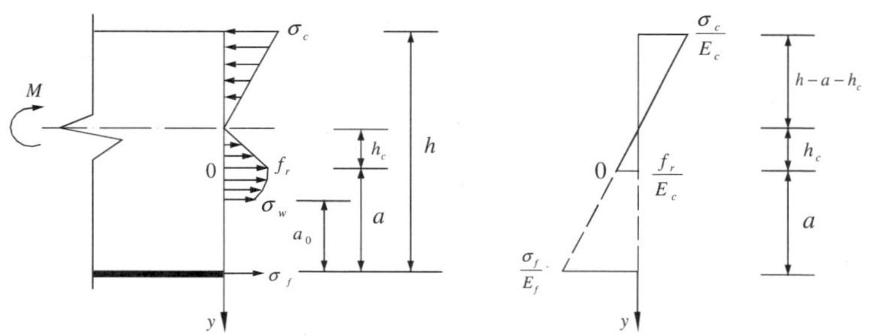

(a) distributions of stresses of mid-span (b) distributions of stresses of mid-span

Fig. 2. Stress and strain across beam section at the cracked plane.

The maximum compressive stress σ_c and CFRP sheet stress σ_f can be given

$$\sigma_f = \frac{E_f}{E_c} \cdot \frac{h_c + a}{h_c} \cdot f_r \tag{10}$$

$$\sigma_c = \frac{f_r(h-a-h_c)}{h_c} \tag{11}$$

The expression to calculate the distribution of cohesive force, σ_w, as follows:

$$\sigma_w = f_t[(1+\frac{c_1^3}{w_0^3}w^3)e^{-\frac{c_2}{w_0}w} - \frac{(1+c_1^3)e^{-c_2}}{w_0}w] \tag{12}$$

Where c_1 and c_2 are the material constants, f_t is the uniaxial tensile strength of concrete, w_0 is the maximum crack opening width when the stress becomes zero. According to the equilibrium in the section(Figure 2(a)), the relationship can be expressed by

$$\frac{1}{2}\sigma_c b(h-a-h_c)=\frac{1}{2}f_r bh_c+\int_0^{a-a_0}\sigma_w bdx+\sigma_f \tag{13}$$

Equation (13) can be written as follows:

$$h_c=\frac{(h-a)^2-\frac{2b_f h_f}{b}\cdot\frac{E_f}{E_c}\cdot a}{2(h-a)+\frac{2b_f h_f}{b}\frac{E_f}{E_c}+(a-a_0)[\frac{2}{\beta w_t}(1+\frac{6\alpha}{\beta^3})-\gamma w_t-\frac{2}{\beta^4 w_t}(\alpha\beta^3 w_t^3+3\alpha\beta^2 w_t^2+6\alpha\beta w_t+6\alpha+\beta^3)e^{-\beta w_t}]} \tag{14}$$

in which, $\alpha=\frac{c_1^3}{w_0^3}$, $\beta=\frac{c_2}{w_0}$, $\gamma=\frac{(1+c_1^3)e^{-c_2}}{w_0}$, $w_t=\frac{a-a_0}{a}CMOD$

Substituting Equation (14) into (10), and then into (7), the stress intensity factor $K_f(a)$ can be determined. The stress intensity factor $K_c(a)$ caused by cohesive force $\sigma(x)$ on the FPZ can be calculated by the following expression:

$$K_c(a)=\int_0^{a-a_0}\frac{2\sigma(x)}{\sqrt{\pi a}}F\left(\frac{x}{a},\frac{a}{h}\right)dx \tag{15}$$

$$F\left(\frac{x}{a},\frac{a}{h}\right)=\frac{3.52\left(1-\frac{x}{a}\right)}{\left(1-\frac{a}{h}\right)^{3/2}}-\frac{4.35-5.28\frac{x}{a}}{\left(1-\frac{a}{h}\right)^{1/2}}+\left[\frac{1.30-0.3\left(\frac{x}{a}\right)^{3/2}}{\sqrt{1-\left(\frac{x}{a}\right)^2}}+0.83-1.76\frac{x}{a}\right]\cdot\left[1-\frac{a}{h}\left(1-\frac{x}{a}\right)\right] \tag{16}$$

For determining the distribution of $\sigma(x)$ in Equation (15), one has to know the distribution of *CMOD* along FPZ and the relation of *CMOD* versus the effect crack length *a*.

If *CMOD* is measured, $w(x)$ can be obtained using Equation (17) and the effect crack length *a* can be calculated using Equation (18):

$$w(x)=CMOD\left\{\left(1-\frac{x}{a}\right)^2+\left(1.081-1.149\frac{a}{h}\right)\left[\frac{x}{a}-\left(\frac{x}{a}\right)^2\right]\right\}^{1/2} \tag{17}$$

$$CMOD=\frac{6PLa}{h^2bE_c}[0.76-2.28\frac{a+H_0}{h+H_0}+3.87(\frac{a+H_0}{h+H_0})^2-2.04(\frac{a+H_0}{h+H_0})^3+0.66(\frac{h+H_0}{h-a})^2] \tag{18}$$

Substituting Equation (17) into (12), and then into (15), the stress intensity factor $K_c(a)$ can be calculated.

3. Calculating Results of K_R-curves

In this section the proposed method is used to calculate the K_R-curves of the concrete beams reinforced with CFRP, whose height is 200mm and a_0/h are 0.2、0.3、0.4 respectively. The material properties of the plain concrete are as follows: f_r=3.23MPa, f_c=50.28MPa, E_c=29.45GPa. The modulus of elasticity of the CFPR is 240GPa, the maximum tensile strain of CFPR is 0.00171 and the cross section area of CFRP A_f=$b_f h_f$=8.53mm^2.

Table 1. Geometry of the three-point bending beams.

Nos. of specimens	Width b mm	Depth H mm	Span L mm	Initial notch a_0 mm	Unstrengthened specimen	CFRP-strengthened specimen	
					P_{max} kN	P_{1max} kN	P_{2max} kN
2021	150	199.5	600	39	10.03	11.17	11.46
2031	150	198	600	58.5	6.65	9.23	11.47
2041	148	199.5	592	79	5.24	8.04	11.77

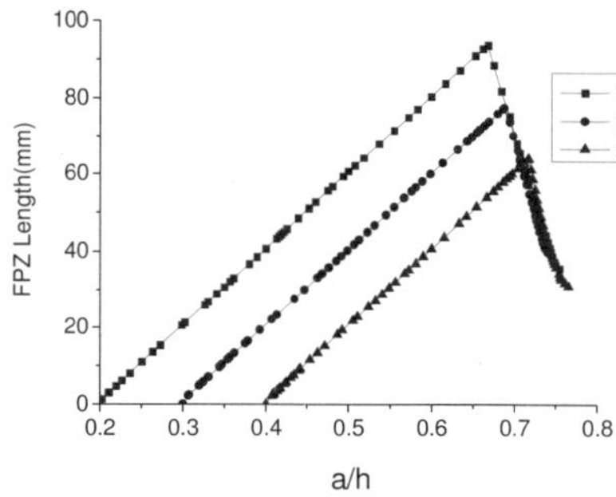

Fig. 3. FPZ lengths with different a_0/h.

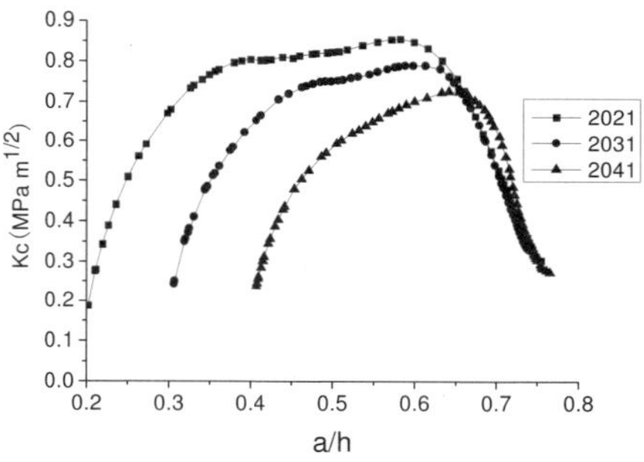

Fig. 4. K_c curves with different a_0/h.

Figure 3 shows the FPZ variety under different a_0/D during the whole crack propagation. Taking CFRP-strengthened No.2021 specimen as an example, when a/h<0.4, the K_c curve increases rapidly. When 0.4< a/h <0.6, the increase of K_c curve is very slow. When a/h >0.6, resistance decreases sharply. Figure 4 shows the K_c curves caused by cohesive force, it is demonstrated that the variety of FPZ length has a obvious effect on the results.

4. Conclusion

Considering the contribution of CFRP as a closing force, a method is proposed to predict the K_R curves of the concrete beam reinforced with CFRP. With *P-CMOD* curves measured from tests, the K_R-curves of the concrete beams are calculated. The study has shown that the FPZ length varies from zero to the maximum, and then decrease. This variety has significant effect to the K_c curves. Furthermore, The contribution of CFRP is more notable than that of the plain concrete after crack extension. The effect of a_0/h is not remarkable for K_f curves.

References

1. H. W. Reinhardt and S. L. Xu: *Eng. Fract. Mech.*, 1999, **64**, 563–587.
2. X. Z. Hu and F. H. Wittmann: *J. Mat. in Civ. Eng.*, 1990, **2**, 15–23.

EFFECT OF LOADING RATE ON REINFORCED CONCRETE MEMBER

MIN LI †

Faculty of infrastructure Engineering, DaLian university of Technology
Dalian, Liaoning, China

HONGNAN LI

Faculty of infrastructure Engineering, DaLian university of Technology
Dalian, Liaoning, China

In order to research the strain rate effect on reinforced concrete members during earthquake loading, based on Timoshenko beam theory and fiber model theory, considering strain rate effect of materials, a new beam element of reinforced concrete member is presented. Dynamic properties of reinforced concrete members subjected to monotonic loading at different loading rates that might be experienced during earthquakes are simulated by the nonlinear finite element program. The influences of loading rate on loading capability and stiffness of reinforced concrete member and the distribution of strain rate at high loading rate are investigated. The loading capability increases with increasing loading rate, the stiffness is not significant influenced by loading rate, the distribution of strain rate changes with time. The simulation results agree well with the test results.

Keywords: Reinforced concrete member; Strain rate; Shear deformation; Timoshenko beam theory; Fiber model.

1. Introduction

Reinforced concrete structure possibly suffers severe earthquake ground motions in its active service, the strain rate at critical sections may be as high as $10^{-1}s^{-1}$. Mechanical and deformative behaviors of structural material at higher strain rate will be different from which at static loading. A large number of research efforts have been devoted to the effect of strain rate on material properties. Properties of concrete under dynamic loading were studied by Bischoff et al,[1] Malvar et al.[2] Strain rate effect on reinforcing steel was studied by M. Li and H. N. Li.[3]

Compared to study on strain rate effect of material, much less available information is available on rate sensitivity of structural members subjected to

† Min Li, email: cnlm225@126.com

earthquake-type loading. Some of the available studies were reported by Bertero et al,[4] Kulkarni S M and Shah S P.[5]

In order to simulate the strain rate effect on reinforced concrete members during earthquake loading, based on Timoshenko beam theory and fiber model theory, considering strain rate effect of materials, a new beam element of reinforced concrete member is developed. Dynamic properties of reinforced concrete members subjected to monotonic loading at different loading rates that might be experienced during earthquakes are simulated by the nonlinear finite element program.

2. Dynamic Constitutive Behavior

2.1. *Rate-dependent compressive stress-strain curves of concrete,*[6]

$$f_{cd} = \begin{cases} k_1 k_2 f_c \left[\dfrac{2\varepsilon}{k_1 \varepsilon_0} - \left(\dfrac{\varepsilon}{k_1 \varepsilon_0} \right)^2 \right] & \varepsilon \le k_1 \varepsilon_0 \\ k_1 k_2 f_c \left[1 - z(\varepsilon - k_1 \varepsilon_0) \right] \ge 0.2 k_1 k_2 f_c & \varepsilon > k_1 \varepsilon_0 \end{cases} \tag{1}$$

$$k_1 = 1 + \frac{\rho_s f_{yh}}{f_c} \tag{2}$$

$$k_2 = \frac{f_{cd}}{f_c} = 0.022 \ln \frac{\varepsilon}{\varepsilon_0} + 0.9973 \tag{3}$$

where, f_{cd} and f_c = dynamic and quasi-static concrete compressive strength, respectively; ε_0 =quasi-static peak strain; ρ_s =volumetric ratio of the hoop reinforcement to concrete core; z =a parameter; f_{yh} = quasi-static yield strength of transverse reinforcement; ε and ε_0 = strain rate and quasi-static strain rate, respectively ,and $\varepsilon_0 = 3.0\times10^{-5}$ s^{-1} in this paper.

2.2. *Rate-dependent tensile stress-strain curve of concrete*

$$\begin{cases} f_{td} = E\varepsilon & \varepsilon < \varepsilon_{tfd} \\ f_{td} = 0 & \varepsilon \ge \varepsilon_{tfd} \end{cases} \tag{4}$$

$$\frac{f_{td}}{f_t} = 0.0714 \ln \frac{\varepsilon}{\varepsilon_0} + 0.9883 \tag{5}$$

Where, f_{td} and f_t = dynamic tensile strength and quasi-static tensile strength; ε_{tfd} = dynamic tensile peak strain ; E = elastic module.

2.3. *Rate-dependent stress-strain curve of steel*

Ideal elastic-plastic model is used, and the relation of dynamic yield strength land quasi-static yield strength is

$$\frac{f_{sd}}{f_s} = 0.0124 \ln \frac{\varepsilon}{\varepsilon_0} + 0.9832 \tag{6}$$

3. Fiber beam element of RC member

3.1. *Basic assumptions of fiber beam element*

1. Plane sections remaining plane
2. No slippage between reinforcing steel and concrete
3. Constitutive model of fiber is 1D constitutive function of material
4. The torsion deformation and geometry nonlinear are not considered

3.2. *Timoshenko beam theory*

Timoshenko beam theory assumes that plane sections remaining plane, while it isn't necessary that the plane is perpendicular to the axis of beam. This theory considers shear deformation.

4. Finite Element Method

The fiber element, including shear deformation and strain rate effect of concrete and steel, is added into Feappv which is a finite element analysis program.

5. Example

5.1. *Outline of the experiment*

A serial of reinforced concrete beams were tested under static and dynamic loading by Kulkarni and shah in 1998,[5] One beam in each pair was tested at a piston velocity of 0.00071cm s^{-1} and the other at a piston velocity of 38cm s^{-1}, the test was controlled by displacement. A pairs of beams are simulated in this paper. A sketch of the dimension of the beam section is shown in Fig.1, The parameters of the beam are such; concrete compressive strength is 46.2 MPa, steel yield strength is 518 MPa, distance between support is 1672 mm, ratio of shear span to effective depth of section is 5.0.

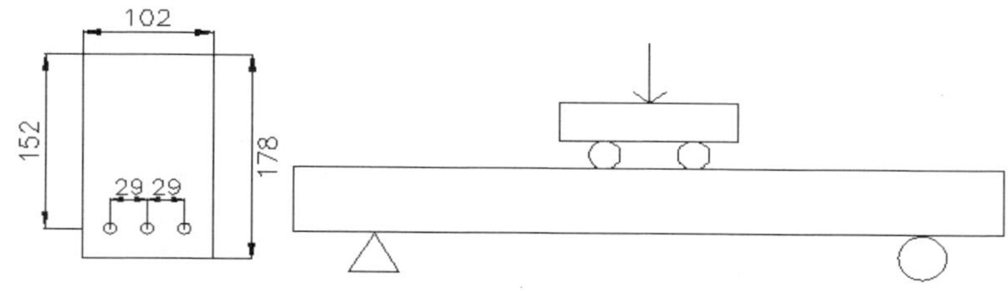

Fig. 1. Sketch of beams in tests (unit: mm).

5.2. *Simulation*

The test is simulated using the fiber beam element model of reinforced concrete which can consider the loading rate in this paper. The load-deflection curves of the beam at different loading rate are shown in Fig.2. As shown in this figure, the loading capability of the beam increases with increasing loading rate, the stiffness is not significant influenced by loading rate, the simulation curves agree well with the experimental curves in prepeak portions of load-deflection curve.

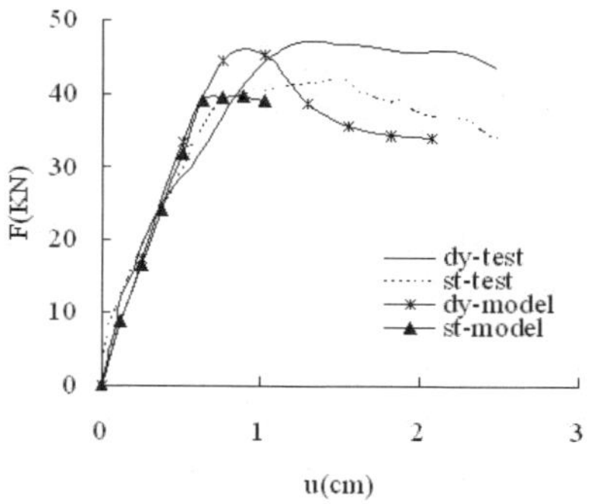

Fig. 2. Load-deflection curve at different loading rate.

5.3. *Strain rate distribution law*

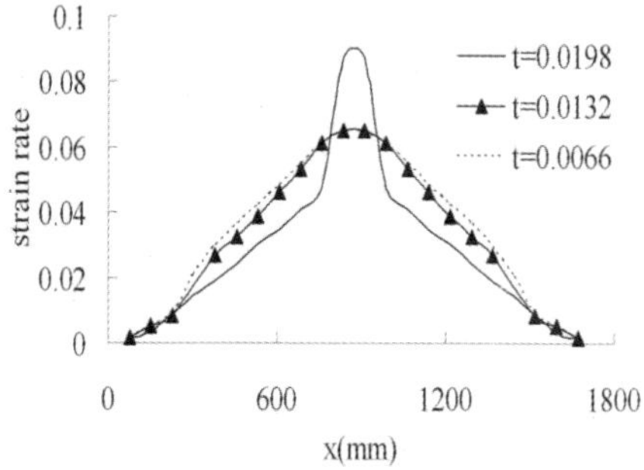

Fig. 3. Distribution of strain rate of extreme edge in compression zone along axis.

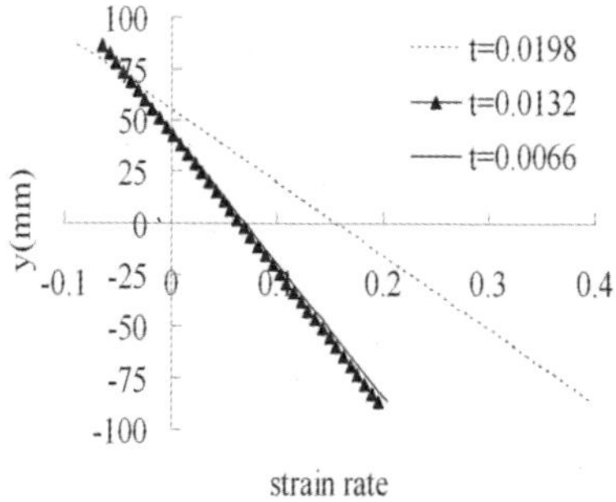

Fig. 4. Distribution of strain rate along section.

Distribution of strain rate of the beam is investigated at dynamic loading. Fig.3 shows the distribution of strain rate (unit is s^{-1}) of extreme edge in compression zone along axis at different time. It appears that strain rate is bigger in mid-span section than in support section, and distribution of strain rate along axis is similar at 0.0132s and at 0.0066s, strain rate becomes bigger suddenly in mid-span section at 0.0198s. It is because that displacement increment in mid-span is bigger than it in other position of the beam in the same time interval, and the beam yields at 0.0198s.

The distribution of strain rate (unit is s^{-1}) in mid-span along height of section at different time is shown in Fig.4, upper zone is in compression and lower zone is in tension. It shows that distribution of strain rate is in agreement with the plane-section assumption. The compression zone becomes smaller with load increasing, and the tension zone is just opposite.

6. Summary

Based on Timoshenko beam theory and fiber model theory, considering strain rate effect of materials, a new fiber element of reinforced concrete member is developed. Dynamic properties of reinforced concrete member subjected to monotonic loading at different loading rates that might be experienced during earthquakes are simulated by the nonlinear finite element program. The loading capability increases with increasing loading rate, the stiffness is not significant influenced by loading rate, the distribution of strain rate changes with time. The simulation results agree well with the test results and the new fiber element is effective.

References

1. B. H. Bischoff and S. H. Perry: *Material and structure.*,1991, 24, 425-450
2. L. Malvar, Javier and C. A. Ross: *ACI material journal.*,1998, 96, (6), 735-739.
3. M. Li and H. N. Li. *China Civil Engineering Journal.*, 2010,43,(4), 70-75 (in Chinese)
4. V. V .Bertero, D. Rea and S .Mahin: *Proceedings of fifth word conference on earthquake engineering* , Rome, Italy,1973,2, 1461- 1470.
5. S. M. Kulkarni and S. P Shah. *ACI structural Journal* .,1998, 95,705-715.
6. P. Soroushian, k. B. Choi and A. Alhamad: *ACI journal*, 1986, 83, 251-259.

SEISMIC RESPONSE OF CONCRETE STRUCTURE REINFORCED WITH SUPERELASTIC SHAPE MEMORY ALLOY

DI CUI[†]

Key Laboratory for Prediction & Control on Complicated Structure System of the Education Department of Liaoning Province, Dalian University, Dalian 116622, China

HONG-NAN LI

Faculty of infrastructure Engineering, DaLian university of Technology Dalian, 116024, China

PING GUAN

Key Laboratory for Prediction & Control on Complicated Structure System of the Education Department of Liaoning Province, Dalian University, Dalian 116622, China

Shape memory alloys (SMAs) have been used as an innovative material for seismic retrofits of buildings and bridges. This paper focus on the seismic response and the self-recentering capability of concrete structures reinforced with superelastic SMAs. An eight-story SMA reinforced concrete (RC) structure is established by replacing the rebar of RC structures with superelastic SMAs. And the structural response of SMA RC structure under different earthquake records is analyzed by using the finite element analysis program ANSYS. The results show that the displacement response of SMA RC structure is less than that ordinary RC structure. Moreover, the residual displacement of SMA RC structure is less than that of RC structure, which demonstrates that SMA RC structure has good self-recentering capabilities.

Keywords: Shape memory alloy; Superelasticity; Concrete structure; Seismic response.

1. Introduction

Superelastic shape memory alloy (SMA) can undergo large amount of inelastic deformations under loading and recovery its original shape after unloading. In addition, SMA exhibits high energy dissipating capability, outstanding low-and high-cycle fatigue and good corrosion resistance. Therefore, superelastic SMA has drawn much attention in civil engineering in recent years. The utilization of SMA to improve earthquake resistance and vibration control of buildings [1-3] and

[†] Di Cui, email: cuidip@163.com

bridges[4-6] has become an important research topic. Furthermore, SMA devices are also used in rehabilitation of historical structures [7]. Since SMA is very expensive, most researches focus on using SMA devices, only very limited research work has been carried out on the use of SMA as reinforcements of structures. Alam et al. [8] investigate the dynamic performance of an eight-story reinforced concrete (RC) frame building reinforced with steel and superelastic SMA in its beam-column joints. The results show that SMA RC frames are able to recover most of its post-yield deformation, even after a strong earthquake.

In this paper, a self-damping and self-recentering structural system was built by replacing the rebar of the columns in RC structure with superelastic SMAs. Then the seismic responses of the SMA RC structures are investigated by using finite element analysis program ANSYS. According to the calculation results, the capability of vibration control and self-recentering of SMA RC structure is discussed.

2. Model of Structures

To investigate the capability of self-damping and self-recentering of SMA RC structure, an eight-story structure is designed in this study. The story height is 1m and the span of frame is 1.5m. For comparison purpose, an ordinary RC structure is also designed as a reference, the dimension and the reinforcement ratio of which are the same with those of SMA RC structure. For ordinary RC structure, the reinforcements in beams and columns are all steel wires; for SMA RC structure, the reinforcements in beams are still steel wires, while the reinforcements in columns are superelastic SMA wires. The mechanical properties of SMA wires and steel wires can be found in reference 9. The dimension of cross section of beams is 120mm×60mm. The details of cross section of columns are shown in Fig.1.

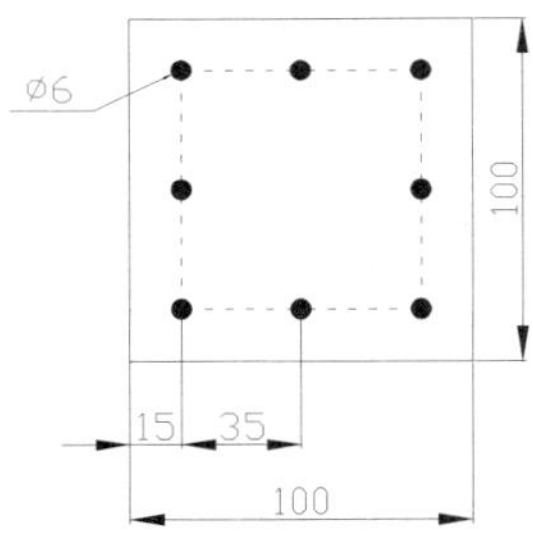

Fig.1. Details of cross section of column (Unit: mm).

In this study, the members (beam and column) of structures are assumed to be made of a homogenous material whose stress-strain diagram is coincidence with that of steel or SMA. The stiffness of the equivalent model can be calculated by the equation

$$EI = E_r I_r + E_c I_c \tag{1}$$

where E, E_r, and E_c are the elastic modulus of the assumed homogenous material, the reinforcements and concrete, respectively. Then the yield stress of the material in the equivalent model can be obtained. Consequently, the constitutive relationship can be determined. The seismic responses of structures are analyzed by using the finite element analysis program ANSYS. The SOLID185 of ANSYS is used to model the beams and columns of the frame structure. The MASS21 of ANSYS is used to model the weight of the structure.

3. Seismic Responses Analysis in Time Domain

The seismic responses of SMA RC structure and RC structure are analyzed in the time domain. The Qian'an, El centro and Ninghe earthquake records are selected as the excitations, which represent different site conditions. The peak accelerations of earthquake records are scaled to 0.4g ($g=9.8m\ s^{-2}$).

The displacement time history curves and the acceleration time history curves of top story under Qian'an earthquake record are shown in Fig.2 and Fig.3. From the figures, it can be seen that the dynamic response of top story under Qian'an earthquake record are reduced in the whole time history under the influence of SMA wires.

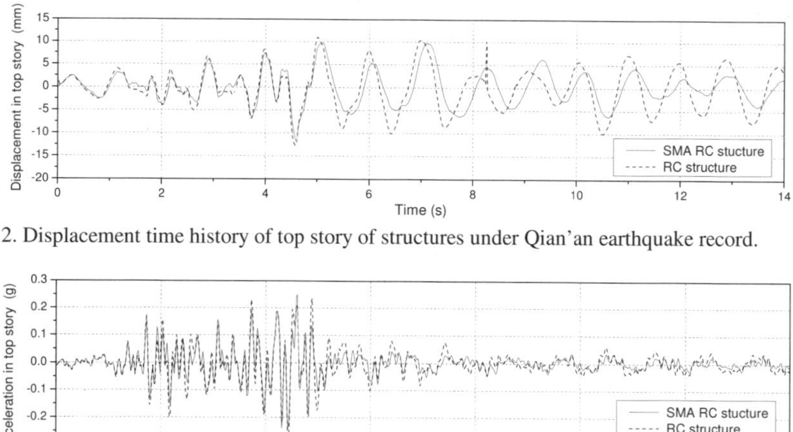

Fig.2. Displacement time history of top story of structures under Qian'an earthquake record.

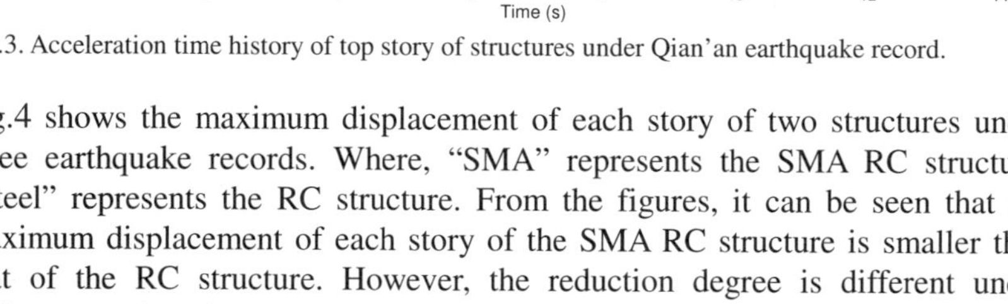

Fig.3. Acceleration time history of top story of structures under Qian'an earthquake record.

Fig.4 shows the maximum displacement of each story of two structures under three earthquake records. Where, "SMA" represents the SMA RC structure, "Steel" represents the RC structure. From the figures, it can be seen that the maximum displacement of each story of the SMA RC structure is smaller than that of the RC structure. However, the reduction degree is different under different earthquake records. The maximum acceleration of each story of two structures under three earthquake records is shown in Fig. 5. Obviously, the maximum accelerations in some stories of the SMA RC structure are larger than those of the RC structure under El centro and Ninghe earthquake records. It is demonstrates that site conditions can influence the seismic response of structures.

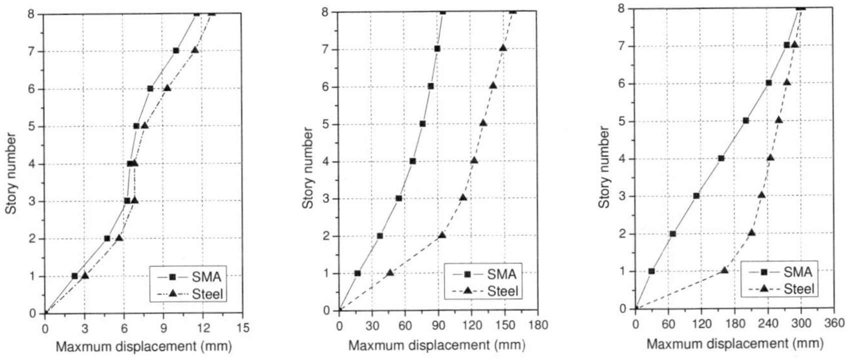

(a) Qian'an (b) El Centro (c) Ninghe

Fig.4. Maximum displacement of each story of structures under three earthquake records.

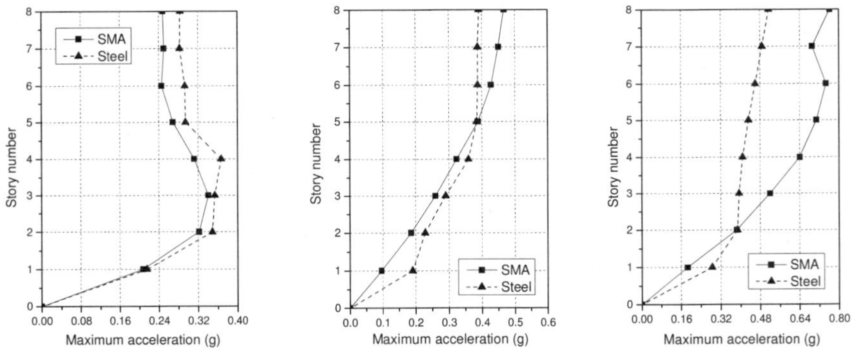

(a) Qian'an (b) El Centro (c) Ninghe

Fig.5. Maximum acceleration of each story of structures under three earthquake records.

4. Self-Recentering Capability of Structures

Table 1 lists the residual displacement of structures of the 1st, the 5th and the 8th stories after earthquake. It is obvious that the residual displacements of the SMA RC structure under different earthquake are all less than those of the RC structure, which reveals the excellent self-recentering capability of the SMA RC structure compared to the RC structure.

Table 1. Residual displacement of structures.

Story number	Residual displacement (mm)					
	Qian'an		El Centro		Ninghe	
	SMA	Steel	SMA	Steel	SMA	Steel
1	0.03	0.36	0.01	41.93	0.34	115.33
5	0.16	0.24	0.07	110.37	1.26	166.49
8	0.22	0.17	0.09	130.73	1.52	189.22

Note: "SMA" represents the SMA RC structure; "Steel" represents the RC structure.

5. Conclusion

In this paper, the seismic responses and the self-recentering capability of SMA RC structure are analyzed. The following conclusions can be drawn: (1) SMA RC structure has less horizontal displacement responses; (2) The acceleration response of SMA RC structure is larger than that of RC structure in some cases, which is due to the increasing of the SMA's elastic modulus when SMA enters into stress-induced martensite stage; (3) The residual displacement of SMA RC structure is less than that of RC structure, which demonstrates that SMA RC structure has good self-recentering capabilities.

Acknowledgments

This work has been supported by the NSFC (Grant No. 50328807, Grant No. 51008040 and Grant No. 50908027).

References

1. M. Dolce, D. Cardone and R. Marnetto: Proceedings of SPIE, 2001, 4330: 238-249.
2. E. Effendy, W. I. Liao, G. Song, Y. L. Mo and C. H. Loh.: ASCE Conference Proceedings, Earth and Space, March, 2006, Houston, U.S.A.
3. E. Choi, K. T. Yang, Tae G H, *et al.* : ESOMAT 2009, 07005.
4. R. DesRoches, M. Delemont: Engineering Structures, 2002, **24**: 325-332.
5. Wilde K, Gardoni P, Fujino Y. Engineering Structures, 2000, **22**: 222-229.
6. E. Choi, T. H. Nam, and B. S. Cho, Canadian Journal of Civil Engineering, 2005, **32**: 957-967.
7. M. Indirli, M. G. Castellano, P. Clemente, *et al.*: Proceedings of SPIE, 2001, 4330: 262-272.
8. M. S. Alam, M. Nehdi, M. A.Youssef: Smart Structures and Systems, 2009, **5**, (5): 565-585.
9. H. N. Li, D. Cui and G. B. Song: Earth and Space, March, 2006, Houston, U.S.A.

ANALYSIS OF THREE-DIMENSIONAL MICROSCOPIC STRESS DISTRIBUTIONS AT A FREE EDGE OF A UNIDIRECTIONAL CFRP LAMINATE

KEITA GOTO and TETSUYA MATSUDA†

Department of Engineering Mechanics and Energy, University of Tsukuba
1-1-1 Tennodai, Tsukuba, 305-8573, Japan

In this study, the distributions of microscopic stresses at a free edge of a unidirectional carbon fibre-reinforced plastic laminate (CFRP laminate) are analysed three-dimensionally, based on a homogenisation theory. For this, the mathematical homogenisation theory for elastic materials is reconstructed for unidirectional CFRP laminates with free edges using a traction-free boundary condition. The present theory is then applied to the analysis of microscopic stress distributions at a free edge of a unidirectional carbon fibre/epoxy laminate subjected to in-plane uniaxial tension. It is shown that microscopic shear stress noticeably occurs at the free edge, especially at the interface region between a carbon fibre and its surrounding epoxy matrix.

Keywords: CFRP laminate; Free Edge; Microscopic Stress; Homogenisation.

1. Introduction

Carbon fibre-reinforced plastic laminates (CFRP laminates) are now regarded as some of the most important engineering materials, because of their excellent specific strength and specific stiffness. These properties contribute to the weight reduction of high-end engineering products such as aircraft, spacecraft and automobiles. This leads to a significant improvement in energy efficiency, resulting in the reduction of environmental load.

In general, CFRP laminates have extremely heterogeneous microscopic structures compared to conventional metallic materials. This can cause complex distribution of microscopic stress and, in some cases, high stress concentration in laminates. In particular, it is of great importance to examine the stress distribution at the free edges of laminates, where high shear stress concentration is apt to occur. Such stress concentration generally takes place at the lamina/lamina interfaces (interlaminar) and the fibre/matrix interfaces around the free edges due to the mismatch of their material constants. This is sometimes called the 'edge-effect', and can bring about microscopic failures of the

† Corresponding author, email: matsuda@kz.tsukuba.ac.jp

laminates, which may result in the macroscopic fracture of the laminates themselves. It is therefore necessary to know the microscopic stress distribution at the free edges of CFRP laminates.

In this study, the homogenisation theory for elastic materials[1] is reconstructed for unidirectional CFRP laminates, which are the most basic CFRP laminates, with free edges using a traction free boundary condition. Using the proposed method, the microscopic stress distribution and stress concentration at the edge of a unidirectional carbon fibre/epoxy laminate are examined.

2. Theory

2.1. *Homogenisation theory for unidirectional CFRP laminates with free edges*

Consider a unidirectional CFRP laminate subjected to a macroscopic uniaxial load in the $y_2 - y_3$ plane (Fig. 1). This laminate is reinforced in the y_1-direction, and has a finite length in the direction, i.e. it has the free edges as illustrated in Fig. 1. On the other hand, the laminate is assumed to be infinite in the y_2- and y_3-directions. The carbon fibres are arranged squarely in the $y_2 - y_3$ plane for simplicity.

For the above-mentioned laminate, a unit cell Y has been defined as indicated in Fig. 1. Describing the microscopic distributions of stress in Y as $\sigma_{ij}(\boldsymbol{y})$, and an arbitrary variation of the perturbed displacement field defined in Y as $\delta u_i^{\#}$, the equilibrium of σ_{ij} can be expressed as

$$\int_Y \sigma_{ij}\delta u_{i,j}^{\#} dY - \int_\Gamma \sigma_{ij} n_j \delta u_i^{\#} d\Gamma = 0, \tag{1}$$

where Γ denotes the boundary of Y, n_i indicates the unit vector outward normal to Γ, and $(\ )_{,j}$ signifies the differentiation with respect to y_j. In the above equation, the second term of the left-hand side, i.e. the boundary integral term, vanishes because of the following reasons: First, on the top and bottom, and front and rear boundary facets of Y, σ_{ij} and $\delta u_i^{\#}$ distribute Y-periodically, while n_i takes opposite directions on the opposite boundary facets. Thus, the boundary integral term in Eq. (1) becomes zero on these boundary facets. This is the same situation as in the usual homogenisation theory which deals with infinite periodic materials. In contrast, on the left and right boundary facets, the above situation holds as it is with respect to the y_2- and y_3-directions, whereas it does not hold in the y_1-direction because the Y-periodicity in this direction no longer exists. However, regarding the y_1-direction, σ_{ij} is zero because the left and right boundary facets are the free edges. The boundary integral term on these facets, therefore, also becomes zero. Consequently, the second term of the left-hand side in Eq. (1) vanishes, resulting in

$$\int_Y \sigma_{ij} \delta u^{\#}_{i,j} dY = 0 . \tag{2}$$

This resulting equation has the same form as that obtained from the conventional homogenisation theory.[1] Thus, the field of microscopic stress σ_{ij} in Y and the relation between macroscopic stress Σ_{ij} and macroscopic strain E_{ij} of the laminate are derived in the same forms:[1]

$$\sigma_{ij} = c_{ijpq}(\delta_{pk}\delta_{ql} + \chi^{kl}_{p,q})E_{kl} , \tag{3}$$

$$\Sigma_{ij} = \left\langle c_{ijpq}(\delta_{pk}\delta_{ql} + \chi^{kl}_{p,q}) \right\rangle E_{kl} , \tag{4}$$

where c_{ijkl} denotes the elastic stiffness of the fibres or matrix, δ_{ij} indicates Kronecker's delta, and $\langle\ \rangle$ designates the volume average in Y defined as $\langle \# \rangle = |Y|^{-1} \int_Y \# dY$, in which $|Y|$ signifies the volume of Y. Moreover, χ^{kl}_i in Eqs. (3) and (4) denotes the characteristic function determined by solving the following boundary value problem:

$$\int_Y c_{ijpq} \chi^{kl}_{p,q} \delta u^{\#}_{i,j} dY = -\int_Y c_{ijkl} \delta u^{\#}_{i,j} dY . \tag{5}$$

2.2. *Semiunit cell and substructure method*

Consider half of the unit cell as illustrated in Fig. 2, which hereafter is referred to as a semiunit cell Y. A close look at Fig. 2 reveals that the internal structure of the laminate has a point-symmetry with respect to the centre of the left boundary facet of Y, C. Consequently, the distribution of χ^{kl}_i also satisfies the point-symmetry with respect to this point. Using the point-symmetry as a

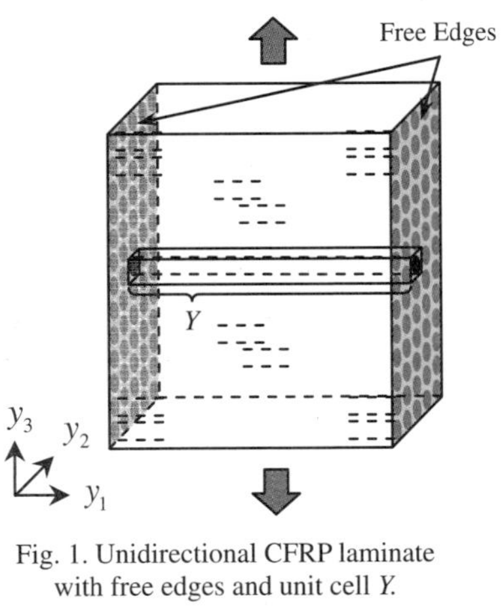

Fig. 1. Unidirectional CFRP laminate with free edges and unit cell Y.

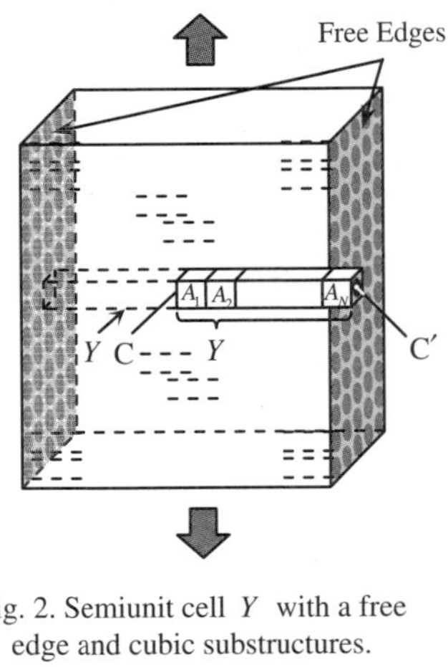

Fig. 2. Semiunit cell Y with a free edge and cubic substructures.

boundary condition on the left boundary facet, Y instead of Y can be employed as the domain of analysis, leading to the following boundary value problem with respect to Y :[2]

$$\int_Y c_{ijpq} \chi_{p,q}^{kl} \delta u_{i,j}^{\#} dY = -\int_Y c_{ijkl} \delta u_{i,j}^{\#} dY \,. \tag{6}$$

Next, the semiunit cell Y mentioned above is divided into cubic substructures[3] A_i $(i = 1, 2, \ldots, N)$ as shown in Fig. 2, and then the boundary value problems for the individual substructures in a finite element discretised form are derived as follows:[4]

$$\boldsymbol{K}_A \boldsymbol{\chi}_{A_i}^{kl} = \boldsymbol{F}_A^{kl} \,, \; (i = 1, 2, \ldots, N) \,, \tag{7}$$

where $\chi_{A_i}^{kl}$ denotes the nodal vector of characteristic function in A_i. Then, the components of $\chi_{A_i}^{kl}$ are divided into two parts, $\chi_{A_i}^{kl(\Omega)}$ and $\chi_{A_i}^{kl(\Gamma)}$, which represent the characteristic functions at the internal and the boundary nodes of A_i , respectively. Then, the boundary value problems for A_i , Eq. (7), are rewritten into the following equation:

$$\begin{bmatrix} \boldsymbol{K}_A^{(\Omega)} & \boldsymbol{K}_A^{(\Omega\Gamma)} \\ \boldsymbol{K}_A^{(\Gamma\Omega)} & \boldsymbol{K}_A^{(\Gamma)} \end{bmatrix} \begin{Bmatrix} \boldsymbol{\chi}_{A_i}^{kl(\Omega)} \\ \boldsymbol{\chi}_{A_i}^{kl(\Gamma)} \end{Bmatrix} = \begin{Bmatrix} \boldsymbol{F}_A^{kl(\Omega)} \\ \boldsymbol{F}_A^{kl(\Gamma)} \end{Bmatrix}, \tag{8}$$

and $\chi_{A_i}^{kl(\Omega)}$ can be expressed as

$$\boldsymbol{\chi}_{A_i}^{kl(\Omega)} = \left(\boldsymbol{K}_A^{(\Omega)}\right)^{-1} \left(\boldsymbol{F}_A^{kl(\Omega)} - \boldsymbol{K}_A^{(\Omega\Gamma)} \boldsymbol{\chi}_{A_i}^{kl(\Gamma)}\right). \tag{9}$$

The elimination of $\chi_{A_i}^{kl(\Omega)}$ from Eq. (8) using the above equation yields

$$\bar{\boldsymbol{K}}_A^{(\Gamma)} \boldsymbol{\chi}_{A_i}^{kl(\Gamma)} = \bar{\boldsymbol{F}}_A^{kl(\Gamma)}, \tag{10}$$

where $\bar{\boldsymbol{K}}_A^{(\Gamma)}$ and $\bar{\boldsymbol{F}}_A^{kl(\Gamma)}$ are expressed as follows:

$$\begin{aligned} \bar{\boldsymbol{K}}_A^{(\Gamma)} &= \boldsymbol{K}_A^{(\Gamma)} - \boldsymbol{K}_A^{(\Gamma\Omega)} \left(\boldsymbol{K}_A^{(\Omega)}\right)^{-1} \boldsymbol{K}_A^{(\Omega\Gamma)}, \\ \bar{\boldsymbol{F}}_A^{kl(\Gamma)} &= \boldsymbol{F}_A^{kl(\Gamma)} - \boldsymbol{K}_A^{(\Gamma\Omega)} \left(\boldsymbol{K}_A^{(\Omega)}\right)^{-1} \boldsymbol{F}_A^{kl(\Omega)}. \end{aligned} \tag{11}$$

Finally, Eqs. (10) are assembled into one equation, which is a boundary value problem with respect to just the boundary nodes of all substructures:[4]

$$\boldsymbol{K}^{(\Gamma)} \boldsymbol{\chi}^{kl(\Gamma)} = \boldsymbol{F}^{kl(\Gamma)}. \tag{12}$$

3. Analysis

Using the present method, microscopic stress distributions at a free edge of a unidirectional carbon fibre/epoxy laminate subjected to 0.5% macroscopic elastic strain in the y_3-direction were analysed. The number of substructures in the semiunit cell Y was set at $N = 30$, which was large enough to avoid interaction of the free edges. Then, each substructure was discretised into eight-node isoparametric elements (4320 elements and 5005 nodes). The volume fraction of fibres (56%) and the material constants were the same as those in the previous studies.[5,6]

Figures 3(a)–(c) show the distributions of microscopic stresses σ_{11}, σ_{12} and σ_{31} in the substructure A_{30}. The deformed shape of A_{30} is also depicted in the figures in which the displacement is magnified 30 times. First, it is seen from the figures that the microscopic stresses exhibit considerably complex distributions in the vicinity of the free edge. Further observation of Fig. 3(b) and (c) reveals that the microscopic shear stresses σ_{12} and σ_{31} significantly occur around the fibre/matrix interface. Especially, the maximum value of σ_{31} is 13.6MPa, which reaches more than 30% of the macroscopic tensile stress (40.9MPa). Such shear stresses, however, almost disappears on the left boundary facet of A_{30}, indicating the local influence of the free edge. On the other hand, it is found from Fig. 3(a) that considerably high compressive stress σ_{11} takes place in the fibre adjacent to the matrix near the free edge, and that this locally-concentrated compressive stress also disappears on the left boundary facet of A_{30}. These results indicate that the free edges in CFRP laminates can strongly affect the microscopic stress distributions in their vicinity.

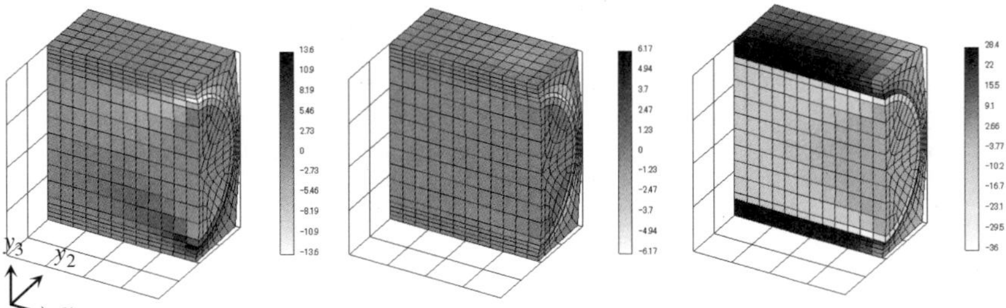

(a) (b) (c)

Fig. 3. Distributions of microscopic stresses; (a) σ_{11}, (b) σ_{12} and (c) σ_{31} in A_{30} at $E_{33} = 0.5\%$.

References

1. E. Sanchez-Palencia: 'Non-Homogeneous Media and Vibration Theory', Lecture Notes in Physics, No.127; 1980, Berlin, Springer-Verlag.
2. N. Ohno, T. Matsuda and X. Wu: *Int. J. Solids Struct.*, 2001, **38**, 2867–2878.
3. O.C. Zienkiewicz and R.L. Taylor: 'The Finite Element Method', 5th edn; 2000, Oxford, Butterworth-Heinemann.
4. T. Matsuda, D. Okumura, N. Ohno and M. Kawai: *Int. J. Solids Struct.*, 2007, **44**, 8274–8284.
5. T. Matsuda, N. Ohno, H. Tanaka and T. Shimizu: *JSME Int. J., Ser. A*, 2002, **45**, 538–544.
6. T. Matsuda, N. Ohno, H. Tanaka and T. Shimizu: *Int. J. Mech. Sci.*, 2003, **45**, 1583–1598.

THE DYNAMIC PROPERTY AND LIQUEFACTION ANALYSIS OF EARTH DAM

DEYONG WANG

School of Naval Architecture, Ocean and Civil Engineering, Shanghai Jiao tong University. 800 Dong chuan Road, Shanghai, P.R. China

XIANQI LUO AND XIURUN GE

School of Naval Architecture, Ocean and Civil Engineering, Shanghai Jiao tong University. 800 Dong chuan Road, Shanghai, P.R. China

Earthquake liquefaction is one of the most important subjects to earth dam. The paper presented the dynamic analysis on the earth dam of Wangqingtuo reservoir; the dynamic characteristics parameters are obtained by field investigation and triaxial test. A cyclic elasto-plastic constitutive model according to liquefied soil which based on Biot's double-phase fluid-saturated porous media dynamic coupling consolidation theory and finite element program LIQCA are used to simulate the seismic liquefaction for the dam foundation soil, and the deformation and liquefaction characteristics for the dam are educed in this paper. Based on the character of earth dam of Wangqingtuo reservoir, the suitable anti-liquefaction measure is presented.

Keywords: Seismic liquefaction; Liquefaction property; Finite element simulation; Anti-liquefaction measures.

1. Introduction

Many earthquakes occurring these years and also made enormous destroy due to soil liquefaction, such as the Great Alaskan earthquake[1] in USA in 1964, the tangshan earthquake in 1976 and the Wenchuan earthquake in 2008 in china. Therefore, the great attention had been paid to the sandy soil liquefaction, especially to earth dam.

Over last decades, the study on the soil liquefaction has focused on the assessment of liquefaction potential[2], evaluation of the deformations and settlement induced by the liquefaction[3], clay content influence on the liquefaction and the liquefaction-induced construction failure, etc. To the seismic response of earth dam, pseudo static method [4] is the main method in the seismic code, despite of safety factors greater than one, caused doubts in utilizing of this method. Nonlinear dynamic response analysis methods arouse much interest to many researchers [5-6].

Wangqingtuo reservoir is a regulative and on-line emergency water supply reservoir of the Chinese national south to north water diversion mid line project. It is about 20 kilometers away from Tianjin city. So it is necessary that the special study on the liquefaction in the dam foundation should be pursued. Based the dynamic behavior of silt which is acquired by laboratory tests, this paper simulate the seismic liquefaction of an earth dam by seismic coupled analysis method [7] with cyclic elasto-plastic model[8].

2. Field Test and Dynamic Tests

The relatively small amount of silt particles (<75 μm) and plastic index present in the samples suggests that it is low plastic silt soil. Standard penetration test and surface wave test are conducted beside the core hole.

The resonant column triaxial tests and cyclic triaxial tests are carried for the dynamic shear modulus G0 and the damping ratio D of unsaturated silts. Experiment research results indicate that dynamic shear modulus of silts decrease with the increase of strain; damping ratio of silts increase with strain.

The cyclic triaxial tests on saturated and undisturbed silts are conducted for the cyclic liquefaction resistance.

3. Dynamic Analysis of Earth Dam

3.1. *Coupled analysis of liquefaction method*

Considering the practical engineering, two-dimensional effective-stress-based finite element program called Coupled Analysis of Liquefaction (LIQCA) based on coupled Biot theory is used to simulate the earthquake response of the earth dam. FEM-FDM coupled analysis succeeds in reducing the degrees of freedom in the descretized equations.

3.2. *Typical section and dynamic parameters*

Considering the reservoir is a plain reservoir of earth dam, Based on the data from the soil tests, the typical two-dimensional dam section (Fig.1) model is simulated in the numerical analysis.

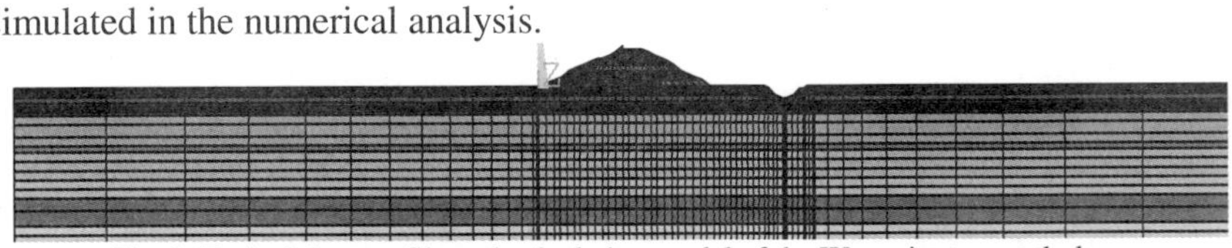

Fig. 1. Typical dam profile and calculation model of the Wangqingtuo earth dam.

In virtue of these dynamic tests, the soil dynamic properties, such as dynamic shear modulus (G), the damping ratio (D) obtained via the resonant column test and the dynamic shear stress ratio (τ/σ_3) by the vibration liquefaction test.

3.3. *Earthquake wave motion*

Consider that the reservoir be working when earthquake strikes the place where the reservoir lies, earthquake intensity of VII is the design earthquake. According to the Chinese national code, a peak ground acceleration of 0.15g is used to evaluate the liquefaction potential analysis. Tang Shan earthquake acceleration wave record (Fig.2) is chosen as the input ground motion, and also the Kobe wave (correspond to VIII earthquake) was adopted for checking.

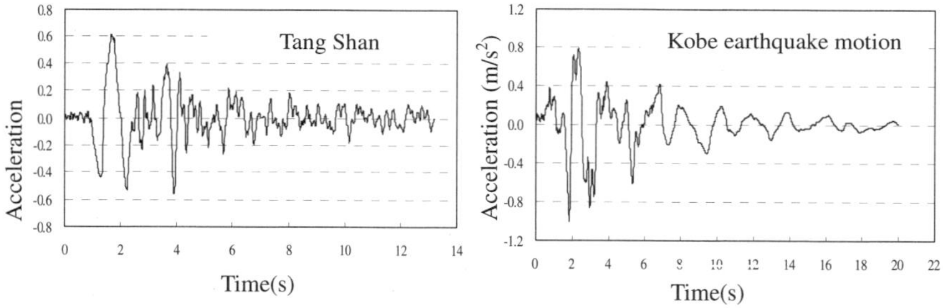

Fig. 2. The earthquake motion of Tang shan earthquake and Kobe earthquake.

3.4. *Results of the dynamic analysis*

Three cases are discussed in the paper. The first case is that calculate with Tang Shan earthquake motion when the level of dead water is 5.4m. The second case is that calculate with Tang Shan earthquake motion when the design flood level is 12.4m. The third case is that calculate with Tang Kobe earthquake motion when the design flood level is 12.4m. Seismic cyclic elasto-plastic model was adopted to dam body material and surface soil, Ramberg-Osgood model was adapted to soil below liquefiable layer under seismic load in the elastic-plastic dynamic response analysis. Accelerations were computed at the nodes and pore pressures at the centroid of elements.

Figure 3 is partial picture of the distribution of excess pore water pressure ratio of case 2 and 3. Because the effective overburden pressure under the dam is stronger, the serious liquefaction areas are dam heel and dam toe. Of course the liquefaction is more serious when adopted Kobe earthquake motion (case 3).

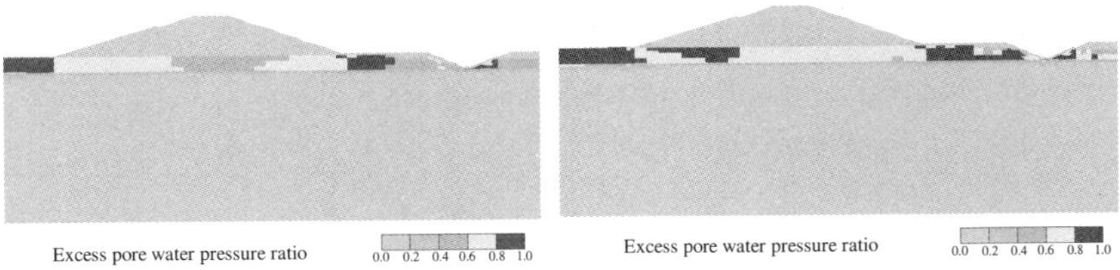

Case 2 Case 3

Fig. 3. Excess pore water pressure ratio under earthquake.

The deformation is evaluated for the below critical cross section to predict the potential non-uniform settlement and level displacement for the significant deformation of the dam due to the earthquake shaking (fig. 4). Node 1, node 2 and node 3 are three representative nodes, the maximum horizontal displacement is 0.002m and the vertical settlement is 0.096m in case 2, The maximum horizontal displacement is 0.028m and the vertical settlement is 0.159m in case 3.

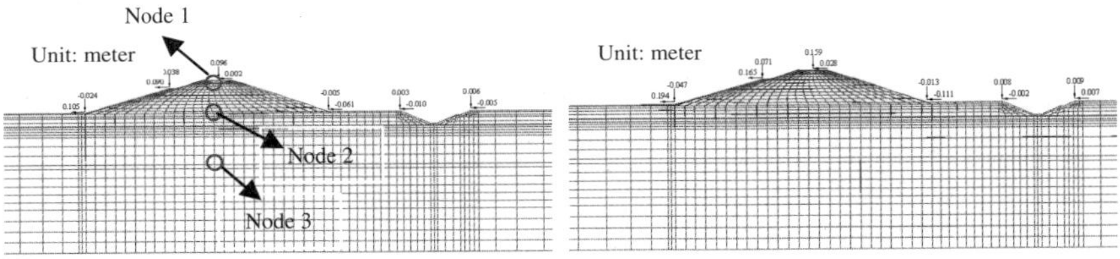

Case 2 Case 3

Fig. 4. Segment of displacement deformation under earthquake.

And also horizontal displacement and acceleration curves versus time of case 2 were showed in Fig. 5. From the displacement time curve and acceleration time curve, the horizontal displacement, horizontal displacement and acceleration got bigger gradually from bottom to the top, which maybe due to amplification reaction of earthquake to soft soil.

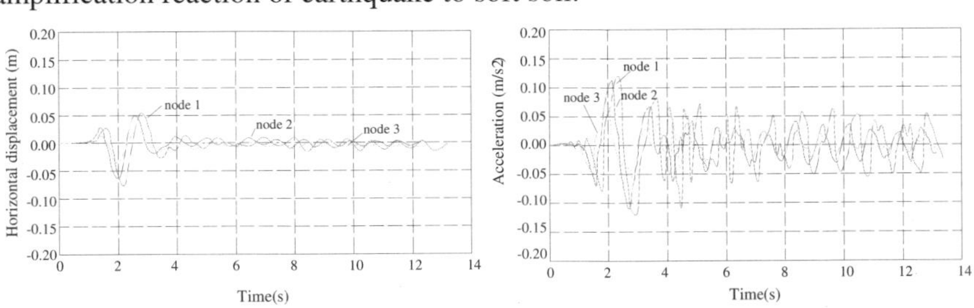

Fig. 5. Horizontal displacement and acceleration curves versus time data of case 2.

4. Conclusion and Discussion

Based large numbers of field and laboratory tests, the characteristics of liquefaction and deformation for Wangqingtuo earth dam foundation are analyzed by elasto-plastic seismic finite element simulation, and shows that the influence characteristics and the influence regularity which reservoir water level changes and seismic intensity bring to the liquefaction and deformation of Wangqingtuo earth dam foundation.

The density of soil and the effective overburden pressure are two important factors to the liquefaction of the earth dam under earthquake; the design team is advised to improve the density of site by dynamic compaction, also weighted platform should be built up on the downstream of the dam. Of cause, drainage is very important, seepage proof wall and seepage intercepting ditch could be adopted to earth dam.

References

1. Seed, H.B., Lee, K.L., Idriss, I.M. and Makadisi, F.I. The Slides in the San Fernando Dams during the Earthquake of February 9, 1971 – ASCE, *J of the Geotechnical Engineering Division*, GT7, pp. 651-688
2. Ishihara k. and Yoshimine M. Evaluation of settlements in sand deposits following liquefaction during earthquakes. *Soils and Foundations*.**32** (1), pp. 173-188.
3. Risheng Piao, P.E.; Arlan H. Rippe, Barry Myers. Earth dam liquefaction and deformation analysis using numerical modeling, 2006-*ASCE*, Geocongras
4. Lei Fu Xiang, wu Zeng .Seismic rehabilitation and analysis of Chaohe earth dam, *Earthquake Engineering and Engineering Vibration*, 2005 4(2):265-273
5. Clough R W， Chopra A K. Earthquake stress analysis in earth dams. *Journal of the English Mechanics Division， ASCE*， 1996， **92**(2)： 197-212
6. Mourad Zeghal, Deceased, and Ahmed M. Evaluation of the Nonlinear Seismic Response of an Earth Dam: Nonparametric System Identification, *Journal of Earthquake Engineering*, 2009, **13**(3):384-405
7. Huang Y, Yashima A, hang F, Sawada K. Numerical simulation for earthquake liquefaction of soil embankments. In: *Proceedings of the 1st international conference on computational methods-computational methods*. Springer, Dordrecht. 2006: 269-273.
8. Chun-Hsien Wu, and Jeen-Hwa Wang. Vertical ground motion analysis for submerged pore-elastic media with random void ratios. *Computers*.2009, **36**(6): 968-976

IDENTIFICATION THE INTERFACE FRACTURE PARAMETER BY INVERSE ANALYSIS: APPLICATION OF KALMAN-FILTER TECHNOLOGY

BIN LIU [1,2] ZHOU LI HE [2] MIN GE DUAN [2] FEI XU [2,*]

[1] *UAV Research and Development Center, Northwestern Polytechnical University, Xi'an, China*

[2] *School of Aeronautics, Northwestern Polytechnical University, Xi'an, China*

In this paper, Kalman filter technology which originally came from the field of signal control is introduced as an inverse analysis to identify some important material parameters. By performing detailed FEM simulation and using the experimental data of the force and displacement curve of the fiber pull-out test, the Kalman filter recursive process is successfully carried out to identify the values of the shear strength τ and the fracture energy G for the SiC (SCS-6) / Ti material interface. The great advantage of this inverse technology is that the influence of experimental noise can be taken into account. The calculation results show that the rate of convergence to the correct solution depends on the number and the initial estimation of unknown parameters to be identified and the accuracy of the measured data. Our analysis demonstrates that the proposed approach is suitable and capable in the identification of material parameters of the composite interface.

Keywords: Kalman filter; Inverse analysis; Material interface; Shear strength; Fracture energy.

1. Introduction

In the field of solid mechanics and material mechanics, it's very difficult to identify some important material parameters through direct experiment, while other physical quantities which are related to these material parameters can be easily obtained through the test. Some studies on the parameter identification of materials by solving certain static and dynamic inverse problem can be found. A method based on an evolutionary algorithm[1] was proposed to identify the parameter of inelastic constitutive equations. A unified approach in the frame of FEM[2] was presented and the isotropic and anisotropic plastic parameters were identified[3]. The nonlinear damage behavior for the interface in the metal matrix composite was investigated[4]. Based on the Gurson-Tvergaard-Needleman model,

[*] Corresponding author, email: xufei@nwpu.edu.cn

identification of ductile structural material parameters were investigated[5,6] as well. Several applications on the material parameter identification by using kinds of inverse analysis are reported[7]. The Kalman filter is introduced as one inverse technique to identify the parameter of the cohesive crack model.

The common inverse methods include Monte Carlo, simulated annealing, genetic algorithm, improved artificial neural network and so on. Some of these are completely non-linear inverse method with the ability of global search, however they aren't so efficient. Since Kalman solved the data filtering problem by a recursive discrete data technique in 1960, the Kalman filter has been the subject of extensive research and application, particularly in the area of autonomous or assisted navigation. The kalman filter is a set of mathematical equations that provides a recursive means to estimate the state of a process in a way that minimizes the mean of the squared error. The filter is so powerful that it supports estimations of past, present and future states, even when the precise nature of the modeled system is unknown[8]. In this paper, the Kalman filter is applied in a fiber push-out test and the shear strength and the fracture energy for the interface of the SiC(SCS-6)/Ti are obtained successfully.

This paper is organized as follows. The next section explains the fundamental of the fiber push-out test. The basic theory and algorithm of Kalman filter are presented in section 3, Section 4 gives several numerical examples to validate the proposed inverse technique and section 5 summarizes.

2. Fiber Push-out Test

The fiber push-out test is a standard experiment to measure the interface properties of fiber reinforced composite materials. With compression loading on one end of the fiber, the fiber will be pushed out from the other end. During the experiment, the force and the displacement can be measured. The typical force-displacement curve is shown in Fig.2 in the section 4.1. It is clear that the relationship between the force and the displacement is linear at both loading and unloading stages. When the force approaches its maximum value, F_{max}, the shear strength of the interface is figured out by the equation, $\tau = F_{max} / \pi dL$, where d and L are the diameter and the length of the fiber respectively. After the maximum point, the interface of two materials starts to separate and the force decreases. Since the distributions of both tension stress and shear stress are not uniform, the above shear strength is the average value of shear strength of the interface, not a real fracture strength[9]. There are attempts to find the real shear strength of the interface. For example, the push-out test was simulated in FEM method with the cohesive zone model and the shear strength of the interface was obtained[9]. The result of force-displacement curve from the numerical simulation of the obtained shear strength agrees well with that from the test. So in this paper, the force-displacement curve from ref.9 is used as the benchmark.

3. Theory and Algorithm of Kalman Filter

The Kalman filter estimates a process in a feedback control form: the filter estimates the process state at some time and then obtains feedback in the form of measurement. As such, the equations for the Kalman filter fall into two groups: time update equations (predict equations), which are responsible for projecting forward the current state and error covariance estimates to obtain the priori estimates for the next step, and measurement update equations (correct equations), which are responsible for the feedback, i.e. for incorporating a new measurement into the priori estimate to obtain an improved posteriori estimate. Indeed the final estimation algorithm resembles that of a predict-correct algorithm for solving numerical problems as shown in Table 1.

Table 1. Kalman filter principle table (– is priori state estimate，ˆ denote estimate).

Time Update（Predict）		Measurement Update（Correct）	
Project the state ahead		Compute the Kalman gain	
$\hat{x}_{k-1}^{-} = Ax_{k-1} + Bu_{k-1} + w_{k-1}$	(1)	$K_k = P_k^- H_k^T (H_k P_k^- H_k^T + R)^{-1}$	(4)
Project the error covariance ahead		Update estimate	
$P_k^- = AP_{k-1}A^T + Q$	(2)	$\hat{x}_k = \hat{x}_k^- + K_k(z_k - H_k\hat{x}_k^-)$	(5)
Measurement z_k		Update the error covariance	
$z_k = h(x_k) + v_k$	(3)	$P_k = (I - K_k H_k)P_k^-$	(6)

Eq.(1) relates the state at the previous time step k-1 to the state at the current time step k. Eq.(2) is a priori estimate error covariance. Eq.(3) is the actual measurement z_k at the state of x_k. Kalman gain K_k in Eq.(4) presents the function of state to the measurement. The u is an optional control input. The random variables w_{k-1} and v_k represent the process and measurement noise respectively. They are assumed to be independent, white, and with normal probability distributions p(w)~N(0,Q) and p(v)~N(0,R). P^- is the priori estimate of error covariance, and the P_k is the posteriori estimate error covariance. The difference of $(z_k - H_k(\hat{x}_k^-))$ is called the residual, which reflects the discrepancy between the predicted measurements $H_k(\hat{x}_k^-)$ and the actual measurement z_k. A zero residual means that the two are in complete agreement. The H_k presents the partial derivative of h with respect to x, $H_k = \partial h(x_k)/\partial x_k$. The error covariance is then updated by K_k and H_k with Eq.(6).

Assume the process noise $w_k=0$, optional control input u=0 and matrix A is the identity matrix, the predicted state and error covariance are simplified as

$$\hat{x}_{k+1}^- = \hat{x}_k \quad , \quad P_{k+1}^- = P_k \qquad (7,8)$$

The steps to perform the Kalman filter are as follows:

1. Assume the estimated initial state $\bar{x}_0$ and initial error covariance P_0;
2. Perform the FEM calculation. Then use the state estimate step k, x_k, to obtain the measurement prediction $\bar{z}_k = H_k(\hat{x}_k^-)$ and the partial derivative function H_k;
3. Calculate the Kalman Gain K_k;
4. Obtain the residual between the predicted and the actual measurements, then estimate the corrected $\hat{x}_k$ and P_k;
5. Get the prior estimates $\hat{x}_{k+1}^-$ and P_{k+1}^- at step k+1 with Eq.(7) and (8);
6. Find out the maximum value of the updated error covariance P_{k+1}^- and compare with the error limit ε;
7. Repeat the above steps from 2 to 6, when the precision is not reached. Otherwise stop the iteration and the identified state estimate is $\hat{x}_k$.

4. Numerical Examples

4.1. *Case 1*

This is a one parameter identification case. Here, the real shear strength $\tau_{\max}$ is chosen as the only state variable in the previous push-out experiment, i.e. $x_k = \tau_{\max}$. The force is defined as the actual measurement, i.e. $z_k = F_{\max}$. In this calculation, the units of force, shear strength and displacement are N, MPa and mm, respectively. The initial parameters of Kalman filter, *i.e.* initial state variables is $\hat{x}_0 = 300$, initial estimate noise covariance is $P_0 = 9$, the measurement noise covariance is $R = 1$ and the process noise covariance is Q=0

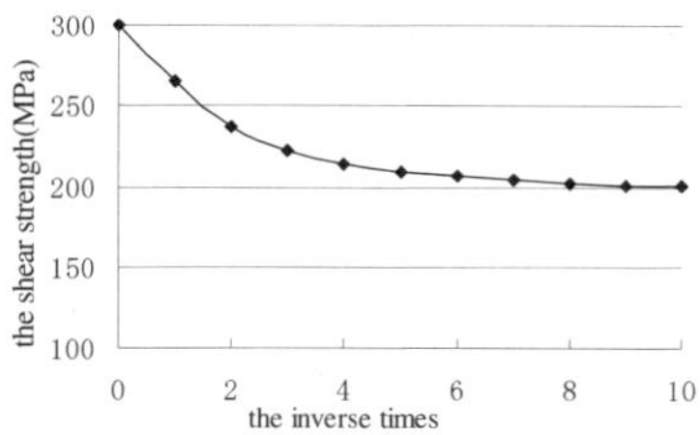

Fig. 1. The inverse analysis of parameter τ.

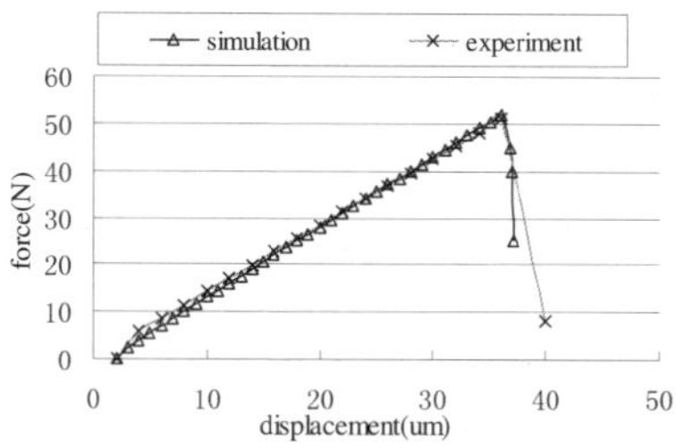

Fig. 2. Comparison between inverse and experimental.

During each step of Kalman filter inverse analysis, the measurement prediction $\bar{z}_k = H_k(\hat{x}_k^-)$ and the partial derivative function H_k are obtained through the FEM simulation on the state variable of x_k. It is found that step by step, the error covariance became smaller and smaller, in this simulation when the maximum value of the updated error covariance matrix is less the error limit of $\min(\lambda_{\max}(P_k)) \le \varepsilon = 1$. The state variable of the shear strength $\tau_{\max}$ is figured out to be 200.8MPa. In this example, totally 10 steps iteration is performed before the limit is reached. The convergence of the shear strength can be found clearly in Fig.1. Fig.2 shows that when the result of $x_k = \tau_k = 200.8MPa$ is selected as the shear strength, the force-displacement curve obtained from the FEM agrees quite well with the experimental result.

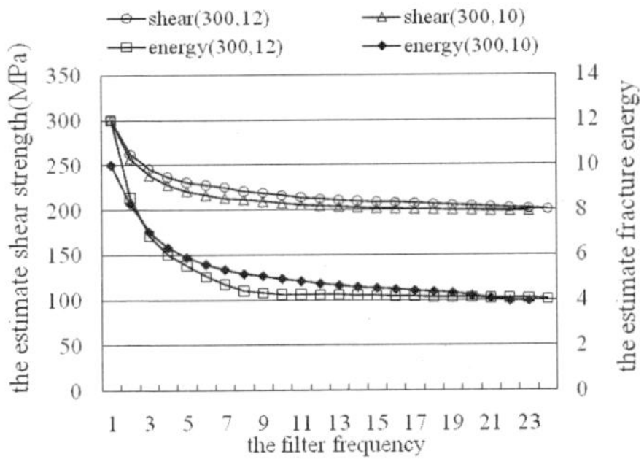

Fig. 3. The inverse analysis of parameter τ and G.

4.2. *Case 2*

This is a two parameters identification case. Different from case 1, both the shear strength and the fracture energy are chosen as the state variables, i.e. $x_k = [\tau, G]^T$. The actual measurements from experiment are the force and displacement of clamp, i.e. $z_k = [F_{\max}, \delta]$. The two kinds of initial values of these state variables are $\hat{x}_0 = [300,10]^T$ and $\hat{x}_0 = [300,12]^T$. The initial estimate noise covariance matrix is assumed as P_0=diag(9,0.1). The measurement noise matrix is R=diag(2.5,2.5) and the process noise is Q=0. Again based on the FEM simulation, the measurement prediction $\bar{z}_k$ and the partial derivative function H_x are obtained by two state variables of $x_k = [\tau, G]^T$. Following the similar algorithm, for the state variables of $\hat{x}_0 = [300,10]^T$, when $\min(\lambda_{\max}(P_k)) \le \varepsilon = 1$ is satisfied, the calculation results after 24 steps iteration are $[200.555, 3.9548]^T$. At the final step the noise covariance are P_{24}= diag(0.9983,0.001). The convergences of parameters τ and G of the two samples are shown in Fig.3. The results show that the force-displacement curves obtained from the above parameters also agree well with the experimental result. Compared with the case 1, it is clear that iteration times increases from 10 to 24 when the identification parameters increase.

4.3. *Case 3*

In order to study the effect of measurement noise, three samples of are considered, which corresponding to the relative error of the measurement about 2%、4.5% and 8% respectively. Following the same algorithm with case 2, the initial identification parameters and results are shown in Table 2.

The data in Table 2 show that the inverse analysis of Kalman filter has the ability to eliminate the noises. Similar results of inverse values can be obtained from different level of noises. However the high level of noise, the more filter times are required.

Table 2. The influence of experimental noise by Kalman filter.

Measurement noise	Initial estimate (τ,G)	Initial error covariance	Filter times	Inverse value (τ,G)	Final error covariance
diag(1,1)	(300,10)	diag(9,0.1)	22	(200.96,3.92)	diag(0.78,0.0005)
diag(5,5)			44	(199.48,4.06)	diag(0.99,0.0015)
diag(16,16)			56	(200.07,3.93)	diag(0.999,0.0035)

5. Conclusion

The study of the identification two parameters of the material interface property through the push-out test of SiC(SCS-6)/Ti fiber reinforced composite shows the feasibility of using the Kalman filter inverse analysis.

1) The convergence of Kalman filter inverse technique is good for both one parameter and more parameters identification.
2) The accuracy of the parameter identification is proved by the comparison with the force displacement curve from the experimental results.
3) The ability of taking the experimental noise into account is a great advantage of the Kalman filter inverse technique.

Acknowledgments

The work described in this paper is supported by the program for 2008 New Century Excellent Talents in University (NCET080454) and the Expertise-introduction Project for Disciplinary Innovation of University (111 project-B07050).

References

1. Tomonari Furukawa and Genki Yagawa: International Journal for Numerical Methods in Engineering, 1997, 40, 1071-1090
2. Mahnken, R. and Stein, E.: Computer methods in Applied Mechanics and Engineering, 1996, 136, 225-258
3. Khalfallah, A. Bel Hadj Salah and H. Dogui: European Journal of Mechanics, A/Solids, 2002, 21, 927-942
4. Lin X.H. Kang, Y.L:Computational materials science, 2005, 32: 47-56
5. C.Alberto,M.Stefano and O.Barbara: International Journal of Fracture, 2000,104(4),349-373.
6. M.Springmann and M.Kuna: Computational materials Science, 2005, 33,544-552
7. G.Bolzon,R.Fedel and G.Maier: Computer Methods in Applied Mechanics and Engineering，2002,191,2847-2871.
8. RE Kalman.;Journal of Basic Engineering, 1960, 82,35-45.

MODELING OF THERMAL STRESS DURING CASTING SOLIDIFICATION PROCESS

DUNMING LIAO† LILIANG CHEN JIANXIN ZHOU RUIXIANG LIU

State Key Lab of Material Processing and Die&Mould Technology, Huazhong University of Science & Technology, Wuhan, Hubei 430074, China

Numerical simulation of thermal stress during casting solidification process can predict crack, residual stress concentration, deformation. Though stress simulation involves too many complex problems such as numerical model, mechanical boundary condition and high thermo-mechanical parameters, it still retains as a difficult hotspot of macro-simulation of casting nowadays. Based on finite difference method (FDM), a 3D FDM/FDM numerical simulation system for temperature and stress analysis during casting solidification process was developed. To verify the system, a standard stress frame model and a practical casting were simulated. Pouring experiments have been carried out in the laboratory. The results of simulation coincided with those obtained from experiment and practical results. When adopting FDM to calculate casting thermal stress, thermal analysis and stress analysis can use the same FD model, which can avoid matching between different models and reduce the errors of temperature load transferring. It makes the simulation of fluid-flow, temperature and stress unify into one model.

Keywords: Casting; Thermal Stress; Numerical Simulation; FDM; Solidification.

1. Introduction

During casting solidification process, many defects occur frequently due to the existence of thermal stress, such as hot tearing, cracking, residual stress concentration and distortion. Modeling of thermal stress of castings during solidification process can provide the following functions: tracking of stress and distortion development during solidification and after shake-out, avoidance of tearing problems and prediction of thermal cracking in casting or permanent molds.[1-4] The process of casting in high temperature is a typical triple nonlinear, i.e. geometry, physics, and boundary nonlinear problem. Since the numerical simulation to 3D thermal stress of casting involves thermo-elastic-plastic-creep theories and high-temperature mechanical properties with temperature dependent nature, the calculation is very complex. Engineering mechanical problems are almost exclusively solved by finite element method (FEM). Many researchers

† Corresponding author, email: liaodunming@hust.edu.cn

have applied integrated FDM/FEM analysis solver, i.e. using FDM to calculate heat transfer problem, while FEM to simulate stress. Some large universal commercial FEA packages are usually adopted, such as ANSYS, Abaqus, MARC, etc.[5] However recent work has applied finite volume methods for stress/strain calculations,[6-8] which virtually bases on FDM.

This paper conducted a study on 3D numerical simulation of casting thermal stress based on FDM. A standard stress frame model and a practical casting were simulated to verify the package.

2. Basic Principle

The basis of the deformation and stress/strain calculation is a transient heat transfer analysis including solidification. In addition, the thermal stress/strain problems are governed by the following field equations:

The equilibrium equations:

$$\begin{cases} \sigma_{ij,j} + F_i = 0 \\ \sigma_{ij} = \sigma_{ji} \end{cases} \qquad \text{(i,j=x,y,z)} \tag{1}$$

The geometric equations (small-strain theory):

$$\varepsilon_{ij} = \frac{1}{2}\left(u_{i,j} + u_{j,i}\right) \qquad \text{(i,j=x,y,z)} \tag{2}$$

The constitutive equations (stress in terms of placements):

$$\begin{cases} \sigma_x = 2G\varepsilon_x + \lambda\theta - \dfrac{\alpha ET}{1-2\nu}, & \tau_{xy} = G\gamma_{xy} \\ \sigma_y = 2G\varepsilon_y + \lambda\theta - \dfrac{\alpha ET}{1-2\nu}, & \tau_{yz} = G\gamma_{yz} \\ \sigma_z = 2G\varepsilon_z + \lambda\theta - \dfrac{\alpha ET}{1-2\nu}, & \tau_{zx} = G\gamma_{zx} \end{cases} \tag{3}$$

In casting process, elasticity, plasticity, the contractions and expansions due to the temperature changes and phase transformations and creep all contribute to generation of strains. So the total incremental strain vector consists of elastic, plastic, thermal and creep components in incremental theory: [12]

$$d\varepsilon_{ij} = d\varepsilon_{ij}^{e} + d\varepsilon_{ij}^{p} + d\varepsilon_{ij}^{T} + d\varepsilon_{ij}^{C} \tag{4}$$

where $d\varepsilon_{ij}$ is total strain increment, $d\varepsilon_{ij}^{e}$ is the elastic strain increment, $d\varepsilon_{ij}^{p}$ is the plastic strain increment, $d\varepsilon_{ij}^{T}$ is the thermal strain increment, $d\varepsilon_{ij}^{C}$ is the creep strain increment. But $d\varepsilon_{ij}^{C}$ was neglected in this study.

3. FDM/FDM Thermal Stress Simulating

This study has applied the thermo-elastic-plastic model to conduct casting stress and strains simulation based on FDM, which is widely used to calculate heat transfer and fluid-flow problems. A FDM/FDM (temperature simulated using FDM, and stress analysis using FDM too) single direction couple approach was adopted. Firstly temperature fields were calculated using FDM, then using the results of temperature fields as thermal load to simulate stress/strain by the same FD model.

4. Application Instance Verification

To verify the system, a standard stress frame sample casting of 0.25% carbon has been simulated. Firstly, calculating the temperature fields using InteCAST,[9] a commercial casting software developed by our research group, the mold was furan resin bonded sand. Then, using the temperature results as thermal load added to thermo-elastic-plastic model to analyze thermal stress using the developed FDM/FDM thermal stress system.

Figure 1 shows the distribution of Y-axis direction stress at the two moments. It shows that at early stage the thin pole bears tensile normal stress whereas the thick pole bears compression stress; while at later stage the situation turns upside-down, the thin pole bears compression normal stress whereas the thick pole bears tensile stress. It is because at early stage, the cool speed of thin pole is much greater than the thin thick pole. The thin pole will contract, but the beam will prevent the contraction of thin pole due to the sand in middle of casting. So the thin pole bears tensile stress. The contraction of thin pole brings compression action to the thick pole, so thick pole bears compression stress. At later stage, the temperature of thick pole drops a lot, but thin pole cools very slowly. The thick pole contracts much, but the sand and beam will prevent its contraction, so the thick pole bears tensile stress at this stage, the contraction of thick pole will compress the thin pole, so thin pole bears compression stress. The behavior predicted by simulation was consistent with the recognized analysis results. Figure 2 demonstrates the distribution of equivalent stress; the equivalent stress of the four hot spots is much bigger than other parts in this figure. Pouring experiments have been carried out in laboratory. The frame casting was poured into furan resin bind sand mold. And Figure 3 shows the photo of real casting, there is a crack shown by the arrow. The results predicted by computation were consistent with the recognized analysis results and experiment result.

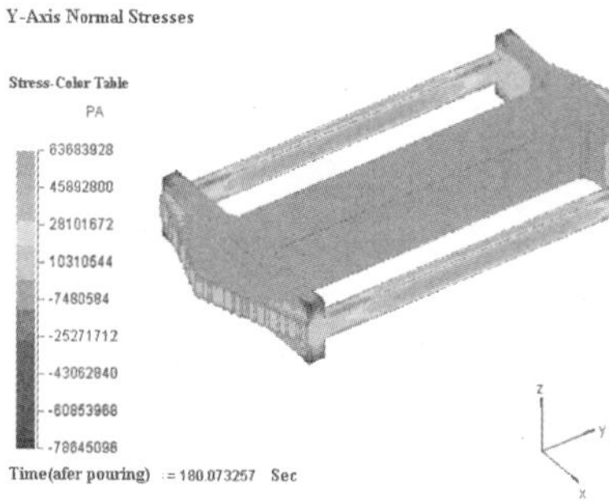

(a) Early stage (180s after pouring)

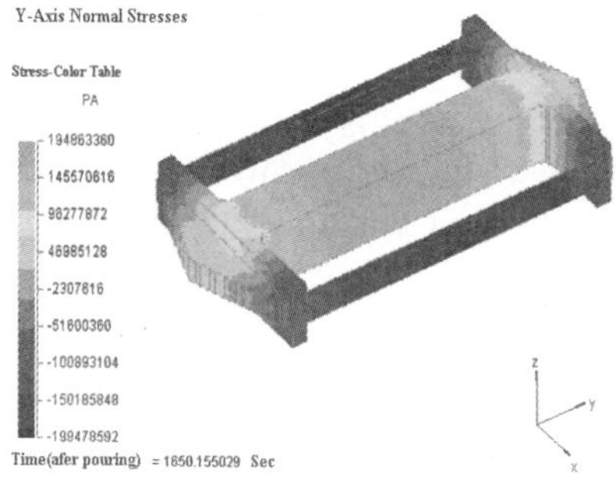

(b) Later stage (1650s after pouring)

Fig. 1. Stress distribution of Y-axis direction at two different moments.

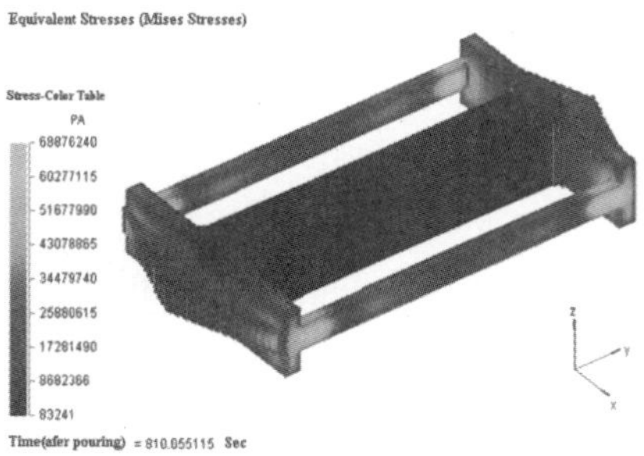

Fig. 2. Distribution of equivalent stress.

Fig. 3. Real casting and defect.

Fig. 4. Model of the head cover castings.

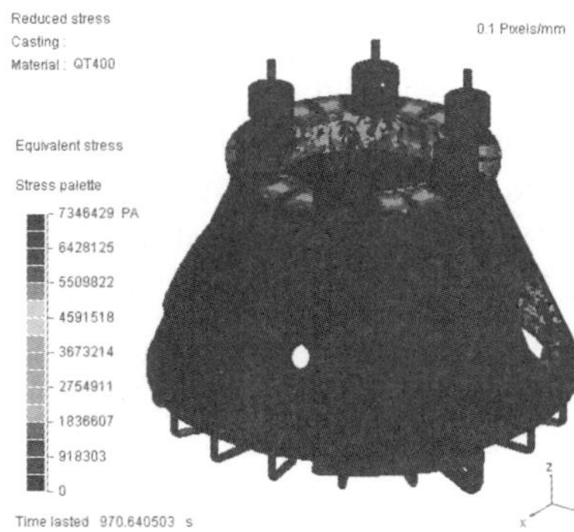

Fig. 5. Equivalent stress distribution of the head cover casting.

A head cover casting of ZG30 from a foundry was selected to perform stress analysis. The temperature fields and stress fields analysis based on a discrete model of approx 6 million elements. Figure 4 shows the 3D model of the head cover casting and its casting process. It has many chills on the surface. Figure 5 displays the distribution of equivalent stress. From Figure 6, it can be found that at the interface between chiller and casting, the equivalent stress is much bigger.

This is because the chills bring dramatic temperature descending; the cooling speed of the interface of casting and chill is much greater than other parts, uneven contractions will happen, which leads to big residual stress. These areas likely come across surface cracking, and the simulated results are accordance with the practical cases approximately.

5. Conclusion

Using FDM, a 3D thermal stress simulation system was developed and calibrated by a standard stress frame model and a practical casting. The results of simulation were in agreement with the experiment result and practical cases. It is very convenient due to the approach where stress and heat transfer analysis using the same FD model, which can avoid matching between different models and reduce the errors of temperature load transferring. It makes the simulation of fluid-flow fields, temperature fields and stress fields unify into one FD model.

Acknowledgments

This work is supported by National Natural Science Foundation of China (50805056); SRF for ROCS, SEM (2009); Fundamental Research Funds for the Central Universities: C2009Z022J.

References

1. Monroe C, Beckermann C: Mater. Sci. and Eng.: 2005,413-414: 30-36.
2. Hatami N, Babaei R, Dadashzadeh M, Davami P: J. of Mate. Proc. Tech., 2008,205(1-3): 506-513.
3. Hattel H, Andersen S and Svoboda JM: Trans. of the Amer. Foundrymen's Soc., Vol. 105 (USA), pp. 31-34, 1998.
4. Liu Baicheng, Xu Qingyan and Xiong Shoumei: Chin. J. of Mech. Eng., 2003(10):53-63.
5. Hannart B, Cialti F and Van Schalkwijk R: Light metals 1994.879~888.
6. Hattel J, Hansen PN and Andersen S: North American Die Casting Association (USA): 99-103, 1993.
7. J.Hattel and P.N. Hansen: Appl.Math. Modeling, 19(4)(1995), 210~243.
8. J. Hattel, J. Thorborg and S. Anderson: In: B. G. Thomas and C.Beckmann, eds. Modeling of Casting, Welding and Advanced Solidification Processes VIII. New York: The Minerals, Metals & Materials Society, 1998. 763~770.
9. Zhou Jianxin, Liu Ruixiang, Chen Liliang, Liao Dunming and Lin Hantong: Int. J. Cast Metals Res., 2002,15(4): 415~419.

PREDICTION OF FORMING LIMIT DIAGRAMS OF DP590 STEEL BASED ON THE M-K MODEL WITH EXPERIMENTAL VERIFICATION

YANSHAN LOU, SEOKBONG KIM, CHANSOO LEE, HOON HUH†

School of Mechanical, Aerospace and Systems Engineering, KAIST, 335, Gwahangno, Daedoek Science Town, Daejeon, 305-701, Republic of Korea

This paper analytically investigates forming limit diagrams (FLDs) of advanced high strength steel DP590 using the Marciniak-Kuczynski model (M-K model) with the strain rate effect. Tensile tests at various strain rates were conducted to obtain the strain hardening and strain rate hardening properties of DP590. The Johnson-Cook model was used to interpolate the experimental data with the strain rate effect. The FLDs of DP590 were then predicted with the M-K model. The material was assumed to be rigid-plastic and subjected to the flow rule based on the Hosford79 yield criterion. Punch stretching tests were conducted to verify the analytical results with different punch velocities to investigate the effect of the forming speed on formability. The analytical model without the strain rate effect underestimated the forming limits of DP590. The computed FLDs with the strain rate effect show good agreement with the experimental result especially when the exponent in the Hosford79 yield criterion is 8 for static forming and 10 for dynamic forming.

Keywords: Forming limit diagrams; strain rate sensitivity; localized necking; sheet metal forming.

1. Introduction

The FLD, first introduced by Keeler and Goodwin[1, 2] in 1969, provides a method for determining process limitations in sheet metal forming and is used to assess the stamping characteristics of sheet metals. The FLD is often used in method planning, tool manufacturing, and in press shops to optimize stamping tools and their geometries. In most cases, FLD can be constructed experimentally. However, the process is labor intensive, material and time consuming[3]. A number of analytical methods have been to compute the FLDs. The popular methods are Hill's local instability model[4] and Swift's diffused instability model[5]. The most famous theory was proposed by Marciniak and Kuczynski[6] in 1967. An initial defect was assumed in a form of groove or trough oriented

† Corresponding author, email: hhuh@kaist.ac.kr

perpendicular to the axis of the maximum principal stress and strain. The effect of yield criteria on the prediction of FLDs was investigated based on the M-K model[7-9].

This paper investigated dynamic FLDs based on the M-K model. The Hosford79 yield function was applied to model yield surfaces of a material. Tensile tests were conducted to obtain the material properties[10]. Punch stretching tests with different punch velocities are employed to obtain FLDs experimentally. The strain hardening and strain rate hardening characteristics were formulated with the Johnson-Cook model. The initial inhomogeneous factor was evaluated by the limited strain in the plane strain condition from experiments. The strain rate effect on the magnitude of FLDs was studied in detail. The calculated FLDs were compared with experimental results in both static and dynamic forming conditions.

2. The Marciniak-Kuczynski (M-K) Model

In the M-K model, an initial defect is assumed in a form of groove oriented normal to the axis of maximum principal stress. The initial imperfection is expressed as the thickness ratio

$$f_0 = t_{b0}/t_{a0}. \tag{1}$$

The value of the initial imperfection f_0 is adjusted so that the predicted limited strain of plane strain (FLD_0) is the same as experimental data. Assuming that the strain ratio outside the groove is a constant

$$\beta_a = d\varepsilon_{a2}/d\varepsilon_{a1}. \tag{2}$$

Two relations between the material inside and outside groove are important to obtain the strain and stress state in these two regions. These two relations are the compatibility and the force equilibrium equations

$$d\varepsilon_{a2} = d\varepsilon_{b2}, T_{a1} = T_{b1}. \tag{3}$$

Where T_{a1} and T_{b1} are forces per unit length normal to the groove direction inside and outside of the groove, respectively.

The Hosford79[11, 12] non-quadratic anisotropic yield function is utilized to describe the yield surface. In the plane stress and planar isotropic conditions, the Hosford79 yield function has the form as follows:

$$|\sigma_1|^a + |\sigma_2|^a + R|\sigma_1 - \sigma_2|^a = (1+R)\bar{\sigma}^a. \tag{4}$$

where a is an exponent which is greater than 2. The value of the exponent was suggested to be 6 and 8 for BCC and FCC metals, respectively.

Table 1. Chemical composition of DP590 steel employed in this study (wt.%).

C	Mn	Si	P	S
0.326	1.7	0.123	0.0162	0.0009

Table 2. Material properties of DP590.

<table>
<tr><th></th><th>A</th><th>B</th><th>n</th><th>C</th><th>$\bar{\varepsilon}_0$</th><th>R_0</th><th>R_{45}</th><th>R_{90}</th><th>R</th></tr>
<tr><td>Static</td><td rowspan="2">381.1</td><td rowspan="2">883.31</td><td rowspan="2">0.502</td><td rowspan="2">0.0341</td><td rowspan="2">0.001</td><td>0.82215</td><td>1.01469</td><td>1.08331</td><td>0.98371</td></tr>
<tr><td>Dynamic</td><td>0.84093</td><td>0.94947</td><td>1.05331</td><td>0.94954</td></tr>
</table>

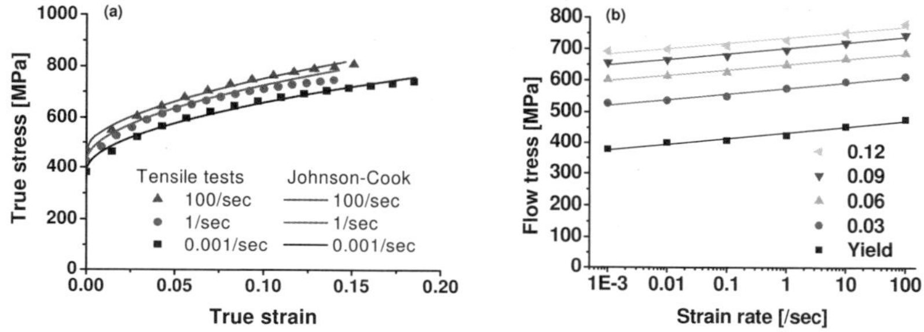

Fig. 1. Fitted material constants in Johnson-Cook model: (a) strain hardening; (b) strain rate hardening.

3. Experiments

3.1. *Uniaxial tensile tests*

The material used in this paper is a dual phase steel, DP590, with the chemical composition as presented in Table 1. The thickness of DP590 steel sheet is 1.2 mm. To predict FLDs with the M-K model, at least two kinds of material properties are required: the stress–strain relation; and the Lankford's coefficient (the R-value). Uniaxial tensile tests were performed along the rolling (0°), diagonal (45°) and transverse (90°) directions. Uniaxial tensile tests at various strain rates were also conducted to obtain the strain–rate hardening coefficient. Tensile tests at low strain rates were carried out using the static tensile machine, INSTRON 5583, while dynamic tensile tests were performed using the high speed material testing machine[10]. The strain rate hardening coefficient is calculated with the yield stresses at different strain rate with the variation of plastic strains and the final strain rate hardening coefficient takes the mean value. Fig. 1 (a) and (b) shows the comparison of the Johnson-Cook model and experiment data as well as the strain rate hardening at different plastic strain.

In the tensile test, circle grids were closely etched on the specimens and the tension process was recorded by a high speed camera to obtain the R-value. The constants in the Johnson-Cook model and R-values for DP590 were summarized in Table 2.

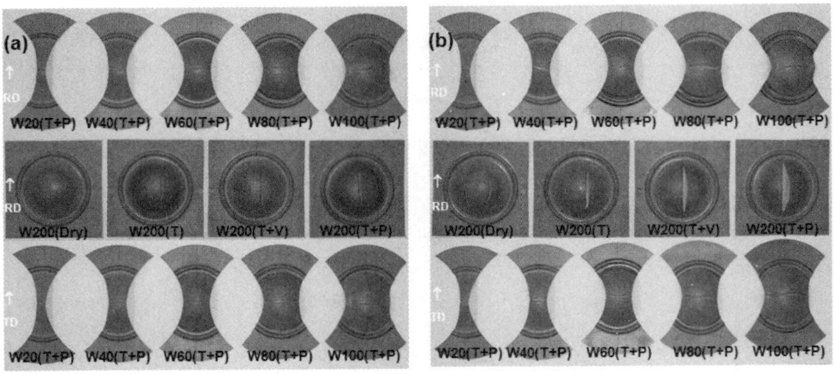

Fig. 2. The specimens of stretching tests with strain rates: (a) static forming; (b) dynamic forming.

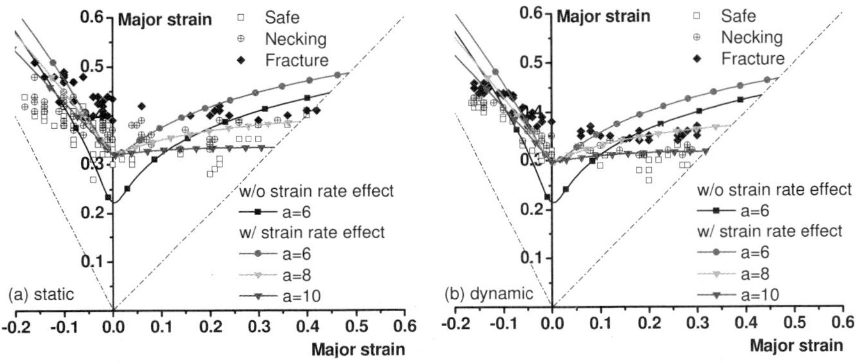

Fig. 3. Comparison of the M-K models and experimental data: (a) static; (b) dynamic.

3.2. *Hemispherical punch stretching tests*

Punch stretching tests were conducted to determine FLDs experimentally. Circle grids were etched on a specimen and the limited strain was calculated by the deformation of the circle after the test. The punch velocity is 20 mm/min for static forming ($\bar{\varepsilon} = 0.0008$~$0.0017$/s) analysis while the forming velocity of dynamic states ranges from 2.5 m/sec to 10 m/sec due to different geometries of specimens ($\bar{\varepsilon} = 65$~109/s). Different sample geometries and lubrication conditions were employed to generate all possible states of stress and strain. The experiment conditions are described in detail in Kim's paper[13]. The tested specimens are shown in Fig. 2.

4. Results and Discussion

The limited strain for plane strain (FLC_0) is of particular importance. Accordingly, the initial imperfect factor is evaluated so that the computed FLC_0

is the same as the experimental results. From the punch stretching tests, the FLC_0 is 0.32 and 0.3 for static and dynamic forming, respectively. The initial imperfect factor f_0 is 0.9932 and 0.9929 in static and dynamic forming conditions, respectively. The FLDs predicted with strain rate effect are higher than that without the strain rate effect because the strain rate inside the groove is higher than that outside the groove and the strain rate hardening effect suppresses the deformation inside the groove. The FLDs predicted with different values of the exponent in the Hosford79 yield function are compared with experimental results in Fig. 3 (a) and Fig. 3 (b) for static and dynamic forming states, respectively. The FLDs predicted without the strain rate effect underestimate the formability around the plane strain condition. The FLDs with the strain rate effect match the experimental data with excellent agreement. In the static forming condition, the predicted FLDs match the experimental results with high accuracy when the exponent a=8 while the M-K model with a=10 produces better results compared with experimental data for dynamic forming.

5. Conclusion

This paper investigated FLDs analytically based on the M-K model and compared the calculated FLDs with the experimental results for DP590 steel. The strain rate effect was considered and shown that the FLDs predicted with the strain rate effect match the experimental results much better than that without the strain rate effect. The FLDs predicted match the experimental results with high agreement when the exponential value of the Hosford79 yield criterion is 8 and 10 for static and dynamic FLDs, respectively.

References

1. S.P. Keeler: *Sheet Met. Ind.*, 1965, **42**, 25–48.
2. G.M. Goodwin: *SAE Paper*, no. 680093, 1968.
3. W.M. Sing and K.P. Rao: *J. Mater. Process. Technol.*, 1993, **37**, 37–51.
4. R. Hill: *J. Mech. Phys. Solids*, 1952, **1**, 19–30.
5. H.W. Swift: *J. Mech. Phys. Solids*, 1952, **1**, 1–18.
6. Z. Marciniak and K. Kuczynski: *Int. J. Mech. Sci.*, 1967, **9**, 609–620.
7. A. Parmar and P.B. Mellor: *Int. J. Mech. Sci.*, 1978, **20**, 387–391.
8. A. Graf and W.F. Hosford: *Metall. Trans. A*, 1990, **21**, 87–94.
9. F. Barlat: *Mater. Sci. Eng.*, 1987, **91**, 55–72.
10. H. Huh, J.H. Lim, S.H. Park: *Int. J. Automot. Technol.*, 2009, **10**, 195–204.
11. R.W. Logan and W.F. Hosford: *Int. J. Mech. Sci.*, 1980, **22**, 419–430.
12. W.F. Hosford: *Proc. 7th NAMRC, Dearborn, MI*, 1979, 191–196.
13. S.B. Kim, H. Huh, H.H. Bok, M.B. Moon: *J. Mater. Process. Technol.*, 2010 (accepted).

NUMERICAL SIMULATION OF RADOME WITH FLUID-STRUCTURE INTERACTION

XU KUN-MEI

School of Astronautics, Beihang University, Beijing, China

HU YONG

AVIC Gas Turbine Establishment, Jiangyou, Sichuan, China

TANG YE

No.25 Institute of the Second Research Academy, CASIC, Beijing, China

For the purpose of prediction of the aerodynamic heating, structure stress, and deformation of the radome for IR seeker, fluid-structure interaction simulation is a good choice to meet the requirement. Unsteady simulation of aerodynamic heating process of radome was conducted. Transient aerodynamic data was obtained through simulation, which include fluid temperature, pressure, velocity, temperature on the two faces of the radome. The CFD results served as boundary condition for the structural analysis. The results showed that the center and edge of the glass would be the risk areas. The numerical simulations also successfully predicted the aerodynamic heating effects on the structure, which is the main reason for the stress and deformation of the structure. The FSI analysis testified that all the parts of the radome meet the requirement of the design. The results provide a valuable reference for the design.

Keywords: Aerodynamic heating; Fluid-structure interaction; Numerical simulation.

1. Introduction

Many guided weapons employ infra-red homing device to locate targets. The device is protected by a radome from aerodynamic heating and high pressure. The radome for IR seeker is made of fragile glass. In the developments of radome for IR seeker, one of the most challenges is the severe aerodynamic heating at nose of the radome in supersonic flight. This could affect the work of the IR seeker, or even damage the device[1,2]. Therefore accurate prediction of aerodynamic heating and structural analysis are required.

Generally, Fluid-Structure Interaction is categorized into two types. One is one way FSI and the other is two way FSI.[3]

One way FSI: The result from a CFD analysis at the fluid-structure interface is applied as a load to the mechanical application analysis. The boundary displacement from the mechanical application is not passed back to the CFD analysis.

Two ways FSI: the results of structural analysis in the mechanical application is transferred to the CFD analysis as a load. Similarly the results of the CFD analysis are passed back to the mechanical application analysis as a load. Two ways FSI is suit for multi-field problems that are highly interactive.

In this work, as the deformation is rather small compare with the radome structure. We only consider a one-way fluid-mechanical coupling.

2. Modeling Approach

2.1. *Fluid dynamics*

The cases presented in this paper involve high speed vehicle. For this reason, the fluid behavior surrounding the vehicle is governed by the Nabier-Stokes equations of compressible flows. In these fluid-structure interaction cases the radome experience significant velocity and altitude changes, therefore, transient flow problems need to be solved. The k -ω turbulence model with a wall law function have been implemented. A pressure-based algorithm based on the PISO algorithm[4] is utilized as the fluid solver.

2.2. *Structure mechanics*

In this work elastic isotropic structures will be considered. The linear elastic material model is used for the problems with small deformations, while the elastic model is applied to structural dynamics tasks with finite deformations and small strains. The structural dynamics will be described using the weak form of the governing equations, which is needed for the application of finite elements methods[5].

2.3. *Fluid-structure interface*

All forces acting on the structure within the fluid-structure interaction have to be taken into account. So the total force F_{struc} is:

$$F_{struc} = F_{struc}^{external} + F_{fluid} \tag{1}$$

where F_{fluid} is determined by fluids solver and $F_{struc}^{external}$ denotes all other applied external forces.

The temperature of the interface is equal to the ambient fluid temperature. The temperature of inner structure is: $t_{inner}=t-\Delta t$.

$$\Delta t = q\delta / \lambda \tag{2}$$

q denotes heat flow density, δ denotes the thickness of the structure, λ is the thermal conductivity.

3. Computational Set-up

3.1. *Structure modeling*

Instead of considering the whole vehicle, in this case, we focus on the radome only while ignore the fuselage. The geometry and mesh of the radome is shown in Figure 1.(a) and Figure 1.(b). The upper structure is sphere, which is consisted of two materials. The windward is made of MgF2, which is rigid and fragile, and can not withstand much force or heat. The base and the back cover are made of TC4, which provides support for the radome and can sustain high pressure and heat flux.

3.2. *CFD modeling*

The commercial CFD code CFX5.0 was chose to calculate the transient flow. The flow field is 500mm×500mm×500mm. The left side boundary is an inlet boundary with set velocity, density and temperature. The rest boundaries are outflow boundaries. The initial run in CFX was performed at Mach 0.1 and altitude 0m. The initial air condition are: T=288K, and P=101325Pa. The Mach number range is from 0.1 to 3.6, and altitude range is from 0 to 5km.

The fluid used for the flow field is air with properties as found in the International Standard Atmosphere definition. All of these properties are altitude dependent.

In this work, it is estimated that the missile accelerate from Mach 0.1 at sea level to Mach 3.6 at 5km in 50 seconds gradually. Of all the heat transfer patterns between the flow and the radome, only the convective heat transfer is considered. The CFD mesh is shown in Figure 1.(c), which neglects the base of the radome. All the mesh contains 375000 cells.

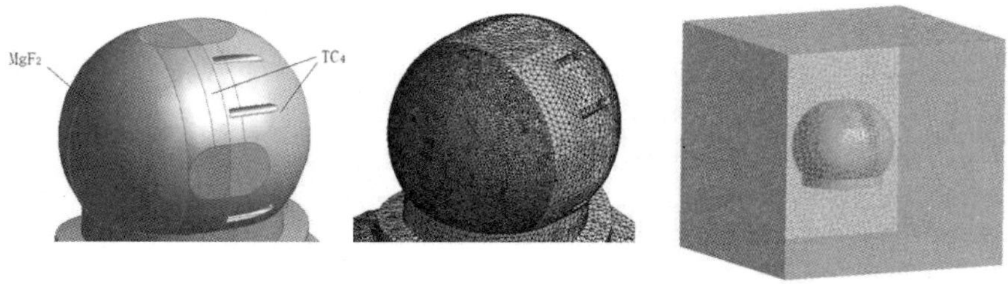

(a) structural model (b) structural mesh (c) CFD mesh

Fig. 1. Structural model and mesh.

4. Result and Discussion

4.1. *CFD results*

Figure 2 is the CFD results at T=50s.

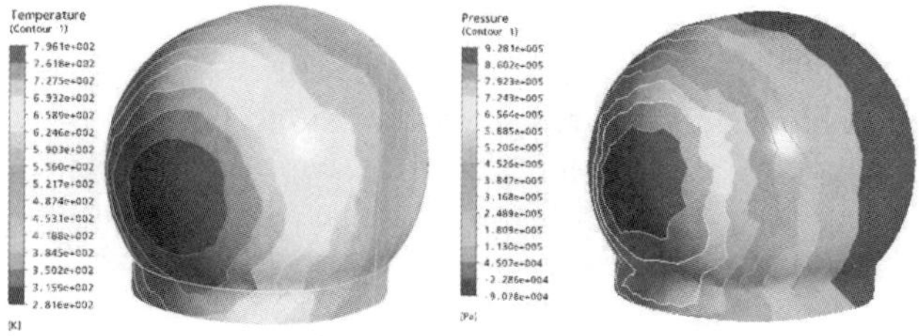

(a) Interface temperature (b) Interface pressure

Fig. 2. CFD results.

At the beginning, as the vehicles fly in subsonic, the temperature of the structure surface raises up gradually. The temperature climbed *dramatically* as the vehicle speed up into supersonic. As the vehicle continues to accelerate in this supersonic flight regime, both of the temperature and pressure reached a maximum of 796K and 0.928Mpa respectively at the center of the radome.

4.2. *FSI results*

Table 1 shows the FSI results from 0s to 50s.

Table 1. Performance results for 0s to 50s.

T (s)	Equivalent stress of TC4 (MPa) and safety factor		Maximum deformation (mm)	Maximum main stress of MgF2 (MPa) and safety factor	
3	394	n>2	0.247	20	n>4
10	381	1.67	0.284	57	1.45
20	380	1.57	0.296	52.4	1.48
30	376	1.55	0.298	54	1.46
40	376	1.54	0.299	55	1.43
50	378	1.53	0.313	57	1.26

Figure 3 and Figure 4 are the FSI results of radome during ascending. The total force of the glass at 3 direction are 468N, 3662N, -21N respectively.

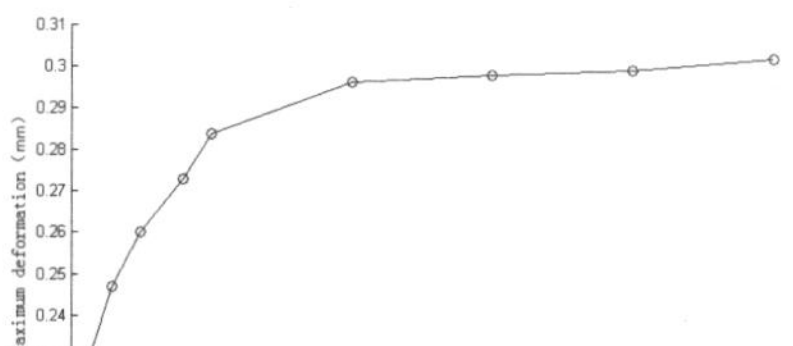

Fig. 3. Maximum deformation.

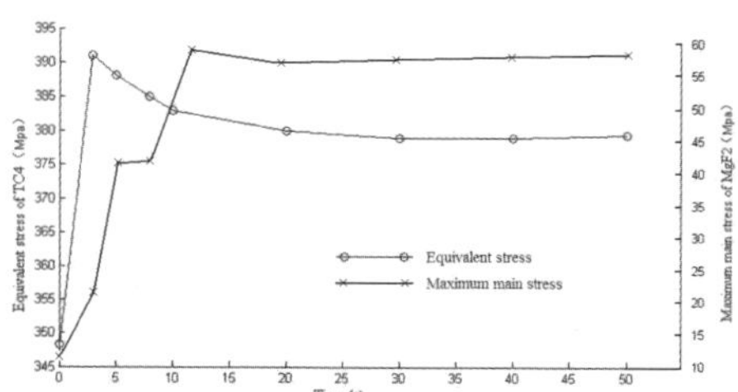

Fig. 4. Stress of each part.

From the result reported in Figure 3, we conclude that the total deformation of the structure increases along with the time span. From 0 to 10s, the deformation increased dramatically as the flight speed raise. It gradually slow down from 10s to 20s, and achieve maximum value of 0.313mm at 50s. The results in Figure 4 illuminated that as the equivalent stress of TC4 and maximum main stress of MgF2 change along with the temperature. The results show that all the parts of the radome meet the requirement of the design.

At the beginning, as the vehicles fly in subsonic, the temperature of the structure surface raises up gradually. The temperature climbed dramatically as the vehicle speed up into supersonic. As the vehicle continues to accelerate in this supersonic flight regime, the stress and deformation of the structure tends to be stable. That's because the temperature of the two sides tends to equal. This again testifies that the pressure only contributes little to the stress and deformation of the structure compare with temperature.

5. Conclusion

The work described in this paper focuses on the application of a FSI algorithm. The goal is to use this algorithm to study radome for IR seeker, thereby providing valid information for the design of the radome. 50 seconds transient simulation of aerodynamic heating process of radome was conducted, which provided boundary conditions for the structural analysis. Through structural analysis, the stress of each part, deformation of the structure, the total force of each part and safety factor are obtained.

The results illuminated that the center and edge of the glass would be the risk areas. The numerical simulations also successfully predicted the aerodynamic heating effects, which is the main reason for the stress and deformation of the structure. The analysis testified that all the parts meet the requirement of the design.

References

1. HE You-jin, ZHANG Peng, PENG Jun, et al. Aerodynamic heating and stress analysis on dome of infrared high speed air to air missiles (in Chinese). Infrared Technology, 2007, 29(7):373-376.
2. Yanping Fan. Structure design and stress analysis of the cooled-cover(in Chinese) [D]. Nanjing: Nanjing University of Science and Technology,2007.
3. Charbel Farhat, Philippe Geuzaine, Gregory Brown and Chuck Harris. Nonlinear Flutter Analysis of an F-16 in Stabilized, Accelerated, and Increased Angle of Attack Flight Conditions. AIAA 2002-1490.
4. Shyy,W. Computational Modeling for Fluid Flow and Interfacial Transport, Elsevier, Amsterdam. The Netherlands, 1994.
5. Galina Sieber. Numerical Simulation of Fluid-Structure Interaction Using Loose Coupling Methods. Fachbereich Maschinenbau an der Technischen Universität Darmstadtzur Erlangung des Grades eines Doktor-Ingenieurs. Genehmigte, 2002.

IN-PLANE STABILITY OF DERRICK OF INCLINED DRILLING RIG

Z. LIANG† G.H. ZHAO F.G. JIANG J.L. TIAN J.J. DENG

Mechatronic Engineering College, Southwest Petroleum University Chengdu, Sichuan, China

Derrick of inclined drilling rig is more prone to compressive-bending instability because the weight of derrick and its accessories has transverse component of force. Here K-type derrick of inclined drilling rig that was top driving was simplified as outstretch beam. Considering influence of shear deformation on stability, mechanical model was established by static method. Through iterating simple-supported part and cantilever part in turn, we obtain deflection curve and moment distribution. Mechanical characters that are influenced by inclination angle of derrick and longitudinal load are analysed. Under the most disadvantageous attitude, stability analysis presents that the derrick of certain home-made rig is still in the elastic state and satisfies the requirement of in-plane stability.

Keywords: Inclined drilling rig; Derrick; Top drive; Lattice beam; In-plane stability.

1. Introduction

Inclined drilling rig whose derrick makes an angle of 45°~90° with horizontal level slants hole from earth surface. Inclined drilling rig is suitable for developing shallow hydrocarbon resources, and can cut down expenses by 50%. On the other hand inclined derrick is more prone to buckling than conventional derrick under longitudinal compressive load. At present most of researches focus on upright derrick and numerical method is usually adopted because of complexity of derrick structure.[1-2] Here K-type derrick of inclined drilling rig is simplified as outstretch beam, and snap-through buckling process is analysed.

2. Mechanical Model

Derrick of inclined drilling rig is simplified as outstretch beam that is sketched in Figure 1a. The beam is loaded with uniform gravity load q', concentrated gravity load G_{DY} of top driver and traveling block and axial compression load P_0, which comes from travelling block and acts on crown block at cantilever end. In order to analyse stability, the derrick is further decomposed into cantilever part BC and

† Corresponding author, email: liangz_2242@126.com

simple-supported part AC that are shown in Figure 1b and 1c. Lateral component of uniform gravity of derrick is described as $q = q'\cos\theta$, and the lateral concentrated force is: $Q = G_{DY}\cos\theta$. Axial forces are assumed invariable longitudinally and denoted by concentrated force P_1 and P_2, which act at beam ends and can be obtained through self-weight dichotomizing. Here K-type derrick is simplified as 4-limb lattice beam-column with tie plates. Figure 1d sketches its cross section. Referring to influence of shearing deformation on compressive bar, equivalent flexural rigidity $E'I'$ of derrick can be derived as:[3]

$$E'I' = \frac{EI}{1 + \dfrac{\pi^2 EI}{(EA_d \sin^2\beta\cos\beta)\cdot(\mu l)^2}}. \tag{1}$$

In which, EI is flexural rigidity of the cross section with respect to imaginary axis i. μ is length factor of derrick. A_d is cross section area of inclined tie plate. β is crossing angle of inclined tie plate and column.

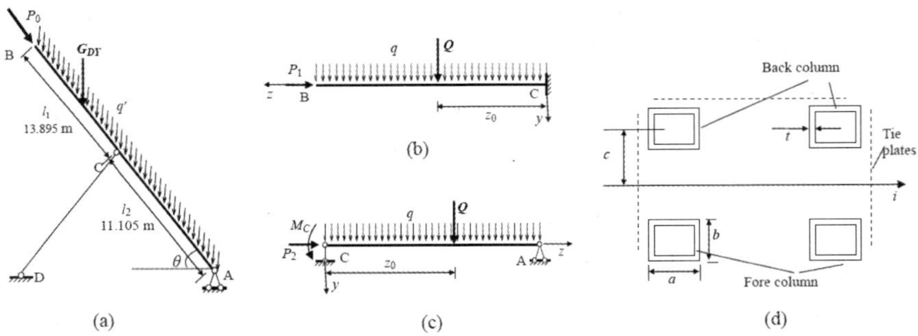

Fig. 1. (a) Sketch of inclined derrick. (b) Cantilever part corresponding to upper segment of derrick. (c) Simple-supported part corresponding to lower segment. (d) Cross section of derrick

Firstly, we establish mechanical model of simple-supported part AC. Figure 1c shows the case that top driver is located in simple-supported part (contrariwise letting Q be zero). For the part AC, differential equation of equilibrium can be derived by moment analyzing of insulator:[4]

$$\begin{cases} y'' + k_2^2 y = \dfrac{qz^2}{2E'I_2'} - \dfrac{1}{E'I_2'}\left(\dfrac{M_c}{l_2} + \dfrac{ql_2}{2} + Q\cdot\dfrac{l_2 - z_0}{l_2}\right)z + \dfrac{M_c}{E'I_2'}, & z \in [0, z_0] \\ y'' + k_2^2 y = \dfrac{qz^2}{2E'I_2'} - \dfrac{1}{E'I_2'}\left(\dfrac{M_c}{l_2} + \dfrac{ql_2}{2} - Q\cdot\dfrac{z_0}{l_2}\right)z + \dfrac{M_c - Q\cdot z_0}{E'I_2'}, & z \in [z_0, l_2] \end{cases} . \tag{2}$$

where $k_2^2 = P_2/(E'I_2')$, and $E'I_2'$ is equivalent flexural rigidity of lower segment of derrick. The boundary conditions are given as expression (3):

$$\begin{cases} y|_{z=0} = 0; \quad y|_{z=l_2} = 0; \\ y|_{z\in[0,z_0]\, and\ z=z_0} = y|_{z\in[z_0,l_2]\, and\ z=z_0}; \quad y'|_{z\in[0,z_0]\, and\ z=z_0} = y'|_{z\in[z_0,l_2]\, and\ z=z_0}. \end{cases} \tag{3}$$

Thus we can get deflection curve of simple-supported part, and further get bending angle φ_c of section C. Here expression of φ_c includes bending moment M_C of section C, which can be obtained by analyzing cantilever part CB. Mechanical model and boundary conditions of cantilever part can be obtained similarly. Based on monodromic conditions of force and deformation at section C of outstretched beam, we can get deformation and internal force by iterating simple-supported part and cantilever part in turn.

3. In-plane Stability Analysis

Here certain home-made derrick of inclined drilling rig is taken as an example. The topmost position of top driver is denoted by $z_{0\max} = 7.1364$m at upper segment, and the maximum hook load is 170T. Other parameters are chosen as: $q' = 1.2748$ Tm^{-1}, $\theta = 45°\sim90°$, $G_{DY} = 11.35$T.

3.1. *Effect of inclination angle on mechanical behavior*

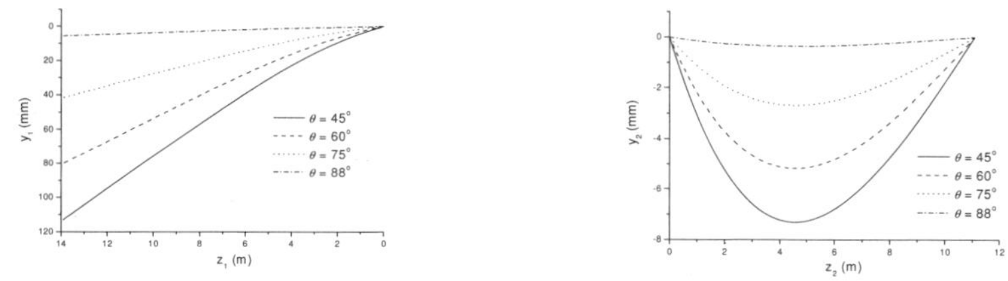

(a) (b)

Fig. 2. Deflection curve of the derrick, (a) Upper segment; (b) Lower segment.

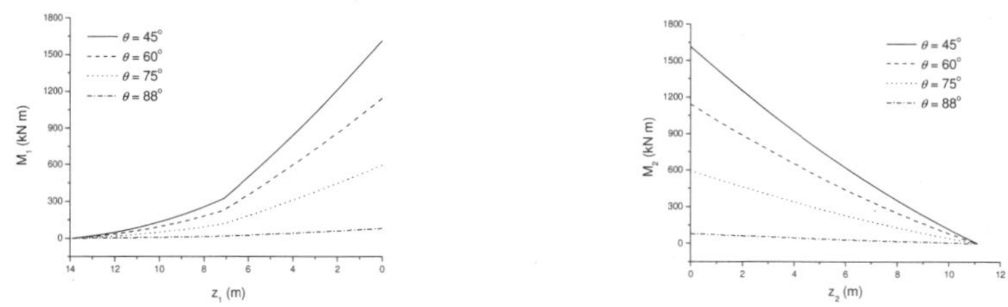

(a) (b)

Fig. 3. Bending moment distribution curve, (a) Upper segment; (b) Lower segment.

Considering that the derrick stands the maximum hook load and top driver is located at the topmost position, we can get influences of inclination angle θ of derrick on its deformation and bending moment. Figure 2 shows deflection curves of derrick for different θ. It's found that lateral deformation of derrick increases along with inclination angle decreasing and the overall deformation is the maximum at $\theta = 45°$. For upper segment of derrick the maximum deflection happens at cantilever end, and for lower segment the section of maximum deflection is not fixed. Lateral deformation of upper segment is much larger than that of lower segment and the maximum value is 112.95 mm at cantilever end. Figure 3 is bending moment distribution for different θ. Larger overall bending moment of the derrick corresponds to less θ, and the maximum bending moment happens at section *C*. Therefore $\theta = 45°$ is the most dangerous inclination angle for derrick to buckling.

3.2. *Influence of axial compressive force on stability*

Snap-through buckling may happen to derrick because that both longitudinal compressive force and lateral force acts on it. Here pulling out of hole (POOH) behavior is analysed by means of increasing weight of drillstring G_{DS}. During POOH, $P_0 = G_{DS}(\sin\theta + \mu\cos\theta)$, where μ is friction coefficient between drillstring and hole wall and chosen as 0.5. Destabilizing processes of derrick for different inclination angles are plotted in Figure 4, in which top driver and traveling block is located at topmost and the curves of the maximum deflection y_{max} *versus* G_{DS} for different inclination angles are presented. As axial compressive force is small, lateral deformation increases slowly along with enhance of compression, and derrick is in stable equilibrium state. When longitudinal compressive force exceeds a certain critical value, which corresponds to compressive-bending buckling, deflection would increase rapidly along with enhance of compression force.

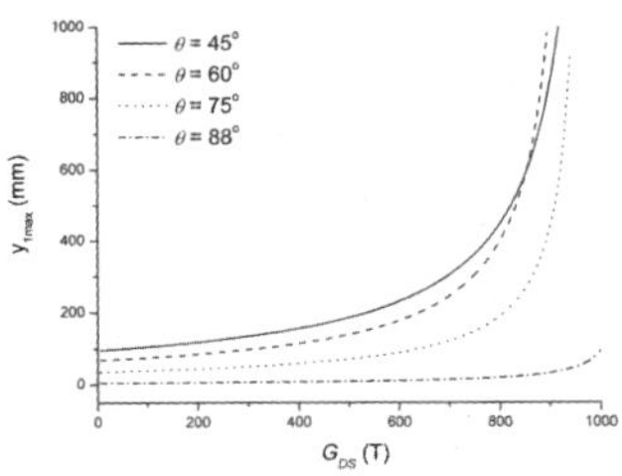

Fig. 4. y_{max} varying with G_{DS} and θ.

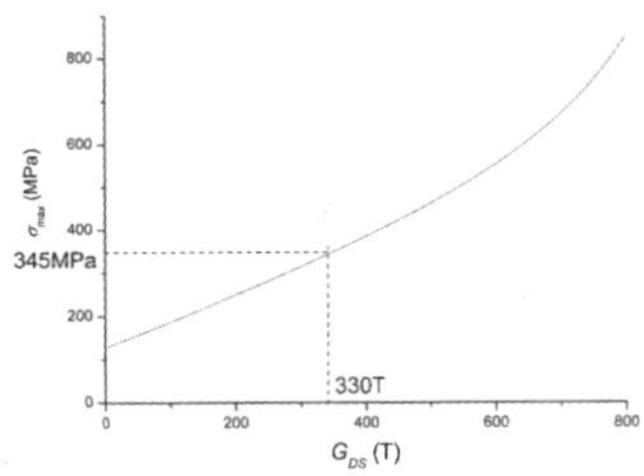

Fig. 5. The max. stress varying with G_{DS}.

Figure 5 shows the maximum stress varying with axial compressive force. Here the inclination angle of derrick is 45° and top driver is located at the topmost position. The derrick is made of steel Q345. From Figure 5 we find that yield happens to the section with the maximum bending moment when G_{DS} reaches 330T. For the inclined drilling rig discussed here, the maximum hook load is only 170T that is far less than 330T. Therefore derrick is still in the elastic state and satisfies the requirement of in-plane stability.

4. Conclusion

Mechanical model of K-type derrick of inclined drilling rig that was top driving was established by static method. The derrick was simplified as outstretch lattice beam-column. By means of iterating upper and lower segment of derrick in turn, snap-through buckling process is presented. It's found that the maximum deflection happens at cantilever end. The state that inclination angle of derrick is 45° and top driver is located at the topmost position is the most prone to buckling. Compressive-bending instability can only occur to upper segment of derrick because that buckling load of cantilever part is much less than that of simple-supported part. Under the condition that axial compressive force is much less than buckling load, some local part of derrick yields. And for the working condition of this home-made rig, derrick satisfies the requirement of intensity and stability.

In this mechanical model some assumptions have been made. For example: lateral displacement of section C is ignored, longitudinal force is regarded as invariable axially, the derrick is simplified as regular 4-limb lattice beam-column and so on. Compared with simulation of Ansys, the conclusions drawn in this paper are in qualitative agreement with the numerical results.

Acknowledgments

This work is supported by Key Laboratory of Oil and Gas Equipment from Ministry of Education (2006STS04) and Science Foundation of SWPU.

References

1. J. Zhu, L.Q. Zou and J.C. Leng: *Oil Field Equipment*, 2008, **37**, (3), 45-48.
2. G.Z. Lu, Q.Y. Jiao: *J. China Agricultural University*, 2007, **12**, (2), 88-91.
3. Z.B. Xia, J. Yao: 'Steel Structures-Theory and Design', 2004, Beijing, China architecture and building press.
4. J. Chen: 'Stability of Steel Structures Theory and Design', 2006, Beijing, Science Press.

TORSIONAL STIFFNESS OF LAMINATED GLASS ELEMENTS IN STRUCTURAL APPLICATIONS — INFLUENCE OF A ELASTO-VISCOPLASTIC IONOMER INTERLAYER ON THE PRE-BREAKAGE BEHAVIOUR

DIETER CALLEWAERT[†]
JAN BELIS, DIDIER DELINCÉ, DELPHINE SONCK, RUDY VAN IMPE
Department of Structural Engineering, Ghent University, Technologiepark 904
9052 Ghent (Zwijnaarde), Belgium

To investigate the mechanical behaviour of laminated glass with the elasto-viscoplastic ionomer interlayer SentryGlas® by DuPont, an experimental program was executed on full-scale glass laminates. The results of this research improved the insight in the complex mechanical behaviour of this stiff interlayer. More specifically, a clear influence of load duration and temperature could be observed for both creep and relaxation during torsion experiments. Namely, within the tested temperature range of 5 °C up to 65 °C, torsional stiffness changed with a factor 3.

Keywords: Laminated glass; Shear transfer; Interlayer; Elasto-viscoplastic; Experimental.

1. Laminated Structural Glass

The last two decades, the use of glass in transparent structural applications has become increasingly successful. This is mainly the result of a growing knowledge of the mechanical behaviour of this brittle material and an amelioration of the available safety concepts, which are necessary for the preservation of the structural integrity of the load-bearing elements. The most applied technique to add some plasticity to a structural glass component consists of combining soft interlayers with the brittle glass sheets. Specifically for this growing market of structural laminated glass, DuPont de Nemours developed the visco-elastic ionomer interlayer material SentryGlas® with enhanced edge stability, stiffness and strength compared to more traditional interlayer materials, such as PVB and EVA [1,2]. Because a thorough understanding of the complicated mechanical behaviour of this ionomer would be highly relevant to industry but has not yet been accomplished for full-scale laminated glass components, a thorough experimental test program has been executed at Ghent University.

[†] Corresponding author, email: Dieter.Callewaert@UGent.be

2. Experimental Programme

At the Department of Structural engineering of Ghent University, both the pre-failure and the post-failure [3,4] influence of SentryGlas® (SG) on the mechanical behaviour of laminated glass is investigated. For the pre-failure research, a combination of torsion and bending experiments on full-scale laminated glass plates is used. The results of the long-duration, torsion creep and relaxation experiments at different temperatures form the base of this paper. The bending experiments are presented and discussed in relation to these torsion results in a second contribution [5].

3. Torsion Experiments

3.1. *Test specimen*

For these tests, 20 laminated samples were made, further referred to as samples A. In Table 1, an overview of their main geometrical characteristics is presented. Because it is generally known that the effective dimensions of products differ from their nominal values and the thickness is highly significant for the calculation of the torsion stiffness because it is raised to the third power, every effective thickness was measured accurately before and after lamination. The following processing of the results is always based on these effective measured dimensions.

Table 1. Nominal dimensions of the test specimens.

Appellation	Number	Type of glass	Length [mm]	Width [mm]	Glass thickness [mm]	SG thickness [mm]
A	20	toughened	1100	360	2 x 6	1.52

3.2. *Experimental setup and procedure*

Based on the experimental research reported by Kasper [6], a torsion test setup was built at Ghent University. Founded on the results of numerical simulations, some modifications of the test setup were implemented [7]. In this setup, a laminated sample is clamped in a fixed support and a hinged support with a lever arm. By applying a dead load or enforcing a certain deformation at the end of this lever arm, the sample is torsed about its longitudinal axis. In Figure 1, the torsion test setup is used for creep (constant torque with dead weight) and relaxation (constant torsional deformation).

Fig. 1. Torsion creep (left) and relaxation (right) experiment.

In a first test cycle, all 20 test specimens were subjected to a relaxation torsion test. More specifically, four relaxation tests of at least 48 hours have been executed at five different temperatures between 5 °C and 65 °C (each time with a temperature interval of 15 °C). These environmental conditions were established in a climatic chamber, containing the entire test setup and the tested samples for at least 60 hours prior to testing.

Subsequently, in a second test cycle, the same twenty specimens were subjected to a torsion creep test. Taking into account the stiffness reduction at increased temperatures, different load levels have been applied at the five different temperatures. This way, after a load duration of 24 hours, all samples had a torsional deformation with a comparable order of magnitude as the relaxation tests. Afterwards, supplementary relaxation tests were performed at intermediate temperatures (10, 15, 25, 30, 40, 45, 55 and 60 °C) to narrow down the temperature interval. With this, a more complete and precise image of the mechanical behaviour of the material could be rendered.

3.3. *Results*

The results of all torsion tests are summarised in Figures 3 and 4. Each curve represents the average of four experiments at a certain temperature. The lightest curves are the results at the lowest temperature (5 °C), while the increasing darkness visualises a raising temperature level (up to 65 °C). All measured loads and deformations have been transformed into a proportional torsional stiffness of the laminate, being the experimental stiffness compared to the theoretical torsional stiffness of a monolithic reference glass plate with the effective dimensions of each sample. The thickness of the monolithic reference glass plate was taken equal to the overall thickness of the laminate. This representation method was chosen, to meaningfully compare the different experiments.

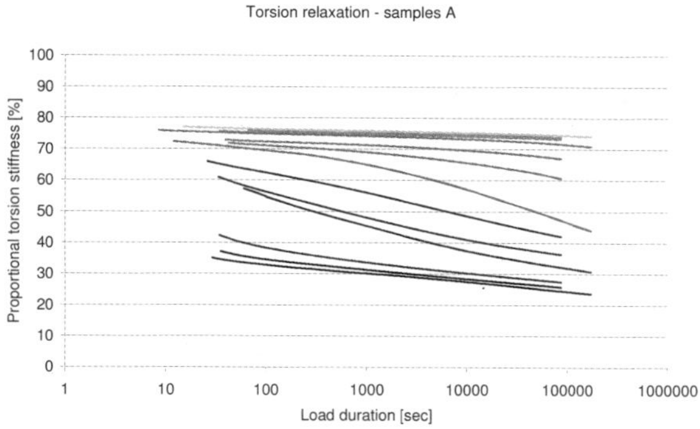

Fig. 2. Proportional torsion stiffness of samples A during relaxation at 5, 10, 15, 20, 25, 30, 35, 40, 45, 50, 55, 60 and 65 °C.

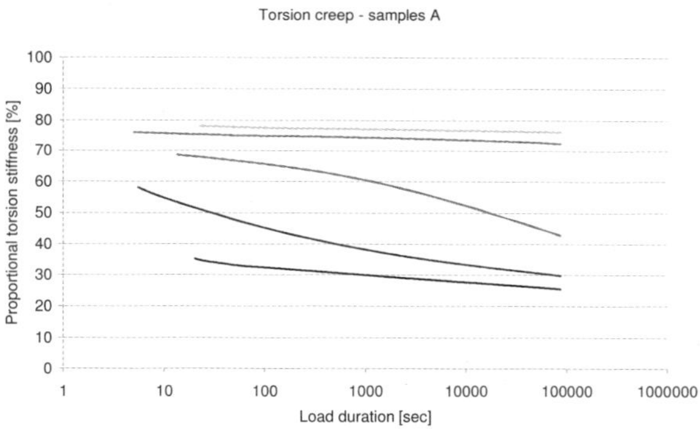

Fig. 3. Proportional torsion stiffness of samples A during creep at 5, 20, 35, 50 and 65 °C.

4. Discussion and Conclusion

Although the shear modulus of the interlayer is at least a factor 100 smaller than the shear modulus of glass, it is clear that the material properties of the visco-elastoplastic ionomer interlayer material extremely influence the overall mechanical behaviour of the laminate also before fracture of a glass plate. By increasing the temperature or the load duration, the stiffness of the interlayer decrease and the proportional stiffness of the laminate reduces significantly.

When comparing the creep tests with the equivalent relaxation tests, the results are almost identical. The proportional torsional stiffness almost reaches the lowest possible limit at the end of the experiments with the highest tested temperature - comparable with the stiffness of two glass sheets with no shear transfer at all.

In Ref. 5, these results are further discussed in comparison with the results of the bending experiments, which were executed simultaneously.

Acknowledgments

This work was supported by the Research Foundation-Flanders (FWO-Vlaanderen. grant nr. 3G018407). Additionally, the authors wish to acknowledge the support of Lerobel for the lamination of all test samples.

References

1. V. Block and P. Davies: Proc. GPD2005, Tampere, Finland, 2005, 118-119.
2. I. Stelzer, S. Bennison and M. Qin: Proc. 3rd Int. Symp. on 'Application of Architectural Glass' (ISAAG), Munich, Germany, 2008, 65-77.
3. D. Delincé, D. Callewaert, W. Vanlaere, J. Belis and J. Depauw: *Int. J. Mod. Phys. B*, **22**, (31-32), 2008, 5447-5452.
4. J. Belis, D. Delincé, D. Callewaert, R. Van Impe and J. Depauw: *Int. J. Mod. Phys. B*, **22**, (31-32), 2008, 5509-5514.
5. D. Callewaert, J. Belis, D. Delincé, D. Sonck and R. Van Impe: Proc. AEPA2010, Wuhan, China – in press.
6. R. Kasper 'Tragverhalten von Glasträgern' PhD. Dissertation, Stahlbau - RWTH Aachen, Germany, 2005.
7. D. Callewaert, J. Belis, R. Van Impe, G. Lagae and W. Vanlaere: Proc. GPD2007, Tampere, Finland, 2007, 118-121.

BENDING STIFFNESS OF LAMINATED GLASS ELEMENTS IN STRUCTURAL APPLICATIONS — INFLUENCE OF A ELASTO-VISCOPLASTIC IONOMER INTERLAYER ON THE PRE-BREAKAGE BEHAVIOUR

DIETER CALLEWAERT[†]
JAN BELIS, DIDIER DELINCÉ, DELPHINE SONCK, RUDY VAN IMPE
Department of Structural Engineering, Ghent University, Technologiepark 904
9052 Ghent (Zwijnaarde), Belgium

To investigate the mechanical behaviour of laminated glass with the elasto-viscoplastic ionomer interlayer SentryGlas® by DuPont, an experimental program was executed on full-scale glass laminates. The results of this research improved the insight in the complex mechanical behaviour of this stiff interlayer. More specifically, a clear influence of load duration and temperature could be observed for both creep and relaxation setups during bending experiments. The stiffness reduction due to the load duration and the temperature raise was most significant for the smaller specimens — with a span of 1m — while for the largest specimen — with a span of almost 3 meter — the reduction remained less than 25% of the monolithic bending stiffness, even after a load duration of 48 hours at 65°C.

Keywords: Laminated glass; Shear transfer; Interlayer; Elasto-viscoplastic; Experimental.

1. Introduction

Because laminated glass becomes increasingly popular in structural applications, both the pre-failure and the post-failure [1,2] influence of a SentryGlas® (SG) interlayer on the overall mechanical behaviour of laminated glass is investigated. The results of the long-duration, full-scale, bending creep and relaxation experiments at different temperatures — executed at the Department of Structural Engineering of Ghent University — form the base of this paper. At the end, these results are also compared to the results of the torsional experiments, presented in an other contribution of the same authors [3].

[†] Corresponding author, email: Dieter.Callewaert@UGent.be

2. Bending

2.1. *Experimental setup*

For the bending experiments, one of the most simple loading conditions was chosen, namely three-point bending. A specific setup was designed in order to be able to perform three-point bending tests about the weak axis with the specimen placed vertically. This way, the self-weight acted about the strong axis of the sample, not causing interfering deflections for the experiment [4]. Figure 1 illustrates the small and large test setups on which the bending experiments have been executed.

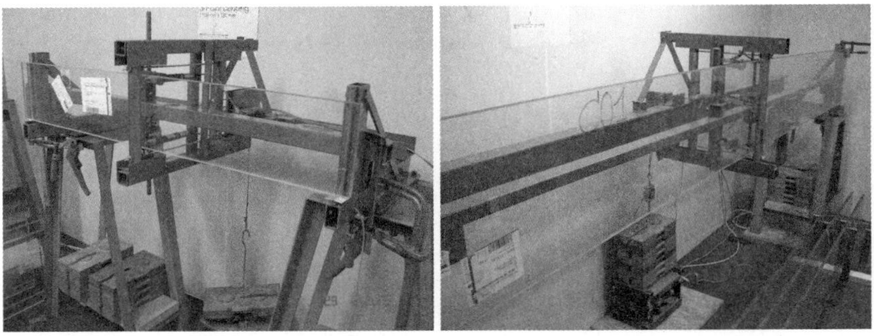

Fig. 1. Small bending creep experiment for test samples B and E with a span of 1050 mm (left) and large bending relaxation experiment for test samples C with a span of 2950 mm (right).

Two different samples lengths had been applied, because preliminary experiments indicated an important influence of the span on the proportional bending stiffness [5]. Additionally, the smallest tests were performed on samples with two different widths (series B and E) to investigate whether a significant edge influence existed. The nominal dimensions of the sample series B, C and E for bending are reported in Table 1. But for all these samples, the effective dimensions were determined as accurately as possible for the further processing of the results.

Table 1. Nominal dimensions of the test specimens.

Appellation	Number	Type of glass	Length [mm]	Width [mm]	Glass thickness [mm]	SG thickness [mm]
B	20	annealed	1100	180	2 x 8	1.52
C	8	toughened	3000	360	2 x 8	1.52
E	20	annealed	1100	120	2 x 8	1.52

2.2. *Experimental procedure*

Simultaneously with the first torsion test cycle [3], creep bending tests with a load duration of at least 48 hours were executed on the samples B and the first four samples of series C at 5, 20, 35, 50 and 65 °C. The dead load on the small and large setup was 150 N and 400 N respectively. During a second cycle with identical temperature levels, similar creep tests were performed on the samples E with a load of 100 N (identical width/load ratio as series B), while the other four samples of series C were subjected to bending relaxation tests. Finally, also more bending creep series were performed on the samples B and the first four samples of series C with a temperature interval of only 5 °C.

2.3. *Results*

All the results of the experiments are summarised in Figures 2-5. In these, a proportional bending stiffness of 100% corresponds to the bending stiffness of a monolithic reference glass plate with the effective dimensions of each sample individually. This overall thickness thus also include the interlayer thickness.

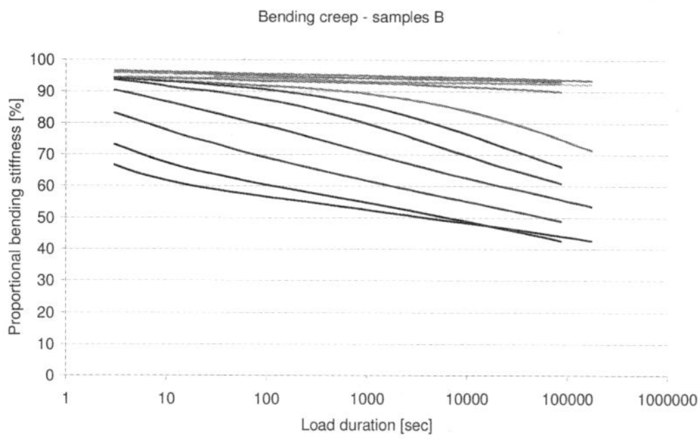

Fig. 2. Proportional bending stiffness of samples B during creep at 5, 10, 15, 20, 25, 30, 35, 40, 45, 50, 55, 60 and 65 °C.

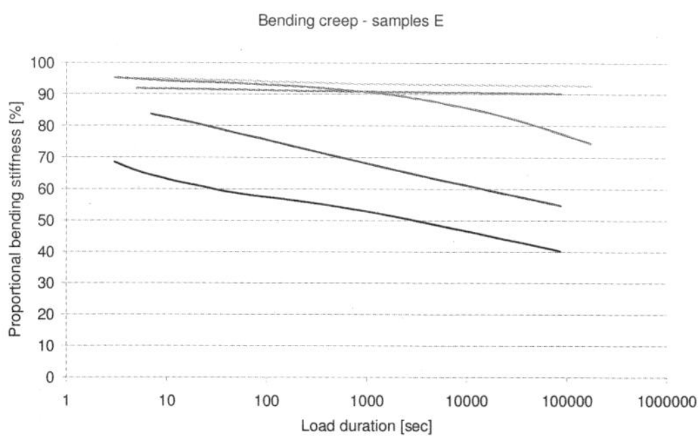

Fig. 3. Proportional bending stiffness of samples E during creep at 5, 20, 35, 50 and 65 °C.

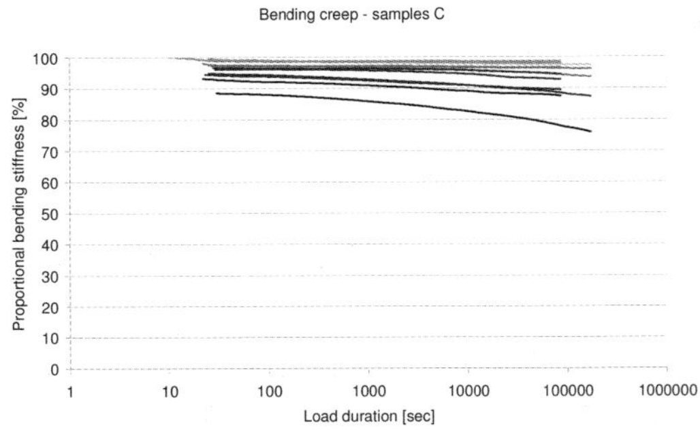

Fig. 4. Proportional bending stiffness of samples C during creep at 5, 10, 15, 20, 25, 30, 35, 40, 45, 50, 55, 60 and 65 °C.

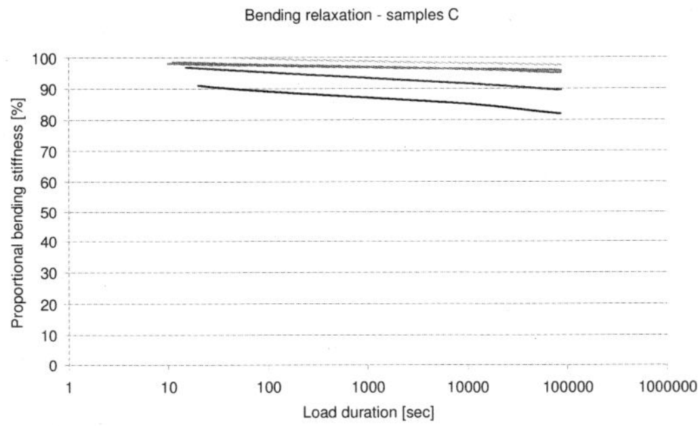

Fig. 5. Proportional bending stiffness of samples C during relaxation at 5, 20, 35, 50 and 65 °C.

3. Discussion

Comparing the creep tests with the equivalent relaxation tests, the results are almost identical. And also for the tests on the samples of series B and E, where only the width of the samples differ, the results show no significant dissimilarity.

When, on the other hand, these bending tests are compared with the torsion tests [3], a major influence of the load type (torsion versus bending) and the geometry (small bending versus large bending) becomes manifest. While the proportional torsion stiffness almost reaches the lowest possible limit at the end of the experiments with the highest tested temperature — comparable with the stiffness of two glass sheets with no shear transfer at all — the decrease is less pronounced for the proportional bending stiffness. But also for the bending experiments, there is a major difference between the small and the large samples with identical temperature condition, load duration and assembly.

4. Conclusion

From all these experiments, it is clear that generalisations between the material properties of a visco-elastic interlayer — determined by the temperature and the load duration — and the overall stiffness of the laminate are not unambiguous. An increase of the span of a plate subjected to bending for instance, increases the proportional stiffness. Also the difference in proportional torsion stiffness compared to the proportional bending stiffness proves that the loading condition of the laminate always has to be taken into account.

It is therefore advisable to directly calculate the stiffness of a laminate for each particular case based on a mechanical material model of the interlayer, rather than searching for a direct connection factor between a certain interlayer stiffness and the overall proportional stiffness of the laminate. This last method might lead to misleading and incorrect assumptions such as the interpretation that some interlayer materials always lead to a quasi monolithic behaviour of the laminated glass, while others are presumed to never be able to transfer any shear between the glass sheets. For instance, if the shear transfer ability of a laminate is established based on the above mentioned large bending test setup, an extrapolation towards the torsional stiffness could be extremely unsafe.

Acknowledgements

This work was supported by the Research Foundation-Flanders (FWO-Vlaanderen. grant nr. 3G018407). Additionally, the authors wish to acknowledge the support of Lerobel for the lamination of all test samples.

References

1. D. Delincé, D. Callewaert, W. Vanlaere, J. Belis and J. Depauw: *Int. J. Mod. Phys. B*, **22**, (31-32), 2008, 5447-5452.
2. J. Belis, D. Delincé, D. Callewaert, R. Van Impe and J. Depauw: *Int. J. Mod. Phys. B*, **22**, (31-32), 2008, 5509-5514.
3. D. Callewaert, J. Belis, D. Delincé, D. Sonck and R. Van Impe: Proc. AEPA2010, Wuhan, China – in press.
4. D. Callewaert, J. Depaepe, K. De Vogel, J. Belis, D. Delincé and R. Van Impe: Proc. 3rd Int. Symp. on 'Application of Architectural Glass' (ISAAG), Munich, Germany, 2008, 51-63.
5. D. Callewaert, D. Delincé, J. Belis and R. Van Impe: Proc. Challenging Glass Conf. on 'Architectural and Structural Applications of Glass', Delft, The Netherlands, 2008, 431-437.

ON THE INELASTIC CONSTITUTIVE EQUATIONS OF PLATES AND SHELLS MADE OF FOAMS

HOLM ALTENBACH[†] & VICTOR A. EREMEYEV

Chair of Engineering Mechanics, Martin-Luther-University Halle-Wittenberg, Kurt-Mothes-Str. 1, D-06120, Halle (Saale), Germany

Within the framework of the shear deformable linear theory of plates and shells we discuss the viscoelastic models of two-dimensional (2D) constitutive equations for the stress resultants (forces and moments). The main goal of the paper focuses on the analysis of the influence of the viscoelastic bulk material behaviour, the through-the-thickness structure of the plate and surface effects on the resultant governing equations. The two-dimensional constitutive equations are represented using the relaxation functions. As an example 2D constitutive equations for plates and shells made of viscoelastic foams are considered. In addition, the functionally graded structures are presented. Finally, we consider the scaling effect in the case of nano-sized foams using the theory of surface stresses.

Keywords: Foam; viscoelasticity; plates and shells; surface tension.

1. Introduction

In the modern engineering there are many two-dimensional models of structures such as plates and shells. The properties of these two-dimensional bodies are described by using proper constitutive equations written for 2D stress and strain measures. Such constitutive equations depend on the behaviour of bulk material, surface effects and the geometry of structure. Here we discuss how 2D constitutive equations inherit the viscoelastic properties of bulk material together with the geometry of 3D body. We consider plates and shells made of polymers or metal foams with non-homogeneous porosity distribution and derive the corresponding constitutive equations. Foams are a very perspective class of materials for modern engineering applications. Metallic and polymeric foams are applied as a material for lightweight structures which are used in civil engineering, in the automotive or aerospace industries since they combine low weight, high specific strength, and excellent possibilities to absorb energy, see [1, 2]. The technical realization is mostly performed as sandwich panels (plates or shells with hard and stiff face sheets and a weak core layer made of foam). The mechanics of viscoelastic isotropic and anisotropic plates is developed in [3-5] among others.

[†] Corresponding author, email: holm.altenbach@iw.uni-halle.de

Within the framework of the linear viscoelasticity we show that 2D relaxation functions of shells have more complicated structures then the relaxation function of the bulk material. In particular, for non-homogeneous shell the spectrum of relaxation time depends not only on the spectrum of bulk material but also on the structure of the shell in the thickness direction. This means 2D structures demonstrate more complicated viscoelastic behaviour then the bulk material. As an example the constitutive equations of viscoelastic plates and shells are presented for functionally graded.

We also discuss the transition to the mechanics of nano-sized foams taking into account the surface stresses. In this case we use the model of surface stresses which generalizes the surface tension in the theory of capillarity. We show how the surface properties of the bulk foam material influence on the inelastic behaviour of plates and shells made of open-cell foams with nano-sized cells.

2. Basic Equations of Linear Theory of Plates and Shells

Let us consider the geometrically and physically linear theory of plates or shallow shells. The equations of motion are formulated as the Euler's laws of dynamics [5]

$$\nabla\cdot\mathbf{T}+\mathbf{q}=\rho\mathbf{u}+\rho\mathbf{\Theta}_1\cdot\boldsymbol{\varphi},\quad \nabla\cdot\mathbf{M}+\mathbf{T}_\times+\mathbf{m}=\rho\mathbf{\Theta}_1{}^T\cdot\mathbf{u}+\rho\mathbf{\Theta}_2\cdot\boldsymbol{\varphi},\tag{1}$$

where $\mathbf{T}$ and $\mathbf{M}$ are the tensors of forces and moments,

$$\mathbf{T}=T_1\mathbf{e}_1\mathbf{e}_1+T_2\mathbf{e}_2\mathbf{e}_2+T_{12}(\mathbf{e}_1\mathbf{e}_2+\mathbf{e}_2\mathbf{e}_1)+T_{1n}\mathbf{e}_1\mathbf{n}+T_{2n}\mathbf{e}_2\mathbf{n},\tag{2}$$

$$\mathbf{M}=M_1\mathbf{e}_1\mathbf{e}_2-M_2\mathbf{e}_2\mathbf{e}_1-M_{12}(\mathbf{e}_1\mathbf{e}_1-\mathbf{e}_2\mathbf{e}_2),$$

$\mathbf{u}$ and $\boldsymbol{\varphi}$ are the vectors of displacements and rotations,

$$\mathbf{u}=u_1\mathbf{e}_1+u_2\mathbf{e}_2+w\mathbf{n},\quad \boldsymbol{\varphi}=-\varphi_2\mathbf{e}_1+\varphi_1\mathbf{e}_2,\tag{3}$$

$\mathbf{n}$ is the unit outer normal vector at the plate surface, $\mathbf{e}_1,\mathbf{e}_2$ are orthonormal vectors in the tangent plane, $\mathbf{q}$ and $\mathbf{m}$ are the vectors of surface loads (forces and moments), $\mathbf{T}_\times$ is the vector invariant of the force tensor, ∇ the nabla (Hamilton) operator, $\mathbf{\Theta}_1$ and $\mathbf{\Theta}_2$ are the first and the second tensor of inertia, ρ is the density (effective property of the deformable surface), the upper index T denotes transposed, and $(\cdots)^{\cdot}$ the time derivative, respectively.

In the case of orthotropic behaviour the following constitutive equations for the stress resultants are valid:

$$\mathbf{T}\cdot\mathbf{a}=\int_{-\infty}^{t}\mathbf{A}(t-\tau)\cdot\cdot\boldsymbol{\mu}\,d\tau+\int_{-\infty}^{t}\mathbf{B}(t-\tau)\cdot\cdot\boldsymbol{\kappa}\,d\tau,\quad \mathbf{T}\cdot\mathbf{n}=\int_{-\infty}^{t}\mathbf{\Gamma}(t-\tau)\cdot\cdot\boldsymbol{\gamma}\,d\tau,\tag{4}$$

$$\mathbf{M}^{\mathrm{T}}=\int_{-\infty}^{t}\boldsymbol{\mu}\cdot\cdot\mathbf{B}(t-\tau)\,d\tau+\int_{-\infty}^{t}\mathbf{C}(t-\tau)\cdot\cdot\boldsymbol{\kappa}\,d\tau,$$

where $\mathbf{A}$, $\mathbf{B}$, $\mathbf{C}$ are fourth-order tensors, and $\mathbf{\Gamma}$ is a second-order tensor. They describe the relaxation functions of the plate, $\boldsymbol{\mu}$, $\boldsymbol{\kappa}$, and $\boldsymbol{\gamma}$ are the tensor of in-plane strains, the tensor of the out-of-plane strains, and the vector of transverse shear strains, respectively. They are given by the relations

$$2\boldsymbol{\mu} = \nabla\mathbf{u} + (\nabla\mathbf{u})^{\mathrm{T}}, \ \boldsymbol{\kappa} = \nabla\boldsymbol{\varphi}, \ \boldsymbol{\gamma} = \nabla\mathbf{u} - \mathbf{a}\times\boldsymbol{\varphi}.$$

Here **a** is the first metric tensor.Equations should be centered and numbered consecutively, as in Eq. (1). An alternative method is given in Eq. (2) for long sets of equations where only one referencing equation number is wanted.

3. Relaxation Functions

The 2D material properties of plates and shells described by the relaxation function **A**, **B**, **C**, and $\boldsymbol{\Gamma}$. In the case of isotropic and symmetric over the thickness plates the effective relaxation tensors have the following form ($\mathbf{B} = \mathbf{0}$)

$$\mathbf{A} = A_{11}\mathbf{a}_1\mathbf{a}_1 + A_{22}(\mathbf{a}_2\mathbf{a}_2 + \mathbf{a}_4\mathbf{a}_4), \ \mathbf{C} = C_{22}\mathbf{a}_2\mathbf{a}_2 + C_{44}(\mathbf{a}_3\mathbf{a}_3 + \mathbf{a}_4\mathbf{a}_4), \ \boldsymbol{\Gamma} = \Gamma\mathbf{a},$$

where $\mathbf{a}_1 = \mathbf{a} = \mathbf{e}_1\mathbf{e}_1 + \mathbf{e}_2\mathbf{e}_2$, $\mathbf{a}_2 = \mathbf{e}_1\mathbf{e}_1 - \mathbf{e}_2\mathbf{e}_2$, $\mathbf{a}_3 = \mathbf{e}_1\mathbf{e}_2 - \mathbf{e}_2\mathbf{e}_1$, $\mathbf{a}_4 = \mathbf{a} = \mathbf{e}_1\mathbf{e}_2 + \mathbf{e}_2\mathbf{e}_1$.

The procedure identification of the relaxation functions for non-homogeneous material was presented in [3,4] using the Laplace transform. The Laplace transform of the components of the relaxation tensors are given by

$$\overline{A}_{11} = \frac{1}{2}\left\langle \frac{\overline{E}}{1-\overline{\nu}} \right\rangle, \ \overline{A}_{22} = \frac{1}{2}\left\langle \frac{\overline{E}}{1+\overline{\nu}} \right\rangle = \left\langle \overline{G} \right\rangle, \overline{C}_{33} = \frac{1}{2}\left\langle \frac{\overline{E}}{1-\overline{\nu}} z^2 \right\rangle,$$

$$\overline{C}_{22} = \frac{1}{2}\left\langle \frac{\overline{E}}{1+\overline{\nu}} z^2 \right\rangle = \left\langle \overline{G} z^2 \right\rangle, \quad \overline{\Gamma} = \lambda^2 \frac{\overline{A}_{44}\overline{C}_{22} - \overline{B}_{42}^2}{\overline{A}_{44}},$$

where λ is the minimal non-zero eigenvalue of the boundary-value problem

$$\frac{d}{dz}\left(\overline{G}\frac{dZ}{dz} \right) + \lambda^2 \overline{G} Z = 0, \quad \left. \frac{dZ}{dz} \right|_{|z|=h/2} = 0.$$

Here G and ν is the shear relaxation function and Poisson ratio of bulk material, respectively, and $\langle(\cdots)\rangle$ denotes the integral over the thickness h.

4. Example of Relaxation Functions. FMG Plate

In this section we consider a FGM plate made of a viscoelastic polymer foam. For the panel made from a porous polymer foam the distribution of the pores over the thickness can be inhomogeneous. Let us introduce ρ_s as the density of the bulk material and ρ_p as the minimum value of the density of the foam. For the description of the symmetric distribution of the porosity we assume the power law

$$V(z) = \alpha + (1-\alpha)\left| \frac{2z}{h} \right|^n, \qquad (5)$$

where $\alpha = \rho_p / \rho_s$ the minimal relative density, n is is the power.

The properties of the foam strongly depend on the porosity and the cell structure. For the polymer foam in [1] the modification of the standard linear viscoelastic solid is proposed. For the open-cell foam and the closed-cell foam the constitutive laws has the form

$$\sigma + \tau_E \sigma = C_1 V^2(z)\left[E_\infty \tau_E \varepsilon + E_0 \varepsilon\right],$$
$$\sigma + \tau_E \sigma = C_2\left[\phi^2 V^2(z) + (1-\phi)V(z)\right]\left[E_\infty \tau_E \varepsilon + E_0 \varepsilon\right]. \qquad (6)$$

Here can be assumed $C_1 \approx 1$, $C_2 \approx 1$, ϕ describes the volume fraction of the solid polymer concentrated near the cell ribs. Usually, $\phi = 0.6 \quad 0.7$. E_∞, E_0, τ_E are material constants of the polymer used in manufacturing of the foam.

From Eqs (6), (7) and using the experimental data presented in [1] one can assume $\nu(t) = \nu = \text{const}$. In this case we have the relations [3]

$$A_{11} = A_{22}\frac{1+\nu}{1-\nu}, \quad C_{33} = C_{22}\frac{1+\nu}{1-\nu}. \qquad (7)$$

A_{22} and C_{22} are given by the relations

$$A_{22} = h\left[\alpha^2 + \frac{2\alpha(1-\alpha)}{n+1} + \frac{(1-\alpha)^2}{2n+1}\right]G(t),$$

$$C_{22} = \frac{h^3}{12}\left[\alpha^2 + \frac{6\alpha(1-\alpha)}{n+3} + \frac{3(1-\alpha)^2}{2n+3}\right]G(t), \qquad (8)$$

for the open-cell foam, and

$$A_{22} = h\left\{\phi^2\left[\alpha^2 + \frac{2\alpha(1-\alpha)}{n+1} + \frac{(1-\alpha)^2}{2n+1}\right] + (1-\phi)\left[\alpha + \frac{1-\alpha}{n+1}\right]\right\}G(t), \qquad (9)$$

$$C_{22} = \frac{h^3}{12}\left\{\phi^2\left[\alpha^2 + \frac{6\alpha(1-\alpha)}{n+3} + \frac{3(1-\alpha)^2}{2n+3}\right] + (1-\phi)\left[\alpha + \frac{3(1-\alpha)}{n+3}\right]\right\}G(t)$$

for the closed-cell foam, respectively. Here we assume that $C_1 = 1$, $C_2 = 1$ and ϕ does not depend on z.

5. Plates made of Nano-sized Foams

Foams demonstrate the so-called size-effect, i.e. dependence of their material properties on the size of cells. For nano-sized foams the size-effect can be more significant then in the case of macro-sized foams. In particular, for the description of material at the nano-scale the theory of surface stresses is widely used [6]. The generalization of surface stresses on the case of viscoelasticity is given in [7]. Let us note the effect of surface viscoelasticity is investigated experimentally for many materials, see for example [8] and can be important for the material with fractal-type surface like foam.

Following [6] we assume that the relaxation function E of a foam strut with the cross-section of the radius R is given by the following relation:

$$E = E\left(1+\chi\hat{l}\right), \quad \hat{l} = l_{in}/R, \tag{10}$$

where E the relaxation function of a bulk material, l_{in} an intrinsic length scale related to the surface property, and χ a non-dimensional function of time describing the viscoelastic of strut surface. Using (10) we propose the following modification of the model of [1] for the nano-sized open-cell foams by the law

$$E_{nf} = E\left(1+\hat{l}\chi_1\right), \; G_{nf} = G\left(1+\hat{l}\chi_2\right), \tag{11}$$

where $E_{\rm nf}$ and $G_{\rm nf}$ are the longitudal and shear relaxation functions of a nanofoam, respectively, χ_1 and χ_2 the relaxation functions related to the surface viscoelasticity [7]. l_{in} is related to the surface effects and is typically in the order of 0.010.1 nm, see [6]. Hence, for foam with nano-size cells one has only one modification for the viscoelastic properties related to surface viscoelasticity properties.

Acknowledgments

The second author was supported by the DFG grant No. AL 341/33-1.

References

1. L. J. Gibson and M. F. Ashby: 'Cellular Solids: Structure and Properties', 2nd edn, 510, 1997, Cambridge, Cambridge University Press.
2. N. Mills: Polymer Foams Handbook Engineering and Biomechanics Applications and Design Guide, 535, 2007, Boston, Butterworth-Heinemann.
3. H. Altenbach and V. A. Eremeyev: *ZAMM*, 2008, 88, 332–341.
4. H. Altenbach and V. A. Eremeyev: *Acta Mech.*, 2009, 204, 137–154.
5. H. Altenbach and V. A. Eremeyev: in 'Cellular and Porous Materials: Modelling–Testing–Application' (ed. H. Altenbach and A. Öchsner), CISM Courses and Lecture Notes No. 521, 167–242, 2010, Wien, Springer.
6. H. L. Duan, J. Wang and B. L. Karihaloo: Theory of elasticity at the nanoscale. In: Adv. Appl. Mech., 2008, vol. 42, Elsevier, pp 1–68.
7. C.Q. Ru: *Appl. Phys. Lett.* 2009, 94, 051,905-1-051,905-3.
8. D. Tranchida, Z. Kiflie, S. Acierno and S. Piccarolo: *Meas. Sci. Techn.*, 2009, 20, 1–9.

EXPERIMENTAL INVESTIGATION OF THE INFLUENCE OF TEMPERATURE ON LOCAL BRIDGING BEHAVIOUR IN LAMINATED GLASS ELEMENTS IN POST-BREAKAGE STATE

DIETER CALLEWAERT, DIDIER DELINCÉ[†],
DELPHINE SONCK, JAN BELIS, RUDY VAN IMPE

Department of Structural Engineering, Ghent University, Technologiepark 904
9052 Ghent (Zwijnaarde), Belgium

The assessment of the post-breakage performances of laminated safety glass elements used in construction needs to take into account the sensitivity to the temperature of the mechanical behaviour and properties of the product, in particular of the interlayer material. After a general problem statement, an overview is given of tests at different scales, with typical observations and results for two different interlayer materials. Through the different presented experimental scales, the mechanisms ruling the bridging behaviour in cracked laminated safety glass are explained, highlighting the difficulty of assessing the mechanical properties to be used in practical design calculations.

Keywords: Laminated glass; Polymer interlayer; Post-breakage behaviour; Experimental.

1. Introduction

To make transparent structural elements in buildings, laminated safety glass products are today undoubtfully one of the best candidates. Laminated glass is an assembly of annealed, or possibly thermally reinforced, flat (float) glass sheets bound together by polymer interlayers.

However, assessing the mechanical performances of laminated glass elements for design purposes is still challenging. This is obviously due to the specific behaviour of glass, which presents a perfect elastic behaviour up to brittle breakage, with good chemical and physical stability. In addition, it is also due to the specific behaviour of the interlayer polymer used, as it is 1) sensitive to temperature and load duration, 2) able to extend to large deformations – compared to usual structural materials such as steel, aluminium, timber and concrete – and 3) sensitive to ageing effects.

An important field of investigation concerns the evaluation of the post-breakage (or post-crack) behaviour, namely the state in which one or more glass sheets are broken and the broken glass pieces are still bonded to the interlayer.

[†] Corresponding author, email: Didier.Delince@UGent.be

This state is occurring before an eventual complete collapse, which has to be avoided or at least delayed. The post-breakage behaviour of laminated glass is highly dependent on the local bridging behaviour, resulting from a combination of the elongation of the interlayer and the delamination from the glass pieces near the cracks. The bridging behaviour is obviously highly dependent on the interlayer and adhesion properties, but there is still a lack of generally accepted assessment methods and corresponding models.[1] Finally the crack pattern, crack propagation and the crushing of the glass in compressive contact zones are also of importance in the global behaviour.

2. Problem Statement

The properties of laminated safety glass products used in glazing units are mainly ruled by impact and dynamic performance requirements (safety of persons, burglar resistance, resistance to explosions,...). However, used as structural elements, the performance requirements are slightly different. As a consequence, besides polyvinyl butyral (PVB), the most widely used interlayer material for laminated safety glass, other stiffer interlayer products have been developed, such as SentryGlas® (SG) by DuPont.

In both cases, the variation of the performances in function of the temperature is not thoroughly investigated, limiting the practical application scope to rules of thumbs and good engineer judgment. Yet, only few experimental results are available to quantify this sensitivity. Therefore, the problem has been experimentally approached by tests at three different scales, detailed in the next paragraphs.

3. Modelling the Interlayer's Mechanical Behaviour

To describe and model the bridging behaviour, thus to characterize the interface behaviour between glass pieces and the interlayer, the mechanical behaviour of both materials has to be known. Indicative values of mechanical properties are given in Table 1.

Table 1. Indicative values of mechanical properties (among others from [1,2,4]).

Property	unit	glass	interlayer (polymer) PVB	SG
Volumetric weight (density) ρ	kg/m³	2,500	1,070	950
Elastic modulus E	N/mm²	70,000	18	300
Tensile strength σ_u / f_u	N/mm²	30-120	> 20	> 30
Deformation at breakage ε_u	%	0.04-0.25	> 250	> 300
Glass transition temperature T_g	°C	~530°C	~10-30°C	~55-60°C

The mechanical properties of the polymer come from uniaxial tensile tests at room temperature[a] achieved on dog-bone samples[2,4] cut from interlayer sheet. Figure 1 shows the general form of the loading curve (nominal stress and strain) of SG and PVB. As explained previously,[3] the mechanical values are merely informative : it was then stated that "simple" uniaxial tensile tests are not sufficient to derive an interlayer material behaviour, especially for SG material, because of the large, non-homogeneous deformation along a part of the loading curve.

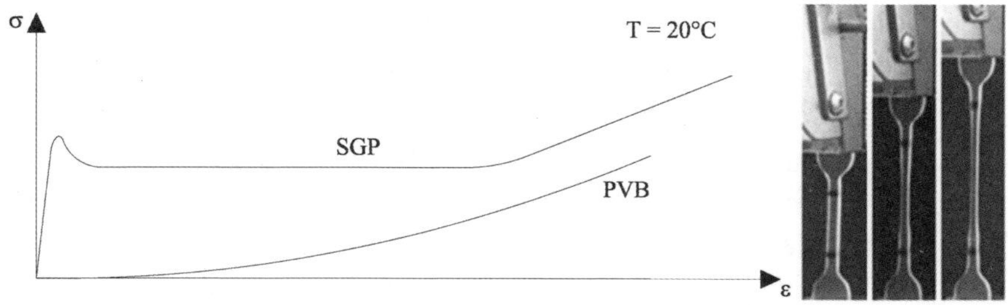

Fig. 1. Schematic loading curves (nominal stress and strain) from uniaxial tensile tests on interlayer samples (left) and non-homogeneous deformation of SG-samples (right).

4. Bridging Behaviour at Room Temperature

In a second phase, the bridging behaviour was investigated on the basis of Through-Cracked Tensile (TCT) tests (Fig. 2), carried out on small samples of laminated glass.[7] From scientific literature, a weak interface behaviour was expected (scheme on the left in Fig. 2) with a steady-state part in the loading curve, at least for PVB-laminates.

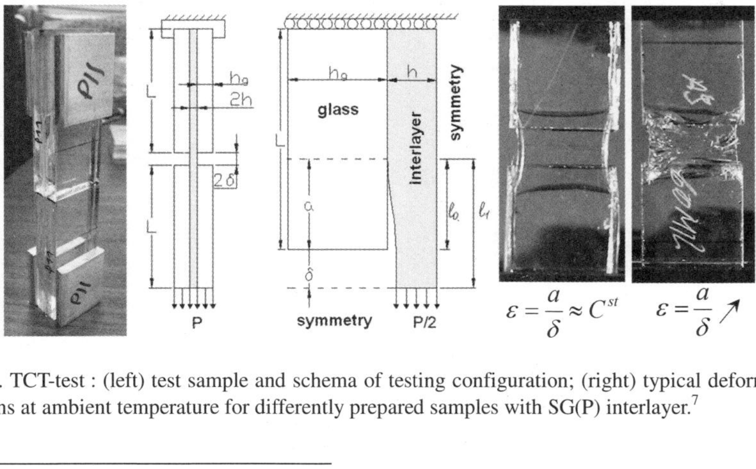

Fig. 2. TCT-test : (left) test sample and schema of testing configuration; (right) typical deformation patterns at ambient temperature for differently prepared samples with SG(P) interlayer.[7]

[a] The considered value of room temperature varies between 20 and 23°C.

During these tests, also other deformation patterns were observed : in some cases, cracks appeared to propagate in the glass sheets as well, and tearing of the interlayer could happen for a small opening of the initial crack, before reaching an eventual steady-state; this happened mainly for the SG(P)-laminates, but also for some PVB-laminates. As a consequence, it does not seem possible to deduce representative values of fracture energy by one single method throughout the whole temperature range. The performed TCT-tests also emphasized the importance of details in the preparation of the test samples, as they clearly influence the adhesion properties.

5. Bending Pre-Cracked Beams at Different Temperatures

From the previous steps, it is clear that deeper experimental research is necessary for modelling the bridging behaviour. Nevertheless, the influence of temperature on the post-breakage behaviour was also investigated by means of further preliminary tests, namely in-plane bending tests on laminated glass beams with SG-interlayer, achieved at three different temperatures (23, 45 and 60°C).[3] The glass sheets were pre-cracked in the middle transversal section of the beam to focus on the bridging behaviour as well.

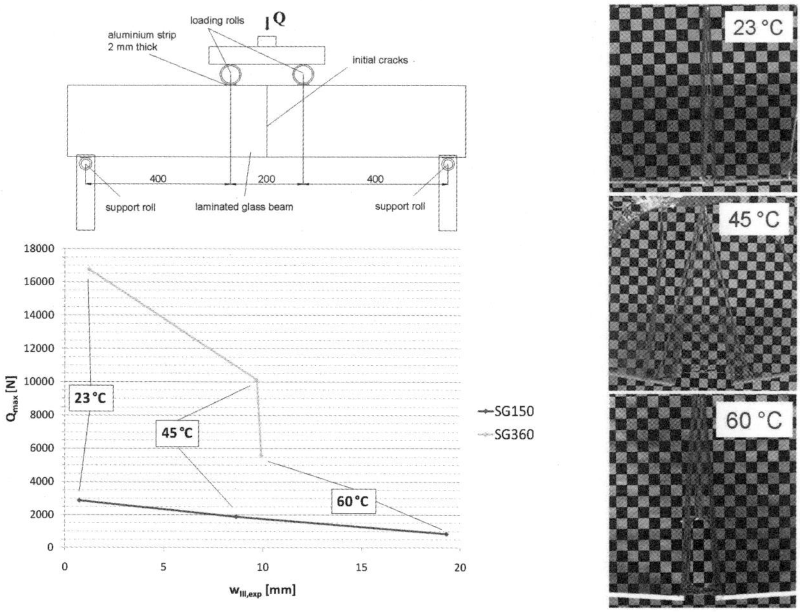

Fig. 3. Bending tests on pre-cracked beams : (left) testing configuration and typical maximal applied force and corresponding deflection; (right) typical deformation patterns.[3]

Different deformation patterns in function of the temperature were observed (Figure 3). At 23°C and 60°C, tearing of the interlayer started for a relatively small opening angle, but with a significative different form of the loading curve, in particular regarding the occurring moment and the value of the maximal load. However, at 45°C, larger delamination lengths generally occurred, with no tearing of the interlayer.

6. Conclusion

The influence of temperature on the post-breakage behaviour of laminated safety glass elements has been investigated experimentally at different scales. From this, it appears clearly necessary to develop and refine the testing methods to investigate and assess the bridging behaviour for different interlayer materials.

Acknowledgments

This work was supported by the Research Foundation-Flanders (FWO-Vlaanderen, grant nr. 3G018407). The authors acknowledge the support of DuPont de Nemours and Lerobel for the test samples, of the Department of Textiles of Ghent University regarding uniaxial tensile tests and TCT-tests, and of J. De Pauw, D. Sonck, F. Galmart and N. Matthijs within their master thesis.

References

1. M. Haldimann, A. Luible and M. Overend: 'Structural Use of Glass', Structural Engineering Documents 10, Zürich, Switzerland, 215, 2008; Int. Asso. for Bridge and Structural Engineering (IABSE), ETH Zürich
2. D. Delincé, D. Callewaert, J. Belis and R. Van Impe: Proc. Challenging Glass Conf. on 'Architectural and Structural Applications of Glass', Delft, the Netherlands, May 2008, 459–467.
3. D. Delincé, D. Callewaert, J. Belis and R. Van Impe: Proc. Challenging Glass 2 on 'Architectural and Structural Applications of Glass', Delft, the Netherlands, May 2010, 407-414.
4. J. Belis, J. Depauw, D. Callewaert, D. Delincé and R. Van Impe: *Eng. Fail. Anal.*, 2009, **16**, (2), 1866-1875.
5. D. Delincé, D. Callewaert, W. Vanlaere, J. Belis and J. Depauw: *Int. J. Mod. Phys. B*, **22**, (31-32), 2008, 5447-5452.
6. J. Belis, D. Delincé, D. Callewaert, R. Van Impe and J. Depauw: *Int. J. Mod. Phys. B*, **22**, (31-32), 2008, 5509-5514.
7. D. Delincé, D. Sonck, J. Belis, D. Callewaert and R. Van Impe: Proc. 3rd Int. Symp. on 'Application of Architectural Glass' (ISAAG), Munich, Germany, October 2008, 41-49.

PREDICTION OF THE PLASTIC DEFORMATION OF CIRCULAR CYLINDRICAL SHELLS

A. DARVIZEH

Department of Mechanical Engineering, Faculty of Engineering, University of Guilan Rasht, P.O. Box 3756, Rasht, Iran

Department of Mechanical Engineering, Faculty of Engineering, Islamic Azad University of Bandar-e-Anzali, Bandar-e-Anzali, Iran

H. RAJABI [†]

Department of Mechanical Engineering, Faculty of Engineering, University of Guilan Rasht, P.O. Box 3756, Rasht, Iran

A study of plastic deformation of circular cylindrical shells subject to dynamic forces is presented. The five governing equations in terms of resultant forces and moments with respect to basic displacement vector components u, v and w are used. Theoretical formulations, based on first-order shear deformation theory (FSDT), take into consideration transverse shear deformation and rotary inertia. The deformation theory of plasticity is employed. Galerkin method is used to convert the partial differential equations (PDEs) to ordinary differential equations (ODEs). The Newmark family of methods is used to solve the system of five coupled second order ordinary differential equations. In order to prove the validity of the presented method and the solving process, the results obtained with the present analysis are compared with a set of available experimental data. Certainly, the aim of this paper is to create a more reliable and precise mathematical model of cylinders to avoid performing several experiments.

Keywords: Plastic deformation; Circular cylindrical shells; Galerkin method; Deformation theory of plasticity.

[†] Corresponding author, email: harajabi@hotmail.com

1. Introduction

In recent years cylindrical shells have found widely used in various engineering applications such as pipeline systems, pressure vessels, aircraft engines, large scale generators and so on. The history of the modeling of the elastic and plastic behavior of cylindrical shells can be summarized as follows: some of the earliest works to study the plastic behavior of cylindrical shells were done by Lee [1] and Batterman [2]. The importance of these articles is in their abundant experimental results. Recently, there are many research studies on the elastic-plastic deformation of shells [3-9]. In all of these articles, the authors have used various methods and assumptions to obtain simplified solutions. However, there is a few works available in the literature pertaining to the plastic deformation of a thick circular cylindrical shell subjected to radial dynamic loading.

There are many technical and industrial applications, in which thick cylinders play an important role, such as aerospace, marine, military, automotive and mining. In many of these applications, embedded devices may be subjected to various dynamic forces. As we know, the dynamic response of structures subjected to dynamic or shock loads is a complex phenomenon [10,11].

In this paper a simple model is presented to investigate the plastic behavior of the shell. The governing shell equations in terms of resultant forces and moments including the transverse shear and rotary inertia are used. The deformation theory of plasticity, which is employed for constitutive equations, leads to acceptable results with relatively less calculation process. Galerkin method is used to convert the partial differential equations (PDEs) to ordinary differential equations (ODEs). Finally, The Newmark family of methods is used to numerically time integration of a system of five coupled second order ordinary differential equations. The results obtained with the present analysis are found to be in good agreement with a set of available experimental data.

2. Governing Equations and Solution Procedure

2.1. *Equations of motion*

Consider a thick-walled circular cylindrical shell made up of a linear orthotropic material. The orthogonal coordinate system (x, ϑ, z) is fixed at the mid-surface of the tube with length L, thickness H, and radius R. x is the axial direction, ϑ is the circumferential direction and z is the radial direction. The deformations of the tube are defined by u, v, w, which are displacements of the point in x, ϑ, z (Fig.1).

The governing equations of motion in the longitudinal, tangential and radial directions in terms of the force and moment resultants are as follows:

$$N_{x,x}+\frac{N_{\vartheta x,\vartheta}}{R}+P_x=I_0\ddot{u}_0+I_1\ddot{\varphi}_x\,,$$

$$N_{x\vartheta,x}+\frac{N_{\vartheta,\vartheta}}{R}+\frac{Q_\vartheta}{R}+P_\vartheta=I_0\ddot{v}_0+I_1\ddot{\varphi}_\vartheta\,,$$

$$Q_{x,x}+\frac{Q_{\vartheta,\vartheta}}{R}-\frac{N_\vartheta}{R}+P_z=I_0\ddot{W}_0, \tag{1}$$

$$M_{x,x}+\frac{M_{\vartheta x,\vartheta}}{R}-Q_x+m_x=I_1\ddot{u}_0+I_2\ddot{\varphi}_x,$$

$$M_{x\vartheta,x}+\frac{M_{\vartheta,\vartheta}}{R}-Q_\vartheta+m_\vartheta=I_1\ddot{v}_0+I_2\ddot{\varphi}_\vartheta,$$

where I_0, I_1 and I_2 are the mass moments of inertia and can be defined as

$$I_0=\int_{-\frac{H}{2}}^{\frac{H}{2}}\rho\left(1+\frac{z}{R}\right)dz,\ I_1=\int_{-\frac{H}{2}}^{\frac{H}{2}}\rho z\left(1+\frac{z}{R}\right)dz,\ I_2=\int_{-\frac{H}{2}}^{\frac{H}{2}}\rho z^2\left(1+\frac{z}{R}\right)dz. \tag{2}$$

The force and moment resultants that act on a shell element are defined as the force and moment per unit length of the shell's mid-surface. The resultants are obtained by integrating the stresses acting on differential area elements of a shell element.

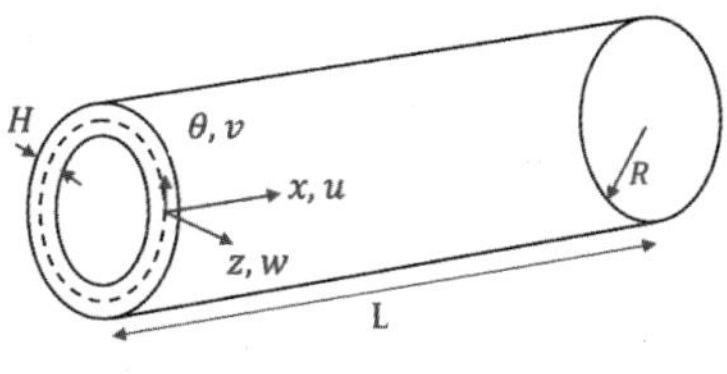

Fig. 1. Circular cylindrical shell in polar coordinate system.

2.2. *Deformation theory of plasticity*

According to the deformation theory of plasticity, the relation between the stress tensor σ_{ij} and the plastic strain tensor ε_{ij} can be written as

$$\varepsilon_{ij}=\frac{3}{2E_s}\sigma_{ij}-\left(\frac{1}{2E_s}-\frac{1-2\upsilon}{3E}\right)\sigma_{kk}\delta_{ij}\,, \tag{3}$$

where E and υ are Young's modulus and Poisson's ration. E_s, which is called secant modulus, is the slop of a line from the origin to any point on the stress-strain curve and can be defined as

$$E_s=\frac{\sigma_e}{\varepsilon_e}, \tag{4}$$

where σ_e and ε_e are the effective stress and strain, respectively.

Now, using Eqs. (3), we can express the constitutive equations into the following form

$$\sigma = Q_{ij}\varepsilon, \tag{5}$$

where $\boldsymbol{Q}_{ij}$ is the plastic stiffness matrix and depends on the material properties of the structure and membrane stresses.

2.3. *Strain-displacement relations*

The strains can be expressed as

$$\varepsilon_x = u_{,x}\,,\ \varepsilon_\vartheta = (v_{,\vartheta} + w)/(R+z)\,, \varepsilon_z = w_{,z}\,,$$
$$\gamma_{\vartheta z} = (w_{,\theta} - v)/(R+z) + \varphi_\vartheta\,,\ \gamma_{xz} = \varphi_x + w_{,x}\,,\ \gamma_{x\vartheta} = u_{,\vartheta}/(R+z) + v_{,x}. \tag{6}$$

As mentioned before, u, v and w are the displacements in the x, ϑ and z directions, respectively.

2.4. *FSDT*

According to the first order shear deformation theory, the displacement fields are of the form

$$u(x,\vartheta,z,t) = u_0(x,\vartheta,t) + z\varphi_x(x,\vartheta,t), \qquad v(x,\vartheta,z,t) = v_0(x,\vartheta,t) + z\varphi_\vartheta(x,\vartheta,t),$$
$$w(x,\vartheta,z,t) = w_0(x,\vartheta,t), \tag{7}$$

where u_0, v_0, w_0, φ_x and φ_ϑ are unknowns to be determined. In fact, u_0, v_0, w_0, φ_x and φ_ϑ are the displacements of a point on the surface $z=0$ and the rotations of transverse normal about its ϑ *and* x axis, respectively.

Substituting Eqs. (7) into Eqs. (6) yield

$$\varepsilon_x = u_{0,x} + z\,\varphi_{x,x}\,, \varepsilon_\vartheta = (v_{0,\vartheta} + z\,\varphi_{\vartheta,\vartheta} + w_0)/(R+z)\,, \varepsilon_z = 0,$$
$$\gamma_{\vartheta z} = (w_{0,\vartheta} - v_0 - z\varphi_\vartheta)/(R+z) + \varphi_\vartheta\,, \tag{8}$$
$$\gamma_{xz} = \varphi_x + w_{0,x}\,, \gamma_{x\vartheta} = (u_{0,\vartheta} + z\,\varphi_{x,\vartheta})/(R+z) + v_{0,x} + z\,\varphi_{\vartheta,x}\,.$$

Substituting Eqs. (8) into Eq. (5), using Eqs. (1) give us a set of partial differential equations in terms of the displacements. Hence, the five governing equations can be expressed in terms of the five basic unknowns (u_0, v_0, w_0, φ_x and φ_ϑ).

2.5. *Galerkin method*

The boundary conditions for simply supported cylindrical shell are

$$x = 0, L:\ v_0 = w_0 = \varphi_\vartheta = N_x = M_x = 0. \tag{9}$$

The displacement field is taken as

$$u_0 = U_0 \cos(\frac{m\pi x}{L}) \cos\left(\frac{n\pi\vartheta}{\varphi}\right), v_0 = V_0 \sin(\frac{m\pi x}{L}) \sin\left(\frac{n\pi\vartheta}{\varphi}\right),$$

$$w_0 = W_0 \sin(\frac{m\pi x}{L}) \cos\left(\frac{n\pi\vartheta}{\varphi}\right), \tag{10}$$

$$\varphi_x = \phi_x \cos(\frac{m\pi x}{L}) \cos\left(\frac{n\pi\vartheta}{\varphi}\right), \varphi_\vartheta = \phi_\vartheta \sin(\frac{m\pi x}{L}) \sin\left(\frac{n\pi\vartheta}{\varphi}\right).$$

The weighted integral statement of the Galerkin method is written as

$$\int_0^{2\pi}\int_0^L (L_{11}u_0 + L_{12}\, v_0 + L_{13}w_0 + L_{14}\varphi_x + L_{15}\varphi_\vartheta + P_x - I_0\ddot{u}_0 - I_1\ddot{\varphi}_x)\delta u_0 dxd\theta = 0,$$

$$\int_0^{2\pi}\int_0^L (L_{21}u_0 + L_{22}\, v_0 + L_{23}w_0 + L_{24}\varphi_x + L_{25}\varphi_\vartheta + P_\vartheta - I_0\ddot{v}_0 - I_1\ddot{\varphi}_\vartheta)\delta v_0 dxd\theta = 0,$$

$$\int_0^{2\pi}\int_0^L (L_{31}u_0 + L_{32}\, v_0 + L_{33}w_0 + L_{34}\varphi_x + L_{35}\varphi_\vartheta + P_z - I_0\ddot{w}_0)\delta w_0 dxd\theta = 0, \tag{11}$$

$$\int_0^{2\pi}\int_0^L (L_{41}u_0 + L_{42}\, v_0 + L_{43}w_0 + L_{44}\varphi_x + L_{45}\varphi_\vartheta + m_x - I_1\ddot{u}_0 - I_2\ddot{\varphi}_x)\delta \varphi_x dxd\theta = 0,$$

$$\int_0^{2\pi}\int_0^L (L_{51}u_0 + L_{52}\, v_0 + L_{53}w_0 + L_{54}\varphi_x + L_{55}\varphi_\vartheta + m_\vartheta - I_1\ddot{v}_0 - I_2\ddot{\varphi}_\vartheta)\delta \varphi_\vartheta dxd\theta = 0.$$

Substituting Eqs. (10) into Eqs. (11) and performing the integration, we obtain the following matrix equation

$$[M]\{\ddot{U}\} + [K]\{U\} = \{P\}, \tag{12}$$

where [M], [K] are the inertia and stiffness matrices and {U}, {P} are the displacements and load vectors. The coefficients of matrix K are very complicated and not presented here.

The Newmark family of methods is used in the present study to numerically time integration of the system of five coupled second order ordinary differential equations. The Newmark's parameters are chosen to be β=1/4 and δ=1/2. By setting these parameters, it results in the constant acceleration scheme, which is desirable because of its second order accuracy and non-dissipative nature.

3. Numerical Example

Consider a thick-walled circular cylindrical shell with simply supported boundary conditions. The dimensions of the studied tube are L = 3.25 [inch], R = 0.967 [inch] and H = 0.064 [inch] and it is made of an aluminium alloy having Young's modulus E = 10200×10^3 [lbf/in^2], density ρ = 0.084 [lb/in^3], elastic Poisson's ratio ν = 0.33, yield stress σ_0 = 21000 [lbf/in^2] and a hardening modulus E_h = 27500 [lbf/in^2].

The uniaxial stress-strain curve expressed by

$$\sigma_e = \sigma_0 + E_h \varepsilon_e. \tag{13}$$

The force acting on the cylinder is

$$P = Ae^{-Bt} sin^2 \omega t. \tag{14}$$

where A, B and ω are 1.64×10^8, 14630 and 1.885×10^5, respectively. The results are shown in Fig. 2. In this graph, the theoretical displacement curve is compared with the results reported by Al-Hassani et al. [12].

It is seen that a good agreement stands between the theoretical results and experimental data. Fig. 3. illustrates profiles of the bulge process at specific points in time.

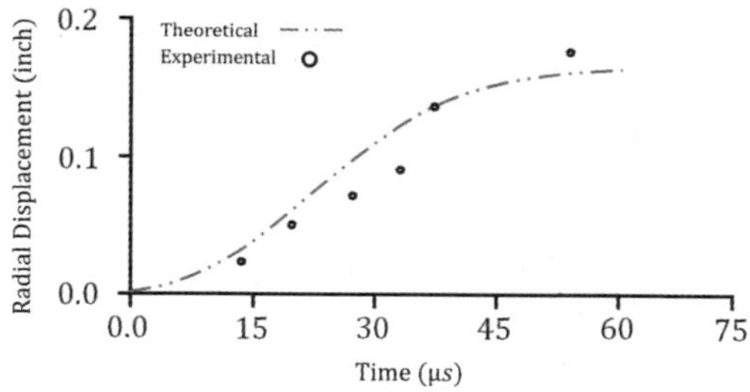

Fig. 2. Theoretical and experimental variation with time of radial expansion for aluminium tube.

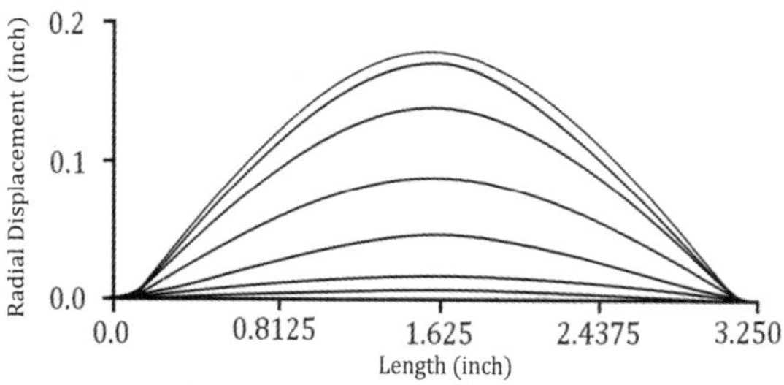

Fig. 3. Calculated profiles at specific points during forming process.

4. Conclusion

A thick shell theory with first order shear deformation is used to analyze the expansion of an aluminium tube. The constitutive equations are based on the deformation theory of plasticity. The material is assumed to be isotropic hardening. Comparison of theoretical and experimental results shows that the presented model correctly predicts the plastic deformation. Therefore, this analysis is suitable for quick and accurate design purposes.

References

1. L. N. H. Lee, 'Inelastic buckling of initially imperfect cylindrical shells subjected to axial compression', *J. Aero. Sci.*, 1962, 29, 87–95.
2. S. C. Batterman, 'Plastic buckling of axially compressed cylindrical shells', *AIAA J.*, 1965, 3, 316–325.
3. D. Bushnell, 'Plastic buckling of various shells', *J. press. Vessel Technol.*, 1982, 104, 51–73.
4. N. Tutuncu and S. J. Winckler, 'Stresses and deformations in thick-walled cylinders subjected to combined loading and a temperature gradient', *J. Reinf. Plast. Comp.*, 1993, 12, 198–209.
5. A. Loghman and M. A. Wahab, 'Loading and unloading of thick-walled cylindrical pressure vessels of strain-hardening material', *J. Press. Vessel Tech.*, 1994, 116, 105–109.
6. J. Zielnica, 'Application of the J_2 plastic flow theory in stability analysis of an elastic–plastic open sandwich cylindrical shell under shear forces and external pressure', Short Commun., Math. Mech. GAMM, 2000, 80, 333–334.
7. M. Shakeri, M. R. Eslami and A. Daneshmehr, 'Dynamic analysis of thick laminated shell panel with piezoelectric layer based on three dimensional elasticity solution', *J. Comp. Struc.*, 2006, 84 , 1519–1526.
8. M. R. Saviz, M. Shakeri and M. H. Yas, 'Electroelastic fields in a layered piezoelectric cylindrical shell under dynamic load', *J. Smart Mat. Struct.*, 2007, 16 , 1683–1695.
9. A. Darvizeh, M. Darvizeh, H. Rajabi and A. Rezaei, 'Analytical analysis of thick-walled tubes coated with piezoelectric layers under dynamic loading', Asian Pacific Conf. on Materials and Mechanics, Yokohama, Japan, 2009.
10. H. Gharababaei and A. Darvizeh, 'Experimental and Analytical Investigation of Large Deformation of Thin Circular Plates Subjected to Localized and Uniform Impulsive Loading', *Mech. Base. Des. Struct. Mach.*, 2010, 38, 171 – 189.
11. H. Gharababaei, N. Nariman-zadeh and A. Darvizeh, 'A Simple Modelling Method for Deflection of Circular Plates under Impulsive Loading using Dimensionless Analysis and Singular Value Decomposition', *J. Mech.*, 2010, 26, 355-361.
12. S. T. S. Al-Hassani, J. L. Duncan and W. Johnson, 'On the parameters of the magnetic forming process', *J. Mech. Eng. Sci.*, 1974, 16, 1–9.

PLASTIC COLLAPSE OF STEEL WATER TOWERS

WESLEY VANLAERE[†], DELPHINE SONCK, RUDY VAN IMPE

Department of Structural Engineering, Ghent University, Technologiepark 904
9052 Ghent (Zwijnaarde), Belgium

ARNE JANSSEUNE, WOUTER DE CORTE

Department of Civil Engineering, University College Ghent, Schoonmeersstraat 52
9000 Ghent, Belgium

The aim of this contribution is to verify an existing design rule for steel liquid-filled conical shells by performing numerical simulations. Questions were raised regarding the reliability of the design rule as early yielding comes into play if an axisymmetric imperfection is located near the lower rim where the stresses are elevated. In our study, full nonlinear numerical simulations are performed with a realistic weld depression as imperfection shape. This weld depression is placed at the most detrimental location near the lower rim. The results of the simulations show that for cones with a moderate relative slenderness the design rule indeed overestimates the buckling strength, but that the error is not pronounced and that the design rule may be considered as still acceptable.

Keywords: Plastic buckling; Imperfections; Liquid-filled conical shells.

1. Introduction

In a typical steel water tower, the water is stored in a conical steel reservoir (Fig. 1a). The weight of the liquid causes compressive stresses in meridional direction and tensile stresses in circumferential direction. The largest meridional stress appears at the base of the cone. These large stresses can trigger the failure of the cone in two different ways. First, the elevated compressive stresses can lead to elastic buckling of the thin-walled cone near the base. The second possible cause of failure is the plastic collapse of the cone if the combination of the meridional and circumferential stresses leads to yielding of the shell wall. This failure phenomenon is dominant for cones with a relatively thicker wall. For cones with an intermediate wall thickness, interaction between the elastic and plastic failure phenomenon appears. This is called elastic-plastic buckling.

For these liquid-filled conical shells, design rules can be found in the 5th Edition of the ECCS European Design Recommendations. These design rules are

[†] Corresponding author, email: Wesley.Vanlaere@gmail.com

based on an elaborate series of experiments performed by Vandepitte *et al.* on liquid-filled conical shells. Although most of the experiments led to elastic buckling of the cone, the design rule can also be used to capture the (elastic-)plastic failure phenomenon.[1-3]

The goal of the present study is to verify this design rule by performing numerical simulations with the finite element package ABAQUS. In the recent past, an initial study verified the design rule for the case of elastic buckling. This study showed that for some geometries – where elastic buckling was expected – the cone failed prematurely due to early yielding near the lower rim. This early yielding was not captured by the present design rule, leading to an overestimation of the failure load.[4-5]

In this contribution, the results of numerical simulations for cones with a relatively thick wall are compared with the design rule. For these cones, plasticity is expected to play a dominant role and the fear exists that the design rule could be very unsafe for these cones.

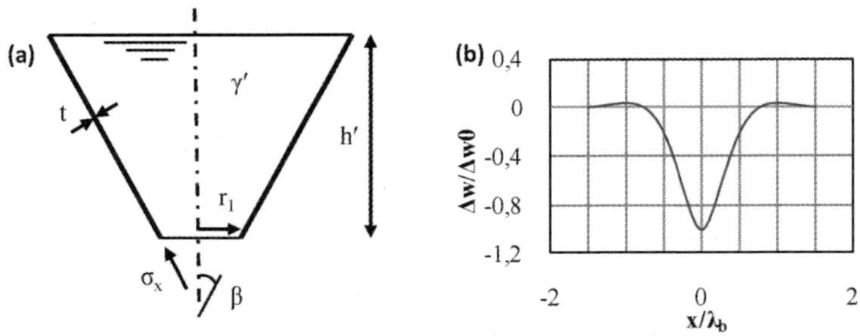

Fig. 1. a) The cone parameters. b) The shape of the weld depression.

2. The Design Rule

Due to space limitations, the design rule as described in the 5th Edition of the ECCS European Design Recommendations[1], can't be repeated here. For this paper, it is important to realize that the design load is characterized by the dimensionless reduction factor χ. This reduction factor is a function of the dimensionless relative slenderness ratio λ. A low relative slenderness ($\lambda \leq \lambda_0$) indicates that the shell will fail due to plastic collapse; a moderate relative slenderness ($\lambda_0 \leq \lambda \leq \lambda_p$) indicates that the shell will fail due to elastic-plastic buckling; a large relative slenderness ($\lambda_p \leq \lambda$) indicates that the shell will fail due to elastic buckling. If $\lambda \leq \lambda_0$, the reduction factor remains constant; if $\lambda_0 \leq \lambda$, an increase of the relative slenderness leads to a decreasing reduction factor.

3. The Numerical Simulations

3.1. *General information*

In a recent past, a limited number of finite element simulations was performed for liquid-filled conical shells with $\lambda/\lambda_p \cong 1$. In these simulations, geometrical nonlinearity, material nonlinearity and imperfections were taken into account. The amplitude of the imperfections were according to quality class C. The first eigenmode of the perfect cone was taken as imperfection shape. This shape is axisymmetrical and is characterized by a limited number of waves in meridional direction. For these investigated geometries, elastic buckling was expected to be the dominant failure phenomenon. However, some of the simulations showed that the cones failed due to early yielding, entailing a lower failure load than predicted by the design load. This phenomenon can be explained by the increase of the von Mises stresses, which is caused by the change in shell curvature due to the outward oriented geometrical imperfections.[4-5]

In this contribution, similar simulations are performed for cone geometries with smaller values of λ/λ_p. For these geometries, plasticity is expected to be important even without the unavoidable presence of geometrical imperfections. This implies that it is possible that the design rule becomes very unsafe when large imperfections are present. For these simulations, the eigenmode is no longer used as imperfection shape. Instead, the more realistic weld depression type A is chosen (see Fig. 1b):[7]

$$\Delta w = \Delta w_0 \cdot e^{-(\pi x/\lambda_b)} \cdot \left(\cos\frac{\pi x}{\lambda_b} + \sin\frac{\pi x}{\lambda_b} \right) \tag{1}$$

In this equation x is the meridional distance from the weld centre, λ_b is the linear bending half-wavelength, Δw_0 is the amplitude of the equivalent geometrical imperfection and Δw is the deviation of the perfect shape (outward is taken positive). Again, the amplitude is taken according to quality class C. In the numerical model, the weld depression is situated at the most detrimental location near the lower rim. An example of the determination of such a location is given in Fig. 2a. In this graph, the ratio of x_0/λ_b is given on the horizontal axis, with x_0 the meridional coordinate of the weld centre. An x_0 value of zero corresponds to a weld centre on the lower rim of the cone. On the vertical axis, the ratio of the reduction factor predicted by ABAQUS and the reduction factor predicted by the design rule is given. The lowest buckling strengths are obtained with the imperfection located near the lower rim.[8]

In the numerical model, the lower rim is simply supported (boundary condition BC1f) and the upper rim is free to deform (boundary condition BC3).

The steel has an elastic, perfectly plastic behavior. The yield stress f_{yk} is taken equal to 240E6 Nm^{-2}, the Young's modulus E equal to 200E9 Nm^{-2} and the Poisson's ratio equal to 0.3. In the numerical simulations, the specific weight of the liquid γ' is increased until failure occurred.

4. Results and Discussion

The results of the numerical simulations are summarized in Fig. 2b. More information on the studied geometries can be found in the master thesis of Gheyle.[8] On the horizontal axis in Fig. 4, the theoretical ratio λ/λ_p is given. On the vertical axis, the deviation between the numerical ABAQUS result and the ECCS design rule is given in terms of percentage:

$$\text{deviation} = \frac{\chi_{ABAQUS} - \chi_{ECCS}}{\chi_{ECCS}} \times 100 \tag{2}$$

Deviation values lower than zero indicate that the design rule leads to an unsafe prediction of the resistance of the liquid-filled conical shell. The graph shows that for ratios $0.5 \leq \lambda/\lambda_p \leq 1.2$ the design rule indeed overestimates the failure load. The overestimation is however limited to about 6%, which is still acceptable.

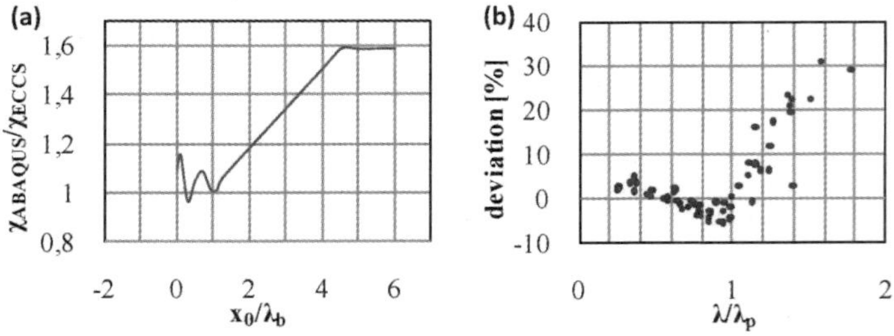

Fig. 2. a) The effect of the location of the weld depression on the failure load. b) The results of the numerical simulations.

Although it was expected that the design rule would become increasingly unsafe with decreasing values of λ/λ_p, this seems not to be the case. The reason can be found in the dimensionless amplitude of the equivalent geometrical imperfection $\Delta w_0/t$ that is taken into account in the numerical simulations. This amplitude $\Delta w_0/t$ is a function of λ/λ_p. A decrease of λ/λ_p leads to a decrease of the dimensionless amplitude $\Delta w_0/t$. This means that for smaller slendernesses the geometrical imperfection of quality class C causes a smaller increase of the von Mises stresses and as a consequence a minor effect of early yielding.[6,8]

5. Summary

In this contribution, an existing design rule for liquid-filled conical shells was verified. Questions were raised regarding the reliability of this design rule, especially for cones where plasticity is a dominant factor in the failure phenomenon. In this study, the failure load of the cones was determined with finite element simulations in which geometrical nonlinearity, plasticity and geometrical imperfections were taken into account. As imperfection shape, the realistic weld depression is used. This weld depression was located at the most detrimental location. The results of the simulations showed that the design rule indeed overestimates the failure load for geometries with a moderate relative slenderness ratio, but the error made is restricted to about 6%, which may be considered as still acceptable.

References

1. G. Lagae, W. Guggenberger and W. Vanlaere in : 'Buckling of Steel Shells, European Design Recommendations', 5th edn, Chapter 14; 2008, ECCS.
2. D. Vandepitte, J. Rathé, B. Verhegghe, R. Paridaens and C. Verschaeve in: 'Buckling of Shells', 375-399;1982,Berlin, Springer.
3. R. Paridaens, D. Vandepitte, G. Lagae, J. Rathé and A. Van den Steen: Proc. Int. Colloq. on 'Stability of Plate and Shell Structures', Ghent, Belgium, April 1987, 425-430.
4. G. Lagae, W. Vanlaere and R. Van Impe: in 'Hommages á René Maquoi', (ed. Jean-Pierre Jaspart, Jacques Rondal), 167-176 ; 2007, Liège, Université de Liège.
5. W. Vanlaere, G. Lagae, W. Guggenberger, W. and R.Van Impe: Proc. 5th Eur. Conf. on 'Steel and Composite Structures', Graz, Austria, September 2008, 1515-1520.
6. CEN: EN 1993-1-6 (Eurocode 3: Design of steel Structures – Part 1-6: Strength and Stability of Shell Structures), European Committee for Standardization, Brussels, Belgium, 2007.
7. J.M. Rotter: in 'Buckling of Thin Metal Shell's (ed. J.G. Teng, J.M. Rotter), 42-87; 2004, London, Spon Press.
8. R. Gheyle: 'Investigation of the structural behavior of steel water towers' (in Dutch), Masterthesis, Ghent University, Ghent, Belgium, 2010.

A DISCUSSION ON ELASTOPLASTIC MODEL OF DOUBLE YIELD SURFACES

XIANQI LUO

School of Naval Architecture, Ocean and Civil Engineering, Shanghai Jiao tong University. 800 Dong chuan Road, Shanghai, P.R. China

DEYONG WANG

School of Naval Architecture, Ocean and Civil Engineering, Shanghai Jiao tong University. 800 Dong chuan Road, Shanghai, P.R. China

The nonlinear model and elastoplastic model are widely used on the constitutive model of soil. Shen Zhujiang's elastoplastic model of double yield surfaces has been widely used to carry out stress strain analysis of high earth dams and other soil engineering structures. The former is represented based on the dilatancy and contraction of soil. In Shen Zhujiang's elastoplastic model, the 2-Dimension-elastoplastic matrixes are unsymmetrical, but its 3-Dimension-elastoplastic matrix is symmetric. Based on that, the 3-Dimension-elastoplastic matrix of the model is detailed deduced. The result shows that the 3-Dimension elastoplastic matrix of the model is also unsymmetrical like as its 2-Dimension elastoplastic matrix. And the rationality of its 3-Dimension-elastoplastic matrix is discussed.

Keywords: Elastoplastic model of double yield surfaces; dilatation; elastoplastic matrix; constitutive model.

1. Introduction

In the constitutive model of soil, the elastic nonlinear model and elastoplastic constitutive model are commonly used. It is assumed that all of the displacement is elastic in the elastic nonlinear model; by changing the elastic constants the nonlinear is behaved. The elastoplastic constitutive model can be divided to two types based on their research method. One is directly determining the experiential formula from specific test data, such as yield function, hardening laws; the other is deriving the yield function from the physical concept of energy, and electing hardening laws to establish the relationship between stress and strain. Shen Zhujiang's elastoplastic model of double yield surfaces (6, 7, 8, and 9) is the first one. Cambridge elastoplastic model [Cambridge K. H. Roscoe] (1) and Lade elastoplastic model [P. V. Lade & J. M. Duncan] (2-5) are the second one.

Shen Zhujiang's elastoplastic model of double yield surfaces is suggested based on the dilatancy and shearing contraction of soil. The elastoplastic matrix is unsymmetrical for the plane strain and axial symmetry，base on these, the 3-D elastoplastic matrix of elastoplastic model of double yield surfaces is derived in this paper, And also the elastoplastic matrix in the elastoplastic model of double yield surfaces is unsymmetrical.

2. Shen Zhujiang's Elastoplastic Model of Double Yield Surfaces

Shen Zhujiang summed up great deal practice, absorbed the advantages of Duncan nonlinear model and Cambridge model, and advanced an elastoplastic model of double yield surfaces. It has overcome many shortcomings such as generalized Hooke's Law can not considered the anisotropy characteristics caused by the stress of dilatancy, and shearing contraction of soil, introduced the concept of the plastic coefficient instead of the concept of hardening and continually used the parameters of well-known Duncan nonlinear model to calculate.

Plastic strain increment added elastic strain increment could obtain the total strain increment $\{d\varepsilon\}$, and then from inverse calculation the stress increment can be written as $\{d\sigma\}=[D_{ep}]\{d\varepsilon\}$. In the three-dimensional problems (9),

$$[D_{ep}]=\begin{bmatrix} d_{11} & \cdots & d_{16} \\ & & \\ d_{61} & \cdots & d_{66} \end{bmatrix}$$

3. A discussion of Shen Zhujiang's Elastoplastic Model of Double Yield Surfaces

From the elastoplastic model of double yield surfaces, the elastoplastic matrix is unsymmetrical for the plane strain and axial symmetry. But its elastoplastic matrix is symmetrical for the three-dimensional matters. Commonly, the three-dimensional form can be degraded to two-dimensional form with the same theory and model. However, there may be some contradiction between two-dimensional elastoplastic matrix and three-dimensional elastoplastic matrix in the elastoplastic model of double yield surfaces.

For three-dimensional problems, the three-dimensional elastoplastic matrix from Refs. (10–11) are also different. The formula is as following:

$$d_{45}=-Qs_{xy}s_{yz}/q^2, d_{46}=-Qs_{xy}s_{zx}/q^2, d_{56}=-Qs_{yz}s_{zx}/q^2 \quad \text{(Ref. 10)}$$

$$d_{45}=-Ps_{yz}/q-Qs_{xy}s_{yz}/q^2, d_{46}=-Ps_{zx}/q-Qs_{xy}s_{zx}/q^2, d_{56}=-Ps_{zx}/q-Qs_{yz}s_{zx}/q^2 \quad \text{(Ref. 11)}$$

4. Derivation of Elastoplastic Model of Double Yield Surfaces

The yield function of double yield surfaces is ($f_1 = p^2 + r^2 q^2 \quad f_2 = q^s / p$). Parameter r is ratio of the long axis and the short-axis; s is power number.

$$p = \frac{1}{3}(\sigma_1 + \sigma_2 + \sigma_3) \qquad q = \frac{1}{3}[(\sigma_1 - \sigma_2)^2 + (\sigma_2 - \sigma_3)^2 + (\sigma_3 - \sigma_1)^2]^{1/2}$$

The total strain increment is divided into elastic strain part $\Delta\varepsilon_{ij}{}^e$ and plastic strain part $\Delta\xi_{ij}{}^p$, and the plastic strain increment is divided into two parts, each part corresponds to one of the yield surface as $\Delta\varepsilon_{ij} = \Delta\varepsilon_{ij}{}^e + (\Delta\varepsilon_{ij}{}^p)_1 + (\Delta\varepsilon_{ij}{}^p)_2$.

Using orthogonal flow rule, then, $\Delta\varepsilon_{ij} = \Delta\varepsilon_{ij}{}^e + A_1 \Delta f_1 \dfrac{\partial f_1}{\partial \sigma_{ij}} + A_2 \Delta f_2 \dfrac{\partial f_2}{\partial \sigma_{ij}}$ where, A_1 and A_2 are the plastic coefficient respective to the yield surface f_1 and f_2. It is a non-negative number, and only when its yield surface is determined as unloading or neutral loading, it is equivalents to 0.

Assuming $v = \varepsilon_1 + \varepsilon_2 + \varepsilon_3$ is volume strain, and $\gamma = \frac{2}{3}[(\varepsilon_1 - \varepsilon_2)^2 + (\varepsilon_2 - \varepsilon_3)^2 + (\varepsilon_3 - \varepsilon_1)^2]^{1/2}$ is octahedron shear strain,

$$\{\sigma\}^{\mathrm{T}} = \{\sigma_x \quad \sigma_y \quad \sigma_z \quad \tau_{xy} \quad \tau_{yz} \quad \tau_{zx}\} \quad \{\varepsilon^0\} = \{\varepsilon_x \quad \varepsilon_y \quad \varepsilon_z \quad \gamma_{xy} \quad \gamma_{yz} \quad \gamma_{zx}\}$$

$$\{\varepsilon\}^{\mathrm{T}} = \{\varepsilon_x \quad \varepsilon_y \quad \varepsilon_z \quad \varepsilon_{xy} \quad \varepsilon_{yz} \quad \varepsilon_{zx}\} = \left\{\varepsilon_x \quad \varepsilon_y \quad \varepsilon_z \quad \frac{1}{2}\gamma_{xy} \quad \frac{1}{2}\gamma_{yz} \quad \frac{1}{2}\gamma_{zx}\right\}$$

$$X = [(\sigma_1 - \sigma_2)^2 + (\sigma_2 - \sigma_3)^2 + (\sigma_3 - \sigma_1)^2]^{1/2} \;;\; Y = [(\varepsilon_1 - \varepsilon_2)^2 + (\varepsilon_2 - \varepsilon_3)^2 + (\varepsilon_3 - \varepsilon_1)^2]^{1/2}$$

$$\Delta v = \Delta\varepsilon_1 + \Delta\varepsilon_2 + \Delta\varepsilon_3 \qquad \Delta v = \frac{\Delta p}{K} + A_1 \Delta f_1 \frac{\partial f_1}{\partial p} + A_2 \Delta f_2 \frac{\partial f_2}{\partial p}, \quad \text{Where, } K = E/(3(1-2\mu))$$

$$\Delta\gamma = \frac{\partial\gamma}{\partial\varepsilon_1}\Delta\varepsilon_1 + \frac{\partial\gamma}{\partial\varepsilon_2}\Delta\varepsilon_2 + \frac{\partial\gamma}{\partial\varepsilon_3}\Delta\varepsilon_3$$

$$= \frac{2}{3}\frac{1+\mu}{E}\frac{\Delta}{Y}[\varepsilon_1(2\sigma_1 - \sigma_2 - \sigma_3) + \varepsilon_2(2\sigma_2 - \sigma_1 - \sigma_3) + \varepsilon_3(2\sigma_3 - \sigma_2 - \sigma_1)] + \frac{2}{3}\left(A_1 \Delta f_1 \frac{\partial f_1}{\partial q} + A_2 \Delta f_2 \frac{\partial f_2}{\partial q}\right)\frac{1}{XY}[\varepsilon_1(2\sigma_1 - \sigma_2 - \sigma_3) + \varepsilon_2(2\sigma_2 - \sigma_1 - \sigma_3) + \varepsilon_3(2\sigma_3 - \sigma_2 - \sigma_1)]$$

$$\Delta q = \frac{\partial q}{\partial\sigma_1}\Delta\sigma_1 + \frac{\partial q}{\partial\sigma_2}\Delta\sigma_2 + \frac{\partial q}{\partial\sigma_3}\Delta\sigma_3 = \frac{1}{3X}[\sigma_1(2\Delta\sigma_1 - \Delta\sigma_2 - \Delta\sigma_3) + \sigma_2(2\Delta\sigma_2 - \Delta\sigma_1 - \Delta\sigma_3) + \sigma_3(2\Delta\sigma_3 - \Delta\sigma_2 - \Delta\sigma_1)]$$

It is worthy to illuminate that from Ref. 10 and 11 the next equation is obtained.

$$\Delta\gamma = \frac{\Delta q}{G} + \frac{2}{3}\left(A_1 \Delta f_1 \frac{\partial f_1}{\partial q} + A_2 \Delta f_2 \frac{\partial f_2}{\partial q}\right), \qquad G = \frac{E}{2(1+\mu)} \tag{1}$$

In fact, the establishment of equation (1) is under the next condition.

$$\frac{\sigma_i}{X} = \frac{\varepsilon_i}{Y} \quad (i=1、2、3) \tag{2}$$

Or, $\dfrac{\sigma_i}{\varepsilon_i} = \dfrac{X}{Y} = K(x)$ (i=1、2、3); or else, if $\dfrac{\sigma_i}{\varepsilon_i} = K_0$, ($i$=1、2、3), then $\dfrac{X}{Y} = K_0$. So the establishment of Eq. 2 is just a special case.

The next is that the 3-D elastoplastic matrix of elastoplastic model of double yield surfaces is derived.

From $\Delta f_1 = 2p\Delta p + 2\gamma^2 q\Delta q, \Delta f_2 = -\frac{q^s}{p^2}\Delta p + s\frac{q^s}{pq}\Delta q$ then

$$\Delta v = \frac{\Delta p}{K} + A\Delta p + C\Delta q \qquad \Delta\gamma = \frac{\Delta q}{G} + \frac{2}{3}(B\Delta q + C\Delta p) \tag{3}$$

Where, $A = 4p^2 A_1 + \frac{q^{2s}}{p^4}A_2, B = 4\gamma^2 q^2 A_1 + \frac{s^2 q^{2s}}{p^2 q^2}A_2, C = 4\gamma^2 qpA_1 - \frac{sq^{2s}}{p^3 q}A_2, K = \frac{E}{3(1-2\mu)}$

$\Delta v^e = \frac{\Delta p}{K}$, $\Delta v^P = A\Delta p + C\Delta q$, $\Delta\gamma^e = \frac{\Delta q}{G}$, $\Delta\gamma^P = \frac{2}{3}(B\Delta q + C\Delta p)$

Let, $$\Delta e_{ij} = \Delta\varepsilon_{ij} - \frac{1}{3}\Delta vI \quad s_{ij} = \sigma_{ij} - pI \tag{4}$$

If the Prandtl-Reuss's flow rule is adopted in π plane, then,

$$\frac{d\varepsilon_x{}'^p}{s_x{}'} = \frac{d\varepsilon_y{}'^p}{s_y{}'} = \frac{d\varepsilon_z{}'^p}{s_z{}'} = \frac{d\varepsilon_{xy}{}'^p}{\tau_{xy}{}'} = \frac{d\varepsilon_{yz}{}'^p}{\tau_{yz}{}'} = \frac{d\varepsilon_{zx}{}'^p}{\tau_{zx}{}'} = d\lambda$$

Where, $d\lambda = \frac{1}{2}\frac{d\gamma^p}{q}$ $\Delta e_{ij}{}^P = \frac{1}{3}(B\Delta q + C\Delta P)\frac{s_{ij}}{q}$ $\Delta e_{ij}{}^e = \Delta\varepsilon_{ij}{}^e - \frac{1}{3}\Delta v^e\delta_{ij} = \frac{\Delta s_{ij}}{2G}$

So $$\Delta e_{ij} = \Delta e_{ij}{}^e + \Delta e_{ij}{}^P = \frac{\Delta s_{ij}}{2G} + \frac{1}{3}(B\Delta q + C\Delta p)\frac{s_{ij}}{q} \tag{5}$$

From $s_{ij}=\sigma_{ij}-p_I$, $(\sigma_1-\sigma_2)^2 + (\sigma_2-\sigma_3)^2 + (\sigma_3-\sigma_1)^2 = 3s_{ij}s_{ij}$ then $q = \frac{1}{\sqrt{3}}(s_{ij}s_{ij})^{1/2}$,

or $s_{ij}{}^2 = 3q^2$, $\Delta q = \frac{1}{3}\frac{s_{ij}}{q}\Delta s_{ij}$ or $s_{ij}\Delta s_{ij} = 3q\Delta q$, and from (5),

$$s_{ij}\Delta e_{ij} = \frac{s_{ij}\Delta s_{ij}}{2G} + \frac{1}{3}(B\Delta q + C\Delta p)\frac{s_{ij}{}^2}{q} = \frac{3q\Delta q}{2G} + (B\Delta q + C\Delta p)q \text{ , } \Delta q = \frac{\frac{2}{3}\left(\frac{G}{q}s_{ij}\Delta e_{ij} - CG\Delta p\right)}{1 + \frac{2}{3}BG} \tag{6}$$

Substitute (6) in (3), Δp is solved as,

$$\Delta v = \left(\frac{1}{K} + A\right)\Delta p + C * \frac{\frac{2}{3}\left(\frac{G}{q}s_{ij}\Delta e_{ij} - CG\Delta p\right)}{1 + \frac{2}{3}BG} \qquad \Delta p = K_p\Delta v - K_q\frac{s_{ij}}{q}\Delta e_{ij} \tag{7}$$

Where, $K_p = \left(\frac{1}{K} + A - \frac{\frac{2}{3}C^2 G}{1 + \frac{2}{3}BG}\right)^{-1}, K_q = \frac{\frac{2}{3}CG}{1 + \frac{2}{3}BG} * K_p$

Substitute (7) in (6), Δs_{ij} is solved as,

$$\Delta s_{ij} = 2G\Delta e_{ij} - \frac{2}{3}G(B\Delta q + C\Delta p)\frac{s_{ij}}{q} = 2G\Delta e_{ij} + M\frac{s_{ij}s_{hk}}{q^2}\Delta e_{hk} - N\frac{s_{ij}}{q}\Delta v \tag{8}$$

Where, $M = \frac{\frac{2}{3}CG}{1 + \frac{2}{3}BG}K_q - \frac{4}{9}\frac{BG^2}{1 + \frac{2}{3}BG}, N = \frac{\frac{2}{3}CG}{1 + \frac{2}{3}BG}K_q$

From $\Delta\sigma_{ij} = \Delta\varepsilon_{ij} + \Delta q\delta_{ij}$, $\Delta\varepsilon_{ij} = \Delta e_{ij} + \frac{1}{3}\Delta v\delta_{ij}$, the next equation can be obtained.

$$\Delta\sigma_{ij} = 2G\left(\Delta\varepsilon_{ij} - \delta_{ij}\frac{\Delta v}{3}\right) + M\frac{s_{ij}s_{hk}}{q^2}\left(\Delta\varepsilon_{hk} - \delta_{hk}\frac{\Delta v}{3}\right) - N\frac{s_{ij}}{q}\Delta v + \delta_{ij}K_p\Delta v - \delta_{ij}K_q\frac{s_{hk}}{q}\left(\Delta\varepsilon_{hk} - s_{hk}\frac{\Delta v}{3}\right)$$

Elastoplastic matrix can be shown as $\{\Delta\sigma\} = [D_{ep}]\{\Delta\varepsilon^0\}$. Thus it is unsymmetrical and different from Ref. 10-11.

5. Conclusion

Shen Zhujiang's elastoplastic model of double yield surfaces is advanced based on the dilatancy and shearing contraction of soil. A discuss of Shen Zhujiang's three-dimensional elastoplastic model of double yield surfaces is conducted. Through the derivation of elastoplastic model of double yield surfaces, the 3-Dimension elastoplastic matrix of the model is also unsymmetrical like as its 2-Dimension elastoplastic matrix.

References

1. Rosco, K. H. And Burland, J. B, *on the generalized stress-strain behavior of 'wetclay'*, engineering plasticity, Cambridge University, 1968.
2. Lade, P. V., in *Elasto-plastic stress-strain theory for cohesionless soil with curve yield surfaces*, Int. J. Soil and Structures, 1977, vol. 13, pp. 1019-1035.
3. Duncan J.M. & Chang C.Y., in *nonlinear analysis of stress and strain in soil*, Journal SMFE, 1970, Vol. 96, NO SM5, p.1629.
4. Duncan J.M. et al, in *Strength, stress-strain and bulk modulus parameters for FEM of stresses and movements in soil masses*, Report No. UCB GT/80-01, Univ. of California, Berkeley, 1980.
5. Naylor, J.M. et al, in *Stress-strain law for soils*, Developments in soil Mechanics, 1978.
6. SHEN Zhujiang, in *Reasonable form of Elastic-Plastic Stress-strain relations on soil*, Chinese Journal of Geotechnical Engineering, 1980, 6, pp. 11~19.
7. SHEN Zhujiang, in *Stress-strain Modes of three-yield surface on soil*, Chinese Journal of Solid Mechanics, 1984, 6(2), pp. 163~174.
8. SHEN Zhujiang, in *Elastic-Plastic analysis of Consolidation Deformation on soft clay ground*, Science in China Series A, 1985, 11, pp. 1049~1060.
9. ZHANG Weimin, SHEN Zhujiang, in *Three-Dimensional elastoplastic FEM analysis of concrete faced rockfill*, Chinese JOURNAL OF HYDRAULIC ENGINEERING, 1992, 4, pp. 75~78.
10. YI Chunlou, in *Study on static and dynamic Analysis Method high Rookfill Dams on Deep Overburdens- Study on State of static and dynamic Working*

Performance of Yele Rockfill Dam. Nanjing, Dissertation, Hohai University, 1995.

11. LI Guoying, in *Study on new structure form and its Design Method on faced rockfill dam*, Dissertation, Nanjing Hydraulic Research Institute, 1994.

THE NUMERICAL SIMULATION ON THE CRACKS OF THE CONCRETE COVER UNDER HOOP REINFORCEMENT CORROSION AND EXPANSION

QIU ZHAOGUO

School of Science, Northeastern University, Shenyang, Liaoning Province, China

ZHANG FENGPENG[†]

College of Resources and Civil Engineering, Northeastern University, Shenyang, Liaoning Province, China

LIANG LI

College of Resources and Civil Engineering, Northeastern University, Shenyang, Liaoning Province, China

Based on the plastic damage theory, Finite Element Models are set up which are about the cracking problem of the rusted hoop reinforced concrete cover. The corrosion ratio is researched at the moment of cracking. The relations between thickness of concrete cover and diameter of hoop reinforcement are studied at critical state. And the cracking paths of concrete cover due to further corrosion expansion of hoop reinforcement are studied after critical state. The results indicate that 3-4cracks are generated in the cracking of the concrete cover. Three of them are main cracks, others are secondary cracks.

Keywords: Sample hoop reinforcement; concrete cover; corrosion expansion rate; cracking path; plastic damage.

1. Introduction

The durability damage and injury of Reinforced Concrete Structures have always been quite an important problem for civil engineers, the propagation theory of cracks and damage caused by reinforcement corrosion expansion is also the research focus of this field[1]. Based on elastic-plastic mechanics, the theoretical model of reinforced concrete corrosion cracking critical corrosive ratio has been established[2-5], and the analytical solution of reinforced concrete corrosion cracking critical point is also presented. However, the theory is based on the assumption that the thickness of concrete cover is much greater than reinforcement diameter and the reinforcement corrosion expansion is uniform. It

[†] Zhang Fengpeng, email:zhangfengpeng@126.com

can't be accurate suitable for the situation of hoop reinforcement. Hoop reinforcement corrosion experiences three stages[6]. (1) The stage of free expansion: gap between reinforcement and concrete is just filled with corrosion products. (2) The stage of compressive expansion: the inner surface of concrete cover begins to format cracks. (3) The stage of crack propagation: the external surface of concrete cover begins to format cracks. Dong Rongzhen[7] etc. point out that mortar layer between reinforcement and concrete is filled with the initial corrosion products freely. The thickness of filling layer is about 20-50μm while other researchers point out that it is 1-20μm[6,8].

2. Calculation Assumptions

In view of the complexity of concrete, all the factors can not possibly be taken into account. Based on the theories and experiments of predecessor, it is assumed that:

Assumption 1: The side of hoop reinforcement far from concrete cover is not corroded, another side near form concrete cover is corroded and expanded, and its boundary is elliptical. The corrosion and expansion direction is along the radial direction.

Assumption 2: In the state of non-corrosion, the value of gap between reinforcement and concrete is a constant. The corrosion products don't diffuse internally in the process of corrosion and expansion.

3. Finite Element Model and Basic Theory

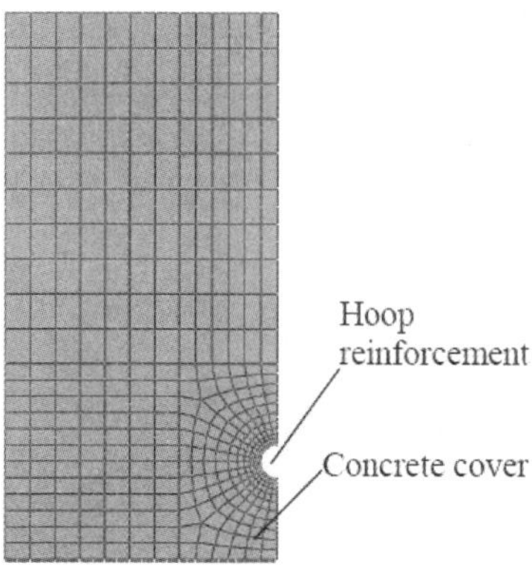

Fig. 1. Finite element model.

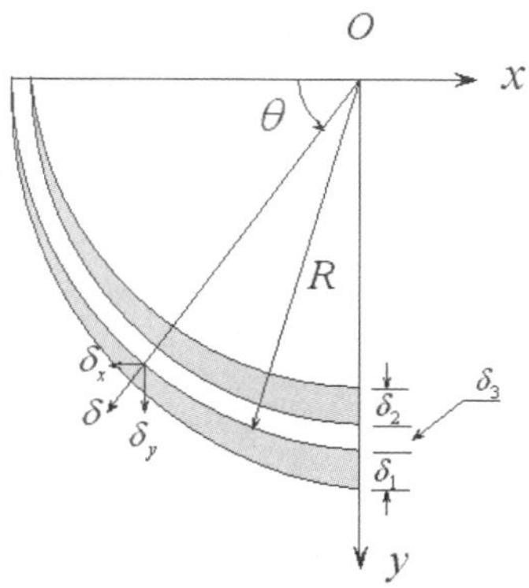

Fig. 2. Displacement loading pattern.

Based on the knowledge of elastic-plastic mechanics, the problem caused by the corrosion and expansion of reinforced concrete can be simplified to the problem of plane strain. The finite element model is built using Symmetry shown in Figure 1.

In the model, the gauges of hoop reinforced are $\phi 8$, $\phi 10$, $\phi 12$, $\phi 14$ and $\phi 16$ respectively.

Li Yonghe and Yuan Yingshu[1,6] etc. systematically analysis the variation characteristics of hoop reinforcement corrosion shape by experiments and point out that hoop reinforcement corrosion mainly occurs on the side close to the concrete cover and the corrosion contour is an Elliptic. Neglect the amount of corrosion in dorsal side of concrete cover and use displacement loading pattern, that is each node is added displacements along x and y(δ_x, δ_y).

$$\begin{aligned} \delta_x &= -R\left(R+\delta_{\max}\right)\sqrt{\frac{1}{\left(R+\delta_{\max}\right)^2+R^2\tan^2\theta}}+R\cos\theta \\ \delta_y &= R\tan\theta\left(R+\delta_{\max}\right)\sqrt{\frac{1}{\left(R+\delta_{\max}\right)^2+R^2\tan^2\theta}}-R\sin\theta \end{aligned} \tag{1}$$

Here: $\delta_{\max}=\delta_1+\delta_3$

Displacement loading pattern is given by Figure 2.

Corrosion expansion coefficient: $\xi=(\delta_1+\delta_2+4\delta_3)/\delta_2$

Corrosion ratio: $\eta=\delta_2/(2R)$, then: $\eta=(\delta_1+4\delta_3)/\left[2R(\xi-1)\right]$

4. Result and Discussion

4.1. *Crack-growth trajectory*

For example, concrete cover is 4cm and the diameter of hoop reinforcement is 8mm. Crack-growth trajectory is shown by Figure 3.

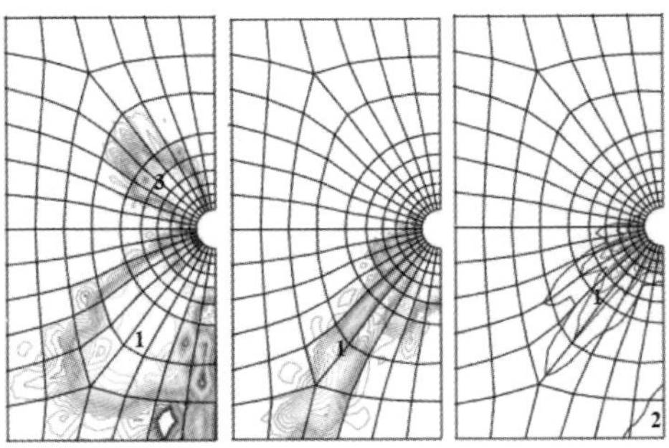

a b c

Fig. 3. Active yield flag at integration points and plastic strain at integration points.

Here, a is active yield flag at integration points at this time $\delta_1=50\mu m$, b is active yield flag at integration points at this time $\delta_1=100\mu m$, c is plastic strain at integration points at this time $\delta_1=100\mu m$.

From active yield flag at integration points and plastic strain at integration points, it can be seen that crack 1 is the most important crack, propagates from inner surface, the angle between crack 1 and vertical direction is about 30°. Crack 2 is the secondary crack, propagates from external surface, along the vertical direction. Crack 3 is the most unimportant crack, stops developing after crack1 and crack 2 is full developed.

4.2. *The effects of hoop reinforcement diameter and concrete cover*

When corrosion expansion reach a critical value, the initial crack will be generated in concrete cover firstly, then with the increase of hoop reinforcement corrosion expansion depth the surface crack begins to generate. It is assumed that $\delta_3 = 15\mu m$, corrosion expansion ratio $\xi = 2$ [9].

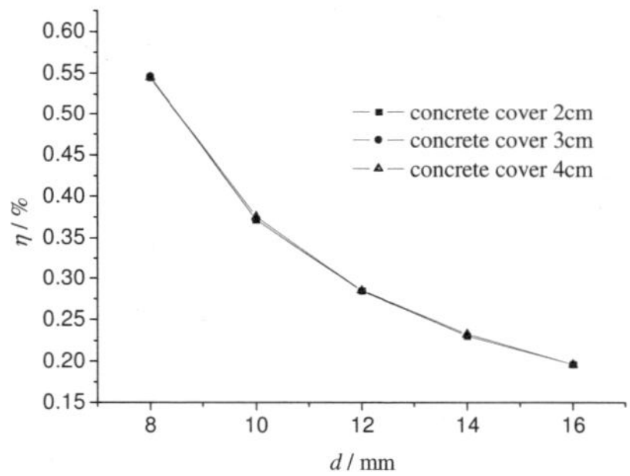

Fig. 4. The initial internal crack.

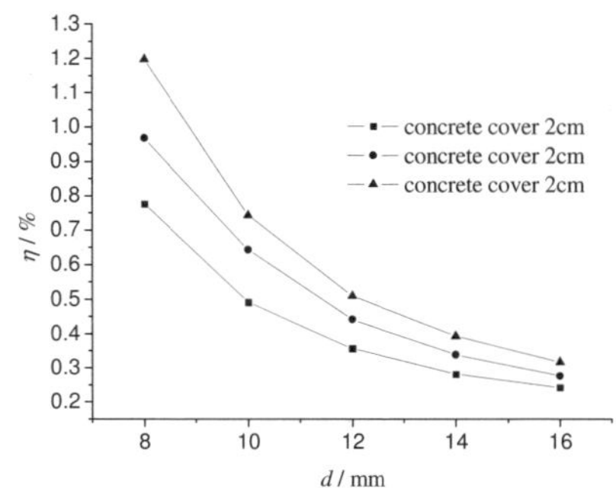

Fig. 5. The initial surface crack.

Figure 4-5 show that the generation of initial crack has little relationship with the concrete cover depth. Corrosion ratio decreases with increasing hoop reinforcement diameter when initial crack begins to generate. When surface crack generates, corrosion ratio η decreases with increasing hoop reinforcement diameter.

5. Conclusion

Numerical models are set up which is applied to research the initial internal cracks and initial surface crack.

A new displacement loading pattern is given in this paper, which is different from previous works.

From this paper the following conclusions can be drawn:

1. Two oblique cracks propagate from inner surface, and one vertical crack propagates from external surface.
2. The generation of initial cracks has little relationship with the concrete cover depth.

Acknowledgments

This work was financially supported by the National High-Tech Research and Development Program ("863" program) of China (No.2007AA06Z107), the Natrual Science Foundation of China (No.50974031), the National Key Technology R&D Program of China (No.2008BAB34B02), the Program for New Century Excellent Talents in Chinese universities (No.NCET-07-0163), and the Fundamental Research Funds for the Central Universities (No.90401006, No. 90405008).

References

1. Y. H. Li, X. R. Ge: *China Civil Eng. J.*, 2003, 36(2):1-5.
2. X. H. Wu: *Mech. in Eng.*, 2007, 29:67-70.
3. Y. X. Zhao, W. L. Jin: *Shuili Xuenao*, 2004, 35: 97-101.
4. Y. S. Cheng, J. Wei, X. H. Luo: *J. Wuhan Univ. of Tech.*, 2007, 29(2):51-53.
5. J. J. Zheng, X. Z. Zhou, C. Q. Li: *Shuili Xuenao*, 2004, 35:62-68.
6. ZHAO Yu_xi, JIN Wei_liang: *Shuili Xuenao*, 2005, 36:939-945.
7. Y. S. Yuan, Y. S. Ji, Y. J. Mu: *China Civil Eng. J.*, 2007, 40(7):5-10.
8. C. Andrade, C. Alonso, F. J. Molina. *Mater. & struct.* 1993, 26:453-464.
9. Y. P. Liu, R. E. Weyers. *Mater. J. (ACI)*, 1998, 95(6): 675-681.

NONLINEAR ANALYSIS OF FRP REINFORCED RC SLAB CARRYING CAPACITY

LIU JUN[†]

School of Sciences, Northeastern University, Shenyang, Liaoning Province, China

ZHANG FENGPENG

College of Resources and Civil Engineering, Northeastern University, Shenyang, Liaoning Province, China

LI QING

School of Architecture and Civile Engineering, Shenyang University of Techology, Shenyang, Liaoning Province, China

Based on the plate theory, a high-order nonlinear FEM model is developed for the coupling of shear/flexure with in-plane deformation of the RC slab strengthened with FRP. In this model, a special plate layer element is built, Darwin-Pecknold Constitutive Model is chosen by modification .Using the FEM model, the carrying capacity of FRP reinforced two-way RC slab is analyzed, the stress distributions of glue layer are obtained and the numerical results are in good agreement with the experimental data. Calculated results show that carrying capacity of FRP reinforced RC slab is about 114% higher than that of un-strengthened RC slab and concrete failure accounts for the disruption of RC slab strengthened with FRP. In addition, the effects of thickness of FRP strip on carrying capacity are discussed.

Keywords: Carrying capacity; FRP; two-way RC slab; debonding; non-linear FEM model; progrssive failure.

1. Introduction

Fiber reinforced plastics(FRP) materials have been recognized as new innovative materials for concrete rehabilitation and retrofit[1-3]. Compared to the studies on reinforceded beam, the researches on FRP-reinforced RC slab are less. The reasons are due the following aspects: First, the experiment of slab (especially two-way RC slab) is more complex than that of beam; second, the mechanical behavior of one-way RC slab is similar to beam and the two-way RC slab is often be designed as one-way RC slab in each direction respectively using

[†] Liu Jun, email: liujunneu@21cn.com

maximum moment method. Sometimes the research results of beam are used directly to slab, but some differences still exist between them.

For two-way FRP-strengthened RC slab, some theoretical analysis and experiments are done by Limam, O.[4-5], but the theoretical analysis is limited to linear analysis. In this paper, based on the plate theory, carrying capacity of two-way FRP-strengthened RC slab is analyzed by a high-order nonlinear FEM method, the results are compared with the experimental results, in addition, the effects of length and width of FRP on carrying capacity are discussed.

2. Analysis Model

2.1. *Basic hypothesis*

1. Deformation of reinforced concrete slab is on the base of plane section assumption. Steel bars contact concrete tightly, result in no slippage effect.
2. Only Shear deformation is considered in glue layer. The bond between FRP and concrete is perfect before loading.
3. Only tensile deformation occurs in FRP.

2.2. *Geometry model*

In this finite element model, a 4-node element is adopted. Every node has 7 degrees of freedom including five concrete mid plane displacements and two other displacements of FRP. Shape function for 4-node plate element, 4-node membrane element and linear bar element are used for concrete, FRP and Steel bar respectively. Node displacements in steel bar are not independent that can be figured out by using concrete mid plane displacements and basic hypothesis 1

2.3. *Constitutive relation of materials*

Property of steel under compress is considered the same as that under tension. Bilinear model is adopted to describe the material nonlinear property. The constitutive relations of steel are as follows

$$\sigma = E\varepsilon \qquad \varepsilon \le \varepsilon_y$$

$$\sigma = \sigma_y + (\varepsilon - \varepsilon_y)\frac{\sigma_u - \sigma_y}{\varepsilon_u - \varepsilon_y} \qquad \varepsilon_y \le \varepsilon \le \varepsilon_u \tag{1}$$

Where ε_y , ε_u are yield strain and ultimate strain of steel respectively, σ_y , σ_u are yield stress and ultimate stress of steel respectively.

Consider the brittleness of glue layer. The brittle constitutive relation is used before debonding.

In FRP reinforced RC plate, concrete is under biaxial stress state and nonlinear elastic model is adopted in this paper. According to Darwin and Pecknold's suggestion, Equivalent tangent modulus of concrete can be calculated from the following equation[6]

$$E_i = \frac{d\sigma_i}{d\varepsilon_{iu}} = \frac{E_0[1-(\frac{\varepsilon_{iu}}{\varepsilon_{ic}})^2]}{[1+(\frac{E_0}{E_s}-2)(\frac{\varepsilon_{iu}}{\varepsilon_{ic}})+(\frac{\varepsilon_{iu}}{\varepsilon_{ic}})^2]^2} \quad (2)$$

It is noticed that Eq. (2) is incorrect when main stress in direction 1 is near zero. So, in this paper, this kind of stress state is predigested as unidirectional compression by considering is E_1 equal to E_2. In this way, better numerical results can be obtained.

3. Calculation Example Analysis

A two-way RC slab simply supported on four sides is analyzed in this section and the size of the slab is 70mm×1150mm×1650mm. With the interval of 30mm, steel bars which have diameter of 6mm are placed in x and y direction and the thickness of concrete protective cover is 17mm. For purpose of comparison with experimental results, paper [5] is consulted for the definition of geometric sizes and material parameters adopted in following numerical analysis. In this example, an concentrated force is added in the centre of the slab.

With the interval of 150mm, Carbon FRP (CFRP) of 50mm width is bonded to the lower surface of RC slab in two direction. The parameters of CFRP, concrete and steel are listed in Tab1 and Tab2.

Table 1. The Characteristics For CFR.

thickness/mm	tensile Strength/Mpa	Maximum tensile strain.	elastic modulus /GPa
1.4	2800	1.6%	163

Table 2. The Characteristics For Concrete And Steel.

	concrete	steel
elastic modulus /Gpa	25	
Poisson ratio	0.2	
compressiv strength /MPa	30	
ultimate compressive strain	0.0022	
yield stress/MPa		400
yield strain		0.002

On the base of symmetry, a quarter of slab is analyzed by dividing into 13×9 elements. Furthermore, concrete is divided into 10 layers in order to simulate the failure process of RC slab layer by layer.

Using the FEM program compiled according to the model build in this paper, carrying capacity of RC slab strengthened with FRP can be obtained .The relationship between the load and central deflection is illustrated in Figure 1. From Figure 1, it is obviously seen that numerical results are in good agreement with Liman's experimental results.

It is found that differentiating from RC beam strengthened with FRP, debond of glue layer and rupture of FRP occurs rarely in RC slab strengthened with FRP. The reason is that shear stress of glue layer and tensile stress of FRP are both less.

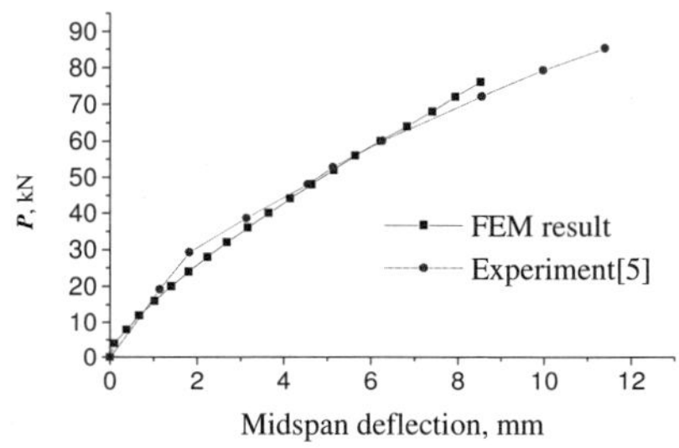

Fig. 1. Load-deflection curve of two-way slab strengthened with CFRP.

The load-deflection curves of the two-way RC slab strengthened and un-strengthened are both illustrated in Figure 2 and the contrast of two curves indicates that the carrying capacity of RC slab strengthened with FRP is about 114% higher than that of RC slab un-strengthened.

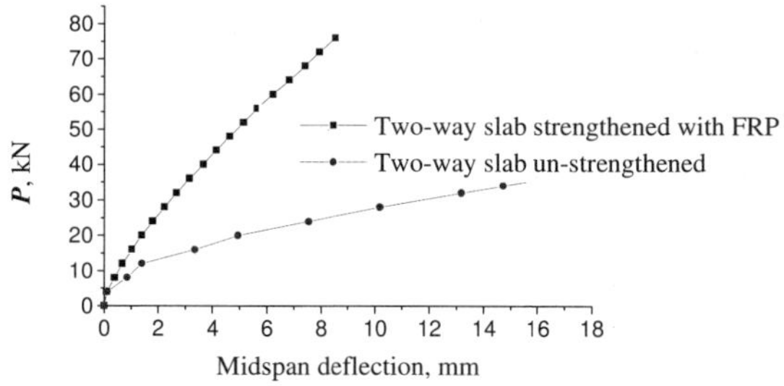

Fig. 2. Load–deflection curves of two-way slab strengthened and un-strengthened.

Figure 3 shows three load-deflection curves of slabs strengthened with three kinds of CFRP strip which have thicknesses of 1.0mm, 1.4mm and 1.8mm respectively. From Figure 3, it is noticed that carrying capacity improve with increase of thickness of CFRP strip.

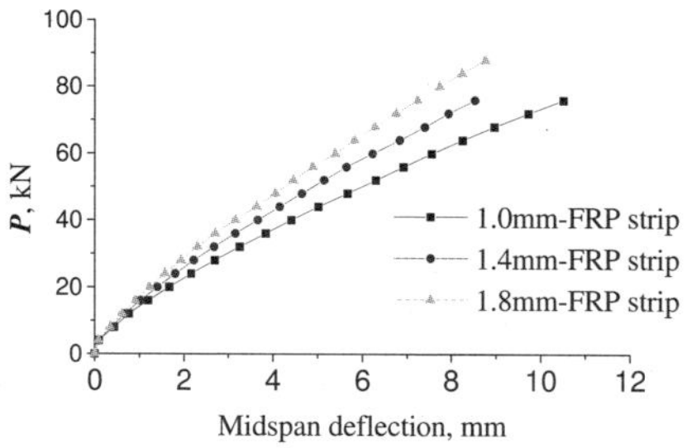

Fig. 3. Load-deflection curves of two-way slab strengthened with CFRP of different thickness.

4. Conclusion

The carrying capacity of FRP reinforced bi-directional RC plate was analyzed using the nonlinear FEM model built in this paper. The numerical results are in good agreement with Liman's experimental data. Therefore the FEM model is effective to predict carrying capacity of FRP reinforced two-way RC plate.

Numerical results show that by strengthened with CFRP, the carrying capacity of RC plate is improved greatly and instead of lack of glue layer strength, concrete failure accounts for disruption of RC slab strengthened with CFRP

It is found that carrying capacity is raised gradually with the increase of width of FRP.

Acknowledgments

This work are supported by the Fundamental Research Funds for the Central Universities (Grant No.90405008), and the National Key Technology R&D Program of China (No.2008BAB34B02).

References

1. V. M. Karbhari, F. Seible: *Appl. Compos. Mater*, 2000, 7, 95-124.
2. J. F. Chen, G. J. Teng: *Construction and Building Meterials*, 2003, 17, 27-41.
3. Z. Wu, H. Yuan, H. Niu: *ASCE J. Eng. Mech.*, 2002, 128, 562-573.
4. O. Limam, G. Foret: *Compos. Struct.*, 2003, 467-471.
5. O. Limam, V. T.Nguyen : *Eng. Struct.*, 2005,27,841-845.
6. X. L. Lv, G. F. Jin, X. H. Wu: in 'Nonlinear finite element theory and application of reinforced concrete structure', 1st edn, 249, 1997, Shanghai, Tongji University Press.

THE TIME DEPENDENCE RESEARCH OF BUCKLING FOR DELAMINATION WITH ELASTIC BRIDGE IN LAMINAR COMPOSITE MATERIAL

ZHANG XIAOCHUN

The Center of Intelligence Transportation System, Southeast University
Nanjing, China
email:zxc01@263.net

LIU SHUSHU

The Center of Intelligence Transportation System, Southeast University
Nanjing, China

Laminar composite materials widely used in engineering usually emerge delaminations because of manufacturing technique and other factors. Delaminations will take place buckling. The particular circumstance buckling is often related to time under constant load. In this paper, the buckling time-delay characters of a circular delamination with elastic bridge were studied by considering laminar composite material as viscoelastic material. The relationship between critical pressure and time of delamination buckling was gained. The effect of cohesion between layers on delaminations time-delay buckling was discussed. Theoretic analysis showed that the creeping buckling of viscoelastic delamination is related to mechanic performances of material, delamination dimension and cohesion between layers. The critical pressure is not a constant value and is related to time. After a period of time, time-delay buckling occurs to delaminations under constant load, and layers cohesion affect the shape of buckling surface if the load is heavy enough.

Keywords: Time dependence; delamination; buckling; composite material.

1. Introduction

The defects of composites which are caused by manufacturing technique and other factors, such as interfacial crack or delamination, have attracted people's attention. These disadvantages of composites not only destroy the integrality of materials' structure and reduce the mechanical property of local structure, but also affect other performances, such as the energy absorbency of built-up beam.

The 3D fibers composite is widely used in order to overcome these disadvantages. The transverse fibers on the interfacial crack and delamination in 3D fibre composites can restrict bucking and growth of the crack, and reduce the loss in overall strength of composites. Though lateral fibers constrain delamination, delamination buckle can happen in certain loads. Cox discussed the buckling problem of one dimensional penetrated delamination bridged by fibers, and derived the important character of the delamination buckling [1]. Siping Li used the energy variational method to solve the problem of circular delamination bridged by linear fibers, and derived the character radius of the delamination buckling and the calculation formula of the critical load [2,3]. Li further analyzed post-buckling and quasistate growth of a circular delamination. [4]

In the technical application and scientific research, time dependence of delamination buckling cannot be ignored, which is always a hidden danger in project structure. In the analysis of viscoelasticity lamina buckling, buckling critical load is related to time. Under a certain load, time delay buckling occurs to the lamina [5]. The buckling rupture is always in time correlation under the constant load. On the one hand mechanical property is reduced by aging of materials. On the other hand the material is rheologic owing to the change of condition such as temperature. So the research on time dependence of delamination buckle in laminar composites can not only evaluate the rupture condition of laminar composites structure and forecast the time of unstability, but also can guide the production and construction.

Laminar composite is regarded as viscoelasticity in this paper. The character of time delay buckling of circular delamination will be studied. Circular delamination radius is a, thickness is d from surface layer and radial pressure applied to the edge. The relationship between critical pressure of delamination buckling, deformation shape parameter and time will be given. The affect of layers cohesion/fibers bridged for shape of buckling surface will be discussed.

2. Governing Equation of Viscoelasticity Composites Delamination Buckling

A circular delamination with radius a subjected to uniformly distributed pressure p along its edge is considered. Let the deflection of the delamination under axi-symmetric buckling at t time be w(r, t). We assume restoring traction q(r) is a linear function of deflection

$$q(r) = \alpha w(r,t) \tag{1}$$

Where α denote equivalent elastic constant of bridge by fibers.

Let the constitutive equation of materials is of linear elastico-viscous property

$$\sigma_{ij} = \delta_{ij}\lambda(t) * d\varepsilon_{kk}(t) + 2G(t) * d\varepsilon_{ij}(t) \tag{2}$$

Where $G(t)$, $\lambda(t)$ denote relaxation function, "*" denote Stieltjes convolution, δ_{ij} denote Kronecker symbol, then

$$\begin{aligned}
\sigma_{rr} &= -x_3[\lambda(t) * d(\varepsilon_{rr} + \varepsilon_{\theta\theta}) + 2G(t) * d(\varepsilon_{rr})] \\
\sigma_{rr} &= -x_3[\lambda(t) * d(\varepsilon_{\theta\theta} + \varepsilon_{rr}) + 2G(t) * d(\varepsilon_{\theta\theta})] \\
\sigma_{r\theta} &= 0
\end{aligned} \tag{3}$$

Equation of geometry at t time is

$$\varepsilon_{rr} = -zw_{,rr}, \qquad \varepsilon_{\theta\theta} = -z(\frac{1}{r}w_{,r}) \tag{4}$$

The differential equation of bridged delamination axisymmetric buckling referencing flexure theory of sheet

$$D(t) * d\left[(\frac{d^4}{dr^4} + \frac{2}{r}\frac{d^3}{dr^3} - \frac{1}{r^2}\frac{d^2}{dr^2} + \frac{1}{r^3}\frac{d}{dr})w(r,t)\right] + p(\frac{d^2}{dr^2} + \frac{1}{r}\frac{d}{dr})w(r,t) + \alpha w(r,t) = 0 \tag{5}$$

Where

$$D(t) = \frac{h^3}{12}[\lambda(t) + 2G(t)] \tag{6}$$

Where p (>0) denotes radial pressure on the edge, h denotes thickness of delamination, $D(t)$ denotes the bending stiffness related to time of the delamination sheet.

Delamination deflections meet the clipped branch boundary condition and delamination cannot embed in under layer ancestral body. So that

$$w(a) = 0, \quad \left.\frac{dw}{dr}\right|_{r=a} = 0, \qquad w(r) > 0 \tag{7}$$

The deflection w(r, t) of delamination at t time can be given in the form of separation of variable between distortion and time, namely

$$w(r,t) = \sum_{m=1}^{\infty} w_m = \sum_{m=1}^{\infty} \eta_m(t) w_m(r) \tag{8}$$

Substituting Eq.8 into Eq.5 leads to

$$\frac{(\frac{d^4}{dr^4}+\frac{2}{r}\frac{d^3}{dr^3}-\frac{1}{r^2}\frac{d^2}{dr^2}+\frac{1}{r^3}\frac{d}{dr})w_m(r)}{-(\frac{d^2}{dr^2}+\frac{1}{r}\frac{d}{dr})w_m(r)-\frac{\alpha}{p}w_m(r)}=\frac{p\eta_m(t)}{D(t)*d\eta_m(t)} \tag{9}$$

Namely

$$\frac{(\frac{d^4}{dr^4}+\frac{2}{r}\frac{d^3}{dr^3}-\frac{1}{r^2}\frac{d^2}{dr^2}+\frac{1}{r^3}\frac{d}{dr})w_m(r)}{-(\frac{d^2}{dr^2}+\frac{1}{r}\frac{d}{dr})w_m(r)-\frac{\alpha}{p}w_m(r)}=k \tag{10}$$

$$\frac{p\eta_m(t)}{D(t)*d\eta_m(t)}=k \quad (k \text{ is a constant}) \tag{11}$$

From Eq.10, constant k is only related to delamination constraint condition and fiber bridge parameter, which reflect the state of delamination distortion. We define it as the delamination deformation shape parameter.

3. The Time-dependence Analysis of Delamination Shape Parameter

From Eq.11, one can find

$$k\int_0^t D(t-\tau)\dot{\eta}_m(\tau)d\tau = p\eta_m(t) \tag{12}$$

In order to derive the nontrivial solution manipulation in Eq.12, doing Laplace manipulation to the above equation leads to

$$[p-ks\overline{D}(s)]\overline{\eta}_m(s)=0 \tag{13}$$

$\eta_m(t)$ is constantly nonzero, so $\overline{\eta}_m(s)$ is constantly nonzero. From the condition of nonzero, one can find

$$ks\overline{D}(s)=p \tag{14}$$

The equation of sheet buckling in image space is derived. Then the time s_β in image space when the sheet buckles under the certain load p_β can be derived, and

$$s_\beta = \overline{Q}^{-1}\left(\frac{p_\beta}{k}\right) \tag{15}$$

Where $\overline{Q}^{-1}$ is the inverse function of $\overline{Q}(s)$, and $\overline{Q}(s) = s\overline{D}(s)$. Do Laplace inverse transformation to Eq.15. Let $p_\beta = pH(t)$, which means considering the sudden and post-constant load. $H(t)$ is the Heaviside function. After transforming equation (15) back to time phase, one obtains

$$k = \frac{p_\beta}{D(t)} \tag{16}$$

Eq.16 shown that delamination deformation shape parameter for viscoelasticity material is changed with time when given load p_β. For any moment t_β, $k = p_\beta / D(t)$ is constant.

4. The Relation between Critical Pressure and Time

From the Eq.10 one can find

$$\frac{d^4 w_m(r)}{dr^4} + \frac{2}{r}\frac{d^3 w_m(r)}{dr^3} + (k - \frac{1}{r^2})\frac{d^2 w_m(r)}{dr^2} + (\frac{1}{r^3} + k\frac{1}{r})\frac{dw_m(r)}{dr} + k\frac{\alpha}{p} w_m(r) = 0 \tag{17}$$

Reference [3] given the zero dimension solution in image space, and the critical pressure is

$$p_\beta^{cr} = \frac{40}{a^2} + 9.52\times10^{-2} a^2 - 4.96\times10^{-4}\sqrt{\frac{2.6\times10^9}{a^4} - 8.516\times10^6 + 1.16\times10^4 a^4}$$

Where

$$a = \sqrt[4]{\frac{\alpha}{D}}a, \quad p = \frac{p}{\sqrt{D\alpha}}, \quad r = \frac{r}{a}$$

Mathematical transformation above equation, obtain

$$k = \frac{40}{a^2} + 9.52\times10^{-2}\frac{\alpha a^2}{D} - 4.96\times10^{-4}\sqrt{\frac{2.6\times10^9}{a^4} - 8.516\times10^6\frac{\alpha}{D} + 1.16\times10^4\frac{\alpha a^4}{D^2}}$$

From the Eq.16, critical pressure and delamination deformation shape parameter can be given

$$\left.\begin{aligned} k &= \frac{40}{a^2} + 9.52\times10^{-2}\frac{\alpha a^2}{D(t_\beta)} - 4.96\times10^{-4}\sqrt{\frac{2.6\times10^9}{a^4} - 8.516\times10^6\frac{\alpha}{D(t_\beta)} + 1.16\times10^4\frac{\alpha a}{D(t_\beta)^2}^{4}} \\ p_\beta^{cr} &= \frac{40}{a^2}D(t_\beta) + 9.52\times10^{-2}\alpha a^2 - 4.96\times10^{-4}\sqrt{\frac{2.6\times10^9}{a^4}D^2(t_\beta) - 8.516\times10^6 D(t_\beta)\alpha + 1.16\times10^4\alpha a^4} \end{aligned}\right\} \tag{18}$$

It is indicated that delamination shape parameter and critical pressure would change non-linearly with time owing to the characters of viscoelasticity composites changing with time. For general viscoelasticity solid, let $\mu(t) = \mu_0 = const$. Consulting reference [5], Eq.6 can be given as

$$D(t) = \left(\frac{h^3}{12}\right)\left(\frac{1}{1-\mu_0}\right)2G(t)$$

$$2G(t) = g_0 + \sum_{i=1}^{n} g_i e^{-t/\tau_i}$$

Then

$$D(t) = \left(\frac{h^3}{12}\right)\left(\frac{1}{1-\mu_0}\right)\left(g_0 + \sum_{i=1}^{n} g_i e^{-t/\tau_i}\right) = A_0 + \sum A_i e^{-t/\tau_i} \tag{9}$$

Where A_0, A_i and τ_i are constants related to materials and delamination thickness h. Substituting Eq.19 into Eq.18. Referencing relaxation function of HDPE[5], let $D(t) = 386 + 1286\times e^{-t/4.13} + 165\times e^{-t/136}$ (mMPa), where $\mu = 0.35$, time unit is hour. Let thickness $h = 10mm$, radius $a = 50 \sim 100mm$, and elastic stiffness of fiber bridged $\alpha = 100\,\text{MPa}$. Choosing delamination radius, time, stiffness of fiber bridged as variables to plot curve of delamination shape parameter and critical pressure as following Fig.1~Fig.4.

Fig.1 and Eq.18 shows the critical pressure of delamination with viscoelasticity feature will reduce nonlinearly with time and the critical pressure is less with radius enlargement. The creep buckling of viscoelastic delamination may be emerged such states as the following:

(1) After loading, if load is large enough, the buckling takes place at once.

(2) If load isn't large to buckle at once, the buckling of delamination will happen on the time t_β after loading due to the viscoelastic feature of the material. This state can be defined as delay buckling.

(3) If load is less than a certain load, the viscoelastic delamination will not lose its stability.

Fig.2 shows the critical pressure of delamination buckling will increase with stiffness (α) of fiber bridged enlargement. This indication that fiber bridged contribute to delamination stability.

Fig.3 is the curve of shape parameter versus time for a=56mm (α=100MPa). Fig.4 is the curve of shape parameter versus time for difference radius a. They show that shape of delamination will change with time. Shape change violently in initial stage of load and hold certain shape ultimately.

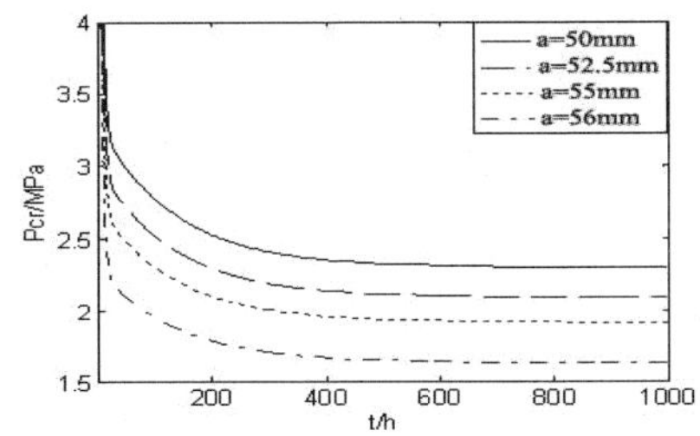

Fig. 1. The curve of critical pressure versus time for difference radius a (α=100MPa).

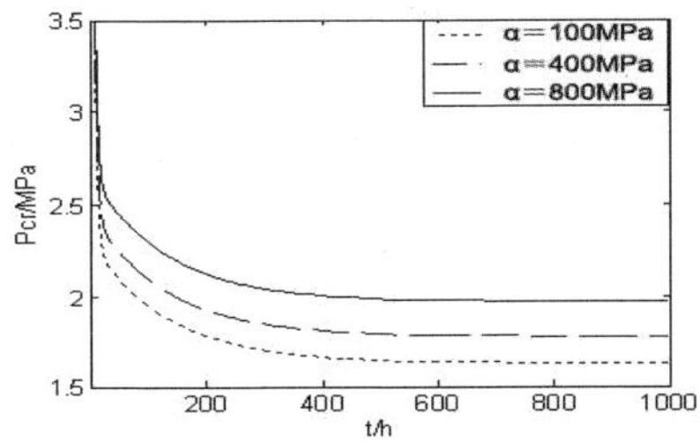

Fig. 2. The curve of critical pressure versus time for difference stiffness of fiber α (a=56mm).

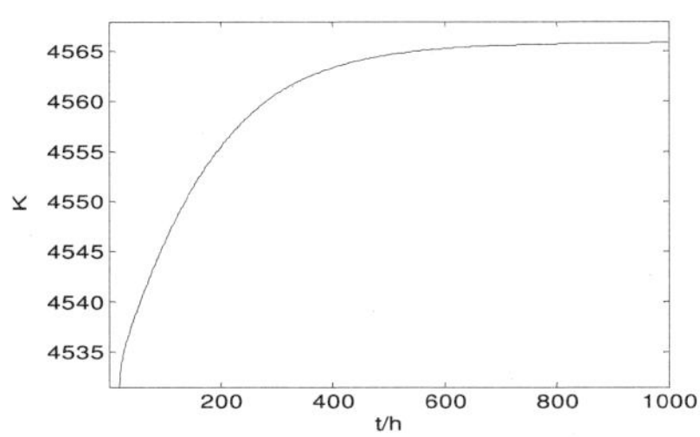

Fig. 3. The curve of shape parameter versus time for a=56mm (α=100MPa).

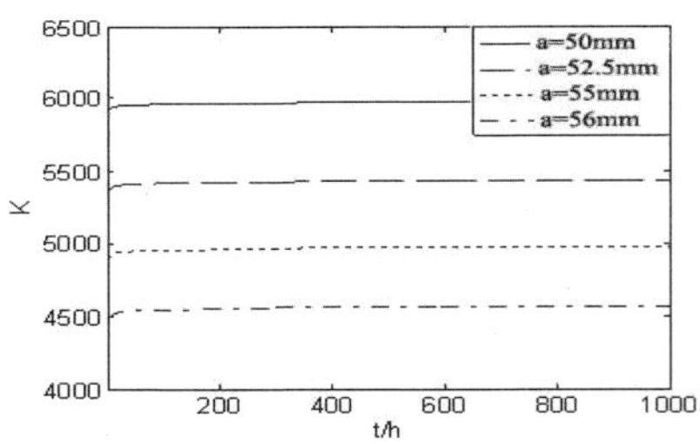

Fig. 4. The curve of shape parameter versus time for difference radius a (α=100MPa).

5. Summary

The analysis show that the buckling of delamination with fiber bridged is time-dependent due to the viscoelastic feature of the material. The relationship between critical pressure and time of delamination buckling was gained. The effect of cohesion between layers on delaminations for time-delay buckling was discussed. The value of critical pressure on delamination is a range relates to time. Under given load, delay buckling will occur. But shape saltation problems were not discussed and need to make further research.

References

1. B.N.Cox: J.Composite Materials. Vol. 17(1996), p. 331
2. S.P.Li,Y.Y.Huang,Y.T.Hu,D.W.Li: Eng. Mechanics. Vol.15(1998), p.127 (in Chinese)
3. S.P.Li:Theoretical Analysis on the Problems of Delamination Bridged by Lateral Fibers in 3D Composites, Ph.D thesis,Huazhong University of Science and Technology, Wuhan(1997)
4. S.P.Li, Y.Y.Huang,Y.T.Hu:ACTA Mechanica Solida Sinica. Vol.10(1997), p.212
5. T.Q.Yang, X.C.Zhang,Q.G.Gang: ACTA Mechanica Sinica.Vol.32(2000), p.319 (in Chinese)

EFFECT OF LOADING RATE ON DYNAMIC CHARACTERISTICS OF REINFORCED CONCRETE FRAME COLUMN

DEBIN WANG†

Faculty of Infrastructure Engineering, Dalian University of Technology
Dalian, Liaoning, China

HONGNAN LI

Faculty of Infrastructure Engineering, Dalian University of Technology
Dalian, Liaoning, China

The response for the reinforced concrete frame columns under quasi-static load and dynamic load with high strain rate are analyzed by finite element software ABAQUS. The influence of loading rate on loading capability, failure modes of reinforced concrete column are investigated, and the simulation results are compared with the test results. The results of high strain rate are compared with those of quasi-static to study the effects of strain rate on dynamic behavior of reinforced concrete columns. Results show that the dynamic behavior of reinforced concrete columns changed greatly when the strain rate effect were taken into account, and the effect of strain rate were different for different axial compression ratio of reinforced concrete columns. When reinforced concrete columns subjected to dynamic load, there was an increase in load carrying capacity. Therefore, the strain rate effect should be considered properly in the seismic analysis of reinforced concrete structures.

Keywords: Reinforced concrete columns; Strain rate; Axial compression ratio; Loading rate.

1. Introduction

Numerous rate relevant constitutive models have been successfully applied to the nonlinear analysis of reinforced concrete structures under dynamic loading. Tedesco developed the rate relevant constitutive model of concrete by MTS and SHPB tests[1]. Basing on the thermodynamic theory, Ragueneau & Gatuingt established plastic damage constitutive model for concrete considering strain rate effect under dynamic loading[2]. It was found that the deformation of reinforced bars deduced and the strength increased under dynamic loading by Yu and so on, and the rate relevant constitutive model of reinforced bar was proposed[3]. Soroushian proposed constitutive model of steel and mechanics property via dynamic loading experiment in different strain rate[4]. Li hongnan & Limin

† Debin Wang, email: wdb1215@163.com

researched mechanics property of steel by MTS facility, and proposed rate relevant constitutive model of steel[5].

In conclusion, although researchers have studied some dynamic properties of concrete and steel, the dynamic behavior of reinforced concrete members researched is limited due to the limitation of capacity of research facilities. In this paper, based on finite element software ABAQUS, the dynamic property of reinforced concrete column account the effect of strain rate is discussed.

2. Finite Element Model

In this paper, both HC4-8L16-T6-0.1P and HC4-8L16-T6-0.2P in literature [6] were accurately simulated, and axial pressure ratio were 0.1 and 0.2, respectively. Fig. 1 shows the specimen dimension in this study.

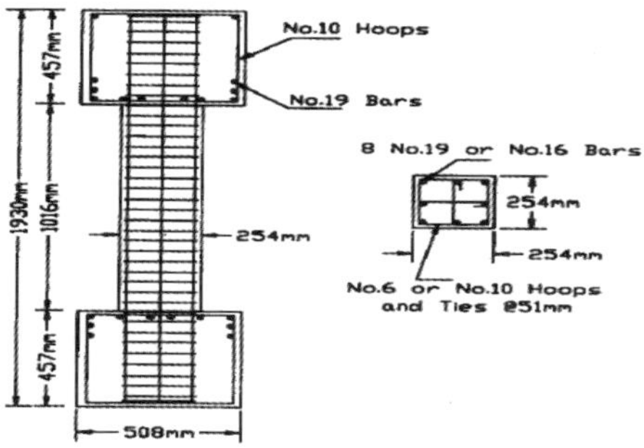

Fig. 1. Specimen dimension.

Finite element model was established by ABAQUS. The nonlinear material model for the reinforced concrete is composed of models for characterizing the behavior the concrete and the reinforcing bars, correspondingly, both concrete constitutive model and reinforcing steel model were concrete damaged plasticity model and linearly elastic-perfectly plastic model, and element types were C3D8R and T3D2, respectively[7]. In order to avoid the stress concentration, the concrete further extended in both top and bottom of reinforced concrete column. The load pattern is that the specimens were simulated under one horizontal loading at the top of column.

3. Strain Rate Relevant Constitutive Model

This paper takes into account both strain rate relevant constitutive model of concrete and strain rate relevant constitutive model of reinforcing steel.

3.1. *Strain rate relevant constitutive model of concrete*

The expression of concrete model is depicted in the following by Bathe and Ramaswamy[1],

$$\frac{\sigma}{\sigma_c} = \frac{(E_0/E_s)(\varepsilon/\varepsilon_c)}{1 + A(\varepsilon/\varepsilon_c) + B(\varepsilon/\varepsilon_c)^2 + C(\varepsilon/\varepsilon_c)^3} \tag{1}$$

where σ = uniaxial stress, σ_c = maximum uniaxial compressive stress, E_0 = uniaxial initial tangent modulus, E_s = uniaxial secant modulus, ε = uniaxial strain, ε_c = uniaxial strain corresponding to σ_c and A,B,C are acquired by test.

In the tension region, the stress-strain curve is showed in the following by Reinhardt(1984)[1],

$$\frac{\sigma}{f_t} = 1 - \left(\frac{\varepsilon}{\varepsilon_f}\right)^k \tag{2}$$

where f_t = uniaxial tension stress, ε_f = ultimate uniaxial tension strain, the parameter $k = 0.31$.

The dynamic–increase factor is the function of strain rate in both the compression region and the tension region, and the expressions were depicted in Formula.3 and Formula.4, respectively[1].

$$\begin{cases} \dfrac{\sigma_{cd}}{\sigma_{cs}} = 0.758 \log \varepsilon - 0.289 \le 2.5, \quad \varepsilon > 63.1 s^{-1} \\ \dfrac{\sigma_{cd}}{\sigma_{cs}} = 0.00965 \log \varepsilon + 1.058 \ge 1.0, \ \varepsilon \le 63.1 s^{-1} \end{cases} \tag{3}$$

$$\begin{cases} \dfrac{\sigma_{td}}{\sigma_{ts}} = 2.929 \log \varepsilon + 0.814 \le 6.0, \ \varepsilon > 2.32 s^{-1} \\ \dfrac{\sigma_{td}}{\sigma_{ts}} = 0.1425 \log \varepsilon + 1.833 \ge 1.0, \ \varepsilon \le 2.32 s^{-1} \end{cases} \tag{4}$$

where σ_{cd} = dynamic compressive stress, σ_{cs} = quasi-static compressive stress, σ_{td} = dynamic tension stress, σ_{ts} = quasi-static tension stress, ε = strain rate.

In this paper, the compressive strain takes into account the effect of strain rate, too. The expression[1] is given by:

$$\frac{\varepsilon_{od}}{\varepsilon_{os}} = 1.08 + 0.112 \log \varepsilon + 0.0193 (\log \varepsilon)^2 \tag{5}$$

3.2. *Strain rate relevant constitutive model of reinforcing steel*

The stress-strain relation of reinforcing steel is depicted linearly elastic-perfectly plastic model. The expression of dynamic-increase factor of reinforcing steel is described in the following[5],

$$\frac{\sigma_{yd}}{\sigma_{ys}} = 1 + c_f \lg\frac{\varepsilon}{\varepsilon_0} \tag{6}$$

$$c_f = 0.1709 - 3.289\times10^{-4}\sigma_{ys} \tag{7}$$

where σ_{yd} and σ_{ys} are yield stress under dynamic loading and under quasi-static, respectively; ε is strain rate; ε_0 is quasi-static strain rate, and $\varepsilon_0 = 1.0\times10^{-5}s^{-1}$; c_f is a function of σ_{ys} (Mpa).

4. Finite Element Simulation

4.1. *Comparison between simulation results and experiment dates*

The axial load ratios were 0.1 and 0.2 for two specimens, respectively. The load-displacement curves were shown in Fig. 2 and Fig. 3,

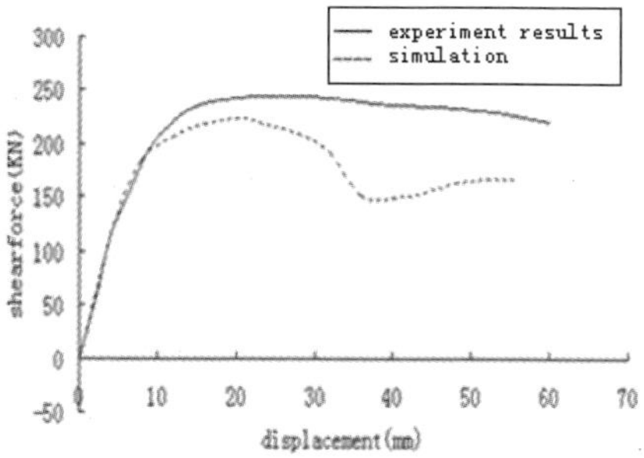

Fig. 2. Axial load ratio 0.1.

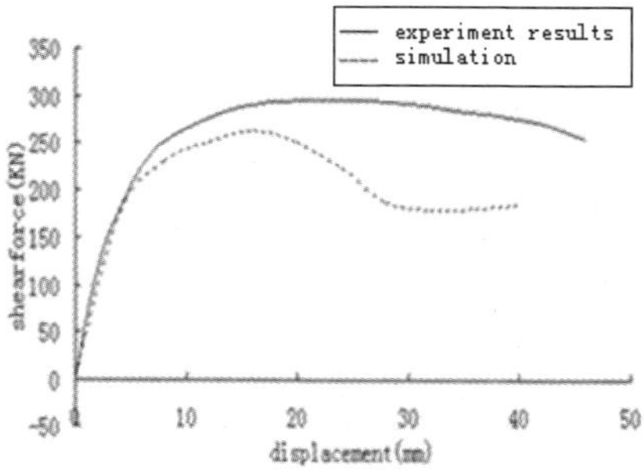

Fig. 3. Axial load ratio 0.2.

In the above pictures, we can see that the analytical results show reasonable correspondence with the experimental results.

4.2. *Comparisons of the dynamic analytical results and the quasi-static analytical results*

To investigate the dynamic mechanical properties of reinforced concrete columns under different loading rate, two specimens were simulated by using the above mentioned constitutive model. This method is executed by introducing the response of strain rate. The strain rate is $1.0\times10^{-2}s^{-1}$ in this paper. The analytical results used ABAQUS were shown in Fig. 4 and Fig. 5,

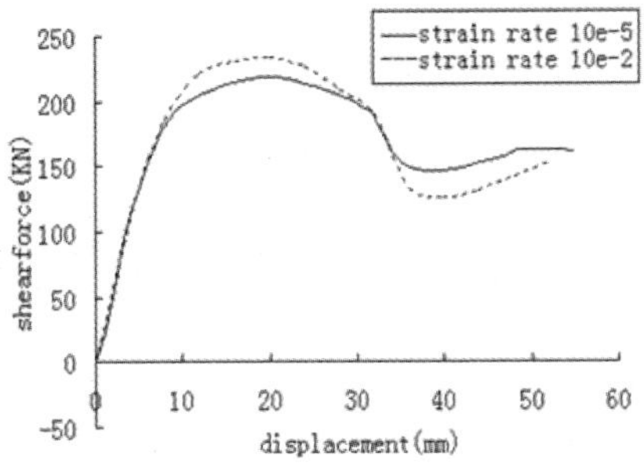

Fig. 4. Axial load ratio 0.1.

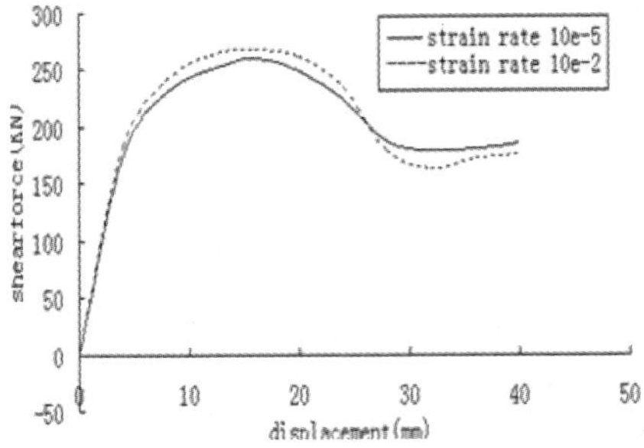

Fig. 5. Axial load ratio 0.2.

Under dynamic loading, however, the higher resistance due to the rate effect was found in the load-displacement curve. In other words, the higher loading rate leads to higher resistance of reinforced concrete column. At the same time, the peak resistance of reinforced concrete column enhanced with the decrease of axial load ratio in the same loading rate.

5. Conclusion

The reinforced column possesses higher resistance ability under higher loading rate, and the shape of the load-displacement curve under dynamic loading is similar to that under the quasi-static loading. The peak resistance of reinforced concrete column enhanced with the decrease of axial load ratio in the same loading rate.

References

1. Tedesco. J.W., Ross. C.A: Journal of Pressure Ves-sel Technology, 1998, 120(4): 398.
2. Frederic Ragueneau，Fabrice Gatuingt: Computers and Structures, 2003, 81, 1287–1299.
3. Haidong Yu, Yongjin Guo, Kaizhi Zhang, Xinmin Lai: Computational Materials Science , 2009, 46, 36–41.
4. Parviz Soroushian. Soroushian P, Choi k B: Journal of structural Engineering, 1987, 113(4): 663-672.
5. M. Li, H. N. Li: *China Civil Engineering Journal*, 2010, 43, (4), 70–75. (in Chinese).
6. Yan Xiao: Journal of structural Engineering, 1998, 111(3):241-251.
7. 'ABAQUS analysis user's manual', 2003, American, ABAQUS INC.

ENERGY-ABSORBING PERFORMANCE OF NESTED TUBE SYSTEMS UNDER COMPRESSION LOAD

CHEN ZONG AND XUE PU†

School of Aeronautics, Northwestern Polytechnical University. 127 Youyi Xilu, Xi'an, Shannxi 710072, P.R.China

This study aims at studying the performance of three nested tube systems under compression load. The tube systems are studied and compared with a single tube. The results show that the proposed nested systems can be specialized designed for different environments, and are effective in absorbing energy by its large plastic deformation, so as to reduce damages for structures.

Keywords: Tube; Nested structure; Plastic deformation; Energy absorption.

1. Introduction

Energy absorption component can absorb impact energy by its plastic deformation, so as to protect the people and structures behind. In recent years, many energy absorption components have been studied. The energy-absorbing components should satisfy some special requirements[1]: ① should utilize plastic deformation, as its major energy-absorption mechanism; ② the peak force must be kept below the threshold that would cause damage or injury; ③ deformation stroke should be long, stable, repeatable and reliable ; ④ light weight and low cost.

Tubes can be used as efficient energy absorbing structures, with light weight, high efficiency in energy absorbing, and low cost. They have been widely used in various engineering structures. Reid and Reddy examined the energy absorption characteristics of tubes with side constraint.[2,3] Morris proposed a nested tubes system and demonstrated its effectiveness under quasi-static loading.[4] The same group studied experimental its energy absorption under lateral loading by drop weight.[5] In this study, three different tube systems are proposed and compared with a single tube. Their energy-absorbing performance is investigated.

† Corresponding author, email: p.xue@nwpu.edu.cn

2. Tube Nested System

In this study, nested tube systems are proposed as shown in Fig. 1, (a), (b) and (c) have the same height. The diameters of the three tubes, named tube A, tube B and tube C, respectively, are d1=30.0mm, d2=20.0mm and d3=10.0mm; the thickness of each tube is 3.0mm and the length of each tube is 50mm.The material of the tubes is mild steel with its Young's Modulus E=225GPa. The linear strain hardening model is used in the plastic stage, namely,

$$\sigma=\sigma_0+E_p\varepsilon \tag{1}$$

Where $\sigma_0 = 450\text{MPa}$, $E_p = 1600\text{MPa}$, $\rho = 7800\text{kg/m}^3$.

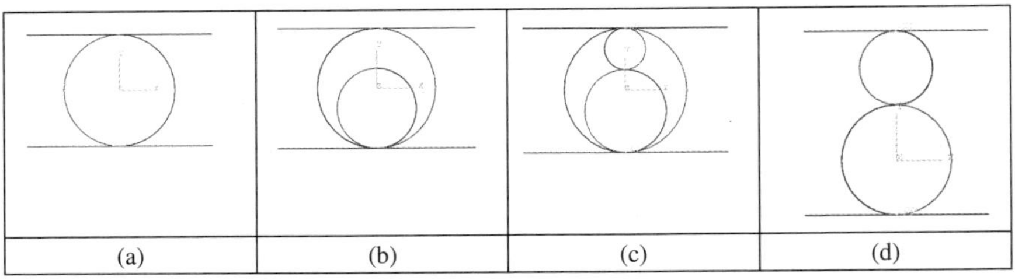

Fig. 1. Four kinds of different tube systems: (a) single-tube system--ST (b) dual-tube system—DT1 (c) triple-tube system--TT (d) dual-tube system—DT2.

3. Results

The four different tube nested systems were applied in the compression load. The pressed distance is approximately 60% of the height of each system. Their reaction forces at the bottom and the deformation were obtained numerically in this study as shown in Fig. 2 and Fig. 3.

It can be seen that DT1 and TT systems can provide a relative higher force plateau than the other two, while DT2 system can provide a longer deformation with a lower force plateau. The curves for both DT1 and DT2 systems have two stages because tube A deforms first under the load, and then tube B deformed after contacted so that the curve jumps to a larger value.

According to plastic theory, a rigid plastic single tube pressed by two rigid plates can be seen as a four-hinge structure, the load-deflection theory equation is:

$$P=\frac{2Yh^2L}{d\sqrt{1-(\delta/d)^2}} \tag{2}$$

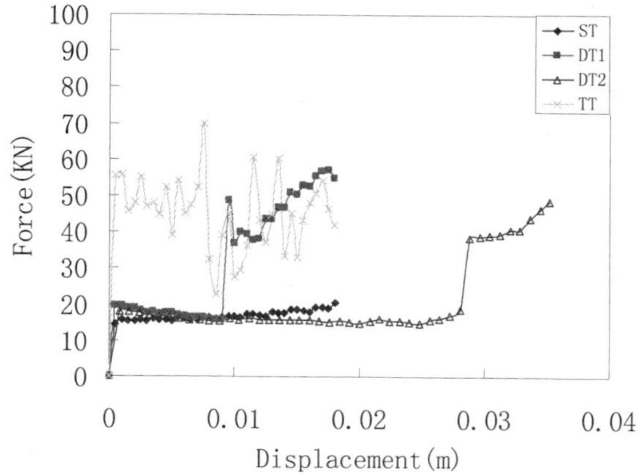

Fig. 2. Curves of reaction force vs. top displacement.

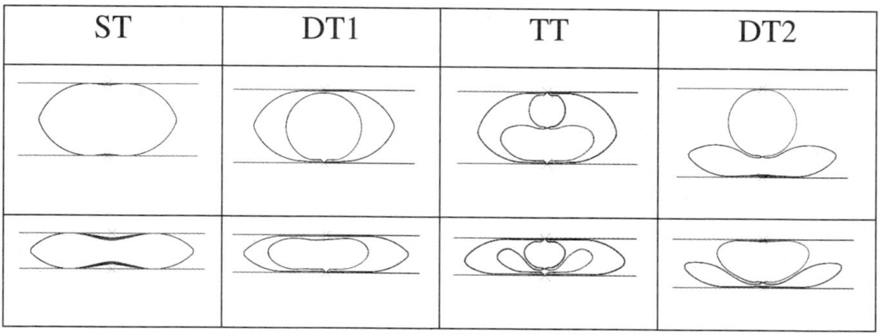

Fig. 3. Deformed configurations of four tube systems.

Y should be yield stress multiplied by $2/\sqrt{3}$. L, h, d, U are tube length, wall thickness, diameter and displacement of top plate respectively. According to equation (1), the reaction force---displacement of ST, DT1, DT2 and TT systems can be evaluated. For example, the reaction flat force of DT1 system is:

$$P_2=\frac{2Yh_1^2L}{d_1\sqrt{1-\left(\frac{\delta}{d_1}\right)^2}}+\Delta\left[\delta-\left(d_1-d_2\right)\right]\frac{2Yh_2^2L}{d_2\sqrt{1-\left[\frac{\delta-\left(d_1-d_2\right)}{d_2}\right]^2}} \tag{3}$$

where

$$\Delta\left[\delta-\left(d_1-d_2\right)\right]=\begin{cases}0, & \delta-\left(d_1-d_2\right)<0\\ 1, & \delta-\left(d_1-d_2\right)>0\end{cases}$$

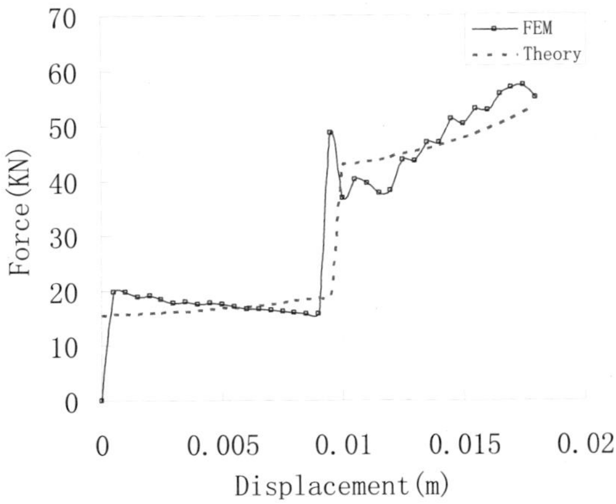

Fig. 4. Comparison between curves by numerical simulation and empirical formula of DT1 system.

Figure 4 shows the comparison between the theoretical prediction and the finite element results. It can be seen that the predicted force agrees with the finite element results very well.

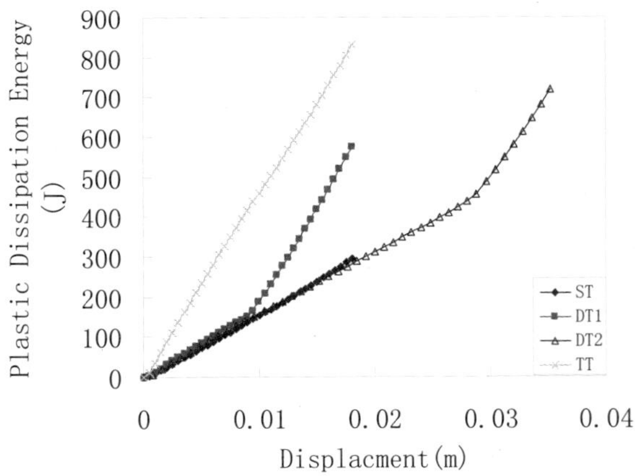

Fig. 5. Curves of plastic dissipation energy vs. top displacement.

Figure 5 shows the curves of plastic dissipation energy vs. displacement. TT system, with a higher force plateau, can absorb more energy per unit deformation than the other systems. Before the deformation reaches 10mm, ST, DT1 and DT2 systems have the same plastic dissipation energy. For DT1 and DT2 systems, the absorbed energy per unit deformation increases due to the deformation of tube B, but DT2 has a much longer displacement than DT1.

From above description, it can be concluded that DT2 system can provide a long deformation with a low reaction force, so it can be used under a weak impact load; TT system and DT1 system can provide a relatively high reaction force and energy absorption per unit deformation, so they can be used under a relatively strong impact load. TT system and DT1 system possess almost the same force plateau, but energy absorption of the TT system is much higher than the DT1 system. Therefore, it is the most effective in energy absorption.

4. Conclusion

This paper proposed three different tube nested systems. Their energy absorption performance was investigated and a comparison was made with a single tube under compression. The results show that tubes connected in series can provide a large deformation and low force plateau, so it performs well in a weak impact condition; If the space is limited, the nested tubes can perform well in a strong impact with its high specific energy-absorption capacity. TT system and DT1 system possess almost the same force plateau, but energy absorption of the TT system is much higher than the DT1 system. Therefore, it is the most effective in energy absorption.

Acknowledgments

Financial support from NSFC (NO. 10872168) and special fund of civil aviation from COSTIND are gratefully acknowledged.

References

1. Lu, G.X, and Yu, T.X: in ‘Energy Absorption of Structures and Materials’ 2003, Cambridge, Woodhead Publishing .
2. Reddy, T.Y. and Reid, S. R: Lateral compression of tubes and tube-systems with side constraints.1979, 21,187-99.
3. Reddy, T.Y. and Reid S.R: Phenomena associated with the crushing of metal tubes between rigid plates.1980, 16, 545-62
4. Morris, E. Olabi, A.G. and Hashimi, M.S.J: Lateral crushing of circular and non-circular tube systems under quasi-static conditions. 2007, 191, 132-135.
5. Olabi, A.G., Morris, E., Hashimi, M. S. J., and Gilchrist, M.D: Optimized design of nested circular tube energy absorbers under lateral impact loading. Mechanical Sciences. 2008, 50, 104-116.

Part II
Materials Processing Engineering

CAUSES AND PREVENTIVE MEASURES FOR CRACKS IN ANCHORED ENDS OF PRESTRESSED CONCRETE HOLLOW SLAB BEAMS

YANG ZHONG†
The Department of Civil Engineering, Ludong University, Yantai, Shandong 264025, China

LIU AIQUN
The Department of Civil Engineering, Ludong University, Yantai, Shandong 264025, China

YIN YIHUI
Institute of Structural Mechanics, China Academy of Engineering Physics, Mianyang, Sichuan 621900, China

Based on construction experiences, test results and active criteria, the causes breeding cracks in anchored ends of prestressed concrete hollow slab beams were analyzed and considered as smallness of dimensions of bearing plates, unreasonableness of beam pre-camber, local compression to the concrete in the anchored ends and largeness of the reaction forced by obtuse angle supports of oblique slabs. The effects of cracks on bearing capacity and normal service were charecterized and related prevention measures were proposed. At last, differences between current and old codes were emphasized in order to ensure safety during construction.

Keywords: Bridge engineering; prestressed concrete; crack adjacent to anchor; prevent measure.

1. Introduction

During construction and application of prestressed concrete beam bridges, lateral, longitudinal and inclined cracks may come into being in different locations of the beam ends which affect the endurance and carrying capacity of the bridges. Making clear the causes inducing the cracks and taking preventive measures can ensure the safety and endurance of the bridges.

---

† Liu Aiqun, email: ytlaq@126.com

2. Analysis of Cracking Causes of the Anchor

The bridge discussed is located in a linear part of a road and ranges about from stakes K20+595.67 to K20+751.71 in Weifang, Shandong. The bridge is 56.04m in length and intersects a river at an angle of 105°. The bridge is made in a form of central one with 4 spans of 13.0m in length and its upper structure consists of prestressed concrete hollow slab beams simply-supported. Each span consists of 7 beams of 1.24m in width. The design load is highroad II. The concrete used is C40, the steel cords are of $7\phi^8 15$ and the anchor tool is of XYM. During construction, cracks were found at the ends of 13 beams, the cracks in different beams were different from each other, the most cracks on a single beam were of 6, and the most width of the cracks was 0.3mm. The cracks were located at the obtuse angle positions, started at 3~6mm behind the anchor slabs along the beam lateral surfaces and prolonged 6~20cm toward the centers. Vertical cracks were 10-20cm away from the slab bottoms. Static load tests indicated that the experimental measured deflections of the central positions were 76.2% of the calculated ones, the measured stresses on the lateral surfaces behind obtuse angle anchors of the slab ends is 58~64% of the theoretical ones, which both were less than the designed values. It was judged that the cracks occurred during tension and by contra positive treatments, the bridge could be put into use.

2.1. *Smallness of the sizes of pads*

The sizes of pads were too small so that the stresses in the concrete below the anchor pads exceeded the designed values and the concrete fractured. According to the formula $\gamma_0 F_{\mathrm{ld}} \leq 0.9(\eta_{\mathrm{s}}\beta f_{\mathrm{cd}} + k\rho_{\mathrm{v}}\beta_{\mathrm{cor}} f_{\mathrm{sd}})A_{\mathrm{ln}}$ in JTG D62-2004 Design codes for steel-concrete and prestessed concrete highway bridges and culverts, the compression stresses in concrete for several anchor pads with different sizes were calculated and the results implied that the sizes of the pads should not be less than 200mm×200mm. However, the practical sizes of the pads in the engineering were 180mm×180mm. Obviously, the concrete stresses under the pads would exceed the designed values.

2.2. *Unreasonableness of the beam pre-cambers*

According to convention for a general steel-concrete beam, the pre-camber is set to 20mm in construction, but after tension, the camber height of the beam increases further more 8~10mm and the beam becomes an arched one with the beam ends being the arch feet, so the whole deadweight of the beam concentrates on the ends and therefore increases the stresses in the anchor pads.

bottom of the beam ends is subjected to both level friction force produced by beam body reversly arching and the whole deadweight of the beam body, as a result, the concrete near the beam ends fractures under compressive stresses.

2.3. *Local compression to the concrete in anchor ends*

The concrete below anchor pads was in a condition of local compression, the big longitudinal compressive stress σ_z spreaded from local to the whole pressed surface, and the spreading span is about the same as the sectional heigth h. Around the central key area subjected to high compressive stress, the lateral tensile stress σ_y existed simultaneously and made the concrete fracture longitudinally under local compression. Therefore this area was in high danger and in the area cracks came into being. If one did not focus attention on this area and not perform sufficient mixing round and solidifying, then the concrete strength would be insufficient and the fracture would happen.

2.4. *Largeness of the reaction forced by obtuse angle supports of oblique slabs*

The stresses were complicated near the obtuse angle supports with the inclined cross plates of the simply supported beams. Further more, the reactions forced by the obtuse angle ends supports were larger several times than that forced by supports with orthogonal plates. In addition, the unilateral tension at obtuse angle ends in construction was also an important reason for the fracture. Due to an elliptical anchor bolt hole of 5mm×15mm being preset at an obtuse angle end of a beam which was located at compressed area, the cross section of the concrete was reduced and the concrete under an anchor of an obtuse angle end became empty. Thus the area under an anchor of an obtuse angle end was a place where various adverse factors converged, once disregarded, cracks occurred inevitably in this area.

3. Effects of Cracks on Carrying Capacity and Normal Application

Synthetical analysis to deflection, strain, concrete strength and cracks of a tested beam indicated that the strength and rigidity of the beam satisfied the demand of designed load. In a statical load test, it was observed that the lengths of original cracks did not change and the widths decreased. So it was concluded that the cracks tended to close up when the beam body was subjected to compressive forces, the concrete fractured in elasticity and not in plasticity. Under loads of various levels, none of fresh cracks came into being in the bottom of the tested beam, especially in the central area. These phenomena show also that the entire strength of the beam satisfies design demands.

High strength steel cords were used in the bridge, so the beams were designed as over-ribbed ones. In order to make use of this part of strength, avoid the compressed concrete to fracture ahead of schedule, and increase lateral rigidity, meanwhile cater to the original intent that the bearing capacity of 8cm bridge surface pavement should be taken into consideration together, the waterproof concrete of the bridge surface was increased to C40 which was the same strength level as that of the beam body, the sizes of the steel cords were changed to $\phi 14$ or $\phi 16$.

In the present engineering, the felted prestressed ribs were used and the steel cords were fixed after the filled slurry solidified. These measures decreased the loads acting on the anchor heads. Due to the felted prestressed ribs with anchors, exterior loads did not increase the forces acting on the anchors. A strict check to compressive strength of the concrete near an anchor implied that, at the tension condition, the concrete under anchor pads suffered from the largest force. The largest force was of a sever disbennifit and should be taken as the basis controlling components design. For strength of concrete increases with time, if the concrete does not fracture during prestresses being loaded, then it will not tend to fracture later on.

From above, it can be concluded that the existing cracks in anchor heads generally do not affect further applications and need only simple treatments. However, for some of them, necessary measures should be taken in order to prevent fracture and avoid hidden troubles in further working.

4. Prevention Measures for Cracks at Anchor Ends

According to the analyses in section 3, the following measures should be taken to assure the bridge to be safe and prevent it to fracture.

Roughen wall surfaces of anchor bolt holes and mend them well by use of epoxy-bonded sanded slurry with high strength($\geq$ M5). Saw off extra parts of anchor bolts and enhance shakeproof baffling blocks. Fill up existing cracks with epoxy to keep the steel cords from rust.

Make forces acting on the 4 supports uniform and any of the supports not to come away.

The cracks existing areas in anchor ends are subjected to both longitudinal compressive forces and vertical support reactions acted by anchor heads. When the support reactions increase or an overloaded vehicle transits, ribbed steel plates of 400mm$\times$400mm$\times$400mm are expected sticked to the lower surfaces of the beam supports, especially for the fractured areas of the beam end bottoms.

A rubber plate of 1m in length and 2~3cm in thickness can be added to the end bottom of a beam body. When the beam being in tension, the rubber plate produces compressive deformation and the compressed area increases, so the concentrated compressive stresses decreases and the concrete in the end bottom of the beam remains in good condition and does not break. Equivalently, a bevel angle of 20cm along longitudinal direction of a beam and 10cm along vertical can be designed at the end bottom of the beam which can also increase effectively the compressed area of the end bottom of the beam in tension.

Enhance the lateral connections between beams and make loads from heavy and overweight vehicles distribute uniformly. In the longitudinal spans of 1.23m from beam ends the gaps between adjacent plates is only 1cm, fill them carefully with epoxy-bonded sanded slurry well. More over, fill the central reamered gaps with small stone-filled concrete in which the stones are not less than C50. Add a steel rib of 40cm in length and $\phi 16$ along the lateral direction on the upper part of the bridge to enhance the lateral rigidity and improve the active loading distribution.

After spanned, the deflection and cracks should be checked by stages, focuses should be on beam bottoms near the centers of the beams and lower bottoms of anchor pads, as well as the area near supports. The main observing stages include the followings: the moment accomplishing spanning the bridge, the moment completing paving the bridge surface concrete, the moment fulfilling paving the bridge surface bitumen layer, one time per 10 days in the previous 3 months using, and then one time per month.

Span each beam and measure its deflection in test field according to the condition of designed supports. Once abnormal phenomena happen with one beam the causes have to be made clear, otherwise, the beam should be disused.

The concrete area A_{cor} ranging indirect ribs should approach to the calculated bottom area A_b locally compressed, and satisfies $A_b > A_{cor} > A_l$, where A_l is the locally compressed concrete area. The depth of indirect ribs should not be less than the shortest side length of compressed sections or the least diameter of components. For spiral ribs, the spiral should be in the form of a bugle with small upper part and big lower part, so that the ribs are located in the boundary area of a highly compressed core zone and play fully a role in crack proof.

References

1. Designing institute for highway planning of Chinese ministry of communications. JTG D60-2004 General codes of highway ridge and culvert design[S]. Beijing: people's communications press, 2004.

2. Designing institute for highway planning of Chinese ministry of communications. JTG D62-2004 Design codes for steel-concrete and prestessed concrete highway bridges and culverts[S]. Beijing: people's communications press, 2004.
3. The first highway engineering bureau of Chinese ministry of communications. JTJ041-2000 Technique codes for construction of highway bridges and culverts[S]. Beijing: people's communications press, 2000.
4. China construction science academy. GB50010-2002 Design codes for concrete structures[S]. Beijing: Chinese construction industry press, 2002.
5. Xu Youlin. Understanding and application of design codes for concrete structures. Beijing: China construction industry press, 2002.
6. Lin Tongyan. Prestressed concrete structure design (3rd edition) [M]. Beijing: Railway industry press, 1983.

NEW IMPROVEMENTS IN THE MULTI-STEP INVERSE FEM FOR THE FAST FORMABILITY PREDICTION OF MULTI-STEP AUTOBODY STAMING PROCESS

BINGTAO TANG[†], XIAOYANG LU

Institute of Engineering Mechanics, Shandong Jianzhu University, Fengming Rd Lingang Development Zone
Jinan, 250101, China
State Key Laboratory of Materials Processing and Die & Mould Technology, Huazhong University of Science and Technology,
Wuhan, 430074, China

In the paper, the multi-step inverse finite element method (FEM) has been introduced to improve the accuracy of simulation in sheet metal stamping. The multi-step inverse FEM can be used to obtain the strain/thickness distribution and shape of blank in the intermediate configurations. An efficient walk-through point location algorithm is used to deal with contact searching problem and restrain the movement of corresponding nodes of intermediate configurations. A constitutive equation based on pseudo deformation theory of plasticity is proposed in order to preserve the computational efficiency of inverse FEM. The presented algorithms are applied to a 2-step cylinder cup deep drawing product. The numerical results compared with explicit dynamic solver LS-DYNA3D confirm its validity in formability prediction of intermediate shapes and final workpiece.

Keywords: Sheet metal forming; Inverse FEM; Constitutive equations.

1. Introduction

There are mainly two approaches for sheet metal forming simulation: an incremental approach and an inverse FEM. [1] The multi-step inverse FEM has been developed recently in order to consider the loading history and keep the simplicity and efficiency of one-step inverse FEM. Ideal forming theory by Majlessi,[2] axisymetric shell elements by Lee and Cao,[3] Pseudo-Inverse Approach by Li and Guo,[4] linear mapping technique by Lee and Huh,[5] sliding constraint surface by Kim,[6] modified arc-length search method by Huang,[7] double section curve expanding method by Tang have been developed.[8] These improvements of multi-step inverse FEM still show weak stability or time cosuming towards complex drawing cases.

[†] Corresponding author, email: tbtsh@hotmail.com

In the paper, developments are proposed to improve the reliability and efficiency of multi-step inverse FEM.

2. Constraint Strategies on Sliding Constraint Surfaces

The sliding constraint surface is introduced to guide the movement of nodes on intermediate configurations in the multi-step FEM. In this study, a point location algorithm, walk-through algorithm, is presented to locate the point quickly.

Consider a triangular element named ABC (in counter clockwise, Fig.1), and the area of the triangle is defined as Δ_{ABC}.

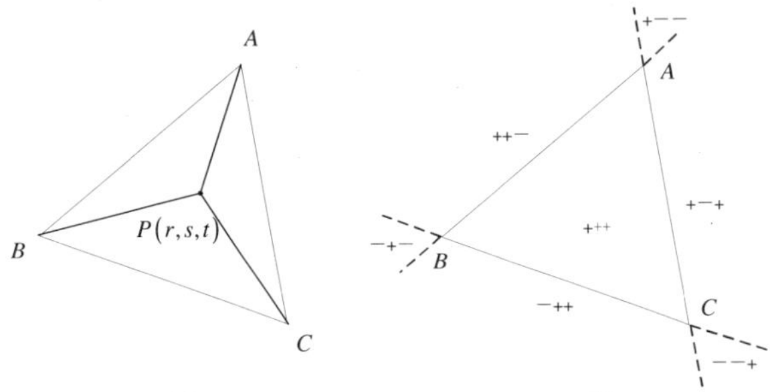

(a) (b)

Fig. 1. Barycentric coordinates and their signs: (a) point in a triangular; (b) signs of barycentric coordinates.

Suppose if the nodes A, B, C are counter clockwise (CCW) arranged, $\Delta_{ABC} > 0$, otherwise it is negative. The barycentric coordinates (r, s, t) of point P with respect to the triangle ABC can be expressed by the ratio of two triangular areas.

$$r = \frac{\Delta_{PBC}}{\Delta_{ABC}}, \quad s = \frac{\Delta_{PCA}}{\Delta_{ABC}}, \quad t = \frac{\Delta_{PAB}}{\Delta_{ABC}} \tag{1}$$

r, s and t satisfy the equation $r + s + t = 1$.

The procedure can be summarized below:

1. Compute the barycentric coordinates (r, s, t) of the query point P with respect to the current triangle element ABC.
2. Searching path walk through the edge that represents by the lowest of the barycentric coordinates. In case Fig.1(b), $r(PBC) < 0$, $s(PCA) > 0$, $t(PAB) > 0$, cross over the edge BC.
3. Stop when all the barycentric coordinates are positive, $r(PBC) > 0$, $s(PCA) > 0$, $t(PAB) > 0$.

3. Constitutive Equations for Multi-step FEM

In the paper, in order to preserve the computational efficiency of inverse FEM, a constitutive equation based on pseudo deformation theory of plasticity is proposed.

Fig.2 shows the uniaxial stress-strain curve for multi-step forming process. For the continuity of deformation from flat blank to final workpiece, the secant modulus E_{s1} can be obtained in a simple form based on Fig.2.

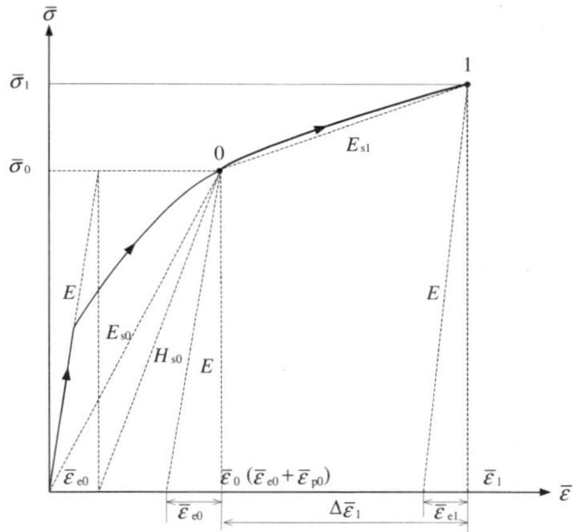

Fig. 2. Uniaxial tress-strain curve for multi-step forming process.

$$E_{s1}=\frac{\bar{\sigma}_1-\bar{\sigma}_0}{\bar{\varepsilon}_1-\bar{\varepsilon}_0}=\frac{\bar{\sigma}(\bar{\varepsilon}_1)-\bar{\sigma}(\bar{\varepsilon}_0)}{\bar{\varepsilon}_1-\bar{\varepsilon}_0} \tag{2}$$

Thus the stress tensor of state ‘1’ can be evaluated

$$\boldsymbol{\sigma}_1=\boldsymbol{\sigma}_0+E_{s1}\mathbf{P}^{-1}(\varepsilon_1-\varepsilon_0) \tag{3}$$

To determine the stress and strain of state ‘1’, the procedure can be summarized as:

1. Compute the displacement, strain and stress of state ‘0’: $\mathbf{u}_0,\varepsilon_0,\boldsymbol{\sigma}_0$
2. Compute displacement and strain increment
   $\Delta\mathbf{u}_1=\mathbf{u}_1-\mathbf{u}_0\ \ \Delta\boldsymbol{\varepsilon}_1=\boldsymbol{\varepsilon}_1-\boldsymbol{\varepsilon}_0\ \ \Delta\bar{\varepsilon}_1=\bar{\varepsilon}_1-\bar{\varepsilon}_0$
3. If $\bar{\sigma}_1<\sigma_s$, the deformation is elastic, the stress increment can be expressed as $\Delta\boldsymbol{\sigma}_1=\mathbf{D}_e\Delta\boldsymbol{\varepsilon}_1$, where $\mathbf{D}_e$ is the elastic matrix
4. If $\bar{\sigma}_1\geq\sigma_s$, the deformation is plastic, the stress increment is $\Delta\boldsymbol{\sigma}_1=E_{s1}\mathbf{P}^{-1}\Delta\boldsymbol{\varepsilon}_1$, where $\mathbf{P}$ is the anisotropic matrix
5. If $\Delta\bar{\varepsilon}_1\geq 0$, it is a loading process. The stress tensor of state ‘1’ can be obtained $\boldsymbol{\sigma}_1=\boldsymbol{\sigma}_0+\Delta\boldsymbol{\sigma}_1=\boldsymbol{\sigma}_0+E_{s1}\mathbf{P}^{-1}\Delta\boldsymbol{\varepsilon}_1$
6. If $\Delta\bar{\varepsilon}_1<0$, it is a unloading process. The stress tensor of state ‘1’ will become $\boldsymbol{\sigma}_1=\boldsymbol{\sigma}_0+\Delta\boldsymbol{\sigma}_1=\boldsymbol{\sigma}_0+\mathbf{D}_e\Delta\boldsymbol{\varepsilon}_1$

The procedure will continue until the convergence criterion for the global iterations are achieved.

4. Numerical Validations

In the paper, a 2-step deep drawing of cylinder cup is adopted to confirm the validity of the proposed developments.

The section line of the axial symmetric final cup for benchmark test is illustrated in Fig.3. The material properties and the process parameters for the analysis of the cylinder cup drawing are: stress-strain curve $\bar{\sigma} = 729.09(0.01345 + \bar{\varepsilon})^{0.1657}$ MPa, elastic modulus $E = 200$ GPa, Poisson's ratio $\nu = 0.3$, Lankford value $r = 0.95$, initial sheet thickness $t_0 = 0.70$ mm, friction coefficient $\mu = 0.13$, blankholder force $F_b = 15.0$ KN.

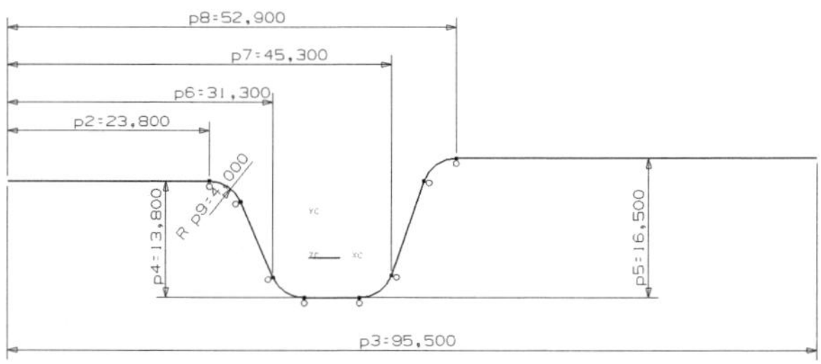

Fig. 3. Dimensions of 2-step forming workpiece.

Figs.4 show the thickness distribution obtained from 2-step inverse analysis and incremental analysis. Fig.4 (a) shows the thickness distribution of intermediate configuration with 1-step inverse analysis method. The comparison shows well agreement between the two results. They both witness the thinning in the area around punch radius and thickening in the area under the binder. The maximum discrepancy in thickness distribution is about 0.03mm. Fig.4 (b) shows the thickness distribution for the final workpiece obtained from 2-step inverse analysis and is compared with that of incremental analysis. The comparison demonstrates that the results of 2-step inverse analysis are in better agreement with those of incremental analysis than those of one-step inverse analysis. However, the thickness distribution from 2-step inverse analysis in the wall and flange region still disagrees with that of incremental analysis, as shown in Fig.4 (b). The increment of the number of step is required to consider more accurate deformation path and the boundary conditions, and improve the accuracy of inverse analysis.

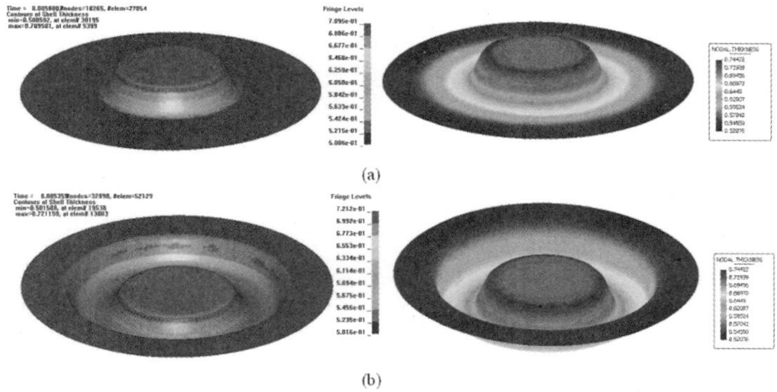

Fig. 4. Comparisons of thickness distribution: (a) from flat blank to intermediate sheet; (b) from flat blank to intermediate sheet and then final workpiece.

5. Conclusion

An efficient walk-through point location algorithm is used to deal with contact searching problem and restrain the movement of corresponding nodes. A pseudo deformation theory of plasticity based constitutive equation is proposed. Comparisons of numerical simulation results between the incremental and multi-step inverse FEM verify the feasibility of the proposed developments.

Acknowledgments

The Project was supported by State Key Laboratory of Materials Processing and Die & Mould Technology(09-15), by the Scientific Research Foundation for Outstanding Young Scientists of Shandong Province(BS2009ZZ011), Science and Technology Program Foundation for Shandong Higher Education Institute (J09LD09) and National Natural Science Foundation of China(60873080).

References

1. A. E. Tekkaya: *J. Mater. Process. Technol.*, 2006, 103, 14–22.
2. S. A. Majlessi, D. Lee: *J. Mater. Shape. Technol.*, 1998, 6, 41–54.
3. C. H. Lee, J. Cao: *Int. J Numer Meth. Eng.*, 2001, 50, 681–706.
4. Y. Q. Guo, Y. M. Li: *J. Mater. Process. Technol.*, 2004, 151, 88–97.
5. C. H. Lee, H. Huh: *J. Mater. Process. Technol.*, 1998, 80-81, 76–82.
6. S. H. Kim, H. Huh: *J. Mater. Process. Technol.*, 2001, 113, 779–785.
7. Y. Huang, Y. P. Chen and R. X. Du: *Int. J. Mech. Sci.*, 2006, 48, 591–600.
8. B. Tang, Z. Zhao, J. Chen and R. X. Yu: *Chin. J. Mech. Eng.*, 2006, 42, 211–217. (in Chinese)

AN STUDY ON THE FLEXURAL PERFORMANCE OF RENINFORCED CONCRETE BEAMS CONSIDERING LOADING HISTORY WITH PRE-STRESSED CFRP SHEETS

ZHE LI†

Department of Civil Engineering, Xi'an University of Technology, Jin hua nan lu 5 #
Xi'an, Shanxi, China

GANG LI

Liuzhou Institute of Architectural Design, Ba yi lu 77-1 #
Liuzhou, Guangxi, China

For the effect of non-prestressed carbon fiber reinforced polymer (CFRP) Sheet Strengthening concrete beams to improve bending stiffness and crack resistance is not satisfactory, and it is difficult to totally unloading on the actual reinforcement, so in this paper, ANSYS is used to research the strengthened effectiveness and impact factors of prestressed CFRP reinforced concrete beams considering the secondary load.

Keywords: Pre-stressed CFRP Sheets; Considering Loading History; Amount of the CFRP Sheets; ANSYS.

1. Introduction

In recent years, the technology of reinforced concrete structures strengthened by CFRP develops rapidly and applies widely. While in the actual reinforced projects, it is difficult to uninstall all loads before strengthening, so it is unsatisfied to enhance the flexural stiffness and anti-cracking performance. When the concrete structures are strengthened by common CFRP, it is unsatisfied to enhance the flexural stiffness and anti-cracking performance, in order to reduce and eliminate the stress of CFRP cloth lag and improve utilization of its strength, and the paper introduces the prestressed CFRP to reinforce concrete structures. In this paper, the finite element analysis software ANSYS is used to analyze ordinary reinforced concrete beams with initial moments in the context of unloading, and not only realize the nonlinear simulation analysis in comparison with experiments of reinforced concrete

---

† Corresponding author, email: lizhe009@163.com
Fund:CXY1020(6)

beams strengthened by the prestressed CFRP, but also further study different factors on the reinforcement effects.

2. Finite Element Analysis General Situation

2.1. *Test Overview*

The section size of concrete beams used in tests is b×h=150mm×300mm, span L=3000mm, and the steel and concrete dimensions of which shown in Figure 1.As can be seen from it, the concentrated forces are applied in the three points respectively, and the initial moments also used in this way, the reinforcement of testing components in Table 1.

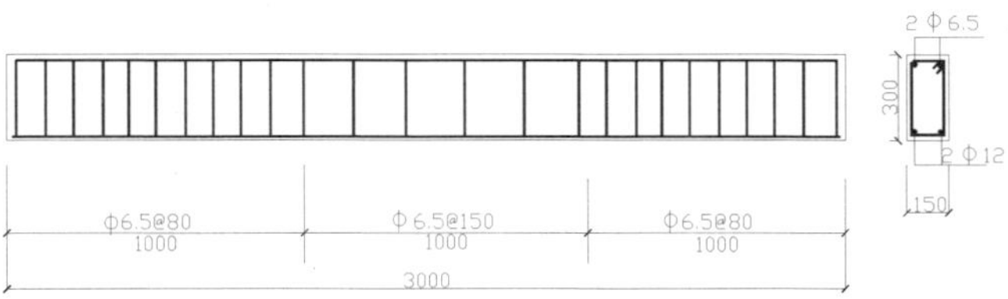

Fig. 1. Dimensions and reinforcement details of specimen.

Table 1. Test loading plan.

Specimen number	M_0(KN·m)	$\beta=M_0/Mu$	F_0(KN)	type and Layers of CFRP
La1-3-15	0	0	15	FTS-CI-30(one layer)
La1-4-15	7.45	0.357	15	FTS-CI-30(one layer)
La1-5-15	12.5	0.597	15	FTS-CI-30(one layer)

2.2. *Basic assumptions*

(1) In the deformation process, the strain distribution of the CFRP, steel and concrete meet the plane-section assumption.

(2) Without regard to the relative slip between steel, CFRP fabric and concrete, there is a reliable bond between CFRP and concrete, and the deformation compatibility is met.

(3) Test beams have sufficient shear strength.

2.3. *Material Properties*

Table 2. Performance index of materials.

Material	Poisson ratio	Modulus of Elasiticity (MPa)	Tensile strength (MPa)	Prism compressive strength(MPa)
Concrete	0.2	2.80×105	2.19	18.54
Steel Bars	0.3	2.10×105	369.6	—
CFRP	0.17	2.45×105	4080	—

In the simulation of test beams, the actual measurement is used as the material strength index. Conversely, for the parameters are not measured it can adopted according to 《Structural load code》[4], the adoption of a specific value, as shown in Table 2.

2.4. *Model building and cell division*

In order to compare with the experimental data, the material real constant, the components geometry and the reinforcement layout entirely consistent with the experimental beams in the model building. Table 2 is the specific test programs of finite element analysis. As the pilot beam geometry and loading programs are symmetrical, only half of the beams can be analyzed in modeling to reduce the computation time without compromising accuracy.

The three-dimensional discrete model is adopted in model building, and the eight-node hexahedral element of solid65 in the concrete. In the calculation, the concrete crushing damage was not taken into account, but the destruction of concrete in tension. Pipe20 unit is used in steel; Shell 41 is adopted in CFRP, using the solid45 unit to set rigid pads which has the same elastic modulus and Poisson's ratio with the concrete material. Finite element model and meshing specific are shown in Fig. 2 and Fig. 3

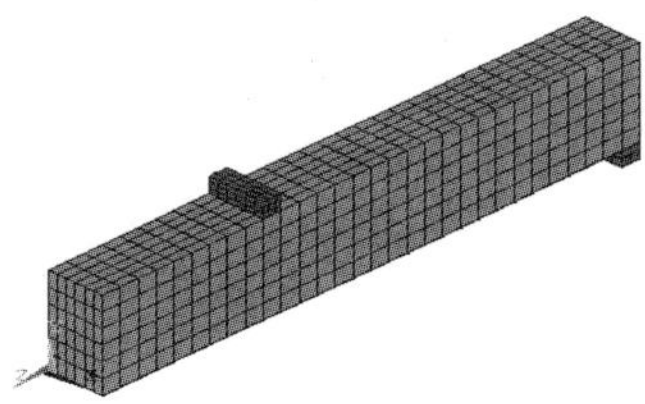

Fig. 2. Unit model of concretes and cushion.

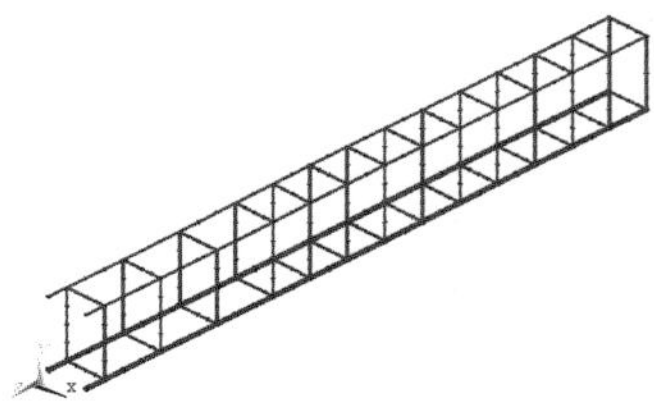

Fig. 3. Unit model of steel bar.

3. Results Analysis

3.1. *Comparison of numerical calculation and experimental results*

As can been see from the Table 3, it is basically consistent with the experimental data of the characteristic values about bearing capacity of reinforced structures calculated by ANSYS finite element program, while there is the error of calculated results and experimental values because of the bond-slip between CFRP and concrete beams which not considered in the model building process and the discrete nature of building materials themselves, but all the works within the scope of permitted, but because in the model building process, it is not considered the bond-slip between CFRP and concrete beams, as well as the discrete nature of building materials themselves, bringing about the error of calculated results and experimental values, but all the works within the scope of permitted. It is reliable that using ANSYS finite element program to simulate flexural capacity of concrete beams strengthened by priestesses CFRP with unloading and able to provide research with relative accurate data, so the method is effective and feasible.

Table 3. Contrast of test results and calculated results.

Specimen number	M_0	β	F_0	CFRP	Results	Mcr	My	Errors (%)	Mu	Errors (%)
La1-3-15	0	0	15	one	Test	14	28.00	—	35.0	—
					Calculate	14.5	28.46	1.64	33.94	-3.03
La1-4-15	7.45	0.357	15	one	Test	10.65	26.65	—	35.65	—
					Calculate	9.91	27.12	1.76	33.68	-5.53
La1-5-15	12.5	0.597	15	one	Test	—	24.15	—	32.15	—
					Calculate	—	25.61	6.04	33.27	3.48

Note: (1) M_0 represent the initial load(KN·m); F_0 represent the Pre-stress of the CFRP(KN);

(2) M_{cr}, M_y, and M_u represent the bending strength as bending strength, yield strength, ultimate strength(KN·m);

(3) $\beta=M_0/M_{u0}$ (M_{u0} represent the ultimate bending moment of the La1-3);

(4) The first number 1 of La1-3 represent amount of the CFRP sheets(the layer of CFRP is one), the second number 3 represent the initial load (M_0=0); the third number 15 represent the Pre-stress of the CFRP(F_0=15KN).

3.2. *Analysis of numerical results*

The influence of different initial moments

Through calculation and analysis, when the amount of CFRP reinforcement is certain, and the priestesses is constant, for the unloading beams strengthened by priestesses CFRP. With the initial moments increase of the reinforcement components increase, yield moments and design moments are decreasing significantly, and as the initial bending moments increase, the decline accelerates obviously.

The influence of different amount of CFRP

As can be seen from calculation and analysis, when other conditions remain unchanged, the bearing capacity of components have risen markedly as the amount of CFRP increases, especially for the design moment and ultimate moment. It mainly because the increase of the amount of CFRP to make more CFRP worked than less amount of it as well as bearing a larger part of tension before destruction in, meanwhile the destruction forms of components result into concrete crushing of compression zone, and the ultimate moment depends on the concrete compressive strength of components, therefore, ultimate bearing capacity has improved greatly .

The influence of different values of pre-stressing

As shown in calculation, for the same components which has the same amount of CFRP and the different initial moments, the carrying capacity of them has improved significantly in the using stage, ultimate moment change very little as the pre-stressed value increases. It is mainly because that, the CFRP can bear the tension in advance and reduce the steel stress and deflection of components effectively by increasing pre-stressed value of it, so as to reduce the steel stress and deflection of components effectively, which is helpful to the bending properties of components, but with the pre-stress value increases, leading to CFRP bears too much tension and causing CFRP damages easily.

4. Conclusion

The initial moment of the components have greater impact on flexural capacity, yield moment and design moment of components decline with the initial moment increase. When the amount of CFRP increases, the bending properties of the components improve significantly, but with the amount of CFRP increases, the failure modes of components result into concrete crushing damage of compression zone, and it is brittle failure. With the amount of CFRP increases, the rise of ultimate strength about reinforced components slows down, it indicates that promotion of ultimate bearing capacity about reinforced omponents is limited. The CFRP can enter the working state in advance, when the pre-stressed value increases, it also avoids the strain lag caused by the initial moment, and increases the normal using loads of components significantly.

References

1. F.F.Peng, SH.Y.Jiang and X.Xie: Building Science Research of Sichuan., 2003，29(2), 39-42.
2. GB50010-2002
3. Wight, Green and Erki: Journal of composites for construction., 2001,11,214- 220.
4. El-Hacha, ASCE, Wight: Journal of Composites for Construction., 2004, 2, 3-13.
5. X.ZH.Lu.Ph.D.Thesis, Tsinghua University, 2004.
6. F.F.Peng, SH.Y.Jiang, W.Fei. Building Science Research of Sichuan., 2003, **29(3)**, 60-63.

MODELING OF SHEAR FRACTURE IN THE SHEET METAL FORMING PROCESSES USING FEM APPROACH

YU SONG†
School of Civil Engineering, Shandong University, Ji'nan 250061, China

ZHAO JUN
School of Mechanical Engineering, Shandong University, Ji'nan 250061, China

In sheet fine-blanking process, severe plastic deformation localized in a narrow shear band near blanking clearance, and ductile fracture present at the final stage. Since the combinations of large nonlinear strain localization, displacement discontinuity and ductile fracture, a new ductile fracture initiation criterion model and elasto-plastic FEM are used to simulate localized severe plastic deformation. The distributions and developing trends of effective strain and damage are investgated, and an experiment also performed to research the forming mechanism and process parameters.

Keywords: Severe plastic deformation; fine-blanking; FEM simulation; Strain localization; Shear band.

1. Introduction

Strain localization always appears in the large plastic deformation of ductile metallic material. During a shearing dominant sheet metal forming process such as blanking, stamping, fine-blanking and piercing, it has been observed that the strain localization phenomena frequently exhibited on the blanked surface. Once the shear band initiates, it would develop and causes the strain intensely, strain localization would improve the capabilities of local plastic deformation, and metal components would be produced in the form of shear fracture. In recent years, numerical simulations of the shearing process of sheet metals have been performed based on the finite element method (FEM) to reduce the time and cost for production and optimize the blanking processes[1-5]. However, there have been few numerical studies which deal with the mechanism of shear fracture or the formation of fracture zone and blanked surface during the blanking process of sheet metals. But, from a manufacturing point of view, the quality and size of fracture zone are often the most important quality parameters. Furthermore, from a numerical point of view, a good description of the crack initiation and propagation requires the development of reliable algorithms. The objective of

† Corresponding author, email: yusong@sdu.edu.cn

the present work is to evolve a methodology to obtain shear fracture initiation criteria and simulate localized severe plastic deformation in sheet blanking using FEM approach. The distributions and developing trends of effective strain and damage, process paramaters affection on fracture are discussed.

2. Modeling of Numerical Simulation

The fine-blanking and its FEM 2D model are shown in Fig.1, parameters of simulation are shown in table 1. Punch, die, v-ring and ejector are defined as rigid bodies, and the sheet metal is defined as elasto-plastic body. The length and height of sheet is 120mm and 5mm respectively. The sheet mechanical properties of the material is shown in table 2. In order to obtain sufficient hydrostatic pressure in shear zone, double v-rings are adopted. The element type is quadrilateral 8-node mixed u-p element, which uses biquadratic interpolation functions to represent the coordinates and displacements, and accurately represents the strain fields in highly large strain analyze. The adaptive remeshing technique based on strain gradient and surface curvature is used to determine the

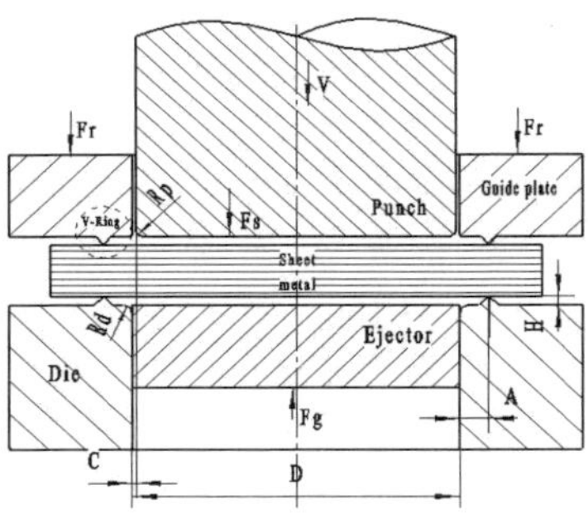

(a)

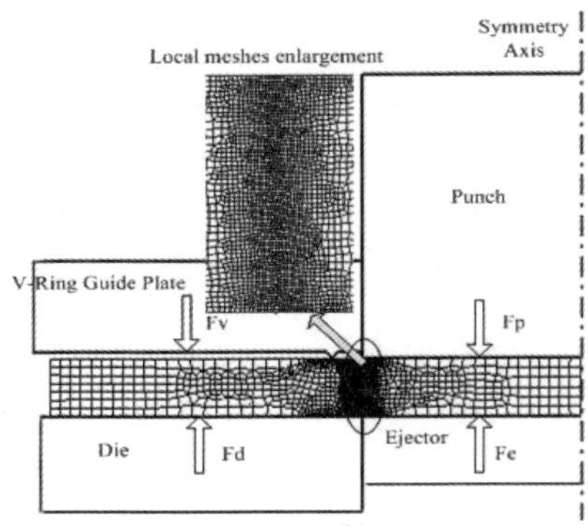

(b)

Fig. 1. Fine-blanking process parameters and FEM model.

Table 1. Process parameters of fine-blanking.

Types of simulation parameters	Parameters
One-side cutting clearance (C)	0.04, 0.14mm
Radius of punch edge (Rp)	0.05mm
Radius of die edge (Rd)	0.15mm
Diameter of inner hole of die (D)	90mm
Height of v-ring (H)	0.8mm
Distance in v-ring - punch (A)	2mm
Vee-ring force (Fr)	250KN
Ejector force (Fg)	125KN
Velocity of punch (V)	30mm/sec
Coefficient of friction (f)	0.1

Table 2. Mechanical properties of sheet metal.

Material type	Elastic-Plastic
Young's modulus (GPa)	210
Poisson's ratio	0.33
Yield strength (MPa)	410
Flow stress (MPa)	$\bar{\sigma} = 480\bar{\varepsilon}^{0.21} + 410$

$$\begin{cases} \int_0^{\bar{\varepsilon}_f} \dfrac{2\sigma_1}{3(\sigma_1 - \sigma_m)} d\bar{\varepsilon} = C, R_\sigma \geq 0.4 \\ \bar{\varepsilon} = C, R_\sigma < 0.4 \end{cases} \quad (1)$$

degree of refinement of meshes. In order to capture shear band phenomenon and reduce the volume loss of elements deletion when cracks development, the edge length of meshes near cutting edges are controlled in 0.01mm. The simulation is divided into two stages: first, the v-ring pressed into sheet metal and ejector applied ejector forces, second, the punch pressed into sheet metal, and plastic shear band formed gradually. In this study, the fracture criterion combined with modified Brozzo criteria[6-8] and effective strain criteria are expressed by equation (1) thought trial-and-error so that the value of the breaking might have the same value as the experiment. C the threshold value of the criterion at the instant of fracture initiation, R_σ =0.4 transition value between tensile fracture and shear fracture for sheet materials. The critical fracture value C=2.4 in this simulation. To verify the numerical simulation results, the actual fine-blanking experiments under the same conditions and material are carried out.

3. Results and Discussion

The development of $\bar{\varepsilon}$ distributions in plastic shear band is illustrated in Fig.2. Plastic deforming zone mainly concentrated near the limited clearance between punch and die, which developed in the direction along the punch edge towards to die edge, and there are no strains outside the shear band which means no plastic deformation happened in those regions. From Fig.2 (a) and (b) we can also see that dramatic plastic deformation occurred in shear band, and the maximum $\bar{\varepsilon}$ developed form 1.56 to 6.87 with penetration of punch from 1mm to 4mm. In Fig.2(c), the maximum $\bar{\varepsilon}$ occurred closed to punch edge and die edge, which are the most dangerous places to induce fracture. In order to observed the severe deformation in shear band of fine-blanking, a photochemical etching method is adopted. The material of specimen has a height of 5mm. Chessboard pattern mesh had been etched on the meridian plane of the specimen. The etched grids with $0.25\times0.25mm^2$ which is fine enough to observe the severe deformation. Then this experiment is performed on SCHMID HSR3200 fine-blanking press, and the process parameters of this experiment are same as those in simulation.

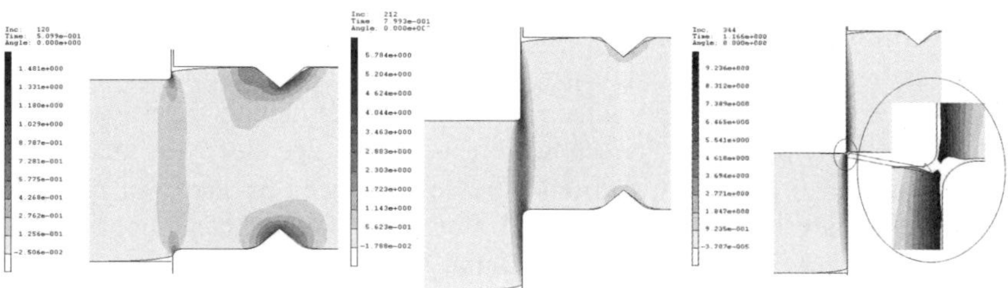

(a) (b) (c)

Fig. 2. Development of shear band when punch penetrated into 15%, 50% and 100% of sheet height.

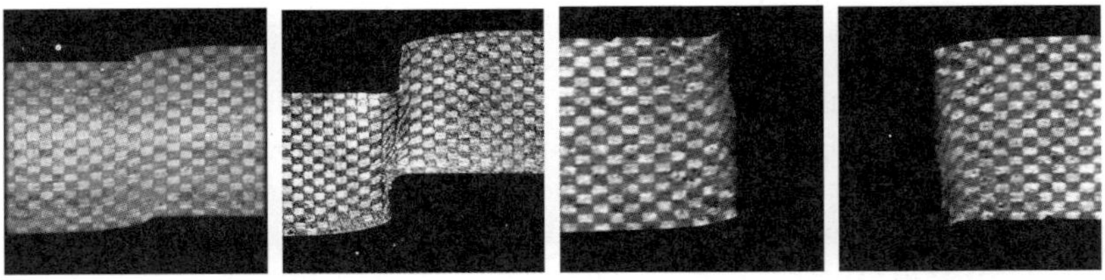

(a) (b) (c) (d)

Fig. 3. Grids deformation when punch penetrated into 15%, 50% and 100% sheet height.

Fig.3 shows grids deformation when punch penetrated into 15%, 50% and 100% of sheet height. At the initial stage of strain localization, the girds in shear band translated and rotated slightly with the shape changing from square to diamond. With the penetration of punch, the girds distorted gradually, and the distorted zone was limited in a narrow shear band. When punch penetrated into 50% of sheet height, the grids are distorted heavily, and the shape shear band is similar as that in simulation. Although grids are so distorted, they are still continuous, and not cut off, which proved that fine-blanking is a continuous plastic deformation process due to the existence of strain localization or shear band.

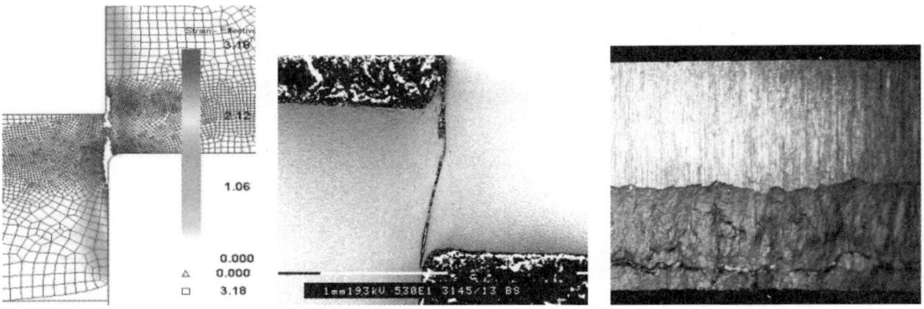

(a) (b) (c)

Fig. 4. (a) position and direction of macro cracks in shear band when punch penetration was 3.2mm. (b) SEM photo (c) actual product side surface.

Hydrostatic stress is the most important factor to influence the plasticity of metal, and negative hydrostatic stress could restrain the onset of fracture. when blanking clearance is 0.04mm which is 0.8% of sheet height, the damage work increased slowly either near punch edge or near die edge slowly at total stages, because the large positive hydrostatic stresses produced by narrower clearance restricted the development of shear band. But when blanking clearance is 0.1mm, the damage work increased slowly at initial stage, but quickly at middle stage because the larger blanking clearance. The damage work near punch edge is larger than that near die edge for the reason of larger positive hydrostatic stress generated by sharp punch edge. In this study, we compare experiments results of the blanked surface on the product and that on the scrap with FEM simulation, as shown in Fig.4. The FEM simulation results are made good agreement with metal experimental results. Although the shear surface and crack positions are slightly different from those of the experimental results on the blanked and the scrap side, the FEM results show the capability of this method for simulating the shear surface, burr and roll-over which usually occurs in fine-blanking process. Fig.4 (a) shows FEM simulate the position and direction of macro cracks developing in shear band of blanking clearance 0.1mm when the punch penetration was 3.2mm. Fig.4 (b) and (c) shows the actual SEM photo and product side surface.

4. Conclusion

Modeling of shear fracture in the sheet metal forming processes and a new ductile fracture initiation are applied to the finite element analysis for fine-blanking process. The modified ductile fracture criteria using relative value of stress variables is made in conjunction with metal experiments and fine-blanking process, good agreement being found to occur. Meanwhile, FEM approach is able to predict the propagation of cracks. The blanked product geometry, such as size of burr, sheared surface and fractured surface in blanking process are successfully simulated. On the basis of these results, it is possible that FEM simulation with shear fracture mechanism can predict the occurrence of defects on a metal forming processes such as blanked surface in the blanking or fine-blanking process.

Acknowledgments

The research work was supported by the National Science Foundation of China (50875156) and National Die and Mold CAD Engineering Research Center.

References

1. D. Brokken, W.A.M. Brekelmans, F.P.T. Baaijens:*J. of Mater. Proc. Tech.*,2003,**103**,51–56.
2. Z H Chen, C Y Tang, T C Lee, et al:*Inte.l J. Mech.l Scie.*, 2002, **44**,1309-1333.
3. Song Yu, Xionglong Xie etc: *J. of Mater. Proc. Tech.*, 2007,**187-188**,169-172
4. A.E. Huespe, A. Needleman , J. Oliver: *Int. J.of Plas.* 2009,**25**,2349–2365.
5. V. Lemialea, J. Chambert P. Picart: *J. of Mater. Proc. Tech.*,2009,**209**,2723–2734.
6. Roland W Lewis, Amir R Khoei: *Int. J. numer. Meth. Eng.*, 2001,**52**, 489-501.
7. Liang Xue, Tomasz Wierzbicki: *Eng. Frac. Mech.*, 2008,**75**, 3276–3293.
8. M Rotiz, Y Lery, A Needleman: *Comp. Meth. Appl. mech. Eng.*1987, **61**, 189-214.

EFFECTS OF MEAN STRESS AND STRESS CONCENTRATION ON HIGH STRENGTH BOLT

ZHOU CONGLING †

School of Mechanical Engineering, Tianjin University of Science and Technology, 1038, Dagu Nanlu, Hexi Dist., Tianjin, 300222, China

LI JIANYU

School of Mechanical Engineering, Tianjin University of Science and Technology, 1038, Dagu Nanlu, Hexi Dist., Tianjin, 300222, China

This study focuses on the effects of mean stress and stress concentration on high strength bolts. The mean stress levels are $0.3\sigma_B$, $0.5\sigma_B$ and $0.7\sigma_B$, respectively. The results show that the fatigue strength increase under the mean loading, but the differences between the loading levels are not so evident; most of the bolts in this study are broken from the bottom of the screw thread, and the cracks initiate from the impurities. Moreover, the stress concentration is obviously reduced as the increase of rounded corner radius at the root of thread and the depth of thread, while the effect of thread pitch is inconspicuous.

Keywords: High-strength bolt; Mean stress; Stress concentration; Finite element analysis.

1. Instruction

High strength bolts are widely used in modern machines, and they are especially important in the modern transport tools, such as cars, trains, airplanes and so on. It is reported that more than 1,000 bolts and 1,500,000 bolts are used in a car and an airplane, respectively[1]. Moreover, with the automobile becoming closer to our daily life, many transport accidents caused by the bolts fatigue failure are reported. Therefore, it is very necessary to study the fatigue properties of high-strength bolts used in automobiles.

Mean stress level is a very important factor that affects the fatigue properties of the high-strength bolts[2, 3]. This study is focused on the effects of different mean stress levels on the fatigue properties of high-strength bolts, as well as how the different pre-process procedures affect the fatigue properties of automobile high-strength bolts. On the other hand, FEM is a very useful method to analyze the stress. In this paper, a simplified model[4, 5] of the bolt is made

† Corresponding author, email: zhoucling@tust.edu.cn

through FEM method to analyze the stress situation during the fatigue test, and then makes a comparison between the FEM analysis results and the fatigue test results.

2. Materials and Experimental Procedures

Materials used in this study are automobile high-strength bolts. Table 1 lists the chemical composition, mechanical properties and the pre-process conditions of the test materials, respectively. Five types of bolts are used in this study, denoted by type 1 to type 5, respectively. The quenching conditions are all at 870°C, ≥5min, and the tempering temperature is 470°C, 445°C, 445°C, 450°C and 460°C, respectively, and the time is longer than 60 minutes. All of the bolts are jointed with nuts made under the same condition: material (45 steel), pitch (1.25mm) and hardness (34 HRC).The fatigue test is realized with a Shimazu-EHF10 machine (±15t, ±25mm) and the mean stress loadings are 0.3, 0.5 and 0.7 times to their tensile strength (σ_B). In the fatigue test, all of the nuts are loaded to the same station under the same load (15Nm) to make sure the fatigue failure initiates and propagates almost at the same position.

Table 1. Chemical compositions, (mass×10^2) %, mechanical properties and process conditions.

Material	C	Si	Mn	P	S	Cu	Ni	Cr	Mo	B	σ_B (MPa)	Processing procedure*
Type1	20	21	81	1.5	2.8	1	1	17	—	0.21	800	T. → H.
Type2	23	18	93	1.5	0.7	1	1	16	—	0.14	900	T. → H.
Type3	34	24	88	2.0	1.3	2	2	14	—	0.15	1000	T. → H.
Type4	41	19	75	1.6	1.5	1	2	101	16	—	1200	T. → H.
Type5	40	19	64	0.8	0.4	1	2	93	22	—	1200	H. → T.

*:T: Thread rolling, H: Heat treatment

3. Experimental Results and Discussion

3.1. *Experimental results*

Fatigue strength The fatigue strength was tested for specimens under the mean stress level of $0.3\sigma_B$, $0.5\sigma_B$ and $0.7\sigma_B$, and the S-N curve under the corresponding mean stress level is shown in Fig.1.

To $0.3\sigma_B$, fatigue limits of type 1 to type 4 are almost the same, while the fatigue limit of type 5 increases 85% to these former 4 types, which is 60MPa higher than that of the other four types. To the case of $0.5\sigma_B$, there is almost no difference between the fatigue limits of types 1, 2 and 3, while the fatigue limit of types 4 and 5 is 10MPa and 40MPa higher, that is, 14% and 57% to the former types. To the case of $0.7\sigma_B$, the fatigue limit of types 1, 3 and 4 are 10MPa lower than that of type 2, while the fatigue limit of type 5 is 40MPa higher. Moreover, compare to the results between the fatigue limits under the three different mean stress values, the highest fatigue limit is that of type 5 under $0.3\sigma_B$ is the highest, and the fatigue limits of type 1, 2 and 4 under $0.7\sigma_B$ are the lowest.

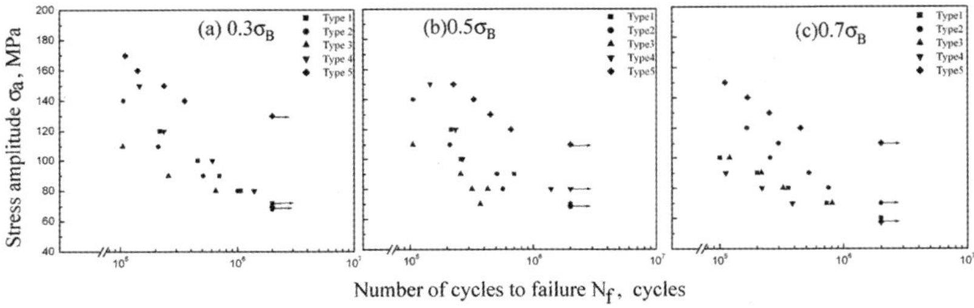

Fig. 1. S-N curves.

Hardness The hardness at the screw thread root is checked by a Vickers hardness tester under the load of 2.94N, and the test results is shown in Fig.2. It is clear that only the surface hardness of type 5 is much higher than the inner part.

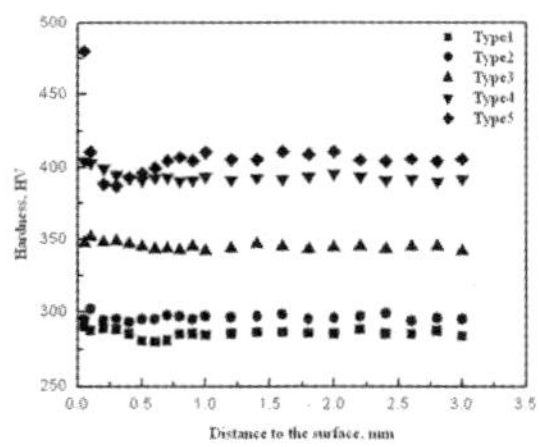

Fig. 2. Hardness on the fractured cross section.

Microstructure behaviors of the fractured cross section A scanning electronic microscope (SEM) is employed to investigate the initiation and propagation behavior of the fatigue crack by observing the fracture surface of the specimen. Based on the fatigue test, it is found that there are two different broken positions on the specimen. Type 1 to 4 breaks at the end of the nuts, while type 5 breaks at the end of the bolt. The representative microstructures of the fractured cross section are shown in Fig. 3. Based on the SEM observation, it is found that the cracks in both the two conditions initiate from the impurities in the specimens, independent on the position where the specimen breaks at.

3.2. *FEM analysis*

A simplified model is made to analyze the stress state on the bolt after the load is on during the fatigue test. Fig. 4 shows a group of the analysis results under the mean stress level of $0.3\sigma_B$, $0.5\sigma_B$ and $0.7\sigma_B$, respectively. The figure clearly shows that the stress level of two places are much higher than the mean value, which means that the bolt will fracture the most easily at this two points in the fatigue test. Moreover, the highest stress for the three cases, $0.3\sigma_B$, $0.5\sigma_B$ and $0.7\sigma_B$ in this group, is 1.21×10^3 MPa, 1.77×10^3 MPa and 2.36×10^3 MPa, respectively. That is, the highest stress becomes much higher with the increase of mean stress value, and it appears at the end of nut that connected to the bolt.

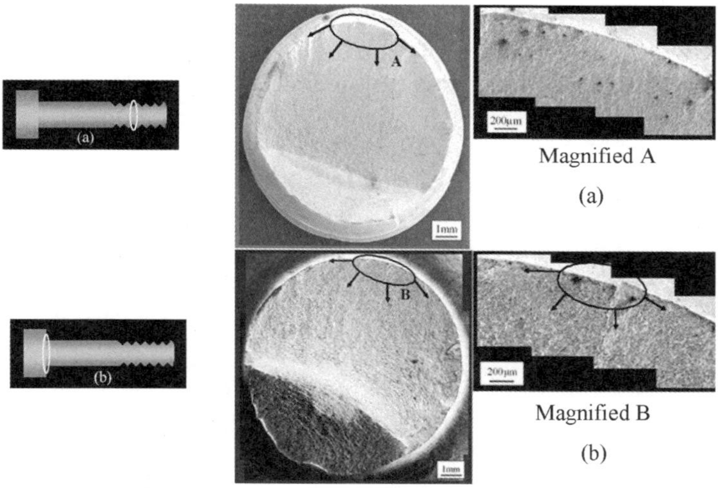

Fig. 3. Microstructure of cross section to the corresponding broken position
(a) σ_a= 90MPa, N= 5.10×10^5cycles (b) σ_a= 140MPa, N= 3.53×10^5cycles.

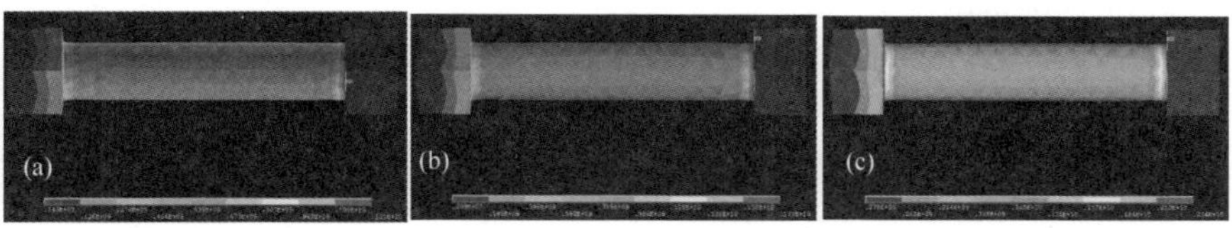

Fig. 4. FEM analysis result.

3.3. *Discussion*

As shown above, the fatigue limit of type 5 is improved much more than the other four types; the fatigue crack initiates and propagates from the impurities in the specimen, independent on where the specimen breaks at; and the surface hardness of type 5 is much higher than the inner part, while it is no much difference for the other four types.

The pre-processing procedure may be a main reason to improve this appearance. From the microstructure observed at the bottom of the screw thread root, it is clear that the thread flow of type 5 is much evident than the other four types. This may be because of the pre-processing procedure for type 5, which is thread rolled after the heat treatment, opposite to the other four types. For type 1 to type 4, the heat treatment after the thread rolling removed the residual stress caused by the thread rolling, which is compressive stress and helpful for the strength improve of the bolts. Moreover, the pre-processing procedure also causes the surface hardness of type 5 much higher than the inner part, because the residual compressive stress may improve the surface hardness.

These may be the main reason why the type 5 breaks at the different position from the other four types of bolts in this test.

4. Conclusion

This paper focuses on the study of mean stress and the pre-processing procedure effects on the fatigue limits of the high stress bolts used in automobiles. The main results obtained are as follows:

(1) The crack initiation and propagation is independent on the mean stress level and pre-processing procedures. All of the cracks initiate and propagate from the impurities in the bolts.

(2) There is no evident strength improvement with the increase of mean stress level, and the fatigue strength of type 5 is the highest under the mean stress of $0.3\sigma_B$.

(3) The pre-processing procedure may be one of the main reasons for the higher strength improvement than the other types.

Acknowledgment

This study is supported by the Talent Fetching Funding of Tianjin University of Science and Technology, 20070401.

References

1. K. Nakahara: *Master Thesis*, 2006.
2. S. Nishida: Failure Analysis in Engineering Applications, 1-20, 1992, Oxford, Butterworth Heinemann Ltd.
3. G. Shi, Y. Shi, Y. Wang, S. Li, H. Chen: Journal of Harbin Institute of Technology, 2005, **37**, 66-69.
4. B. Lu, H. Wang, X. Zhou, G. Wu: Heilongjiang Electric Power, 2004, **26**, 100-102.
5. D. Fang, J. Chen: Development and Application of Materials, 2007, **22**, 37-39.

IMPLEMENTATION AND APPLICATION OF A REAL-TIME CONTROL SYSTEM USING MATLAB/xPC

XIAOFENG HE

School of Mechanical Science & Engineering, Huazhong University of Science & Technology, No. 1037 Luoyu Road, Wuhan, 430074, China

ZHENGYI JIANG

Faculty of Engineering, University of Wollongong, Northfield Road, Wollongong, NSW 2522, Australia

CHRISTOPHER COOK

Faculty of Engineering, University of Wollongong, Northfield Road, Wollongong, NSW 2522, Australia

YINSHUI LIU†

School of Mechanical Science & Engineering, Huazhong University of Science & Technology, No. 1037 Luoyu Road, Wuhan, 430074, China

A real-time control system for rolling mill control such as the Hille 100 experimental rolling mill was developed based on Matlab Simulink, Real-Time Workshop and xPC Target. The Host-Target PCs architecture for the control system was introduced, then the approach for the implementation of data acquisition, real-time control and the associated graphic user interface (GUI), were discussed in detail. Some limitations of the commercial software were also presented. The developed control system has been verified with Hille 100 rolling mill in laboratory, indicating the effective application in rolling mill control.

Keywords: Rolling mill; Data Acquisition; Real-time Control; Matlab/xPC.

1. Introduction

Modern control technologies are being applied in almost every field of industries. In general, a control system design process consists of design, modelling, simulation, implementation and evaluation.[1] Subsequently the design is modified until all the system demands are fulfilled. These processes are often complicated, and need a long-term to be finished. However, with the

† Corresponding author, email: liu_yinshui@163.com

development of computer science, the complicated work is becoming simple. The Matlab software provides such rapid prototype tools for achieving hardware-in-the-loop simulation or constructing a real-time control system.[2-7]

An AGC (Auto Gauge Control) control system for a Hille 100 experimental rolling mill was developed successfully in 2008 by using tools of Simulink, Real-Time Workshop and xPC Target in Matlab 2009a. The goal of the system is to achieve real-time control of the roll gaps, as well as the data acquisition during rolling process. In this study, the detail of how to implement a data acquisition, real-time control and the associated GUI will be presented.

2. Architecture of Real-Time Control System

The developed real-time control system adopted the Host-Target type architecture. The Host PC uses Windows XP as an operating system combined with Matlab 2009a, which contains Simulink 7.3, RTW 7.3 and xPC Target 4.1 toolboxes.

Simulink provides a graphical user interface (GUI) for building models as block diagrams. xPC Target 4.1 supports a serials of I/O hardware in its block library, which can be added to the model.[8] RTW is used to generate code from the block diagram model. Then the code can be compiled to the executable real-time kernel using Microsoft Visual C/C++ or Watcom as a compiler, and downloaded directly to the Target PC via TCP/IP or RS-232 protocol.[2, 9] The Target PC is booted by a disk created from Host PC, and runs the real-time kernel in a DOS environment. The DAQ Card installed on the Target PC will be connected to the device sensors and amplifiers with its AD and DA ports, respectively.

So a great feature of this architecture is to build a bridge between the simulation model and hardware without writing low-level programming language, as well as to ensure the system's real time performance.

3. Implementation of Data Acquisition

Figure 1 gives the roll gap control model on the operation side for the rolling mill. It contains rolling forces data acquisition blocks, which is not only for monitoring process parameters, but also providing data source for calculating and adjusting the input signals of the control system.

Because the measured signals are inevitably coupled with many kinds of disturbance with high frequency in work site, therefore the low pass filters are often adopted for data acquisition. Matlab software provides very powerful Filter Design & Analysis Tool (Fdatool) to design digital filter blocks which can be added to the model. However, the characteristics of different types of filters are different. Taking two typical low pass filters, Butterworth and Equiripple, as example, it can be found by simulation that the Equiripple filter will cause more steady error compared with the Butterworth filter. Therefore, the Butterworth filters were finally selected for the model in Figure 1.

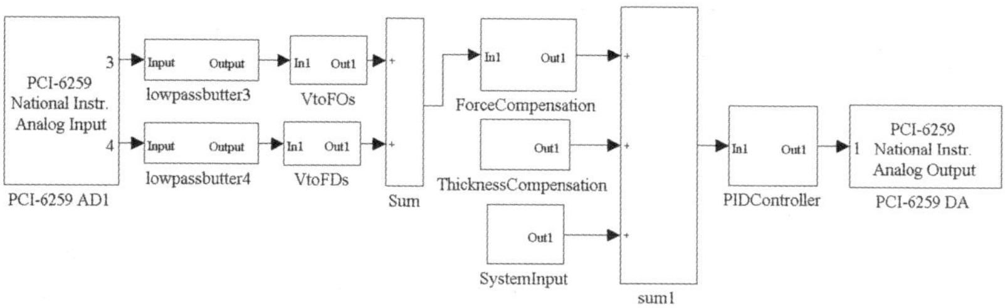

Fig. 1. Block diagram of the roll gap control on the operation side for Hille 100 rolling mill.

4. Implementation of Real-Time Control

The main function of the model in Figure 1 is to send the overall input signals to the PCI-6259 DA port and control the servo valve by a G122-202A1 servo amplifier. The overall input signal is integrated by the preset value, the force compensation value and thickness compensation value. The preset value is composed of the Entry and Desired Strip Thicknesses input through the Host PC by operator, and the Initial Roll Gaps which are measured before rolling.

The force compensation value C_F adopts the following fundamental equation:

$$C_F = \alpha_F dPM/2 \text{ (mm)} \tag{1}$$

where M is the rigidity of the rolling mill, α_F is the force compensation coefficient which can be obtained by test, α_F=0-1. dP is the rolling force variation compared with a base value.

The thickness compensation value C_{TH} is obtained by

$$C_{TH} = (h_{ave} - h_d)((W + M)/(W + M - \alpha_{TH} W)) \text{ (mm)} \tag{2}$$

where h_{ave} is the average exit strip thickness during last rolling process, h_d is the desired exit strip thickness. α_{TH} is the thickness compensation coefficient, α_{TH}=0-1， W is the deformation resistance of the rolled material.

5. Design of Graphic User Interface (GUI)

After modelling, xPC Target Explore tool can be used to load, build and download the model to the Target PC, then run in real-time, as well as modify the blocks' parameters on-line. However, this GUI is not friendly for users.

Figure 2 illustrates the GUI for Hille 100 short strip rolling process designed by using GUIDE layout panel.[10] This GUI has the following main functions.

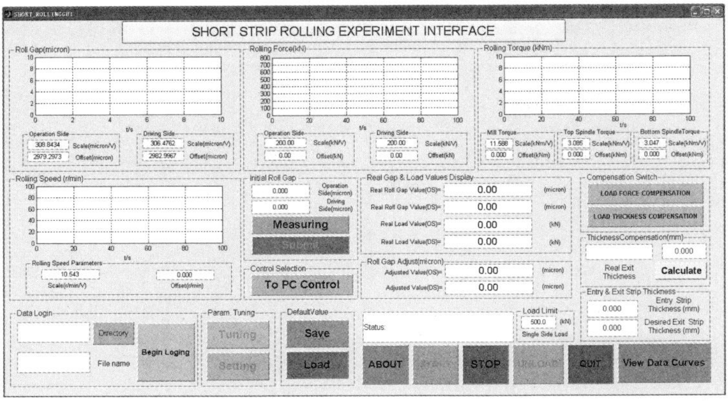

Fig. 2. GUI for Hille 100 short strip rolling process.

1) Measuring and Submit buttons: measure the initial roll gaps of the work rolls, then take their average values and display on the screen.

2) Tuning and Setting buttons: tune the controller parameters, select the mill type and material, and set the sampling parameters, etc.

3) START button: set xPC environment, load and build the model and download it to the Target PC, set the parameters of the Target object, then start the application. When the program is running, the acquired data can be displayed on the screen with curves and numerical values. In addition, the rolling mill can be switched to the PC or Manual control.

4) Begin Logging toggle button: to select if the experimental data is saved or not.

5) View Data Curves button: open another GUI and plot data curves for viewing and analysing the results after experiments. Figures 3 to 4 are the displacement (roll gap), rolling force on the operation side respectively when rolling a hot strip with entry thickness 12mm, reduction 5.95mm. From Figures 3 and 4, it can be seen that when the strip was moved into the roll, the gap increased about 150 μm with the increase of rolling force, and then they kept relatively stable values. When the strip was running out of the work rolls, the roll gap decreased with the rolling force and was recovered to its original state after a short transient process.

6. Limitations of the Matlab/xPC

During using Matlab/xPC, some limitations of the Matlab/xPC were found:

- Not all the xPC Target blocks have been tested to be perfect. For example, the PCI6259 A/D block in the xPC Target Library (Version 4.1) was verified to have bug. When this block was used to achieve multi-channels data acquisition, the correct results could not be obtained until a file named 'xpcionim.c' was replaced with a patch provided by The MathWorks Inc.
- The compatibility between different versions of software is not so good. Some commands' formats from Matlab Help files or its website may not be supported by the using version.

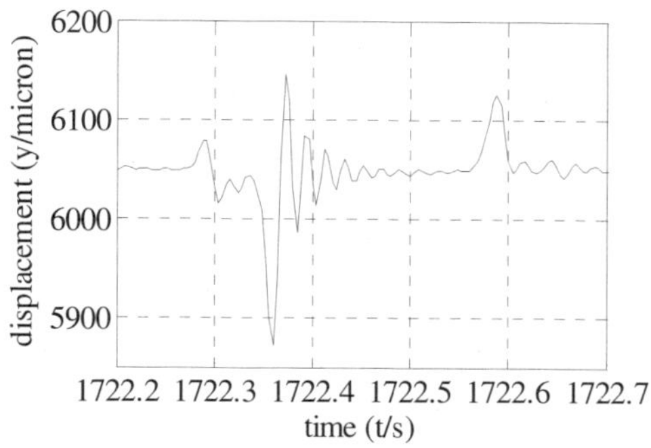

Fig. 3. The displacement (roll gap) curve on the operation side when rolling a hot strip with entry thickness 12mm, reduction 5.95mm.

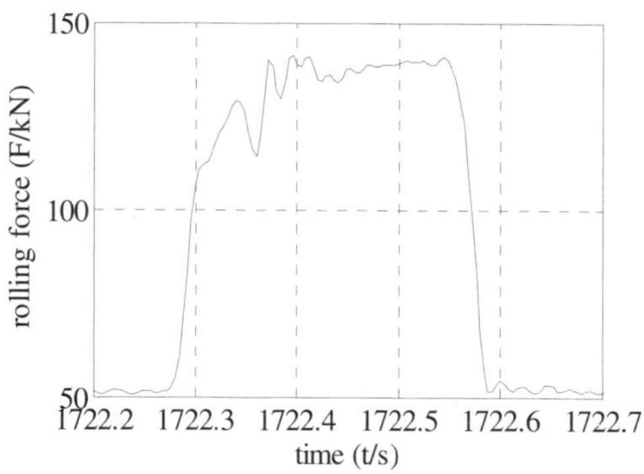

Fig. 4. The rolling force curve on the operation side when rolling a hot strip with entry thickness 12mm, reduction 5.95mm.

7. Conclusion

This paper presents the implementation of a real-time control system for a Hille 100 experimental rolling mill based on Matlab/xPC. The details of constructing data acquisition, real-time control system and the design of the associated GUI are discussed. The developed control system was successfully applied to practice and upgraded the performance of the rolling mill.

Although limitations of Matlab/xPC have been found during the implementation process, they will not handicap to verify that the Matlab/xPC has powerful functions which will make the modelling, simulation and implementation of a real-time control system easy and fast. This work will provide a reference for the rapid implementation of real-time control or monitoring systems not only in steel industries, but also in many other automatic control fields.

Acknowledgments

The work was financially supported by the Faculty of Engineering, University of Wollongong, Australia.

References

1. C. Ridderström, J. Ingvast, J. Wikander: Proceedings 2001 ICRA. IEEE International Conference on 'Robotics and Automation', Seoul, Korea, May 2001, IEEE, 1327-1333.
2. P. S. Shiakolas, D. Piyabongkam: Proceedings of the 40th IEEE Conference on 'Decision and Control', Orlando, Florida USA, December 2001, IEEE, 1348-1351.
3. A. Adgar, C. S. Cox: UKACC Control 2004 Mini Symposia, Bath UK, September 2004, University of Bath, ID-095.
4. Asan Gani, M. J. E. Salami-SMIEEE, Md. Raisuddin Khan: Proceedings of 2003 IEEE Conference on 'Control Applications', Istanbul Turkey, June 2003, IEEE, 538-544.
5. K. H. Low, Heng Wang, Michael Yu Wang: Proceeding of the 2005 IEEE International Conference on 'Automation Science and Engineering', Edmonton Canada, August 2005, IEEE, 345-350.
6. F C Teng: 2000 IEEE International Conference on 'Systems, Man, and Cybernetics', Nashville USA, October 2000, IEEE, 2697-2702.
7. Cheol Hoon Park, Sang-Kyu Choi, Young Su Son, Young Hee Han: 2009 World Congress on 'Computer Science and Information Engineering', Los Angeles USA, March 2009, IEEE, 701-705.
8. 'xPC Target User's Guide', 10, 2002, Natick USA, The MathWorks, Inc..
9. 'Real-Time Workshop User's Guide', 1, 1999, Natick USA, The MathWorks, Inc..
10. 'Creating Graphical User Interfaces', 1, 2008, Natick USA, The MathWorks, Inc..

STUDY ON STRETCHING EFFICIENCY OF HEAVY FORGED SHAFT BASED ON ORTHOGONAL EXPERIMENTAL DESIGN

QINXIANG XIA †, GUANGMING WEI, YULIN HUO

School of Mechanical & Automotive Engineering, South China University of Technology, Guangzhou, 510640, China

SU HUANG, MING LI

Guangzhou Shipyard Company Limited, Guangzhou, 510382, China

During stretching of heavy forged shafts, there are three major factors that affect the stretching efficiency, including the rotation mode, anvil gap and feed length. Based on the research of model simplification, successive blows and rotation control, a reasonable finite element simulation model is established by the numerical simulation software Deform-3D. The effect of the three factors on stretching efficiency is analysed by numerical simulation and orthogonal experiment. The results show that the effect of the rotation mode on stretching efficiency is the maximum, and the effect of the feed length is the minimum.

Keywords: Heavy forged shaft; stretching efficiency; orthogonal experiment; numerical simulation.

1. Introduction

For heavy forgings, multiple forging passes and several blows per pass are involved, the forging billet is fed at a certain distance after per blow and is rotated after per pass, so the stretching process is very time-consuming[1]. Liu presented new forging processes, such as tapered-anvil upsetting and LZ forging method, and the process parameters for stretching large rotor was optimised by numerical simulation[2]. Yang optimised the parameters for forging heavy rotor via orthogonal experiment, such as blank shape, tool width ratio, reduction of thickness and pre-cooling time[3]. Banaszek presented a numerical analysis of the effect of relative feed during stretching of large forgings[4]. Recently, the research on forging of heavy forgings was mainly focused on the optimization of upsetting and stretching.

The parameters influencing the stretching efficiency, including the rotation mode, anvil gap and feed length, are analysed and optimised via orthogonal

† Qinxiang Xia, email: meqxxia@ scut.edu.cn

experiment, and the combination of the parameters is obtained to maximize the stretching efficiency.

During stretching between two flat-anvils, the blank is locally compressed. The deformation can be regarded as local upsetting plus effects of rigid-edge. The flow of metal is constrained by tools and undeformed metal[5]. Therefore, the stretching efficiency is affected by the feed length and reduction of thickness. In addition, the rotation mode, which determines the cross sectional shape of the forging billet, also affects stretching efficiency.

2. Establishment of Finite Element Model

2.1. *Model size simplification*

In finite element model, the mesh size significantly affects the accuracy of simulation result. If meshing the model with the actual size of heavy forgings, the mesh number will be more than 400 thousand, which would consume too long time in simulation. Thus, to simplify the model is indispensable[6].

Fig.1 shows the simulation of stretching by 300 mm width flat-anvils, with a 30% relative reduction of thickness and 300 mm feed length, the same as the anvil width. In Fig.1 (a), after the first blow, the equivalent strain of the region more than 500 mm away from the right end of the forging billet is less than 0.003. In Fig.1 (b), after the second blow, the equivalent strain of the region more than 800 mm away from the right end of the forging billet is less than 0.003. It's concluded that the influence of undeformed region more than 800 mm away from the right end on the deformed region is slight, and can be neglected. Therefore, the length of model can be shortened to 900 mm, three times of the width of anvil.

(a) Equivalent strain after the first blow (b) Equivalent strain after the second blow

Fig. 1. Equivalent strain nephogram.

3. Orthogonal Experimental Design

The According to the three main influence factors on the stretching efficiency, including the rotation mode, anvil gap and feed length, three different values are selected for each factor. Then an orthogonal array experiment for 3 factors and

3 levels is designed, totally 9 testing cases which are simulated by the finite element software Deform-3D.

The initial dimension of blank is Φ650 mm×900 mm, and then is forged into square cross-section with 500 mm on each side by square forging. The process parameters are as follows: the initial temperature of blank is 1200 °C, the initial temperature of upper and lower anvils is 300 °C, the environment temperature is 28°C, the shearing frictional coefficient is 0.3, the heat transfer coefficient between forging billet and anvils is 11 Ns^{-1} mm°C, and the heat transfer coefficient between forging billet and environment is 0.02 Ns^{-1} mm°C. It's assumed that all blank surfaces are free surface without boundary constraint and there exists heat transfer between forging billet and environment.

To ensure the flatness and surface finish of forging, 0.4~0.8 width of the anvil is taken as the feed length[7]. For heavy forged shaft, the feed length is close to 0.75 width of the anvil to increase the stretching efficiency[8]. Therefore, three values of feed length 250 mm, 300 mm and 350 mm, and three values of anvil gap 370 mm, 360 mm and 350 mm are selected. As for the rotation mode, the forging billet is rotated 90° after 1 forging pass, 3 passes and 5 passes.

4. Orthogonal Experimental Design

Table 1 shows the orthogonal simulation results of 9 models, the stretching efficiency is defined as: $\eta=\Delta l/n$; where, n is the blow times, Δl is the elongation of each model.

It is shown that the range of each factor differs greatly from each other, where 1.48 for rotation mode and 0.11 for feed length. Therefore, the rotation mode has the maximum effect on the stretching efficiency, while the feed length has the minimum effect.

The analysis of variance (F-test) results show that the significance of the three influence factors on the stretching efficiency is distinctly different, where the feed length has the smallest significance in results. Thus it is unnecessary to carry out multiple comparisons within 'feed length' group.

Fig.2 shows the effect curve of orthogonal analysis. The variation of effect of the rotation mode is the maximum. The stretching efficiency reaches the highest value when the forging billet is rotated 90° after each forging pass, while reaches the lowest value when the forging billet is rotated 90° after 5 passes. As for the anvil gap, the stretching efficiency reaches the highest when the anvil gap is 350 mm, while reaches the lowest when the anvil gap is 370 mm. With increasing deformation degree, initially, the stretching efficiency increases obviously, then the increment slows down gradually. As for the feed length, the variation of effect is the minimum. The stretching efficiency reaches the highest when the feed length is 350 mm, while reaches the lowest when the feed length is 250 mm. Therefore, the sequence of influence factor on the stretching efficiency from large to small is: rotation mode, anvil gap, feed length.

Table 1. Orthogonal simulation results of stretching efficiency.

Factor / Test No.	Feed length	Anvil gap	Rotation mode	Test results		
	mm	Reduction ratio of thickness % (Anvil gap)	Rotate 90° after n forging passes	Elongation (Δl)	Blow times (n)	Stretching efficiency(η)
Test 1	250	26% (370)	1	168.2	43	3.91
Test 2	250	28% (360)	3	212.8	60	3.55
Test 3	250	30% (350)	5	216.1	64	3.38
Test 4	300	26% (370)	3	202.7	65	3.12
Test 5	300	28% (360)	5	210.5	70	3.01
Test 6	300	30% (350)	1	278.4	56	4.97
Test 7	350	26% (370)	5	204.4	71	2.88
Test 8	350	28% (360)	1	279.9	58	4.83
Test 9	350	30% (350)	3	271.8	79	3.44
Average 1	3.61	3.30	4.57	——	——	——
Average 2	3.70	3.80	3.37	——	——	——
Average 3	3.72	3.93	3.09	——	——	——
Range	0.11	0.63	1.48	——	——	——

F-test shows that multiple comparisons should be carried out within ‘rotation mode’ group. However, Fig.2 shows that the stretching efficiency reaches the maximum when the forging billet is rotated 90° after each forging pass. Therefore, the multiple comparisons among different levels of the rotation mode are not necessary. Since the feed length slightly affects the stretching efficiency, the multiple comparisons among different levels are not necessary either. Therefore, the optimised process parameter combination can be figured out from Table 1, which is 300 mm feed length, 30% of reduction ratio of thickness and 90° rotation after each forging pass in test 6. The maximum stretching efficiency is 4.97%, that is, the average elongation of the forging billet is 4.97 mm per pass.

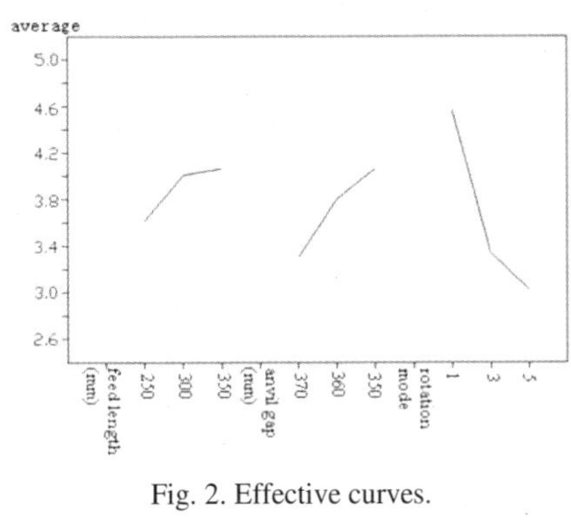

Fig. 2. Effective curves.

The marine propeller shaft of 13,500 ton ship is forged by the optimized process parameters as shown in Fig.3. The forging material is 35# steel, 7.8 ton weight. The finial dimension of the round cross-sectional forging is Φ350 × 9430 mm, the length-to-diameter ratio is 27.

Fig. 3. Heavy forged propeller shaft.

5. Conclusion

1. During stretching of heavy forged shaft, the length of finite element model can be shortened to three times of the width of anvil. The successive blows can be simulated by the function of 'Hammer Multiple Operation'.
2. The sequence of influence factors on the stretching efficiency from large to small is that the rotation mode, anvil gap and feed length.
3. The maximum stretching efficiency is 4.97% on condition of the optimised combination of process parameters: 90° rotation of the forging billet after each forging pass, 350 mm anvil gap and 300 mm feed length. In this way, the average elongation of the forging billet is 4.97 mm per pass.

Acknowledgments

This project is financially supported by Industrial Science and Technology Supporting Project of Guangzhou City (2008Z1-D221) and Major Science and Technology Project of Guangdong Province (2009A080304004).

References

1. Zhiwen Zhang: 'Forging Technology', 65-68; 1988, Beijing, China Machine Press.
2. Zhubai Liu: 'New Plastic Forming Technology and Its Mechanics Principle', 78-87; 1995, Beijing, China Machine Press.
3. D. Y. Yang: The 13th International Forge Master Meeting, Korea, 1997, 135-151.
4. G. Banaszek, P. Szota: Journal of Materials Processing Technology, 2005, **169**, 437-444.
5. Zhenqiang Zhang, Xuanrong Xin, Ting Liu, China Metalforming Equipment &Manufacturing Technology, 2006, **6**, 50-53.
6. Xinbo Lin, Die and Mould Technology, 2000, (3), 75-80.
7. Wei Zhu, Zhiliang Zhang: Computer Simulation, 2005, **22**, (3), 88-91.
8. Zhubai Liu, Liyong Ni, Guohui Liu:New 'Theory and Technology of Large Forgings Deformation', 84-88; 2009, Beijing, China Machine Press.

STORAGE METHOD OF FE EQUATIONS AND ITS APPLICATION IN NUMERICAL SIMULATION OF SHEET METAL FORMING

BAO YIDONG[†], DU GUO KANG, CHEN WEN LIANG, WANG ZHI GUO

College of Mechanical and Electrical Engineering,
Nanjing University of Aeronautics and Astronautics, 29 Yudao Street
Nanjing, 210016, China

A finite element equation always is a linear large sparse matrix equation and solving finite element equations is a key process in the numerical simulation of sheet metal forming. The solving method of equations and storage technology of the stiffness matrix affect the calculate efficiency of FEM numerical analysis significantly. In order to improve the solving efficiency of the self-developed code Quick-Form, the direct solver is replaced by the iterative solver. Based on two definitions of general adjacent node and general nodal adjacent relation, the generation principle of the total stiffness matrix and the distribution rules about nonzero matrix are introduced. An improved compact storage scheme named one dimension array with variational band width is proposed for the iterative algorithm solver, which can save large physical memory and avoid the optimization of nodal numbering. Numerical example demonstrates the higher efficiency and less memory needed of this proposed algorithm.

Keywords: Finite element method; Sparse matrix; Compact storage scheme.

1. Introduction

Recently the finite element method (FEM) is the main method in numerical simulation of sheet metal forming. There are the incremental FEM based on flow theory and the total FEM based on deformation theory in the numerical simulation methods of sheet metal forming.[1-3] Finite element analysis always comes down to solving the linear equations, so the calculation algorithm of linear equations and matrix storage structure will determine the solving efficiency of finite element method.

Generally, the calculation algorithms of large linear equations have two ways: the direct method and the iteration method. The iteration method is usually used to solve large linear equations. The SSOR-PCG (Symmetric Successive over Relaxation- Preconditioned Conjugate Gradient) iteration algorithm is

† Corresponding author, email: baoyd@nuaa.edu.cn

mostly used by many commercial FEM softwares (e.g. ANSYS, COSMOS, ALGOR etc.). [4]

The low calculation efficiency reason in some forming process of big parts simulated by self-developed code QuickForm is investigated. In order to improve the solving efficiency of Quick-Form, the direct solver is replaced by the iterative solver: the SSOR-PCG iteration algorithm. Based on two definitions of general adjacent node and general nodal adjacent relation, the generation principle of the total stiffness matrix and the distribution rules about nonzero matrix are introduced. An improved compact storage scheme named one dimension array with variational band width is proposed for the iterative algorithm solver based on the constant total number of nonzero elements in total stiffness matrix. This method can save large physical memory and avoid the optimization of node numbering. Finally, numerical example result shows efficiency and accuracy of this method.

2. Large-step Static Implicit Algorithm

QuickForm system is a self-developed code based on large-step static implicit algorithm. Large-step Static Implicit Algorithm is a specially designed new simulation algorithm which specializes in sheet metal forming. It uses the uncoupling thought to separate the two processes of bending effect and membrane drawing effect during sheet forming and solve them respectively.[5]

This algorithm consists of two modules: the intermediate configuration prediction module and the stress equilibrium iteration module. The intermediate configuration prediction module: Before incremental step iteration, an approximate method named pre-stressed membrane element is used to achieve the intermediate configuration of sheet in this incremental step. The sheet shape of intermediate configuration depends on the current tooling position. In the contact zones this form is defined by the tools, and in the other zones it is determined by the pre-stressed membrane element; The stress equilibrium iteration module: After the intermediate configuration of sheet is determined, the nonlinear material flow within this intermediate configuration is carried out, which means that the sheet nodes can only slide on this intermediate configuration.

In each incremental step of QuickForm system, a linear large sparse matrix equation $\boldsymbol{K\delta} = \boldsymbol{P}$ ($\boldsymbol{K}$ is global stiffness matrix) will be generated and to be solved by iteration method, so the calculation algorithm of linear equations and storage scheme of global stiffness matrix $\boldsymbol{K}$ will affect the solving efficiency of QuickForm system.

3. Storage Scheme of Global Stiffness Matrix

The old QuickForm system use the direct slover based on the LDLT decomposition algorithm. Considering memory problem, one dimension array with variational band width is already used to storage the total stiffness matrix, but the array of total stiffness matrix still have many zero matrixes. On the other hand, the node number will affect band width of total stiffness matrix, so optimization of nodal numbering should be carried on before assembly element stiffness matrix. The direct solver in old QuickForm system is replaced by the iterative solver which could save much more memory storage based on above reasons.

Firstly, the general adjacent nodes denote all nodes of nodal adjacent elements. If two nodes have general adjacent relation, they are the general adjacent nodes; otherwise they are not the general adjacent nodes. Let $\boldsymbol{LJ}^{i}$ denotes the number aggregate of all general adjacent nodes of node *i*, e.g. the number aggregate of all general adjacent nodes of node2 is $\boldsymbol{LJ}^{2} = \{1,2,3,4,5,6,7,8,10\}$, as in Fig.1.

The terms of element matrices are asigned in total stiffness matrix depend on node numbering. This procedure is called by element assembly. From element assembly process, we can found that:

(1) When two nodes have general adjacent relation, their corresponding matrix in total stiffness matrix is nonzero sub-matrix, otherwise is null sub-matrix filled entirely with zeros;

(2) The amount of nonzero sub-matrices is constant and determined by nodal general adjacent relation, the node number does not matter;

(3) When the node number is changed, the position of nonzero submatrix in total stiffness matrix will be changed also, however the value of nonzero sub-matrix is fixed.

An improved compact storage method of one dimension array with variational band width is proposed based on above characters and the total stiffness matrix is constant during iteration process.

In order to obtain the position information of non-zero submatrix in symetric sparse matrix ***A***, a one dimension integer array named by ***NRNODE*** will be generated to store all general adjacent node numbers of each node in lower triangular matrix of ***A*** by ascending order.

Let *m* are total number of nodal d.o.f., the total stiffness matrix will be a symetric sparse matrix ***A*** ($m \times m$). This improved storage method will use three one dimesion arrays: the first real array is ***AK***, this array will store all non-zero submatrices in lower triangular matrix of ***A*** by row order; the second integer array is ***LMD***, this array will store the main diagonal element address of matrix ***A*** in ***AK*** array; the third integer array is ***NCOL***, the length of ***NCOL*** is equal to ***AK*** and this array will store the column number of ***AK*** element in matrix ***A***.

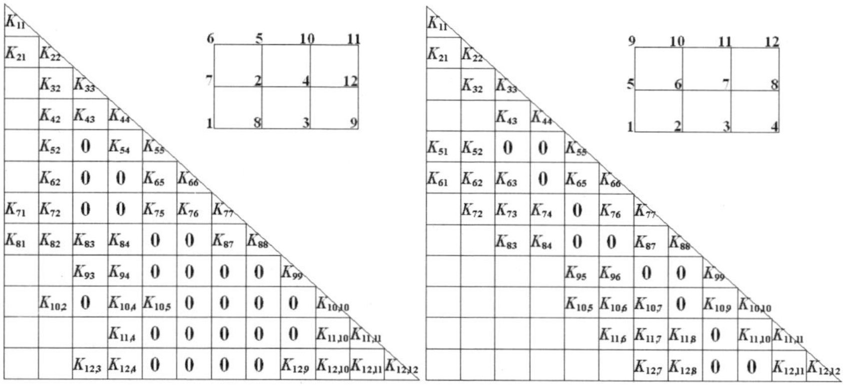

a) Before optimization of node number b) After optimization of node number

Fig. 1. Total stiffness matrix of one demesion array with variational band width.

All element matrices can be assemblied into the one dimension array ***AK*** which will store total stiffness matrix based on above two one dimesion integer arrays: ***LMD*** and ***NCOL***.

4. Numerical Results and Discussion

The test example used to validate the proposed method is the ladder pan as shown in Fig. 2. The computer CPU is Intel Pentium Dual CPU, T2330 1.6GHz and the memory is 1.0 G. The material named DQSK is used and the mechanical properties of this material and test conditions are as follow: the initial sheet thickness = 1.0mm; the average Lankford value ravg = 1.6; the friction coefficient = 0.15; and the flow stress curve of $\bar{\sigma} = 549.03(0.0013 + \bar{\varepsilon})^{0.22}$ (MPa).

Fig. 2 shows the finite element models of tools of ladder pan. The initial blank mesh includes 7 916 nodes and 15 500 triangle elements. This finite element model is simulated by the original QuickForm system with direct slover (LDLT) and the new QuickForm system with iterative slover (SSOR-PCG), repectively. Final thickness distributions from two diffirent methods are shown in Fig. 3. The comparison results show that the iterative solver with the proposed improved compact storage method have almost same accuracy result with the old QuickForm system.

The Comparison of memory occupation and simulation time between the old and new solver is shown in table 1. Firstly, we can find that the old QuickForm system is affected deeply with node number from the comparison result, the occupation memory will be decreased and the calculation efficiency will be increased when the node number of model is optimized. However, the occupation memory of new QuickForm system doesn't change at all between before and after optimization of node number. The affection of simulation efficiency from node number situation could be neglected. Therefore the new QuickForm system doesn't need optimize the node number of model before simulation. Secondly, an improved compact storage method used in new QuickForm system can save much more memory space.

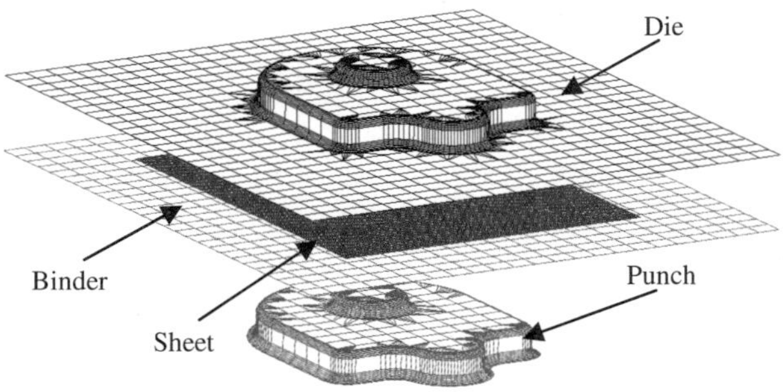

Fig. 2. Finite element models of ladder pan.

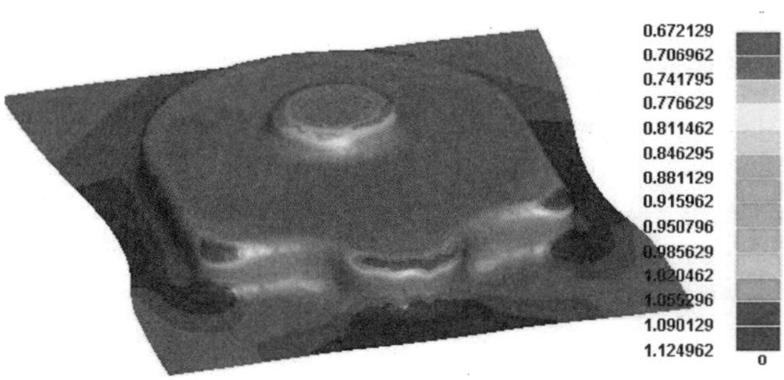

a) Result from old solver

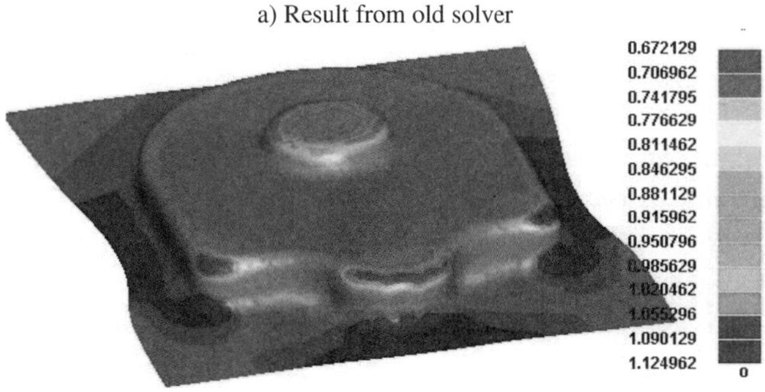

b) Result from new solver

Fig. 3. Final Thickness distribution.

Table 1. Comparison of memory occupation and simulation time between the old and new solver

Comparison items	*Before optimization of node number*		*After optimization of node number*	
	Old solver	New solver	Old solver	New solver
Memory of total stiffness matrix, Kb	130,772	2,017	45,410	2,017
Occupation proportion of total stiffness matrix	94.9%	13.7%	87.1%	13.7%
Total simulation time, s	4642.9	709.2	925.7	691.2
Simulation time of equations calculation, s	4037.0	116.7	335.5	92.9

5. Conclusion

An improved compact storage scheme named one dimension array with variational band width is proposed for the iterative algorithm solver as the important module of self-developed QuickForm system, which can save large physical memory and avoid the optimization of nodal numbering. Comparison results between the old and new QuickForm system have proved that the proposed compact storage method is efficient and accurate.

Acknowledgments

This work was funded by the NUAA Research Funding, NO. NS2010118, the Scientific Research Foundation for the Returned Overseas Chinese Scholars of State Education Ministry and National Nature Foundation (project number: 50805075), these supports are gratefully acknowledged.

References

1. Y.Q. Guo, J.L. Batoz, H. Naceur, et al: *Computers and Structure.*, 2000, **78**, 133–148.
2. S.H. Kim, S.H. Kim and H. Huh: *Journal of Materials Processing Technology*, 2001, **113**, 779–785.
3. G.Y. Li: *Chinese Journal of Computational Mechanics*, 1996, **13**, 253–268.
4. Y.F. Liu, F.Z. Shi, X.Y. Xu: *Journal of Plasticity Engineering*, 2006, **13**, 15–19.
5. W. Kubli, J. Reissner: *Journal of Materials Processing Technology*, 1995, **50**, 292–305.

ESTABLISHMENT OF SPRINGBACK PREDICTION MODEL FOR THE ROTARY-DRAW BENDING OF THIN-WALLED RECTANGULAR TUBE CONSIDERING VARIATION OF YOUNG'S MODULUS

W.Q. DONG, Y.L. LIU*, G.Y. ZHAO, H. YANG

State Key Laboratory of Solidification Processing, Northwestern Polytechnical University, Xi'an, Shaanxi, 710072, PR China

Springback is one of the main defects affecting the forming precision of 3A21 thin-walled rectangular tube in rotary-draw bending process. The material hardening mode and Young's modulus influence the springback prediction greatly, while the Young's modulus changes versus plastic strain rather than a constant. Hence, the variation of the Young's modulus has been taken into account in the springback analysis to improve the springback prediction precision. Based on Von-Mises yield function and Swift isotropic hardening model, a new material constitutive model is developed considering the variation of Young's modulus. Then, a springback prediction model is established using this new constitutive law. Furthermore, the reliability of the model is validated experimentally. Comparisons between simulation results and experimental data show that the prediction model with the new constitutive law is able to predict the springback significantly more accurately.

Keywords: Thin-walled rectangular Tube; Rotary-draw bending process; Young's modulus; Numerical simulation; Springback.

1. Introduction

Springback has a significant effect on the final shape of the product. Therefore, it is crucial to predict springback exactly and effectively for high-precision bending parts[1]. The material constitutive model influences the distribution of the internal stress calculated in the formed part[2]. It is found that the Young's modulus affects springback greatly as a vital elastic parameter in the material constitutive model and it changes versus plastic strain during bending process[3]. Therefore, to improve the springback prediction capability, the variation of Young's modulus with plastic strain should be considered in the numerical simulation.

A great deal of researchers has been dedicated to study springback prediction. Morestin and Boivin[4] investigated that the Young's modulus decreased in the form of a piecewise linear function during plastic deformation

*Corresponding author. Tel.: +86 29 8846 0212 803; Fax: +86 29 8849 5632
E-mail address: lyl@nwpu.edu.cn (Y.L. Liu).

experimentally. Li[5-6] addressed that the springback prediction precision can be improved by using the power-exponent hardening model, and developed a variational law of Young's modulus with plastic deformation of aluminum alloy by tensile tests. Gu[7] studied the springback for NC bending of stainless steel thin-walled round tube with a constant Young's modulus comprehensively. Hence, the objective of the present study is to formulate a new constitutive model considering the variation of Young's modulus with plastic strain. Then, a springback prediction model is established using this new constitutive model. Furthermore, the reliability of the model is validated.

2. Constitutive Model for 3A21 Rectangular Tube

2.1. *Constitutive equations considering the variation of Young's modulus*

The relationship between Young's modulus and plastic strain has been studied and described by a piecewise linear function, as show in Eq.(1).

$$E_{av} = \begin{cases} 60.29 - 12.27\overline{\varepsilon}^{p} / 0.038 & \overline{\varepsilon}^{p} \le 0.038 \\ 48.02 & \overline{\varepsilon}^{p} > 0.038 \end{cases} \tag{1}$$

2.1.1. *Elastic deformation process*

Based on Hooker's law, the relationship between the stress and the strain is stated as,

$$\boldsymbol{\sigma} = \lambda \operatorname{trace}\left(\boldsymbol{\varepsilon}^{e}\right)\mathbf{I} + 2\mu\boldsymbol{\varepsilon}^{e} \tag{2}$$

where λ and μ are Lame constants. Since the variation of Young's modulus versus plastic strain is considered here, the Lame constants can be expressed as the functions of equivalent plastic strain in Eq.(3) and (4),

$$\lambda\left(\overline{\varepsilon}^{p}\right) = \frac{\upsilon E_{av}\left(\overline{\varepsilon}^{p}\right)}{(1+\upsilon)(1-2\upsilon)} \tag{3}$$

$$\mu\left(\overline{\varepsilon}^{p}\right) = \frac{E_{av}\left(\overline{\varepsilon}^{p}\right)}{2(1+\upsilon)} \tag{4}$$

where $\overline{\varepsilon}^{p}$ is equivalent plastic strain.

2.1.2. *Plastic deformation process*

When plastic deformation occurs, the stress state of material satisfies yielding criterion, Von-Mises yielding criterion is adopted here, which can be seen in Eq.(5),

$$F(\sigma_{ij}) = \frac{3}{2}\sigma'_{ij}\sigma'_{ij} - \sigma_{s}^{2} = 0 \tag{5}$$

where σ_{ij} is stress tensor component, $F(\sigma_{ij})$ is a geometrical curved surface in the stress space, $\sigma'_{ij} = \sigma_{ij} - \sigma_m$ is stress deviatoric tensor component, σ_m is average stress, σ_s is the initial yield stress.

According to associated flow rule, plastic strain increment is written as,

$$\Delta\varepsilon^p = \Delta\gamma \frac{\sigma'_{ij}}{\left|\sigma'_{ij}\right|} \tag{6}$$

where $\Delta\gamma$ is plastic multiplier.

Equivalent plastic strain increment is defined in Eq.(7),

$$\Delta\bar{\varepsilon}^p = \sqrt{\frac{2}{3}\boldsymbol{\varepsilon}^p \cdot \boldsymbol{\varepsilon}^p} \tag{7}$$

The relationship between equivalent plastic strain and plastic multiplier $\Delta\gamma$ can be stated as,

$$\Delta\bar{\varepsilon}^p = \sqrt{\frac{2}{3}}\Delta\gamma \tag{8}$$

Isotropic hardening law is usually adopted for the most of work-hardening material. Swift isotropic hardening model is used as the hardening law for 3A21 aluminum alloy and is shown in Eq.(9),

$$\sigma_s(\bar{\varepsilon}^p) = Y\left(\bar{\varepsilon}_0 + \bar{\varepsilon}^p\right)^n \tag{9}$$

where Y is strength coefficient, n is hardening exponent, $\bar{\varepsilon}_0$ is initial pre-strain in the forming process.

Substituting from Eq.(9) into Eq.(5), Von-Mises yielding criterion based on swift isotropic hardening model can be written in Eq.(10),

$$F(\sigma_{ij}) = \frac{3}{2}\sigma'_{ij}\sigma'_{ij} - \sigma^2{}_s\left(\bar{\varepsilon}^p\right) = 0 \tag{10}$$

Thus, the constitutive equations (2) and (10) used in the elastic deformation and plastic deformation respectively are established.

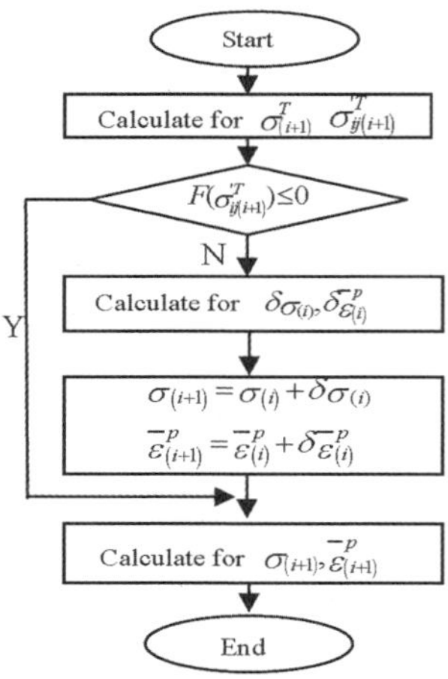

Fig. 1. The flow chart of constitutive model.

2.2. *Integration algorithm of the constitutive equations*

Fig.1 is the flow chart of the integration algorithm of the constitutive equations. The calculating steps are illustrated detailedly as following.

The plastic strain incremental is supposed to be zero, according to Eq.(2), the trail stress in the forming process can be written as,

$$\boldsymbol{\sigma}^{\mathrm{T}}_{(i+1)} = \boldsymbol{\sigma}_{(i)} + \lambda_{(i+1)}\mathrm{trace}\left(\Delta\boldsymbol{\varepsilon}^{e}_{(i+1)}\right)\mathbf{I} + 2\mu_{(i+1)}\Delta\boldsymbol{\varepsilon}^{e}_{(i+1)} \tag{11}$$

where the superscript(i) stands for the variable before stress updating, while$(i+1)$ stands for the variable after stress updating, $\boldsymbol{\sigma}^{\mathrm{T}}_{(i+1)}$ is trail stress, the deviatoric tensor of the trail stress is shown in Eq.(12),

$$\boldsymbol{\sigma}'^{\mathrm{T}}_{ij(i+1)} = \boldsymbol{\sigma}^{\mathrm{T}}_{(i+1)} - \frac{1}{3}\mathrm{trace}\left(\Delta\boldsymbol{\varepsilon}^{e}_{(i+1)}\right)\mathbf{I} \tag{12}$$

Substituting Eq.(11) into Eq.(9), if,

$$F\left(\boldsymbol{\sigma}'^{\mathrm{T}}_{ij(i+1)}\right) = \frac{3}{2}\boldsymbol{\sigma}'^{\mathrm{T}}_{ij(i+1)}\boldsymbol{\sigma}'^{\mathrm{T}}_{ij(i+1)} - \sigma_s^2\left(\bar{\varepsilon}^p\right) > 0 \tag{13}$$

the compatibility condition is satisfied with yielding surface at $t_{(i+1)}$ in the process and is expressed as,

$$\sigma'_{ij(i+1)} = \sqrt{\frac{2}{3}}\left(\sigma^0\left(\bar{\varepsilon}^p_{(i)}\right) + \sqrt{\frac{2}{3}}h_{(i)}\Delta\gamma\right)\mathbf{q} \tag{14}$$

The deviatoric tensor of the trail stress is obtained in Eq.(15),

$$\sigma'_{ij(i+1)} = \boldsymbol{\sigma}'^{\mathrm{T}}_{ij(i+1)} - 2\mu\Delta\gamma\mathbf{q} \tag{15}$$

By solving the equations (14) and (15) simultaneously, $\Delta\gamma$ can be calculated by Eq.(16),

$$\Delta\gamma = \frac{\left(\boldsymbol{\sigma}'^{\mathrm{T}}_{ij(i+1)} : \boldsymbol{\sigma}'^{\mathrm{T}}_{ij(i+1)}\right)^{1/2} - \sqrt{\frac{2}{3}}\,\mathrm{Y}\left(\bar{\varepsilon}_0 + \bar{\varepsilon}^p\right)^n}{2\mu\left(1 + \frac{h_{(i)}}{3\mu}\right)} \tag{16}$$

thus, the state variables are updated by Eq.(17) and Eq.(18) from $t_{(i)}$ to $t_{(i+1)}$,

$$\boldsymbol{\sigma}_{(i+1)} = \boldsymbol{\sigma}^{\mathrm{T}}_{(i+1)} - 2\mu\Delta\gamma\mathbf{q} \tag{17}$$

$$\bar{\varepsilon}^p_{(i+1)} = \bar{\varepsilon}^p_{(i)} + \sqrt{\frac{2}{3}}\Delta\gamma \tag{18}$$

In each increment a backward-Euler method is used to solve for $\Delta\gamma$. At each iteration, the yield condition Eq.(13) is checked for the updated value and the state variable are updated by Eq.(17) and Eq.(18). The integration algorithm of the constitutive equations is realized by Fortran language and implemented into ABAQUS software by user subroutine VUMAT.

3. Establishment of Springback Prediction Model considering Variation of Young's Modulus

3.1. *Establishment of springback model*

A new FE model for the bending process based on literature[8] has been established. However, the variation of Young's modulus versus plastic strain was not taken into account. Hence, the constitutive model built in section 2 is implemented into the FE model. The output files of the results in the bending process are imported into ABAQUS/Standard software, such as the final shape of the rectangular tube, related information of elements, nodes, the stress and strain, etc.

3.1.1. *Material model*

The material property must be redefined in springback calculation because of the application of VUMAT for the bending process in which the variation of Young's modulus has been introduced into the simulation. Springback is assumed to be a pure elastic unloading process, and no reverse yielding occurs. Thus, the constitutive equation (2) for a pure elastic unloading process is carried out by user subroutine UMAT under the ABAQUS/Standard environment.

3.1.2. *Treatment of constrains*

To keep the system energy unchangeable during springback computation, the whole translation and the whole rotary movement of the bent-tube must be constrained. The constrains cannot restrict the real springback deformation. There is rarely deformation at the end A (as shown in Fig.2), thus the constrains are loaded to section A and the predefined stress-strain field is also applied to the tube at the same time.

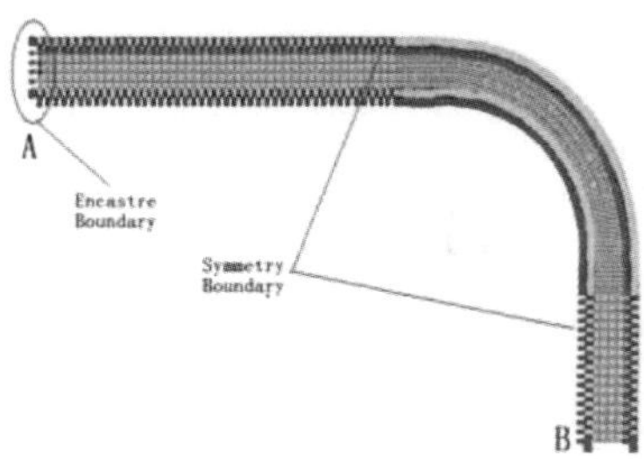

Fig. 2. Constrains on springback model.

3.1.3. *Stabilization of the springback model*

Because the springback computation is difficult to converge, the damping factor is used to control the convergency of the springback model. In the simulation, the damping factor is set to 0.0002 when the ratio of viscous damping energy ALLSD and total strain energy ALLSE is less than 0.01.

3.2. *Validation of the model*

The established springback prediction model is validated experimentally. The experiments are carried out using a W27YPC-63NC microcomputer controlled bender. The material property parameters of 3A21 aluminum alloy are K=236MPa, n=0.16, Poisson ratio ν =0.33, and the expression for Young's modulus is seen in Eq.(1). Experimental and simulation conditions are keep the same.

Fig.3 shows the variational curves of the springback angles with the bending angles under different conditions.

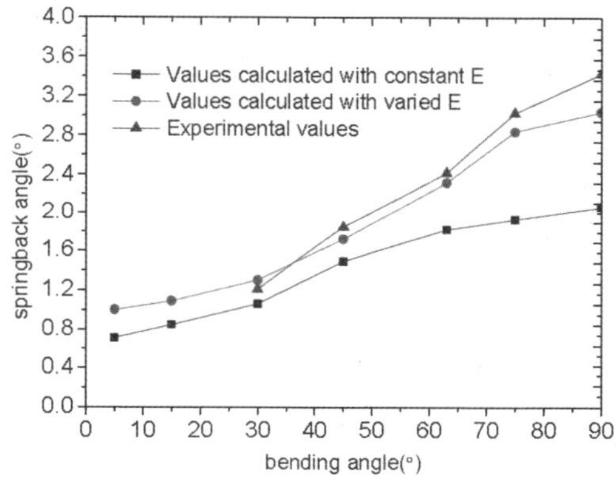

Fig. 3. The relationship between bending angle and springback angle.

It can be seen that the springback angles increase with the increase of bending angles under three conditions. The simulation results with the consideration of the variation of Young's modulus are much closer to the experimental values, and the maximum relative error is only 11.37%; while the simulation results of springback angles with a constant Young's modulus show a significant disagreement with the experimental values when the bending angle is bigger than 30°, the maximum relative error can reach to 39.94% in this condition. It is obvious that a higher prediction precision of the springback angle can be obtained by the consideration of the variation of Young's modulus versus plastic strain. Therefore, the established springback prediction model considering the variation of Young's modulus is valid and reliable. It is an effective aid for springback prediction.

4. Conclusion

(1) The E_{av} of 3A21 aluminum alloy of rectangular tube decreases rapidly with the increase of plastic strain firstly, and the decrease is about 20.23% when the plastic strain is up to 0.038. Then the E_{av} stabilizes at 48.02MPa basically.

(2) Based on Swift isotropic hardening law and Von-Mises yielding criterion, a new constitutive model considering the variation of Young's modulus is built. The integration algorithm of the constitutive equations is realized by Fortran language and implemented into ABAQUS software by user subroutine.

(3) A springback prediction model with the new constitutive model is established under the ABAQUS environment. Furthermore, the reliability of the model is validated by experiments. Comparisons between simulation results and experimental data show that the prediction precision of springback angle is improved significantly with the consideration of the variation of Young's modulus.

Acknowledgments

The research is supported by the National Natural Science Foundation of China (Nos.50575184 and 50975235), NPU Foundation for Fundamental Research (No. NPU-FFR-200809), State Key Laboratory of Materials Processing and Die & Mould Technology, Huazhong University of Science and Technology (No. 08-3).

References

1. A. Taherizadeh, A. Ghaei, D. E. Green, W. J. Altenhof: J. International Journal of Mechanical Sciences., 2009, 51, 314–325.
2. X.C. Li, Y. Y Yang, Y. Z. Wang, J. Bao, S. P. Li: M. Materials science and Technology., 2002, 10, (3), 277–280.
3. S.L. Zang, J. Liang, C. Guo: J. International Journal of Machine Tools & Manufacture., 2007, 47, 1791–1797.
4. F. Morestin, M. Boivin: J. Nuclear Engineering and Design., 1996, 161, 107–116.
5. X.C. Li, Y. Y Yang, J. Bao, S. P. Li: J. Journal of Harbin Institute of Technology, 2000, 32, (5), 54–56.
6. X.C. Li, Y. Y Yang, Y. Z. Wang, J. Bao: J. The Chinese Journal of Nonferrous Matals., 2002, 12, (4), 701–705.
7. R.J. Gu, H. Yang, M. Zhan, H. Li, H. W. Li: M. Computational materials science., 2008, 42, 537–549.
8. G.Y. Zhao, Y.L. Liu, H. Yang, C. H. Lu, R.J. Gu: M. Materials science and EngineeringA, 2009, 499, 257–261.

OPTIMIZATION OF GAS-ASSISTED INJECTION MOLDING PROCESS FOR A HANDLE PRODUCT BASED ON ORTHOGONAL EXPERIMENT

JIANG SHAOFEI†, LI JIQUAN, LU CONGDA

Key Laboratory for Mechanical Manufacture & Automation of Ministry of Education, Zhejiang University of Technology, Hangzhou 310032, China

Based on orthogonal experiment design, this paper present a methodology to optimize process parameters, which include melt temperature, mold temperature , gas injection pressure, gas delay time and gas injection time. L16(45) orthogonal layout incorporate interaction with pairwise orthogonal design is constructed based on Moldflow, and variance analysis is calculated to observe the influence of process parameters. Optimized GAIM process parameters was obtained, in which the measurements were gas penetration volume, airway length and relative thickness. A case study of the handle product by GAIM was resented, and the results verified validity of the methodology.

Keywords: Gas-Assisted Injection Molding (GAIM); optimization; orthogonal experiment design.

1. Introduction

Gas-assisted injection molding (GAIM) was developed to overcome the shortcomings of the conventional injection molding method recent years[1,2]. There are many advantages associated with the process, include material savings, free of sink marks in thick sections and free of warpage in long plates. However, how to get the optimized GAIM process under the condition of the coupling among main process parameters is unintelligible.

The main parameters of gas-assisted injection molding are mold temperature, melt temperature, gas injection pressure, gas delay time and gas injection time, and so on[3-5]. Mold trial is used to obtain the relative perfect technical conditions at present. It will take lots of times of experiments, because of the complex influence of several parameters and their combination in practical production. Thus, it is necessary to test as many times as possible to obtain a better process condition. S.S. Miguel [6] optimized the runner system of injection mold of a back plastic sheet cover of automobile by orthogonal design,

† Corresponding author, email: zjujsf@163.com

to get the more adaptable shape of gas channel.. A. Polynkin[7] established a numerical analysis model of gas-assisted handle by utilizing Short Cut Method, which could perfectly evaluate the shape of gas penetration and the distribution of wall thickness.

The orthogonal design method is one of effective method to reduce the experiment times. Taking a gas-assisted handle as research object, how to obtain the relative perfect GAIM conditions was studied.

2. The Establishment of Analysis Model and the Pretreatment

2.1. *The establishment of analysis model and material analysis*

A handle of some vehicle was taken as the research object, as shown in Fig. 1. In order to simulate GAIM of the handle with spillover cavity, we add the spillover cavity to the model of the handle, as shown in Fig. 2.

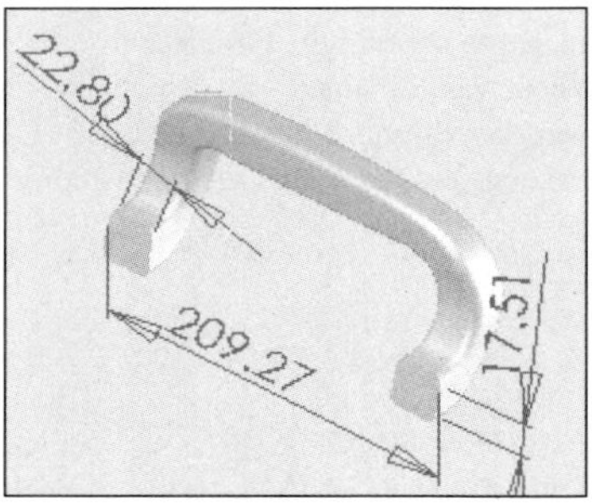

Fig. 1. Model of the handle.

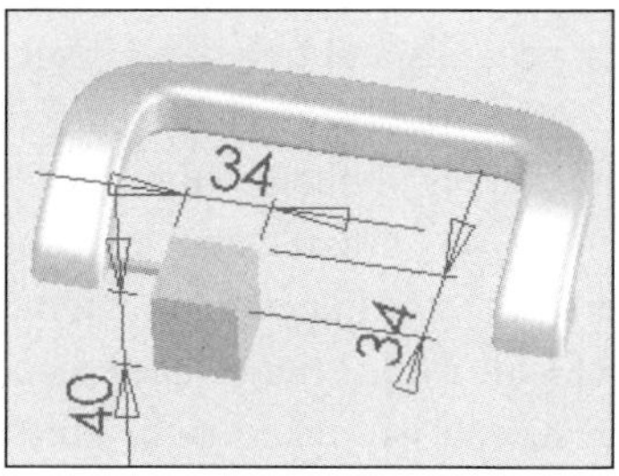

Fig. 2. Model with spillover cavity.

2.2. *Pretreatment of Model*

The model was convert into step(STEP AP203) format and imported to MPI. A grid model was obtained by grid division and runner system established on the base of grid model, as shown in the Fig. 3.

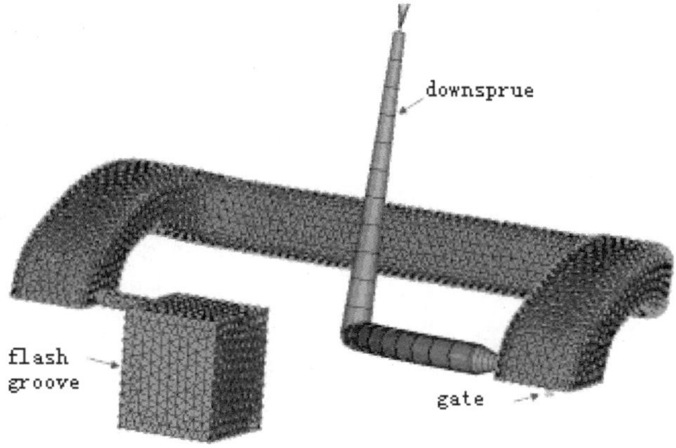

Fig. 3. The rounded analysis model.

3. The Optimization Analysis of Gas Injection Molding Techniques

3.1. *Initialization and the experiment result*

The main parameters of the gas-assisted injection molding are mold temperature, melt temperature, short-shot size of the plastic resin and delay time, and so on. To obtain preferable parameters combination and the influence of them, we got the orthogonal design to study the GAIM of the handle. Five parameters was chosen as the experiment factors, and we place the relative parameter level based on the molding conditions provided by the materials supplier. The experiment factors and factor levels are shown in Table 1. The gas injection time is 10 seconds.

Table 1. Factors and levels setting.

levels	mold temperature A/(℃)	melt temperature B/(℃)	gas injection time C/(%)	gas delay time D/(s)	gas injection pressure E/(MPa)
1	30	225	4	0.5	6
2	40	235	6	1.5	8
3	50	245	8	2.5	10
4	60	255	10	3.5	12

Gas penetration volume is the basis to charge the injection molding quality under the gas-assisted molding conditions, and was chosen as the measurement of gas-assisted molded parts quality. Design of experiments was based on the orthogonal table L16(45). The experiment design and results is shown in Table 2. The calculation formula of the "lager the better" signal-to-noise ratio was taken as the evaluating index of the gas penetration volume and gas channel length, while the calculation formula of "smaller the better" for the relative wall thickness. In the simulation, gas channel length is 320 mm, the wall thickness of the handle is 2.2 mm, and 1.1 mm was chosen when calculate the relative wall thickness.

Table 2. Orthogonal matrix and the result of the evaluating index.

Experiment No	factor level settings					volume of gas penetration (%)	signal-to-noise ratio η_1/dB
	A	B	C	D	E		
1	30	225	4	0.5	6	26.82	28.569
2	30	235	6	1.5	8	28.6	29.127
3	30	245	8	2.5	10	28.38	29.060
4	30	255	10	3.5	12	29.45	29.382
5	40	225	6	2.5	12	28.53	29.106
6	40	235	4	3.5	10	27.22	28.698
7	40	245	10	0.5	8	27.95	28.928
8	40	255	8	1.5	6	27.45	28.771
9	50	225	8	3.5	8	27.81	28.884
10	50	235	10	2.5	6	26.78	28.556
11	50	245	4	1.5	12	29.4	29.367
12	50	255	6	0.5	10	29.16	29.296
13	60	225	10	1.5	10	28.27	29.027
14	60	235	8	0.5	12	28.7	29.158
15	60	245	6	3.5	6	26.48	28.458
16	60	255	4	2.5	8	27.77	28.872

3.2. *Optimization of gas-assisted handle's technique*

We could get a fine parameters combination through the 16 groups datum. In the results, we could divide them into two groups, one of which only take into account of a single index every time while the other synthesis the 16 groups datum of orthogonal experiments to carry on the single index range analysis to get the preferable parameters combination.

sing histogram to reveal the experiments results of the three evaluating index's signal-to-noise ratio in Fig. 4, which could help to find the best parameters combination readily.

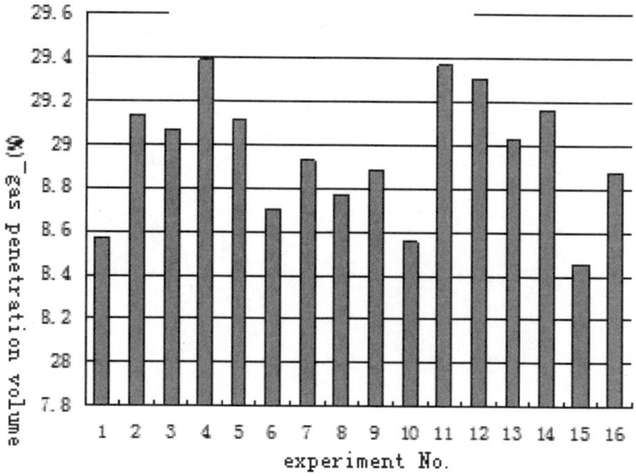

Fig. 4. The signal-to-noise ratio of the gas penetration volume.

We could figure out that the gas penetration is the lager the better from the Fig. 4: the experiment No.4 and No.11 have the largest gas penetration volume. The parameters of experiment No.4 are A1B4C4D4E4, and the definite shape of the gas channel is shown in Fig. 5. The parameters of experiment No.11 are A3B3C1D2E4, and the definite shape of the gas channel is shown in Fig. 6. We can find out that the gas penetration volume is various under the different technical conditions from the 16 different groups datum. From Fig. 5 and Fig. 6, the gas penetration of experiment No.4 and No.11 is satisfied. The final product according to experiment results was satisfying enough for the manufacture, as shown in Fig. 7.

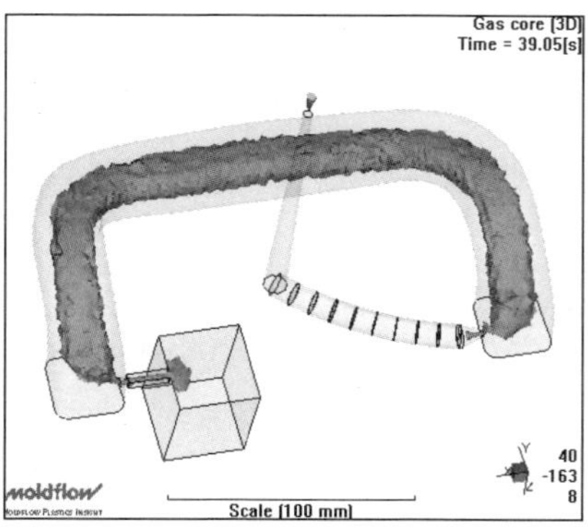

Fig. 5. Gas core of experiment 4.

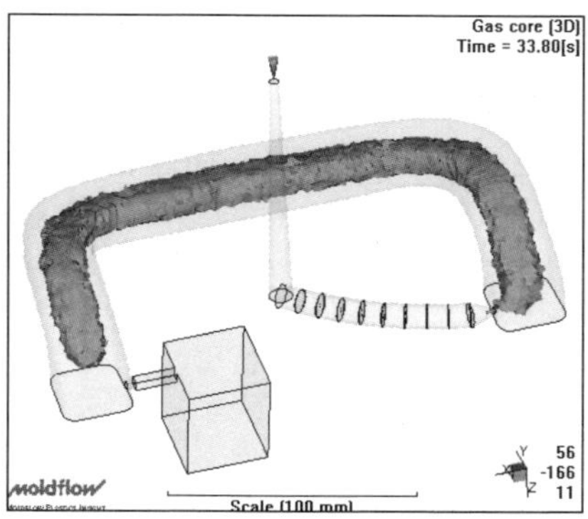

Fig. 6. Gas core of experiment 11.

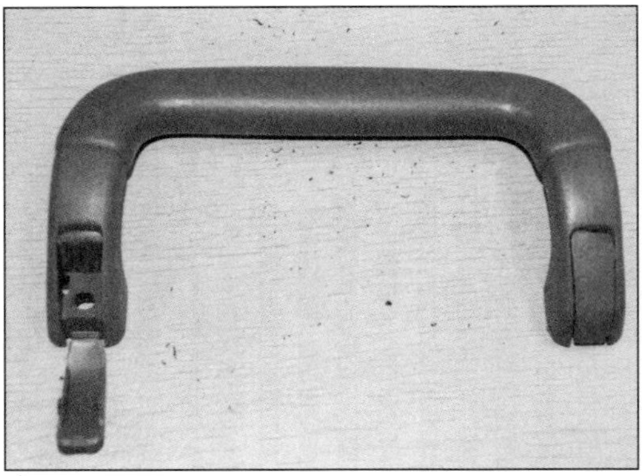

Fig. 7. The concrete injection molding product.

4. Conclusion

This paper introduce orthogonal experiment to study the influence trend to the molded-part quality of the five parameters: mold temperature, melt temperature, gas injection pressure, the delay time and the gas injection time. And the injection molding was optimized to obtain satisfied product. Test proved the method was correct and effective.

Acknowledgments

This work was sponsored by Science and Technology Project of the Education of Zhejiang Province (No.Y200908117), Priority Themes of Major Projects in Zhejiang Province (No. 2008C01059-1), and Science and Technology Projects of Zhejiang Province (No. 2006C21042).

References

1. L Chen, Jianhui Li, Huamin Zhou, Dequn Li, Zhicheng He, Qian Tang: *J. Mater. Process. Technol.*, 2008, 208(1-3): 90-98
2. A.J. Carrillo, A.I. Isayev: *Polym. Eng. Sci.*,2009,49(12):2350-2373
3. N.S. Ong., H. L. Lee, M. A. Parvez: *Adv. Polym. Tech.*, 2001,20(4):270-280
4. Huamin Zhou, Hua Zhang, Dequn Li: *CMES-Comp. M.*, 2009, 46(3): 209-220
5. B. Yang, X.R. Fu, W. Yang, S.P. Liang, N. Sun, S. Hu, M.B. Yang: *Macromol. Mater. Sci.*, 2009, 294(5): 336-344
6. S.S. Miguel, G. Antonio, B. Arasanz, J. Aurrekoetxea, L Aretxabaleta: *J. Mater. Process. Technol.*, 2006,178: 369-378
7. A. Polynkin, J.F.T Pittman, J. Sienz: *Polym. Eng. Sci.*, 2005, 45(8): 1049–1058

THE EFFECT OF ION ASSISTING DEPOSITED TIN/TI COATING ON THE FRETTING FATIGUE BEHAVIOR OF TI-811 ALLOY AT ELEVATED TEMPERATURE

XIAOHUA ZHANG[†]

Corrosion and Protection Research Laboratory, Northwestern Polytechnical University, YouYi West Road 127
Xi'an, Shaanxi, 710072, China

LIUDAO XIN

Corrosion and Protection Research Laboratory, Northwestern Polytechnical University, YouYi West Road 127
Xi'an, Shaanxi, 710072, China

A TiN/Ti composite coating was prepared on a Ti-811 titanium alloy substrate by an ion-assisted arc deposition (IAAD) technique with a view to improving the fretting fatigue resistance of the titanium alloy at elevated temperature. The composition distribution, bonding strength, micro-hardness, ductility, tribological properties, and fretting fatigue resistance at elevated temperature of the coating were investigated. The results indicate that the IAAD technique can be used to prepare a TiN/Ti composite coating with high hardness, good ductility, excellent bonding strength, and high load-bearing capability. The TiN/Ti composite coating can improve the resistance to wear and fretting fatigue of the Ti alloy at 350 °C because of the good toughness and high bonding strength of the IAAD TiN/Ti composite coating with excellent lubrication of Ti particles in the coating.

Keywords: Titanium alloy; Fretting fatigue; Ion-assisted deposition; Elevated temperature.

1. Guidelines Introduction

Because of their low density, high Young's modulus, excellent corrosion resistance, and good thermal stability, titanium alloys have been widely used in manufacturing the fans, compressor disks, and blades of advanced aircraft engines[1]. However, titanium alloys are very sensitive to fretting fatigue (FF) damage[2, 3]. Hence, the development of surface modification techniques to improve the fretting fatigue resistance (FFR) of titanium alloys is vital[4–7]. To date, no surface treatment methods used to improve the FFR of titanium alloys at

[†] Corresponding author, email: yhzhangxh@nwpu.edu.cn

elevated temperature have been entirely satisfactory due to a lack of a clear understanding of the mechanisms of FF as a result of the complexity and great number of influencing factors on FF damage. To improve the FFR of titanium alloys at elevated temperature, it is necessary to devise other effective surface treatment methods.

TiN coating leads to superior abrasion resistance, a low coefficient of friction, high temperature stability, and high hardness. TiN coating is thus widely used in many industries. Research has indicated that the tribological and fretting wear performances of metals may be improved by a TiN coating[8]. However, our research has shown that the FFR of titanium alloys is significantly reduced by chemical vapor deposited (CVD) TiN coating and plasma nitriding[9]. It is difficult to improve FFR by surface treatment methods because the anti-friction and anti-fatigue properties should be improved at the same time. However, the methods for settling the anti-friction and anti-fatigue properties are usually incompatible. The anti-friction property can be remarkably improved by hard coatings or surface hardening treatments. However, the fatigue resistance is often reduced due to the poor ductility of the hard treated surface. As a result, the FFR cannot be improved by traditional surface hardening treatments. To improve the FFR of titanium alloys at elevated temperature, it is necessary to investigate how to exploit processes for preparing coatings to rationally match their strength and ductility by optimizing their structures. Ion-assisted deposition (IAD) is a surface treatment method that effectively combines ion implantation with the physical vapor deposition (PVD) technique. This surface treatment method can be used to prepare coatings with high density and high bonding strength under low-temperature processing conditions[10]. The vacuum cathode arc deposition (VCAD) technique has the advantages of non-line-of-sight, fast deposition rate and the capacity to prepare metal–ceramic composite coatings[11]. Improving the FFR of Ti alloys at elevated temperature by TiN/Ti composite coating using the ion-assisted arc deposition (IAAD) method, which combines IAD with VCAD, has not been reported. In addition, it is valuable to investigate whether the FFR of Ti alloys can be synergistically improved at elevated temperature by a duplex surface modification consisting of SP and IAAD TiN/Ti composite coating.

2. Experimental Procedures

FF specimens and fretting pads were obtained from Ti-811 titanium alloy bars (∅ 16 mm). Ti-811 alloy is almost all α-phase and contains 7.9% Al, 1.0% Mo, 0.99% V, 0.05% Fe, 0.1% C, 0.01% N, 0.001% H, 0.06% O, and balance, Ti. The material was treated by double annealing (910 °C for 1 h, cooled in air, 580 °C for 8 h, cooled in air). The resulting microstructure is an equiaxial α-phase and intergranular β-phase. The mechanical properties of the alloy are: σ_b=931 MPa, $\sigma_{0.2}$=890 MPa, δ=23% and Ψ=46%.

A PLG-100C high-frequency fatigue machine was used to conduct FF tests. A schematic drawing of the FF test setup is shown in Fig.1. The load was set in pull-pull mode. The contact state between the pad and the specimen was flat to flat, with a rectangular contact area of 2 mm × 6 mm. Relative slip between the specimen and pad was introduced by the difference in elastic deformation between them. The contact stress of the pads with the specimen was 85 MPa, which was loaded by using a stressing ring. The span length between the fretting bridge foots was 15 mm. The slip was approximately 36 μm for an axial loading force of 500 MPa in sinusoidal form at 110 Hz with a stress ratio of 0.1. The FF life was taken as the average value of three specimens. An electrical resistance furnace was used for heating, and the temperature was controlled at 350 ± 1 °C using a feedback temperature controller with a rectifier and a K-type thermocouple.

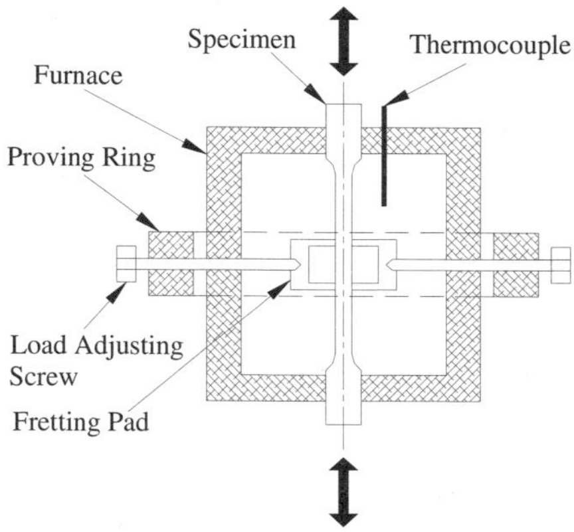

Fig. 1. Fretting fatigue test schematic.

3. Results and Discussion

Fig. 2 compares the FF lives (FFL) of the Ti-811 alloy (labeled BM), the IAAD TiN/Ti coating (labeled TiN/Ti) at 350 °C. The FFL of the Ti-811 alloy at elevated temperature was improved by a factor of 2.4 by the IAAD TiN/Ti coating. The reasons for this may be summarized as follows: The TiN/Ti coating showed good wear resistance, high ductility, and high bond strength. Moreover, the surface contact regime between the TiN/Ti coating and the fretting pads involved the convex peaks contacting the flat surface, which resulted from the softer Ti particles randomly distributed in the hard TiN/Ti coating. The contact area and fretting area were decreased, so the probability of crack initiation was reduced. The softer Ti particles on the harder TiN coating surface provided a good solid lubricating function and absorbed some fretting movement energy. Therefore, the wear resistance of the Ti-811 alloy was significantly improved by the TiN/Ti coating. Simultaneously, the pure Ti interlayer and the mixed Ti and N gradient layer formed a good transition layer between the TiN coating and the

Ti-811 alloy substrate. The transition layer not only resulted in the hardness of the whole coating gradually changing along the coating thickness, but also improved the bond strength of the coating. At the same time, cracks initiated in the TiN coating surface were arrested by this transition layer. Hence, the fatigue resistance of the TiN coating was improved. The internal stress of the TiN/Ti coating was relaxed and the ductility of the coating was enhanced due to the soft Ti particles interfusing in the coating. Fig. 3 compares the SEM morphologies of the fretting fatigue damaged surfaces of the Ti-811 alloy and the IAAD TiN/Ti coating. The FF damaged surface of the Ti-811 alloy is characterized by fatigue detachment. Because the hardness of the Ti-811 alloy is low, its anti-friction property is poor. The surface of the IAAD TiN/Ti coating was only slightly scratched. Simultaneously, the Ti particles on the surface of the TiN/Ti coating were planished. No cracks appeared on the planished surface. Because the TiN/Ti coating had good anti-adherence properties and high wear resistance, the probability of forming propagating cracks was reduced and the FF resistance of the Ti-811 alloy was improved.

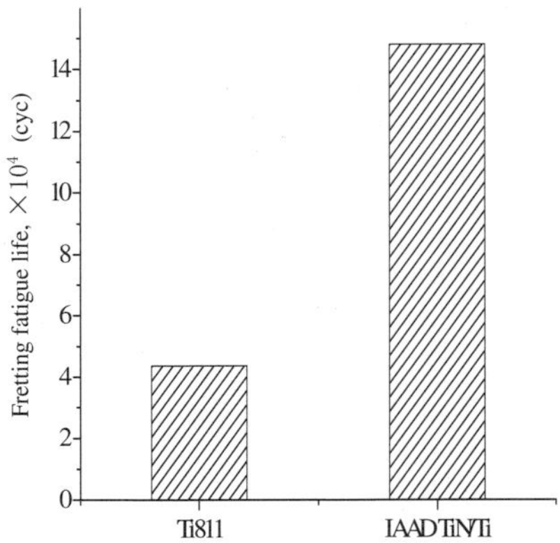

Fig. 2. Fretting fatigue lives of Ti-811 alloy and TiN/Ti coating at 350℃

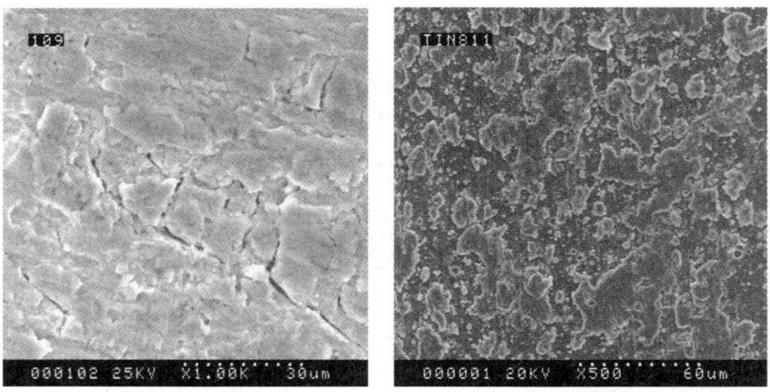

(a) Ti-811 alloy (b) IAAD TiN/Ti coating

Fig. 3. SEM morphologies of FF damaged surfaces of Ti-811 alloy with and without TiN/Ti film

4. Conclusions

In this paper, the effects of the TiN/Ti composite coating prepared by an ion-assisted arc deposition (IAAD) technique on the FFR of Ti-811 titanium alloy at elevated temperature were investigated. This analysis leads to the following conclusions: A TiN/Ti coating was prepared on a Ti-811 titanium alloy substrate by IAAD technique. The TiN/Ti coating exhibited high hardness, excellent bonding strength, high load-bearing capability, good ductility, and a low coefficient of friction, and so the wear resistance of the Ti-811 titanium alloy was significantly improved by the coating at 350 °C. The fretting fatigue life (FFL) of the Ti-811 alloy with the TiN/Ti coating was improved by a factor of 2.4 as compared to the uncoated substrate at 350 °C because of the excellent wear and fatigue resistance and good ductility of the coating.

Acknowledgments

This work was supported by the National Natural Science Foundation of China (No.50671085, No.50771070) and the National High-Tech Research and Development Program of China ("863" Program, 2007AA03Z521) and Fundamental Research Foundation of Northwestern Polytechnical University (JC200902).

References

1. C. W.James, A. S.Edgar: *Acta Material.*, 2003, 51, 5775–5779.
2. D. W. Hoeppner: *Tribol. Int.*, 2006, 39, 1271–1276.
3. A. K.Ray, G. Das and V. R. Ranganath: *Eng. Failure Anal.*, 2004, 11, 613–617.
4. A. L. Hutson, M. Niinomi and T. Nicholas: *Int. J. Fatigue*, 2002, 24, 1223–1234.
5. R. Wei, M. Shankar and H. Jeffrey: *Surf. Coat. Technol.*, 2004, 192, 177–188.
6. D. X. Liu, J. W. He and J. P. Celis: *Surf. Coat. Technol.*, 1999, 116, 234–238.
7. L. Hyukjae, M. Shankar and S. Shamachary: *Mater. Sci. Eng. A*, 2005, 390 227–232.
8. J. H. Sung, T. H. Kim and S. S. Kim: *Wear*, 2001, 250, 658–664.
9. D. X. Liu, H. Chen and J. W. He Jia: *Trans. Met. Heat Treat.*, 2001, 22, 49–54.
10. D.X. Liu, B. Tang and H. Chen: *Chin. J. Nonferrous Metals*, 2001, 11, 454–460.
11. H. Y. Lee, Y. M. Kim: *Surf. Coat. Technol.*, 2005, 193, 303–308.

WEAR CHARACTERISTICS OF TRADITIONAL AND NEW GENERATION BRAKE DISCS

DARIUSZ RUDNIK[†]

Motor Transport Institute, Jagiellonska 80
Warsaw, 03-301, Poland

ANDRZEJ WOJCIECHOWSKI

Motor Transport Institute, Jagiellonska 80
Warsaw, 03-301, Poland

JERZY SOBCZAK

Foundry Research Institute, Zakopianska 73
Cracow, 30-418, Poland

In the research the some problems concerning the disadvantageous effects occurring in cast iron brake discs in operating conditions are presented. The qualitative metallographic investigations of microstructure of materials used in traditional brake discs (cast iron) and modern brake discs (composites reinforced with SiC and fly ash) are presented. The studies of the friction properties of the traditional and composite brake discs are also presented.

It was proved that in the case of not ventilated brake discs only the friction sets consisted of the cast iron brake discs and the investigated friction pads can be used in vehicle brake systems.

The results allow to ascertain that the pressure decreasing in a brake system and in consequence the work temperature decreasing and brake discs wear decreasing will make possible composite brake discs based on Al-alloys to put into use in not high loaded applications, for example in the light passenger cars in a rear axle.

Keywords: MMC brake discs; wear; composites; reinforcement phase; microstructure.

1. Introduction

For several years metal matrix composites (MMC) are developed because of their attractive exploitation properties. The high level of mechanical properties and low weight of products fabricated from MMC are advantageous especially in aerospace and motor industry solutions. Introducing the MMC reinforced by

[†] Corresponding author, email: dariusz.rudnik@its.waw.pl

hard particles or fibers, in total volume or locally, results in improvement of mechanical properties of ready products[1-4].

The results of previous own studies were the premises to further systematic tests of friction sets. This investigations made possible to identify three disadvantageous changes occurring in traditional brake discs in operating conditions: friction over-wear, brake disc cracking as a result of thermal shock (temperature gradient) - when hot brake disc has contact with water and changes geometry of friction elements (brake discs and pads) during operating period.

These negative phenomenon forced the formulation of materials requirements for brake elements as follows: stable shape and dimensions in different operating conditions (temperature, humidity), resistance for water influence, good mechanical properties, in high temperatures too, friction wear resistance, high thermal conduction.

Taking into account the aspects as above, the aim of presented investigations was to assess which friction set – brake disc-brake pads - may be treated as an optimum solution in a vehicle design.

2. Experimental Procedure

The brake discs made from various materials such as: cast iron, metal matrix composites reinforced with SiC and with fly ash (ALFA composites) in chosen friction sets (36 brake discs with various friction pads) were tested. The measurements were carried out by means of KRAUSS apparatus, simulating a work of brake disc-brake pad set. The tests were conducted according with Annex 8 of ECE R-90[5], constant pressure method.

3. Results and Discussion

The typical microstructure of traditional cast iron brake disc consists of flake graphite distributed in pearlite matrix of cast iron.

The microstructures of metal matrix composites reinforced with SiC used in the producing of modern brake discs consists of reinforcing phase (SiC particles) uniformly distributed in Al-alloy matrix consisted of eutectic precipitates distributed in α-solid state. There is visible a remainder of interdendritic arrangement of eutectic precipitates.

The microstructures of metal matrix composites reinforced with fly ash consists of reinforcing phase (fly ash particles of spherical form) distributed in Al-alloy matrix consisted of eutectic precipitates interdendritic distributed in α-solid state. There are also visible a primary Si crystals.

The friction pads used in the tests can be divided to five general groups as follow: pressed with metals powders and sintered (semi – metal) with metals powders and resins for passenger cars and light trucks, pressed for trucks and buses, pressed for sport and racing cars, pressed for MMC brake discs.

The results of friction investigations are presented in Fig. 1-2.

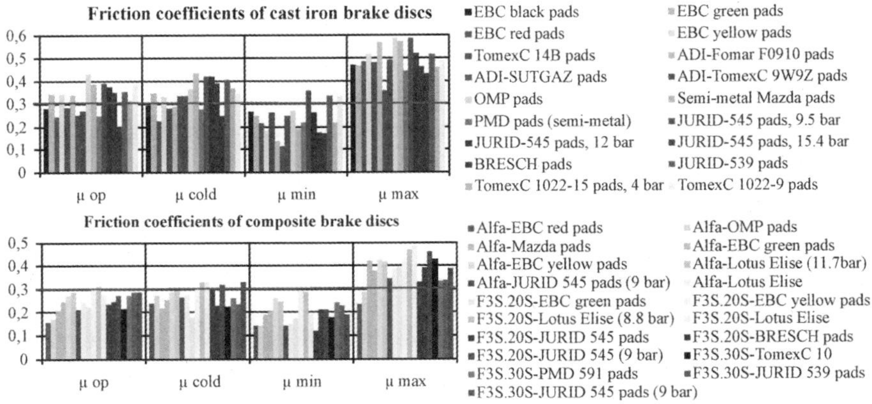

Fig. 1. Comparison of friction coefficients of cast iron and metal matrix composites brake discs with various brake pads.

The meanings of the values shown on diagrams "x" axis are as follow:

- μ_{op} - operating friction coefficient - it is average from values obtained in cycles from 2 to 7. The measurement is performed in every cycle, 1 second after brake starting,
- μ_{max} - maximal friction coefficient - it is the maximum μ value from cycles 1 to 8,
- μ_{min} - minimal friction coefficient - it is the minimum μ value from cycles 1 to 8.

The diagram shown in Fig. 1 concerns the friction coefficients of cast iron brake discs and composite brake discs reinforced with SiC or fly-ash which working with various brake pads of different producers - the types as above.

In agreement with requirements of ECE90 during production of an approved brake pads type, test samples must demonstrate compliance with the values (with proper tolerances) registered for brake pads which were tested during the car homologation test. But according to investigator technical experience we know, that results of μ_{op} over 0.3 in first approximation, are acceptable for use in cars. So the iron brake disc in comparison to MMC disks in this test conditions are better - friction coefficients for more sets are bigger then 0.3.

Next parameter, which is important for efficiency of friction set is temperature, especially fading temperature, which dependent on brake disc and pad materials. Fading causes loss of efficiency when brakes are hot that can occur during repeated or sustained overload of brakes, especially in a case of high load or high speed conditions.

The diagrams shown in Fig. 2 concerns the fading temperatures of cast iron and composite brake discs reinforced with SiC or fly-ash and working with various brake pads of different producers with different pressures in brake system. Fading temperatures are lower for best selected sets of MMC brake disc and pads than for iron discs because of higher thermal conductivity of aluminum reinforced with SiC or fly-ash particles. It is significant that if temperature is higher than 350°C, there is brake pads surface degradation.

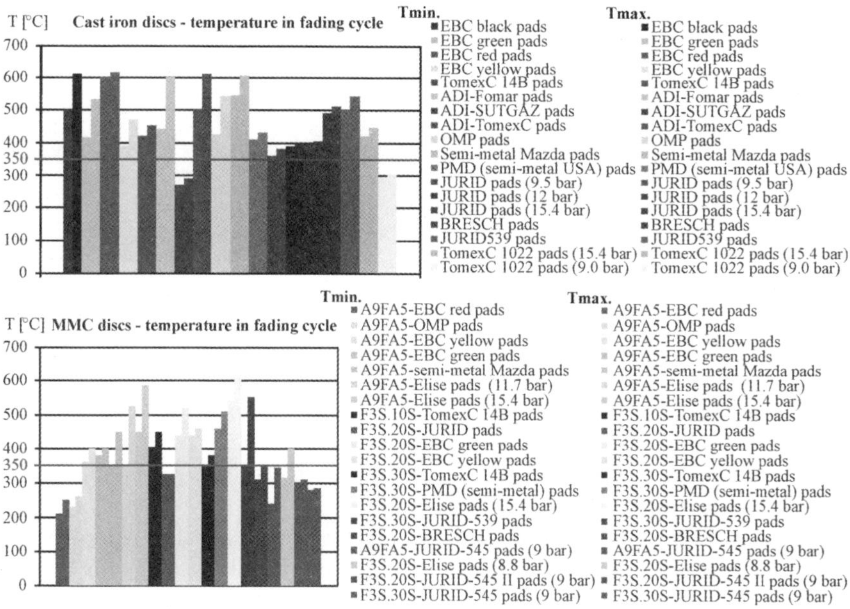

Fig. 2. Cast iron and metal matrix composites discs - temperature in fading cycle.

4. Concluding Remarks

The results make possible to formulate that:

- application of selected brake disc with selected pads in car brake systems depends on friction coefficient which is closely connected with work temperature;
- dynamic investigations on Krauss apparatus showed that if the brake discs are not ventilated only the friction sets composed of the cast iron brake discs and the brake pads can be used in vehicle brake systems,
- low brake efficiency (μ≤0,3) and high work temperature (t≥350°C) make some of the friction sets impossible to use,

- not ventilated composite brake discs are not perform technical requirements and can not be used in the braking system on a front axle in vehicles of M1 and N1 category,
- results proved that if adequate composition of brakes pads and adequate pressure in brake system are used, it is possible to recommend them in car brake system solutions.
- simultaneously based on results of this research and other research[2] it was ascertained that metal matrix composites brake discs are better than cast iron because of their properties such as: lower weight, multiply lower wear, faster heat dissipation, high thermal shock resistance.

Acknowledgments

This work was sponsored by the Ministry of Science and Higher Education of Poland under the project No. O R00 0061 06.

References

1. J. Sobczak: ‘Kompozyty metalowe, ISBN 83-913045-8-2’, Instytut Odlewnictwa - Instytut Transportu Samochodowego, Kraków-Warszawa 2001.
2. A. Wojciechowski, J. Sobczak: ‘Kompozytowe tarcze hamulcowe pojazdow drogowych’, Instytut Transportu Samochodowego, Warszawa, 2001.
3. D. Rudnik, J. Sobczak: ‘Tloki kompozytowe do silnikow spalinowych’, Instytut Transportu Samochodowego, Warszawa, 2001.
4. D.B. Miracle, S.L. Donaldson: ‘ASM Handbook Volume 21: Composites’, 2001, ASM International.
5. Regulation No. 90: ‘Uniform Provisions Concerning the Approval of Replacement Brake Lining Assemblies and Drum Brake Linings for Power-Driven Vehicles and Their Trailers’, 2009, http://www.unece.org/trans/main/wp29/wp29regs81-100.html.

MICROSTRUCTURE VARIATIONS OF CAST IRON AFFECTING ITS MECHANICAL BEHAVIOUR

KRYSTYNA PIETRZAK†

Motor Transport Institute, Jagiellonska 80
Warsaw, 03-301, Poland

The cast iron of different graphite shape was tested with respect of its microstructural and mechanical behavior. The original combinatorial method was applied in the quantitative metallographic assessment of cast iron microstructure. The Forster's elastomat was used for the assessment of the logarithmic vibration decrements, by means of resonance curve (ϑ_{fk}) method. Then the same samples were tested by means of the modified low cycle fatigue method (MLCF) in order to determine the mechanical parameters. It was possible to ascertain that the quantitative relationship between damping capacity and proposed synthetic stereological parameter of graphite form obtained as exponential relation, simultaneously means that the logarithmic vibration decrement exponentially increases with the increase of square root from the quotient of relative graphite area and average chord of graphite precipitates. It was possible to ascertain also that the best part of remaining mechanical characteristics determined by means of (MLCF) method decreases with the increase of stereological parameter $\xi = (S_V/l)^{1/2}$.

Keywords: Cast iron; damping capacity; mechanical properties; microstructure; stereological parameters AEPA 2010.

1. Materials and Methods

The cast iron of different graphite shape was tested with respect of its microstructural and mechanical behavior. The microstructure variations were obtained by the adequate modification of Ti or Sb additions and total evaporation of Mg. The investigations were started from 100% spherical graphite precipitates through the mixed graphite forms (modification by Ti or Sb) up to total evaporation of Mg when the 100 % of flaky graphite precipitates were obtained. The original combinatorial method[1] was applied in the quantitative metallographic assessment of cast iron microstructure. This method bases on the phase quanta theory according to which any microstructure can be treated as an arrangement of its measurable elements in the material matrix. In modern equipment resolutions (like image analyzers) the measureable element means one pixel. In the paper, basing on other results[2], two estimators were determined:

† Corresponding author, email: krystyna.pietrzak@its.waw.pl

"r" as estimator of relative graphite area S_V, where $S_V = 4r$[3, 4] and "m" as estimator of its volume fraction V_V, where $V_V = (m/M)x100\%$[3, 4]. Next the specific computational formulas[5] were used for the determination of geometrical microstructure parameters combined, as the summarizing effect, in the synthetic variable. All needed measurements were carried out in the same conditions using the computer image analyzer CLEMEX joined on line with an optical microscope PMG3. Forster's elastomat was used for an assessment of the logarithmic vibration decrements $(\vartheta_{fk})^2$, by means of resonance curve method.

Then the same samples were tested by means of the modified low cycle fatigue method (MLCF)[6, 7] in order to determine the following mechanical parameters:

- stress limit ($R_{0,02}$, $R_{0,05}$, $R_{0,10}$, $R_{0,20}$),
- ultimate stress strength (R_m),
- accommodation limit (R_a),
- fatigue limit in bending-rotational conditions (Z_{go}),
- elongation (A_5),
- Young's modulus (E: E_0, E_{10}, E_{80}, E_{180}),
- fatigue strength exponent (Basquin's coefficient - b),
- fatigue deformability exponent (c),
- strain-hardening exponent in cycle deformations (n),
- strength cycle coefficient (K'),
- maximal permissible true strain in changing loads (ε_{max}).

The new software algorithm was proposed[7] in the determination of fatigue parameters as above. A comparative verification of the new procedure with bibliographical data confirmed the possibility of the replacement of classical LCF by proposed original MLCF method as especially suitable test in the case of heterogeneous materials microstructure, particularly when it is hard to design Wohler's diagrams as well as to carry on research according to the classical LCF method[7].

2. Results

The microstructures of investigated cast iron are shown in Figure 1 a, b, c (non etched state, illustrating the differences observed between graphite forms) and in Figure 1 d, e, f (etched state, illustrating the similar pearlitic matrix microstructure). On the basis of the quantitative metallographic microstructure assessment it was possible to formulate the synthetic stereological parameter as $\xi = (S_V/l)^{1/2}$, see Table 1, where S_V means the relative area of graphite precipitates and l means the average chord of those. The results concerning calculated correlations between proposed synthetic stereological parameter ξ and determined mechanical characteristics are put together in Table 1.

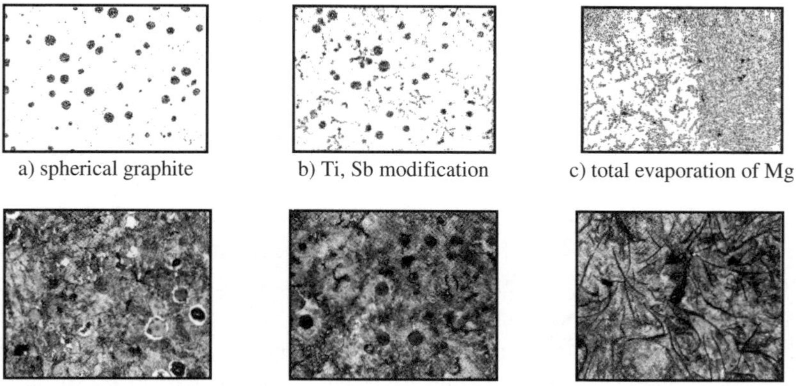

a) spherical graphite　b) Ti, Sb modification　c) total evaporation of Mg

d) spherical graphite　e) Ti, Sb modification　f) total evaporation of Mg

Fig. 1. Forms of graphite precipitates (a, b, c - non etched state) and matrix microstructures (d, e, f - etched state) of the cast iron.

Table 1. Correlation coefficients (and their calculated and theoretical significances) between the synthetic stereological parameter ξ and mechanical characteristics.

		Mechanical properties					
Stereolog. parameter	$(S_V/l)^{1/2}$	$\ln\vartheta_{fk}$	E_0	E_{10}	E_{80}	E_{180}	$R_{0,02}$
Correl. coefficient	ρ	0.9332	-0.8368	-0.8651	-0.9022	-0.7363	0.4441
Calculated signific.	t_o	10.704	6.302	7.112	8.624	3.768	2.044
		Mechanical properties (continue)					
Stereolog. parameter	$(S_V/l)^{1/2}$	$R_{0,05}$	$R_{0,10}$	$R_{0,20}$	R_a	R_m	Z_{go}
Correl. coefficient	ρ	-0.8884	-0.8744	-0.7876	-0.8914	-0.9357	-0.8998
Calculated signific.	t_o	7.980	7.208	4.608	8.110	10.932	8.504
		Mechanical properties (continue)					
Stereolog. parameter	$(S_V/l)^{1/2}$	A5	b	c	n	K'	ε_{max}
Correl. coefficient	ρ	-0.8056	-0.3327	-0.6199	0.6574	0.3547	-0.6907
Calculated signific.	t_o	5.606	1.455	3.257	3.597	1.564	3.938
Number of samples	n	19					
Theor. sig.	t_t	2.110 for α=0,05 2.567 for α=0,02 2.898 for α=0,01 3.965 for α=0,001					

Data completed in Table 1 show the significant correlations (on the significance level: α=0,001) obtained between the best part of mechanical properties and proposed synthetic stereological microstructure parameter

$\xi = (S_V/l)^{1/2}$. It is connected with a fact that all calculated significances (t_o) for: logarithmic vibration decrements (ϑ_{fk}), Young's modulus (E_0, E_{10}, E_{80}, E_{180}), stress limits ($R_{0,02}$, $R_{0,05}$, $R_{0,10}$, $R_{0,20}$), accommodation limit (R_a), ultimate stress strength (R_m), fatigue limit in bending-rotational conditions (Z_{go}), elongation (A_5) meet theoretical significance criterion - t_t = 3.965 for significance level - α=0.001, see Table 1. Results concerning: fatigue deformability exponent (c), strain-hardening exponent in cycle deformations (n) and maximal permissible true strain in changing loads (ε_{max}) meet the analogous theoretical significance criterion t_t = 2.567 for significance level - α=0.01. Only two mechanical parameters fatigue strength exponent (Basquin's coefficient - b) and strength cycle coefficient (K') do not meet theoretical significance criterion even on the significance level - α=0.05. Additionally it was found the best fit for exponential relationship in the case of logarithmic vibration decrements (ϑ_{fk}), see Figure 2a and polynomial quadratic relationships in the case of rest mechanical properties, as exemplum, see Figure 2b.

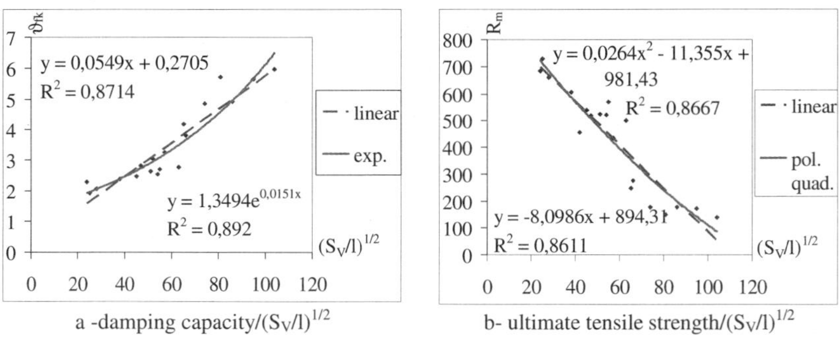

Fig. 2. Quantitative relationships between selected mechanical characteristics (a - damping capacity, b - ultimate tensile strength, c - Young modulus, d - fatigue limit) and synthetic stereological parameter.

Simultaneously it was possible to ascertain that if values of logarithmic vibration decrements (ϑ_{fk}) decrease, the rest mechanical properties increase. This regularity is in agreement with the fact, that susceptibility to damping vibrations is connected with properties of softer phase. In contrary, the ultimate tensile strength and the rest mechanical properties resulting from simple tensile test or resulting from presented MLCF method are connected with the harder phase. In other words, in the case of cast iron, damping capacity is connected with graphite precipitations, but the rest mechanical properties are connected with a matrix of cast iron. Thus, any changes of graphite precipitates morphology will influence on damping capacity and rest mechanical properties. Based on quantitative metallographic assessment, it was possible to demonstrate that synthetic geometrical microstructure parameter - $\xi = (S_V/l)^{1/2}$ describes best any changes of graphite precipitates and correlates significantly with mechanical characteristics, see Table 1.

3. Concluding Remarks

The paper presents the results of non-destructive and destructive mechanical tests as well as the effects of metallographic assessment carried out by means of combinatorial method. It was possible, as final results:

- to formulate the synthetic stereological parameter of graphite form, on the basis of the quantitative metallographic assessment, e.g.: $\xi = S_V/l^{-1/2}$.
- to ascertain that there is the exponential relationship between damping capacity and proposed synthetic stereological parameter of graphite form. This fact simultaneously means that the logarithmic vibration decrement exponentially increases with the increase of square root from the quotient of relative graphite area and average chord of graphite precipitates.
- to ascertain that the best part of remaining mechanical characteristics determined by means of modified low fatigue method linearly decreases (but the best fit was obtained in the case of the polynomial quadratic) with the increase of proposed synthetic stereological variable.

References

1. B. Kęsy: *Metaloznawstwo i Obróbka Cieplna*, 1990, **103-105**, 9-13.
2. K. Pietrzak, J. Ryś: Proc. 3th Int. Conf. on 'Stereology in Materials Science', Kraków-Szczyrk, Poland, October 1990, Paper 21.
3. E.E. Underwood: 'Quantitative Stereology', (ed. Addison-Wesley Publishing Company), Ontario, 1970.
4. J. Ryś: 'Stereology of Materials', (ed. Fotobit Design), 1995, Cracow.
5. K. Pietrzak, A. Klasik, Z. L. Kowalewski, D. Rudnik: *International Journal of Modern Physics B (IJMPB)*, 2008,**Vol. 22, Nos. 31/32**, 5533-5538.
6. M. Maj: 'Criterions of Exploitation Strength of Cast Iron based on Mechanical Properties of Materials', PhD Thesis, AGH University of Science and Technology, Faculty of Foundry Engineering, Cracow, Poland, 1984.
7. M. Maj, W. Moćko, K. Pietrzak, A. Klasik: *Archives of Foundry and Engineering*, 2009, **vol. 9, issue 4**, 129-134.

RESEARCHES ON FLOW BEHAVIOR AND CONSTITUTIVE EQUATION OF A390 ALUMINUM ALLOY

HAITING GUO, XINYUN WANG†, LEI DENG, JUCHEN XIA

State Key Laboratory of Material Processing and Die & Mould Technology, Huazhong University of Science and Technology, 1037 Luoyu Road, Wuhan, 430074, China

Flow behavior of an A390 Aluminum alloy was investigated at temperatures of 290-490 ℃ and at strain rates of 0.001-1 s^{-1}. The true stress-strain curves were obtained from the hot compression tests. The results indicate that the flow stress of A390 aluminum alloy decreases with the increase of temperature and the decrease of strain rate. It is also proved that work hardening and dynamic softening phenomena occur during hot compression. By applying the hyperbolic sine constitutive model, the constitution equation of A390 aluminum alloy was obtained.

Keywords: A390 aluminum alloy; Hot compression; Flow behavior; Constitutive equation.

1. Introduction

A390 alloy is a typical hypereutectic aluminum silicon alloy. Because of its interesting properties such as high wear resistance, low thermal expansion coefficient, good corrosion resistance and castability, it is widely used for piston parts in aviation and automobile industries.[1]

The common forming technology is gravity casting, low pressure casting, squeeze casting and semi-solid forming. However, defects such as pores, peeling and cold shut are generated in the casting parts. Plastic forming technology can effectively avoid these quality problems.[2]

It is necessary to study workability and characterize flow behavior of A390 aluminum alloy during hot deformation. The constitutive equation can describe the stress state during the deformation when the strain, strain rate and temperature change. The establishment of constitutive equation can be obtained by two strategies: one is based on the description of the mechanism of deformation, establishing a physical constitutive equation. The other method depends on macroscopic stress, strain and internal state, establishing a phenomenonal constitutive equation by regression analysis of experimental

† Corresponding author, email: wangxy_hust@163.com

data.[3] Sellars proposed a hyperbolic sine constitutive equation of Arrhenius type including deformation activation energy and temperature.[4]

By introducing Arrhenius type equation, the flow stress σ can be expressed as function of strain rate and temperature. The equation could be coupled into finite element code to obtain distribution of internal state variables. Moreover, it also provides convenience for studying the relationship between deformation condition and the evolution of the material microstructure and properties.[5]

Compression tests were used to investigate the effects of temperature and stain rate on the flow stress of the A390 aluminum alloy. The deformation temperature varied from 290 ℃ to 490 ℃, and an engineered strain rate range of 0.001 to 1 s^{-1} was covered. Constitutive equations which can describe the material deformation behavior are established.

2. Material and Experiment

The material used in this research is A390 aluminum alloy with chemical compositions shown in Table 1.[6]

Table 1. A390 aluminum alloy chemical composition.

Element	Si	Cu	Mg	Fe	Mn	Ti	Zn	Al
wt. %	16~18	4~5	0.45~0.6	≤0.50	≤0.20	≤0.10	≤0.10	bal

Hot compression tests were carried out on Zwick/Roell Z020 universal testing machine, at six temperatures of 290 ℃, 330 ℃, 370 ℃, 410 ℃, 450 ℃ and 490 ℃, and at five strain rates of 0.001 s^{-1}, 0.01 s^{-1}, 0.1 s^{-1}, 0.5 s^{-1} and 1 s^{-1}. The high reductions of the specimens were 50%.

3. Results and Discussion

3.1. *Flow stress behavior*

True stress-strain curves of A390 aluminum alloy compressed at different conditions are shown in Figure 1. It can be seen that flow stress increases rapidly with the increase of strain at the initial stage of deformation, then flow stress reaches a peak value, and as the strain continues to increase, stress slightly reduces.

That is because at the initial stage, work hardening occurs which leads to the deformation resistance increasing. During the later deformation, dynamic restoration occurs which causes decrease of stress.[7]

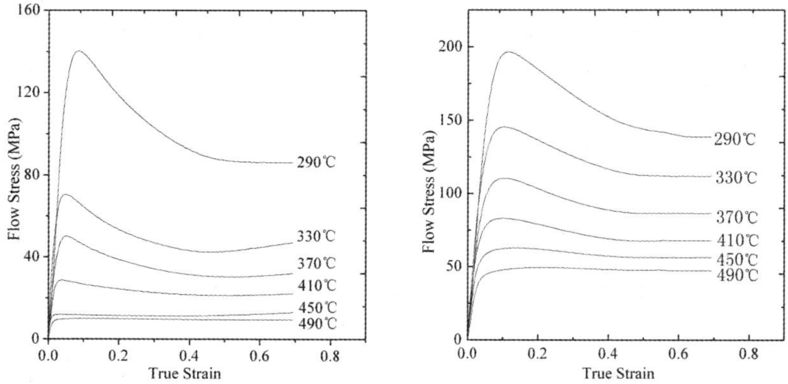

(a) 0.001/s (b) 1/s

Fig. 1. True stress-strain curves of A390 Al alloy deformed at different conditions.

3.2. *Constitutive equation*

It is common to adopt Arrhenius equation to set up the constitutive equation. The relationship of stress σ and strain rate fits the following Eqs. (1), (2) and (3) at low, high and all stress conditions, respectively.

$$\dot{\varepsilon} = A_2\sigma^n \tag{1}$$

$$\dot{\varepsilon} = A_1 \exp(\beta\sigma) \tag{2}$$

$$\dot{\varepsilon} = A[\sinh(\alpha\sigma)]^n \exp(-\frac{Q}{RT}) \tag{3}$$

Where A_1, A_2, A, n, n_1, α and β are constants, Q the activation energy, R the gas constant and T the absolute temperature. α, β and n_1 meet the relationship of $\alpha=\beta/n_1$.[8]

The following equations can be derived from Eq. (1) and Eq. (2);

$$\ln\dot{\varepsilon} = \ln A_1 + n_1 \ln\sigma \tag{4}$$

$$\ln\dot{\varepsilon} = \ln A_2 + \beta\sigma \tag{5}$$

The peak values of stress under different deformation conditions are substituted into Eq. (4) and Eq. (5), so the relationships between $\ln\sigma$ and $\ln\dot{\varepsilon}$, and σ and $\ln\dot{\varepsilon}$ are plotted in Figure 2. The values of n_1 and β are obtained from the slope of the line. The values of the parameters are: n_1= 4.94269, β= 0.10461, $\alpha=\beta/n_1=0.02116$.

Supposing the deformation activation is independent of temperature, Eq. (3) is transformed into Eq. (6) as follows:

$$\ln\dot{\varepsilon} = \ln A - \frac{Q}{RT} + n\ln[\sinh(\alpha\sigma)] \tag{6}$$

The peak values of stress under different conditions are substituted into Eq. (6). The relationship between $\ln\dot{\varepsilon}$ and $\ln\sinh(\alpha\sigma)$ is plotted in Figure 3. The slope of the line is the value of n which is 4.40327. It can be seen that $\ln\dot{\varepsilon}$ and $\ln\sinh(\alpha\sigma)$ fit a linear relationship very well, which proved that the relationship of the stress and the strain rate meet the form of hyperbolic sine. So this kind of function can be used to describe relationship between flow stress and strain rate under all the stress levels. It provides a theoretical basis for controlling the stress levels by controlling strain rate during hot working.

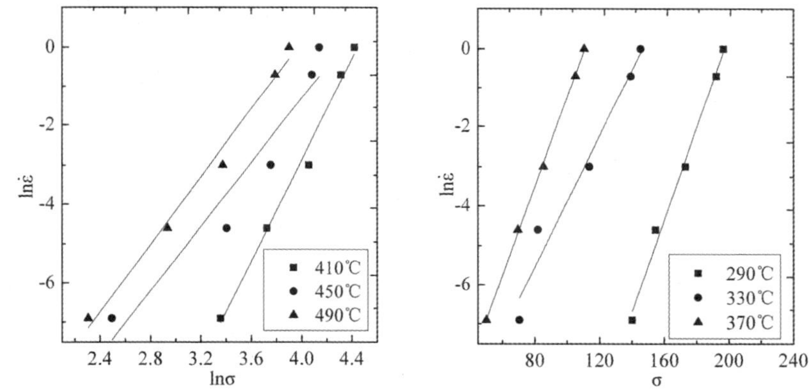

(a) $\ln\sigma-\ln\dot{\varepsilon}$ (b) $\sigma-\ln\dot{\varepsilon}$

Fig. 2. The relationship between peak stress and strain rate.

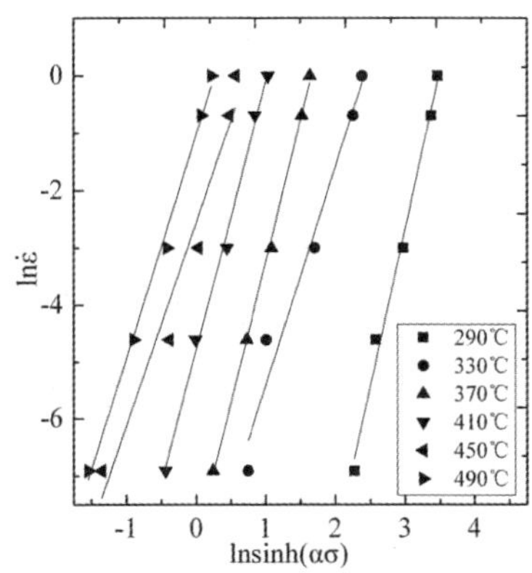

Fig. 3. $\ln\sinh(\alpha\sigma)$ versus $\ln\dot{\varepsilon}$.

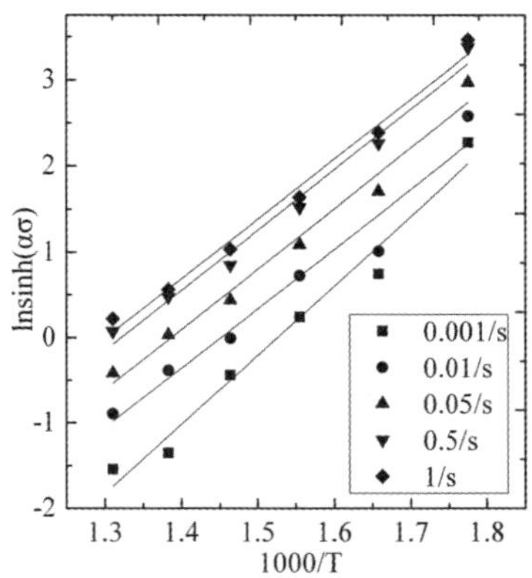

Fig. 4. $\ln\sinh(\alpha\sigma)$ versus $1000/T$.

The Eq. (6) is transformed into Eq. (7) as follows

$$\ln\sinh(\alpha\sigma)=\frac{Q}{nRT}+\frac{\ln\dot{\varepsilon}-\ln A}{n} \tag{7}$$

The relationship between $\ln \sinh(\alpha\sigma)$ and $1000/T$ is plotted in Figure 4. The slope of the line is the value of $Q/1000nR$, so the value of Q is calculated to be 263850.487634 J/mol. Then the value of A is obtained, which is A=1.84E+18.

The constitutive equation of A390 aluminum alloy is expressed as $\dot{\varepsilon} = 1.84\times 10^{18}[\sinh(0.02116\sigma)]^{4.4}\exp(\frac{-263850}{8.3145T})$.

4. Conclusion

Work hardening and dynamic softening phenomena occur during the hot compression deformation. The flow stress decreases with increasing temperature, while increases with increasing the strain rate. The constitutive equation is obtained as $\dot{\varepsilon} = 1.84\times 10^{18}[\sinh(0.02116\sigma)]^{4.4}\exp(\frac{-263850}{8.3145T})$ This study provides constitutive model for the finite element solving of A390 aluminum alloy during hot working.

Acknowledgments

This work was supported by State Key Laboratory of Materials Processing and Die & Mould Technology and Ministry of Industry and Information Technology of China (No. 2009ZX04014-082).

References

1. Y. Zhang, D. Q. Yi, W. X. Li, Z. S. Ren, Z. Q. Zhao and J. H. Zhang: *Trans. Nonferrous Met. Soc Chin.*, 2007, **17**, s413–s417.
2. J. F. Wang, J. P. Xie, Z. X. Liu, J. W. Li, Y. G. Wang and M. X. Wang: *Foundry*, 2005, **54**, 24–27.
3. J. M. Zhou, L. H. Qi and G. D. Chen: *Mech. Sci. Technol.* 2005, **24**, 212–216.
4. C. M. Sellars and T. Mukherjee: *Metal. Master. Trans. B*. 1972, **3**, 953–962.
5. W. X. Wu, Y. Han, H. Zhong, Y. K. Le and H. Zhang: *Chin. J. Nonferrous Met.* 2009, **19**, 1403–1408.
6. W. Li, J. R. Luo, G. D. Lan and Q. H. Liang: *J. Huazhong Univ. Sci. Technol.* 2008, **36**, 92–95.
7. F. Liu, D. B. Shan and Y. Lu: *Mater. Sci. Technol*, 2003, **11**, 55–58.
8. G. Z. Quan, G. H. Zhang, J. Zhou and K. W. Liu: *Trans. Mater. Heat Treat.* 2009, **38**, 23–25.

RESEARCH ON MODEL OF ON-LINE MEASUREMENT FOR VERTICAL RING ROLLING MILL

XIAOKAI WANG

School of Materials and Engineering, Wuhan University of Technolog
Wuhan, 430070, China

LIN HUA†

School of Materials and Engineering, Wuhan University of Technolog
Wuhan, 430070, China
Hubei Key Laboratory of Advanced Technology of Automotive Parts
Wuhan, 430070, China

On-line measurement and process control which has been widely used in forging equipments is rarely applied to the vertical hot ring rolling mill. This paper firstly establishes mathematical model of ring geometric shape and size corresponding to single fixed guide mode. By means of the model, the accurate expression of ring's dimensions is derived. The curved surface which governs the relations of ring's outer radius, feeding amount and rolls' geometric parameters is constructed. The precise and simple function of the ring's outer radius in rolling process is derived based on the surface-fitting method. The prototype vertical ring rolling mill was designed and relative experiment was carried out. The results indicated that the method is useful for on-line measurement and control of vertical hot ring rolling.

Keywords: Vertical ring rolling; On-line measurement; Mathematical model; Surface-fitting.

1. Introduction

Vertical ring rolling (VRR) mill is the common equipment for manufacturing the small and medium size rings such as bearing ring, aircraft turbine ring, gearing ring, etc.[1] Figure 1 shows the VRR principle. Up to now, the horizontal hot ring rolling (HHRR) and cold ring rolling (CRR) have almost realized automatic control. However, to the VRR, it is difficult to realize on-line measurement and process control due to the sharp increase of the ring's outer radius and the limitation of the ring rolling mill's structure.[2] Relative report is also not presented at home and abroad. In the practical production, the VRR process still depends on human's experience, which results in large dimension error and high consumption of raw material.[3] Therefore, this paper studies establishment a

† Lin Hua, email: wxk0919@163.com

measurement model of VRR, and provides basis for on-line measurement and control of VRR.

2. Geometrical Model and Analysis Formula of VRR

Ma et al.[4] provided the geometrical model of VRR, and obtained the ring's center coordinates. However, the solution of the model is very complex and dose not suit to on-line measurement and control of VRR process. For the sake of a more practical on-line measurement method, new geometrical models and calculation methods must be adopted.

During the VRR process, The mandrel roller's location keeps static, so its center can be regard as the origin of the geometrical model's coordinate, as shown in F Figure 2. Guide roller is fixed with an arm which is set at an angle of β degree to the vertical direction. Measuring roller moves at an angle of α degree to the vertical direction. For the convenience of calculation, an auxiliary roller whose radius is equal to the measuring roller is assumed at the contact of driving roller and ring. So that an auxiliary circle can be defined by the center of auxiliary roller (O_5), guide roller (O_3) and measuring roller (O_4). According to the geometrical relations, the ring's center coordinate and outer radius can be calculated if the auxiliary circle's radius is calculated.

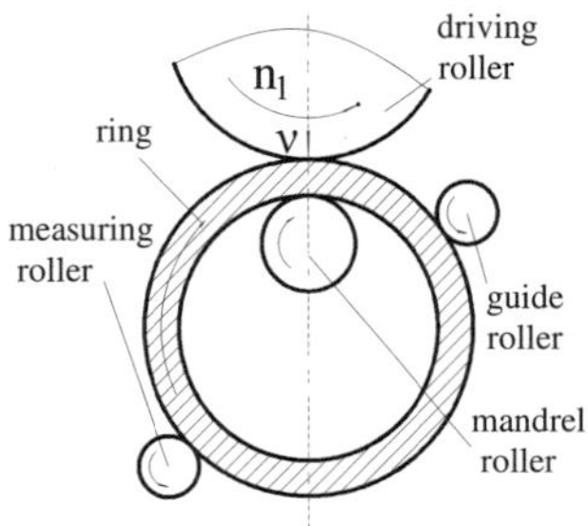

Fig. 1. VRR principle.

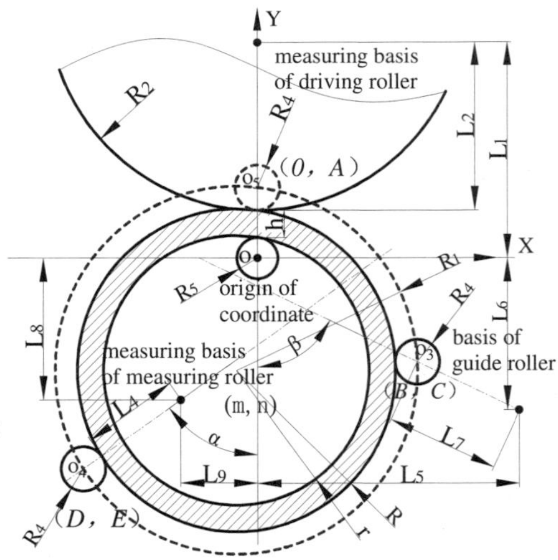

Fig. 2. Geometrical model of VRR.

$(0, A)$、(B, C) and (D, E) in Figure 2 are represented the coordinates of O_5、O_3、O_4, respectively. Center coordinate of ring is defined as (m, n).The following equations set can be obtained based on the geometrical relations.

$$\begin{cases} m^2 + (A - n)^2 = R_1^2 \\ (B - m)^2 + (C - n)^2 = R_1^2 \\ (D - m)^2 + (E - n)^2 = R_1^2 \end{cases} \tag{1}$$

where

$$\begin{aligned} &A = L_1 - L_2 + R_2, \quad B = L_5\text{-}L_7\sin\beta, \\ &C = L_7\cos\beta\text{-}L_6, \; D = -L_9 - (L_4 + R_4)\sin\alpha \end{aligned} \tag{2}$$

where L_2 and L_4 are the displacement of driving roller and measuring roller, respectively. L_1 to L_7 are the dimensions of the VRR mill.
From Eq. (1), it can achieved,

$$\begin{cases} m = \dfrac{B^2C + A^2B + E^2C - BC^2 - BD^2 - A^2C}{2(AB - BD + CE - AC)} \\ n = \dfrac{C^2E + D^2E + A^2D + AB^2 + AE^2 - C^2A - AD^2 - B^2D - DE^2 - A^2E}{2(AB - BD + CE - AC)} \\ R_1 = \sqrt{m^2 + (A - n)^2} \end{cases} \tag{3}$$

We can have every dimension of ring:

$$R = R_1 - R_3, \; h = |O_yO| - L_2 - R_4, \; r = R_1 - R_3 - |O_yO| + L_2 + R_4 \tag{4}$$

3. Numerical Solution of Ring's Geometry

Assume that the parameters about a certain VRR mill as below: R_2=180mm; R_4=18.5mm; R_5=17.5mm; L_1=182mm; L_5=223.5mm; L_6=127.5mm; L_7=96mm; α=55°; β=65°. Pretend that the measuring range of the displacement sensors of driving roller and measuring roller are 160mm and 80mm.Utilizing MATLAB, a surface which represents the relations of auxiliary circle's outer radius versus displacement of driving roller and measuring roller can be plotted as Figure 3.

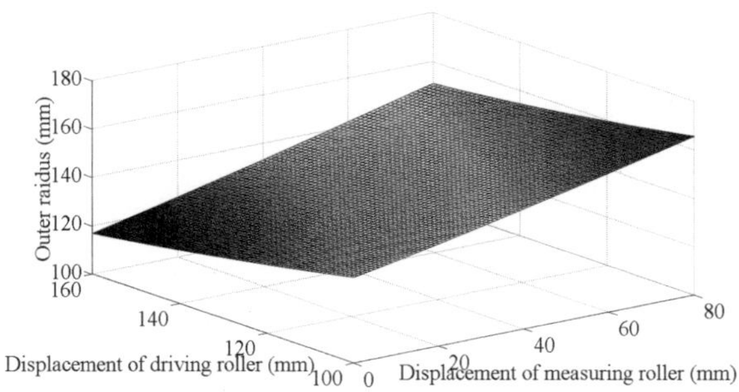

Fig. 3. The surface of outer radius.

From the surface, we can gain the value of auxiliary circle's radius if we know the displacement of driving roller and measuring roller. The ring's radius is the difference between the auxiliary circle's radius and the measuring roller's radius.

4. Simplification of the Geometrical Formula

For some control device, especially the Programmable Logic Controller (PLC), it's too difficult to program tedious formula like Eq.(3). So Eq.(3) must be simplified.

Least-squares surface Fitting method[5] is chose to reduce Eq.(3). Eq.(3) can be denoted as a function $R = f(X,Y)$, where X and Y stand for the displacement of measuring roller and driving roller, Using MATLAB surface -fitting toolbox, Eq.(3) can be reduced to quadratic polynomial as follows.

$$R_{\text{outer radius}} = \text{a+b}X + \text{c}Y + \text{d}X^2 + eY^2 + fXY \tag{5}$$

Where *a, b, c, d, e, f* are the coefficients of Fun.(5).

5. The Development of VRR Pprototype

According to the measurement model, the prototype of D51-350 vertical ring rolling mill was designed as shown in Figure 4. The rectangular section ring workpiece was selected to roll. Figure 4 shows the situation of the VRR experiment.

Fig. 4. The VRR prototype and the ring rolling process.

In order to investigate the quality of the formed ring in detail, the average dimensions of the ring were measured. The results of experiment are summarized in Table 1. It is can be seen that the design dimensions and the experiment result are approximate. It is indicated that on line measurement method is feasible to the VRR process.

Table 1. The results of experiment.

Item	Design dimension (mm)	Experiment result (mm)	Relative error (%)
Outer radius	285	283.8	0.42
Inner radius	200	196.8	1.60
Width	51	51.8	1.56

6. Conclusion

From the foregoing analyses, the following conclusions can be drawn:

1. A new geometrical model of VRR has been found, and the geometrical formula of ring's center coordinate and outer radius has been deduced.

2. A precise and simple function of ring's outer radius in VRR process is derived based on the surface-fitting method. Through analysis the accuracy, the fitting function proves to be feasible.

3. The prototype of D51-350 vertical ring rolling mill has been contrived. The experiment results demonstrate that the on-line measurement method can be used for the VRR process.

Acknowledgments

The authors would like to thank the National 863 Program (2009AA04Z112) and the National Major Science-technology Project (2009ZX04014-074) for the support given to this research.

References

1. L.Hua, X.G. Huang, C.D.Zhu: 'Theory and Technology of Ring Rolling', 2001, Beijing, Mechanical Industry Press.
2. Allwood, Julian M, Tekkaya: *Steel. Res. Int.*, 2005, 2, 111–120.
3. L.Hua, X.S.Mei, X.T.Wu: *Trans. Nonferrous Met. Soc. China.*, 1999, 9, (2), 213–217.
4. S.Z.Ma, G.S.Zhou: Bearing, 1999, 35, 12, 9–12.
5. Y.C.Chen: 'High Level Programming Language about MATLAB', 2004, Beijing, Tsinghua University Press.

RESEARCH ON ANALYTICAL ALGORITHM OF CONTACT PROBLEM USING FINITE ELEMENT ANALYSIS FOR BULK FORMING

ZHONG-JIN WANG†

School of Materials Science and Engineering, Harbin Institute of Technology, No.92, West Da-Zhi Street, P.O. Box 435, Harbin 150001, P.R.China

BIN-XIAN YUAN

School of Materials Science and Engineering, Harbin Institute of Technology, No.92, West Da-Zhi Street, P.O. Box 435, Harbin 150001, P.R.China

Dynamic contact problem between die and billet arises during metal forming processes, the expression method of die shape is an important aspect in handling contact problem. In this paper, we put forward a new way to express the die shape through a high-order algebraic curve. Utilizing the finite element analytical program of contact judgment algorithm based on expression of high-order curve analytical equation, we analyzed the process of local expression of cylinder with dies of curved shapes, and then compared the analytical results with the business software DEFORM-2D to verify the reliability of expression of die shape in high-order curve analytical equation and relevent contact judgment algorithm.

Keywords: Bulk forming; Die shape; Analytical method; Contact judgment.

1. Introduction

In recent years, in order to predict processing results and enhance efficiency, finite element method (FEM) has been applied to simulate metal forming processes. The simulation of metal forming processes involves the expression of die shape, contact searching and judgment, which have a great influence on the results.

Many methods have been developed to express the die shape in finite element analysis, including the dot matrix method[1], the analytical method[1-2], the finite element mesh approximation method[3-4], the parametric surface method[5], and the triangle diamond method[6]. The analytical method, which accurately expresses a simple curved surface, can transform the dynamic contact problem into system of equations to slove. But when it comes to the complex curved

---

† Corresponding author, email: wangzj@hit.edu.cn

surface of a die shape, the precision of the analytical method is low. In this paper, we proposed a method to express the die shape in a high-order algebraic curve, programmed the corresponding contact judgment algorithm, and then compared the analytical results with the business software DEFORM-2D to verify the reliability.

2. Analytical Expression of the Die Shape

The irregular die shape is expressed by curves in a two dimension problem. In order to construct a smooth, homogenous high-order algebraic curve, we consider transition curve:

$$C: F(x,y) = h^n(x,y) + \lambda f(x,y) g(x,y) = 0 \tag{1}$$

where $C_1 : f(x,y) = 0$, $C_2 : g(x,y) = 0$ are two algebraic curves, called source curves, $C_3 : h(x,y) = 0$ is an algebraic curve that intersects with both C_1 and C_2, called control curve, and $n > 1$ is an integer, λ is an arbitrary constant.

It can be deduced that transition curve C intersects with C_1 and C_2 at the same points where the control curve C_3 intersects with C_1 and C_2, and transition curve C has the same normal lines with both C_1 and C_2 at the intersection points respectively.

It is shown that high-order curve analytical equations are influenced by source curve equations, control radius and shape parameter λ. We can get ideal and smooth curves of different shapes through changing these factors above. Under the same control radius and source curve equations, we can also obtain different curve shapes through varying shape parameter λ, which can be used to optimize the mold curve shapes.

3. Analytical Expression of the Die Shape

In finite element analysis, the die is always rigid so the nodes of the billet couldn't penetrate it. If a node of the billet penetrates the die, adjust the node to the surface of the die along the normal line. The contact judgment problem only involves boundary nodes, including liberty boundary nodes and restraint boundary nodes.

3.1. *Contact judgment for liberty boundary nodes*

It is necessary for each liberty boundary node to be judged. Fig. 1. is the sketch map of contact judgment for straight lines and high-order algebraic curves, where n and t are normal vector and tangential vector of the die respectively.

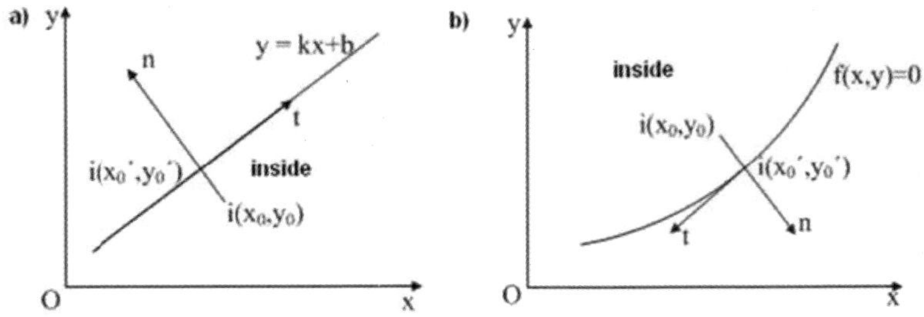

Fig. 1. Sketch maps of contact judgment a) Straight lines b) High-order algebraic curves.

For a straight line, the flow chart is given below:

1. Judge the position of the node i: If $y_0 \le kx_0 + b$, the node i is considered to have contact with the die. Both effects are independent and additive.
2. Adjust the position of the node i: First, we should acquire the normal equation $y = -\frac{1}{k}x + \left(y_0 + \frac{1}{k}x_0\right)$ of $y = kx + b(k \ne 0)$, which goes through the node $i(x_0, y_0)$. Then, solve system of equations $\begin{cases} y = kx + b \\ y = -\frac{1}{k}x + \left(y_0 + \frac{1}{k}x_0\right) \end{cases}$ to obtain (x_0', y_0'), that is the new position of the node i.
3. Establish a local coordinate system at the new position, and apply the constraint.

For a high-order algebraic curve, the flow chart is given below:

1. Judge the position of the node i: The node $i(x_0, y_0)$ is considered to have contact with the die when $y_0 - y \ge 0$, in which y is obtained by solving $f(x_0, y) = 0$ through Newton iteration method.
2. Adjust the position of the node i: First, we should acquire the normal equation of the high-order algebraic curve $f(x, y) = 0$, which goes through the node $i(x_0, y_0)$. Then, solve equations set of the normal equation and $f(x, y) = 0$ to obtain (x_0', y_0'), which is the new position of the node i.
3. Establish a local coordinate system at the new position, and apply the constraint.

3.2. *Contact judgment for restraint boundary nodes*

For the restraint boundary nodes, the nodal forces should be verified after the iteration, the flow chart is given below:

1. Judge the position of the node i : Calculate the normal force at the local coordinate through transition matrix. The node $i(x_0, y_0)$ is considered to be a liberty boundary node when the normal force is more than zero, or it is still a restraint boundary node.
2. Adjust the position of the node i : The position adjustment method for a restraint boundary node is the same as the method of a liberty boundary node, which is mentioned above.
3. Establish a local coordinate system at the new position, and apply the constraint.

4. Numerical Examples

Utilizing the finite element analytical program of contact judgment algorithm based on expression of a high-order curve analytical equation presented in this paper, we analyzed the process of local expression of cylinder with dies of curve shapes. The finite element analysis model is shown in Fig. 2., the billet is an axisymmetric cylinder with $D = 40mm$, $H = 10mm$. Based on the symmetry, the simulation model is right half part of the cylinder, with 512 quadrangle meshes and 561 nodes. The yield stress model of the billet is $\bar{\sigma} = 10\dot{\bar{\varepsilon}}^{0.1}$, friction coefficient between the die and the billet is 0.5, the velocity of the die is 1mm/s and time step is 0.01s .

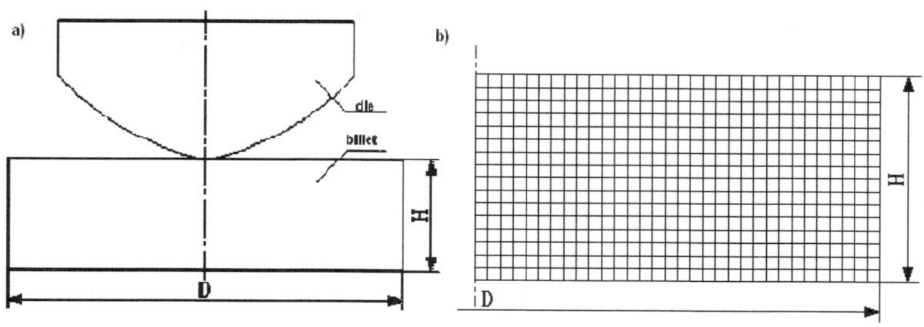

Fig. 2. Finite element model for local expression of cylinder a) Finite element model b) Meshes of the billet.

The parameters of the die shape are shown in Table 1. High-order curve expressions can be calculated:

$$(x^2 + y^2 + 10x - 77.5y + 675)^3 + 0.005(y - 10)(x - 15) = 0 \tag{2}$$

Table 1. Shape parameters of the die.

Endpoint coordinate (mm)	*Control arc radius (mm)*	*Equation of boundary line*	λ
(0,10),(15,17.5)	29.18	y-10=0、x-15=0	0.005

For die shape mentioned above, when the strokes are 2mm and 8mm respectively, the mesh configuration, effective stress and effective strain, which were obtained by the program presented in this paper and DEFORM-2D respectively, are shown in Fig. 3.–Fig. 4. The isoline distributions of effective stress and effective strain are in agreement basicly.

5. Conclusion

This paper introduces a new way to express the die shape in a high-order algebraic curve, and presents contact judgment algorithm based on expression of the high-order curve analytical equations. We analyzed the process of local expression of cylinder with dies of curve shapes, to verify the reliability of expression of die shape in a high-order curve analytical equation and relevent contact judgment algorithm.

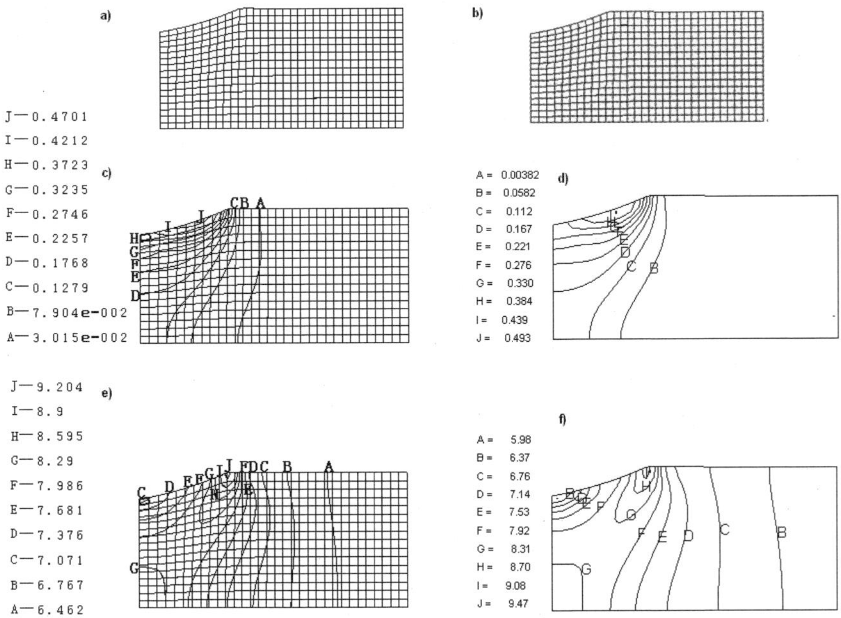

Fig. 3. Numerical results and isoline contours distribution at stroke 2mm a) Mesh (in this paper) b) Mesh (DEFORM) c) Effective strain (in this paper) d) Effective strain (DEFORM) e) Effective stress (in this paper) f) Effective stress (DEFORM).

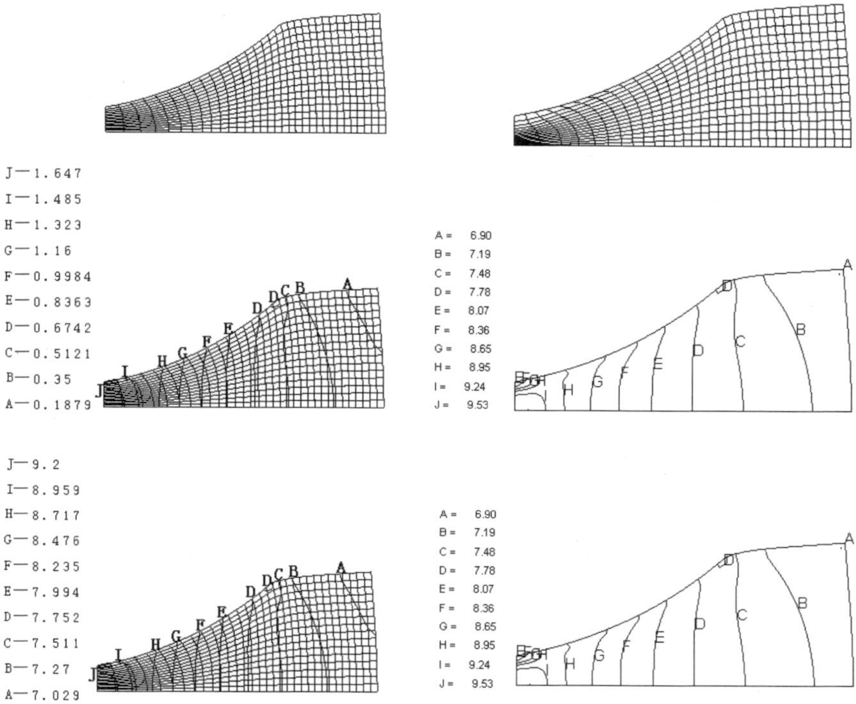

Fig. 4. Numerical results and isoline contours distribution at stroke 8mm a) Mesh (in this paper) b) Mesh (DEFORM) c) Effective strain (in this paper) d) Effective strain (DEFORM) e) Effective stress (in this paper) f) Effective stress (DEFORM).

References

1. X. Zuo, J. Chen, Y. P. Wei, X. Y. Ruan: *Forging and Stamping*, 1997, **6** ,59–61.
2. Y. X. Zhao, W. L. Xu, Z. Q. Lin, X. M. Lai: *Chinese Journal of Mechanical Engineering*, 2001, **37,** (4), 99–102.
3. X. N. Pan, X. C. Su, H. Li, D. Wang: *Coal Mine Machinery*, 2007, **28**, (4), 68–70.
4. M. Q. Wang, Y. M. Zhu, W. X. Liu: *Machinery Design and Manufacture*, 2004, (1), 22–24.
5. D. J. Yoo, I. S. Song, D. Y. Yang: *Int. J. Mech. Sci.* 1994, **36**, (6), 513–546.

AMENDMENT AND RECONSTRUCTION OF EMPIRICAL FORMULAS CALCULATING BURSTING PRESSURE OF SPECIAL SPHERICAL VESSELS(I)

YIN YIHUI [†], LIU XINFU

Institute of Structural Mechanics, China Academy of Engineering Physics Mianyang, Sichuan China

SUN YIN

Committee of Science and Technology, China Academy of Engineering Physics, Mianyang, Sichuan 621900, China

TAN YUN, MEI JUN, LIU YUANDONG AND YU SHAORONG

Institute of Structural Mechanics, China Academy of Engineering Physics Mianyang, Sichuan China

By using several traditional empirical formulas, the bursting pressures of a kind of special spherical vessels were calculated, a comparison between calculated and tested results was performed, and the features of errors distribution were analyzed. Based on the features of errors distribution, referring to the methods constructing traditional empirical formulas and taking the constituent characteristics of the original formulas into consideration, the traditional empirical formulas were amended and reconstructed. Moreover the amended and reconstructed formulas were validated by experimental data. The validation results indicate that the accuracy of the amended formulas is higher than that of the traditional empirical formulas and the accuracy of the reconstructed formulas is further higher than that of the amended formulas, which implies that the amended and reconstructed formulas are practically useful in design of the discussed special spherical pressure vessels, the related amendment and reconstruction methods are reasonable and can be used in researches of analogous problems. In this paper the amendment case was presented. The reconstruction case will be presented in a subsequent paper[1].

Keywords: Spherical pressure vessel; bursting pressure; empirical formula; test.

[†] Corresponding author Yin Yihui, email: yinyh@caep.ac.cn

1. Introduction

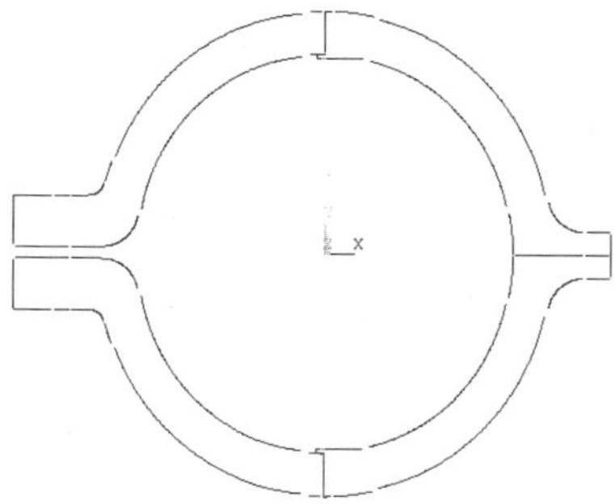

Fig. 1. Special spherical pressure vessel.

During 50 to 70 decades of last century, Faupel studied calculating formulas of bursting pressure of spherical vessels subjected to inner pressure, meanwhile carried out many experiments, and presented empirical formulas of bursting pressure named after his name which have been widely employed in engineering. However, some literatures claim that Faupel's formulas have some shortages. The first is that the errors are still about ±15%, the second is that its calculated results tend to safety when the ratio of yield to strength limits $\eta(=\sigma_s/\sigma_b)$ is small, but its results tend to be danger when η is big[2]. Therefore, in consideration of special spherical pressure vessels, whether Faupel's formulas and other related empirical formulas can be directly applied or not has to be studied. Therefor, the bursting pressure calculation problem for the special spherical pressure vessels as shown in Fig. 1 is studied in this paper. This vessel consists of two half spherical shells welded together along the equator, the top of the right half shell is closed and there is a head with a hole at the top of the left half one. According to different size and material, totally 57 vessels belong to 12 classifications are studied, their basic parameters and mean values of measured bursting pressure are shown in table 1 in which each mean value is obtained through the measured bursting pressure of the same classification vessels.

Table 1. Basic parameters and measured mean bursting pressure of special spherical pressure vessels.

No.	1	2	3	4	5	6	7	8	9	10	11	12
Amount of vessels	4	6	4	3	3	3	3	6	5	6	4	10
Ratio of outer to inner radius K	1.139	1.148	1.154	1.257	1.109	1.114	1.114	1.146	1.159	1.171	1.222	1.199
Yield strength/MPa	258.3	258.3	258.3	258.3	221.9	221.9	376.8	376.8	376.8	376.8	376.8	376.8
Ultimate strength/MPa	574.2	574.2	574.2	574.2	660.0	660.0	723.5	723.5	723.5	723.5	723.5	723.5
Mean value of measured P_b /MPa	112.4	128.7	125.1	220.5	112.8	115.0	143.3	181.0	180.9	201.3	256.5	233.9

Firstly, bursting pressures of the spherical vessels were calculated by using traditional empirical formulas, the calculated and measured data were compared and the error characteristics were analyzed. Then, recurring to the method to establish traditional empirical formulas and using the constituent's nature of traditional empirical formulas, amended several empirical formulas and validated the amended formulas by using experimentally measured data. The validation indicates that the precision of the amended formulas is higher than that of related original formulas.

2. Several Traditional Empirical Formulas and their Errors

Several traditional empirical formulas and their errorsSeveral traditional empirical formulas calculating bursting pressure of spherical vessels subjected to inner pressure and their errors used to calculating the special spherical vessels in table 1 are shown in table 2, where σ_b is material limit strength, K is the ratio of outer radius to inner one, called radii ratio, σ_s material yield strength, $\eta = \sigma_s / \sigma_b$, called ratio of yield to strength limits, n is material hardening index. The following can be seen from table 2:

The errors of the calculated results from Faupel's formulas are negative with a range of -7.6%~-32.3%, i.e. the results tends to be on the small side; and those from Mises' formulas are positive with a range of 9.2%~32.5%, i.e. the calculated results tends to be on the big side. The cases from the formulas in American ASME-III are similar with those from Mises' formulas. The results errors from Vanlterson's formula, Svensson's formula and the Formula based on relation of true stress-strain are all tend to be on the middle and a little bigger than zero, the maximum among them is 14.7%.

Table 2. Traditional formulas and error range of calculated bursting pressure for special spherical vessel.

Name	Expression	Error range	Reference
Faupel's formula	$P_{b_f} = 2\sigma_s(2-\eta)\ln K$	-7.6%~-32.3%	Ref. [3, 4]
Mises' formula	$P_{b_m} = 2\sigma_b \ln K$	9.2%~32.5%	Ref. [3, 4]
Vanlterson's formula	$p_{b_v} = \sqrt{3}\sigma_b \ln K$	-5.4~14.7%	Ref. [3]
Svensson's formula	$p_{b_s} = 2(2/3)^n\left(1-K^{-1}\right)\sigma_b$	-9.6%~10.1%	Ref. [5]
Formula based on relation of true stress-strain	$P_{b_t} = \sigma_b\left(\dfrac{0.433}{n+0.227}\right)\left(\dfrac{2}{\sqrt{3}}\dfrac{e}{n}\right)^n \ln K$	-9.2%~10.2%	Ref. [6]
Formula recommended in ASME-III	$p_{b_{A\text{-}III}} = \dfrac{4}{3}\dfrac{K^3-1}{K^3+1}\sigma_b$	7.7%~30.8%	Ref. [7]

3. Amendment to Traditional Empirical Formulas

3.1. *Amendment by narrowing errors range*

At the beginning, Faupel assumed that bursting pressure of a vessel was between the fully yielding pressure which was $2\sigma_b \ln K$ and at which the wall material fully yielded and the limit strength pressure which was $2\sigma_b \ln K$, and it is controlled by the ratio η of yield stress to limit stress of the material, therefore he took the bursting pressure as $p_b = \eta(2\sigma_b \ln K) + (1-\eta)(2\sigma_s \ln K)$. By simplifying it, he constructed Faupel's formula which is common used in engineering field later on. Here by using the method Faupel constructed Faupel's formula then, and as the errors features of Faupel's formula and Mises' formula in table 2 is concerned, recombine the two formulas and apply further the controlling action of η, the 1st amended formula is constructed as

$$P_{b_1} = \eta P_{b_m} + (1-\eta)P_{b_f} = 2\sigma_s[1+(1-\eta)(2-\eta)]\ln K \tag{1}$$

Note that the following approximate equations satisfy under the condition of K being not very big

$$\ln K \approx 2\frac{K-1}{K+1} \approx \frac{K^2-1}{K^2+1} \approx \frac{2}{3}\frac{K^3-1}{K^3+1}, \quad \text{and } \ln K > 2\frac{K-1}{K+1} > \frac{K^2-1}{K^2+1} > \frac{2}{3}\frac{K^3-1}{K^3+1}$$

Therefore the amended formula (1) can also be expressed as

$$P_{b_1} = 4\sigma_s[1+(1-\eta)(2-\eta)]\frac{K-1}{K+1} \tag{2}$$

3.2. *Amendment by choosing mean value of the calculating values from Faupel's formula and Mises' one*

The data in table 2 show that the experimentally measured values of bursting pressure of the special pressure vessels in table 1 are just at the middles of the calculated values of Faupel's formula and Mises' formula. Based on this, if choosing the middles of the calculated values of Faupel's formula and Mises' formula as bursting pressure of vessels, the 2nd amended formula is obtained as

$$P_{b_2} = \sigma_b[2-(1-\eta)^2]\ln K \tag{3}$$

3.3. *Amendment by reassembling and permuting some continents in existing formulas*

Exchanging the positions of Faupel's formula and Mises' formula in (1), one can get

$$P_b^* = \eta P_{b_f} + (1-\eta)P_{b_m} = 2\sigma_b[(2-\eta)(1+\eta^2)-1]\ln K \tag{4}$$

Again choosing the middle value of the calculated values of (3) and (4) as the bursting pressure of a vessel and using the approximate equation about *K*, the 3rd amended formula is obtained as

$$P_{b_3} = \sigma_b\left(1+\eta^2-\frac{2}{3}\eta^3\right)\frac{K^3-1}{K^3+1} \tag{5}$$

In addition, choosing 1/4 of Faupel's formula, 1/4 of Mises' formula and 1/2 of Vanlterson's formula to sum, similarly using the approximate equation about *K*, the 4th amended formula is obtained as

$$P_{b_4} \approx \sigma_b\left[2-(1-\eta)^2+\sqrt{3}\right]\frac{K-1}{K+1} \approx \sigma_b\left[1+\frac{1}{2}(\sqrt{3}-(1-\eta)^2)\right]\frac{K^2-1}{K^2+1} \tag{6}$$

3.4. *Amendment to the empirical formulas including the terms* $\sigma_b[(K^3-1)/(K^3+1)]$ *or* $\sigma_b(1-K^{-1})$

Through similar method with the above, an empirical formula including $\sigma_b[(K^3-1)/(K^3+1)]$ can be obtained as

$$p_{b_5} = \sigma_b\left(\frac{158}{25}\eta^2-\frac{287}{60}\eta+\frac{239}{120}\right)\frac{K^3-1}{K^3+1} \tag{7}$$

And empirical formula including $\sigma_b(1-K^{-1})$ as

$$p_{b_6} = \sigma_b\left(\frac{114}{13}\eta^2-\frac{333}{50}\eta+\frac{83}{28}\right)\left(1-\frac{1}{K}\right) \tag{8}$$

4. Validation to Amended Empirical Formulas by Practical Examples

The bursting pressures of the special spherical vessels in table 1 were calculated by using traditional empirical formulas such as Faupel's one etc and the amended empirical formulas in section 3, due to space, here only the calculation results of part formulas with related experimentally measured ones were typically shown in table 3. It can be seen from table 3 that compared with experimentally measured data, as Related error mean, Excellent ratio, Good ratio and Standard error are concerned, the amended formulas have higher precision, and as the errors range is concerned, totally speaking, the errors span of the amended formulas is narrower than that of the original ones.

It is worth to mention that due to bursting position being at weldings, the value of radii ratio *K* of the special spherical in table 1 is chosen as the welding depth at the joint ends, not as the wall thickness of the even zone. In table 3, "Excellent ratio %" and "Good ratio %" mean the percentage ratio of the vessels classifications that the absolute value of related error is not bigger than 3% and 5% to total classifications, respectively.

Table 3. Calculated bursting pressures from different formulas with their errors.

No.	Faupel's formula		Mises' formula		Formula based on real stress-strain relation		Formula (5)	
	Pressure /MPa	Error /%	Pressure /MPa	Error /%	Pressure /MPa	Error /%	Pressure /MPa	Error /%
1	103.9	-7.6	149.0	32.5	123.9	10.2	126.0	12.0
2	110.2	-14.3	158.0	22.8	131.5	2.1	133.4	3.7
3	114.6	-8.4	164.3	31.4	136.7	9.3	138.6	10.8
4	183.1	-17.0	262.5	19.1	218.4	-1.0	216.4	-1.9
5	76.3	-32.2	136.5	21.0	113.5	0.6	110.4	-2.1
6	79.7	-30.7	142.5	23.9	118.5	3.1	115.3	0.2
7	120.6	-15.8	156.6	9.3	130.3	-9.1	137.0	-4.4
8	152.2	-15.9	197.6	9.2	164.4	-9.2	172.1	-4.9
9	164.2	-9.2	213.2	17.8	177.3	-2.0	185.2	2.4
10	175.7	-12.7	228.1	13.3	189.7	-5.8	197.7	-1.8
11	223.7	-12.8	290.4	13.2	241.5	-5.8	248.9	-3.0
12	202.2	-13.5	262.5	12.2	218.4	-6.6	230.2	-1.6
Related error /%	-7.6~-32.3		9.2~32.5		-9.2~10.2		-4.9~12.0	
Mean error /%	-15.8		18.8		-1.2		0.8	
Excellent ratio /%	0		0		33.3		58.3	
Good ratio /%	0		0		41.7		83.3	
Standard error σ	27.2		30.3		10.6		7.3	

5. Conclusion

Based on analysis of shortages of existing empirical formulas in calculating bursting pressure of special spherical pressure vessels and characteristic of calculation errors, several traditional empirical formulas were been amended and related new formulas were presented. By using available experimental data, validations to the amended formulas were carried out. The validation results indicate that the accuracy of the amended formulas is much increased, their maximum error decreased to 12.0% from original 32.5%. The amended formulas have simple expressions and are good at calculating bursting pressure of the special spherical pressure vessels discussed in the present paper. Further more, the related amendment methods are reasonable and can be used in researches of analogous problems.

Acknowledgments

The paper is Supported by science and technology development fundation of China academy of engineering physics(No. 2008A0301010).

References

1. Yin Yihui, Liu Xinfu, Sun Yin, etal. Amendment and reconstruction of empirical formulas calculating bursting pressure of special spherical vessels(II), Proceedings of the 10th asia-pacific conference on enginnering plasticity and its applications, Wuhan, China, 2010.
2. Zhu Xuezheng, Chen Guoli. Calculation of explosive pressure for high pressure vessel(in Chinese). Petro-chemical Equipment Technology, 1995, 16 (1): 23-26
3. Shao Guohua, Wei Zhaocan, et al. Super-highpressure vessels(in Chinese), Beijing: Chemical Engineering Press of China, 2002
4. Liu Xinmin, Wei Riyan. Analysis of special structures(in Chinese). Beijing: National defense press of China, 1995
5. Zheng Jingyang, Kuang Jiyong, et al. Study of bursting pressure of multilayered super-highpressure vessels(in Chinese), Chemical Engineering & Machiner, 1994, 21(5): 271-277
6. Edition Committee. Mechanical structure strength, the 19th section in mechanical design (1), in the 4th volume of mechanical engineering handbook(in Chinese), Beijing: Mechanical Engineering Press of China, 1980
7. Codes for boiler and pressure vessels of ASME, the 3rd section, 1971

AMENDMENT AND RECONSTRUCTION OF EMPIRICAL FORMULAS CALCULATING BURSTING PRESSURE OF SPECIAL SPHERICAL VESSELS(II)

YIN YIHUI †, LIU XINFU

Institute of Structural Mechanics, China Academy of Engineering Physics Mianyang, Sichuan China

SUN YIN

Committee of Science and Technology, China Academy of Engineering Physics, Mianyang, Sichuan 621900, China

TAN YUN, MEI JUN, LIU YUANDONG AND YU SHAORONG

Institute of Structural Mechanics, China Academy of Engineering Physics Mianyang, Sichuan China

By using several traditional empirical formulas, the bursting pressures of a kind of special spherical vessels were calculated, a comparison between calculated and tested results was performed, and the features of errors distribution were analyzed. Based on the features of errors distribution, referring to the methods constructing traditional empirical formulas and taking the constituent characteristics of the original formulas into consideration, the traditional empirical formulas were amended and reconstructed. Moreover the amended and reconstructed formulas were validated by experimental data. The validation results indicate that the accuracy of the amended formulas is higher than that of the traditional empirical formulas and the accuracy of the reconstructed formulas is further higher than that of the amended formulas, which implies that the amended and reconstructed formulas are practically useful in design of the discussed special spherical pressure vessels, the related amendment and reconstruction methods are reasonable and can be used in researches of analogous problems. In a previous paper[1] the amendment case was presented, in this paper the reconstruction case was presented.

Keywords: Spherical pressure vessel; bursting pressure; empirical formula; test.

1. Introduction

In reference [1], the errors of several traditional empirical formulas in calculating bursting pressure of the special spherical pressure vessels in Fig. 1 and table 1 which were also expounded there were analyzed and some amended

† Corresponding author Yin Yihui, email: yinyh@caep.ac.cn

empirical formulas were put forward. In this paper, some reconstructed empirical formulas were presented and validated through comparing calculated data by using reconstructed formulas with experimentally measured data. The validation indicates that the precision of the reconstructed formulas is further higher than that of related amended formulas.

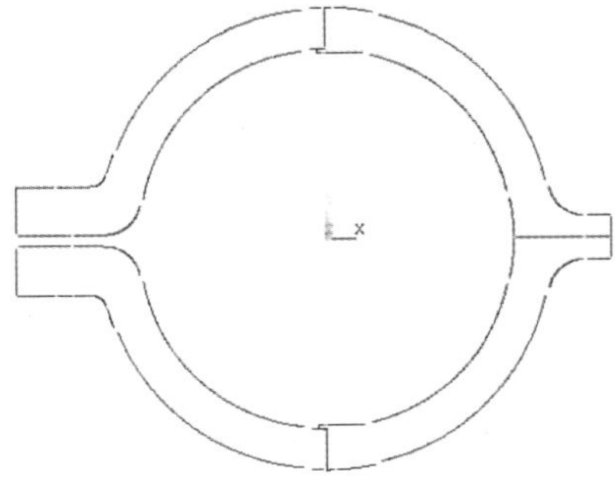

Fig. 1. Special spherical pressure vessel.

Table 1. Basic parameters and measured mean bursting pressure of special spherical pressure vessels.

No.	1	2	3	4	5	6	7	8	9	10	11	12
Amount of vessels	4	6	4	3	3	3	3	6	5	6	4	10
Ratio of outer to inner radius K	1.139	1.148	1.154	1.257	1.109	1.114	1.114	1.146	1.159	1.171	1.222	1.199
Yield strength/MPa	258.3	258.3	258.3	258.3	221.9	221.9	376.8	376.8	376.8	376.8	376.8	376.8
Ultimate strength/MPa	574.2	574.2	574.2	574.2	660.0	660.0	723.5	723.5	723.5	723.5	723.5	723.5
Mean value of measured P_b /MPa	112.4	128.7	125.1	220.5	112.8	115.0	143.3	181.0	180.9	201.3	256.5	233.9

2. Reconstruction of New Empirical Formulas

2.1. *Constitutive form of empirical formulas calculating bursting pressure of spherical vessels and method to reconstructing new empirical formulas*

After checking various empirical formulas of bursting pressure of spherical vessels, it is found that they can be expressed as the following

$$p_b = \psi(\sigma_s, \sigma_b, n) f(K) \tag{1}$$

where the meanings of σ_s, σ_b, n and K are the same as the previous, $\psi(\sigma_s, \sigma_b, n)$ can be named as material properties function of vessels, $f(K)$ can be named as geometrical properties function. If material hardening effect is neglected, then the bursting pressure can be expressed as

$$p_b = \sigma_b \varphi(\eta) f(K) \tag{2}$$

Here as an exploration and example, the reconstruction of (2) is discussed only. Obviously, the key of reconstruction of empirical formulas is to find out concrete expressions of functions $\varphi(\eta)$ and $f(K)$.

Firstly, rewrite Faupel's formula as $p_{b_f} = 2\sigma_b\eta(2-\eta)\ln K$, it can be seen that in this formula $\varphi(\eta) = 2\eta(2-\eta)$ is a quadric multinomial about η without constant term, $f(K) = \ln K$ is a logarithm function about K. By similar analysis to every amended formulas, it is known that in these formulas, the function $\varphi(\eta)$ is either a cubic multinomial about η without constant term or linear term, or a full quadric multinomial; The function $f(K)$ is either a logarithm function about K, or an improper fraction about combination of K^3. According to these characteristics, generally speaking, in formula (10), the function $\varphi(\eta)$ can be supposed as quadric or cubic multinomial about η, $f(K)$ as improper fraction about K or combination of K^3. Here as an example, the case is discussed that $\varphi(\eta)$ is quadric function about η, $f(K)$ is a certain combination of components in existing formulas, i.e.

$$p_b = \sigma_b(\alpha\eta^2 + \beta\eta + \gamma)f(K) \tag{3}$$

where α, β, γ are coefficients to be determined, they can be determined with the help of experimentally measured data. In the following, as an example, the method and process to reconstruct an empirical formula is presented.

Choosing the middle value of the calculation values of the amended formulas $p_{b_5} = \sigma_b[(158/25)\eta^2 - (287/60)\eta + (239/120)](K^3-1)/(K^3+1)$ and $p_{b_6} = \sigma_b[(114/13)\eta^2 - (333/50)\eta + (83/28)](1-K^{-1})$ in Ref. [1] as new bursting pressure, a new formula is obtained

$$p_b = \sigma_b\left[\left(3.16\eta^2 - 2.39\eta + 1\right)\left(\frac{K^2+K-1}{K^3+1}\right) + 2.5\left(3.02\eta^2 - 2.29\eta + 1\right)\right]\left(1 - \frac{1}{K}\right) \tag{4}$$

where the coefficients of the same order terms in two quadric trinomials about η are closed in value, in order to simplify formula constitute and make application convenient, choosing the middle value of the coefficients of the same order terms as the coefficient of the same order term in the new quadric trinomial, simplifying (4) yields

$$p_{b_l} = \sigma_b\left(3.09\eta^2 - 2.34\eta + 1\right)\left(\frac{K^2+K-1}{K^3+1} + 2.5\right)\left(1 - \frac{1}{K}\right) \tag{5}$$

Generalizing (5) yields the following reconstructed formula with a general form

$$p_{b_l} = \sigma_b\left(\alpha\eta^2 + \beta\eta + 1\right)\left(\frac{K^2+K-1}{K^3+1} + 2.5\right)\left(1 - \frac{1}{K}\right) \tag{6}$$

where the coefficients α, β can be determined by using the least square method to fit experimental data. Its process is as follows:

Letting that $\varphi_i(\eta) = \alpha\eta_i^2 + \beta\eta_i + 1$ and $f_1(K) = [(K^2 + K - 1)/(K^3 + 1) + 2.5](1 - K^{-1})$, then according to (14), one can get that $\varphi_i = \varphi(\eta_i) = p_{b_i} / \sigma_{b_i} f_I(K_i)$, $(i = 1, 2, ..., 12)$. Solving this formula by using experimentally measured data p_{b_i} $(i = 1, 2, ..., 12)$, one can solve out φ_i. Let its error be $R_i = \alpha\eta_i^2 + \beta\eta_i + 1 - \varphi_i$, $(i = 1, 2, ..., 12)$, then its square sum is

$$\sum_{i=1}^{12} R_i^2 = \sum_{i=1}^{12} \left(\alpha\eta_i^2 + \beta\eta_i + 1 - \varphi_i\right)^2 = \Phi(\alpha, \beta) \tag{7}$$

According to the least square theorem, coefficients α, β should make Φ be minimum, therefore let

$$\frac{\partial\Phi}{\partial\alpha} = 2\sum_{i=1}^{12} \left(\alpha\eta_i^2 + \beta\eta_i + 1 - \varphi_i\right) \cdot \eta_i^2 = 0,$$

$$\frac{\partial\Phi}{\partial\beta} = 2\sum_{i=1}^{12} \left(\alpha\eta_i^2 + \beta\eta_i + 1 - \varphi_i\right) \cdot \eta_i = 0 \tag{8}$$

From (8), the positive equations to coefficients α, β are reached as

$$\alpha\sum_{i=1}^{12}\eta_i^4 + \beta\sum_{i=1}^{12}\eta_i^3 = \sum_{i=1}^{12}(\varphi_i - 1)\eta_i^2, \quad \alpha\sum_{i=1}^{12}\eta_i^3 + \beta\sum_{i=1}^{12}\eta_i^2 = \sum_{i=1}^{12}(\varphi_i - 1)\eta_i \tag{9}$$

Substituting φ_i and related η_i determined with the help of experimentally measured data into (9) yields the following

$$\begin{cases} 1.04081681\alpha + 2.151592052\beta = -1.818103146 \\ 2.151592052\alpha + 4.512135332\beta = -3.910478383 \end{cases} \tag{10}$$

By solving out α and β from (10) and substituting them into (6), a reconstructed formula is produced as

$$p_{b_l} = \sigma_b \left(\frac{135}{43}\eta^2 - \frac{26}{11}\eta + 1\right)\left(\frac{K^2 + K - 1}{K^3 + 1} + 2.5\right)\left(1 - \frac{1}{K}\right) \tag{11}$$

2.2. *Other reconstructed formulas*

By similar method to construction of empirical formula (11), some empirical formulas were constructed, among them are the following:

$$P_{b_2} = \sigma_b \left(\frac{34}{11}\eta^2 - \frac{138}{59}\eta + 1\right)\left(\frac{K^2 + K - 1}{K^3 + 1} + 2.5\right)\left(1 - \frac{1}{K}\right) \tag{12}$$

$$P_{b_5} = \sigma_b \left(6\eta^2 - \frac{227}{50}\eta + 2\right)\left(\frac{1}{K^3} - \frac{5}{K^3 + 1} + \frac{6}{K^3 + 2}\right)\left(K^3 - 1\right) \tag{13}$$

$$P_{b_7} = \sigma_b \left(\frac{35}{51}\eta^2 - \frac{221}{420}\eta + \frac{34}{155} \right)\left(\frac{1}{K^3} + \frac{6}{K^3+1} + \frac{2}{K^3+2} \right)\left(K^3 - 1\right) \quad (14)$$

$$P_{b_{10}} = \sigma_b \left(\frac{141}{77}\eta^2 - \frac{25}{18}\eta + \frac{38}{63} \right)\left(\frac{2}{K^3} - \frac{8}{K^3+1} + \frac{11}{K^3+2} \right)\left(K^3 - 1\right) \quad (15)$$

3. Validation to Reconstructed Empirical Formulas by Practical Examples

Table 2. Calculated bursting pressures from reconstructed formulas with their errors.

No.	Formula (12)		Formula (13)		Formula (14)		Formula (15)	
	Pressure /MPa	Error /%	Pressure /MPa	Error /%	Pressure /MPa	Error /%	Pressure /MPa	Error /%
1	123.3	9.7	123.0	9.4	122.0	8.6	122.8	9.3
2	130.5	1.4	130.5	1.4	129.1	0.4	130.1	1.1
3	135.4	8.3	135.8	8.5	134.0	7.2	135.2	8.1
4	209.6	-5.0	221.7	0.5	210.4	-4.6	214.8	-2.6
5	111.9	-0.8	111.1	-1.6	111.4	-1.3	111.4	-1.3
6	116.7	1.5	115.9	0.8	116.1	1.0	116.2	1.0
7	141.2	-1.4	139.6	-2.6	139.8	-2.5	140.0	-2.3
8	176.6	-2.4	175.9	-2.8	174.7	-3.5	175.6	-3.0
9	189.8	4.9	189.8	4.9	187.8	3.8	189.1	4.5
10	202.3	0.5	203.3	1.0	200.2	-0.5	202.0	0.4
11	253.5	-1.2	261.5	2.0	250.8	-2.2	256.4	-0.1
12	237.8	1.7	235.0	0.5	235.3	0.6	235.9	0.8
Related error%	-5.0~9.7		-2.8~9.4		-4.6~8.6		-3.0~9.3	
Mean error%	1.4		1.8		0.6		1.3	
Excellent ratio%	66.7		75.0		58.3		75.0	
Good ratio%	83.3		83.3		83.3		83.3	
Standard error σ	6.3		5.7		5.9		5.5	

The reconstructed empirical formulas (12)~(15) were applied to calculating bursting pressure of the vessels in table 1, the calculated results as well as their comparison with related experimentally measures data are shown in table 2. Here, the meanings of "Excellent ratio %" and "Good ratio %" are the same as those in Ref. [1] and the comparison is cross. The method is to divide the test vessels into different groups, use the measured data of a certain group to determine the parameters in reconstructed formulas and further determine the formulas and calculate out bursting pressure by using the formula, and then use the measured data of the other groups to validate the calculated data. The data comparison indicates that, compared with traditional and amended formulas in Ref. [1], the calculation precision of the reconstructed formulas increases further, especially, the errors range becomes narrower and the maximum error is within 10%.

4. Conclusion

Based on analysis of shortages of existing empirical formulas in calculating bursting pressure of special spherical pressure vessels and characteristic of calculation errors, several traditional empirical formulas were reconstructed and related new formulas were presented. By using available experimental data, validations to the reconstructed formulas were carried out. The validation results indicate that compared with the amended formulas in Ref. [1], the accuracy of the reconstructed formulas is further increased, their relative errors range decreased to -3.0%~9.3% from -4.9%~12.0% of amended formulas and their excellent ratio get to 75.0%. The reconstructed formulas have simple expressions and are good at calculating bursting pressure of the special spherical pressure vessels discussed in the present paper. Further more, the related reconstruction methods are reasonable and can be used in researches of analogous problems.

Acknowledgments

The paper is Supported by science and technology development fundation of China academy of engineering physics(No. 2008A0301010).

References

[1] Yin Yihui, Liu Xinfu, Sun Yin, etal. Amendment and reconstruction of empirical formulas calculating bursting pressure of special spherical vessels(I), Proceedings of the 10th asia-pacific conference on enginnering plasticity and its applications, Wuhan, China, 2010

[2] Zhu Xuezheng, Chen Guoli. Calculation of explosive pressure for high pressure vessel(in Chinese). Petro-chemical Equipment Technology, 1995, 16 (1): 23-26

[3] Shao Guohua, Wei Zhaocan, et al. Super-highpressure vessels(in Chinese), Beijing: Chemical Engineering Press of China, 2002

[4] Liu Xinmin, Wei Riyan. Analysis of special structures(in Chinese). Beijing: National defense press of China, 1995

[5] Zheng Jingyang, Kuang Jiyong, et al. Study of bursting pressure of multilayered super-highpressure vessels(in Chinese), Chemical Engineering & Machiner, 1994, 21(5): 271-277

[6] Edition Committee. Mechanical structure strength, the 19th section in mechanical design (1), in the 4th volume of mechanical engineering handbook(in Chinese), Beijing: Mechanical Engineering Press of China, 1980

[7] Codes for boiler and pressure vessels of ASME, the 3rd section, 1971

DEDUCTION OF STRESS AND ELASTIC STRAIN DISTRIBUTION FOR TORSIONAL CIRCULAR SHAFT WITH MINIMUM ENERGY PRINCIPLE

LI ZHIHUI †

Department of Engineering Mechanics, Xi'an University of Technology, 58 Yanxiang Road, Xi'an, Shaanxi, 710054, China

SHI JUNPING

Department of Engineering Mechanics, Xi'an University of Technology, 58 Yangxiang Road, Xi'an, Shaanxi, 710054, China

TANG ANMIN

Department of Engineering Mechanics, Xi'an University of Technology, 58 Yangxiang Road, Xi'an, Shaanxi, 710054, China

The microscopic mechanisms corresponding to elastic and plastic deformation were analysed respectively. According to minimum deformation energy principle, stress and elastic strain distributions for torsional circular shaft in the condition of small plastic deformation were given. The result shows that, the appearance of small plastic deformation does not change the elastic property of material and its elastic strain distribution. That is, in the condition of small plastic deformation, shear stress and shear elastic strain at any point of cross-section for torsional circular shaft are proportional to the distance between the point and axes, and the distributions are the same to which in the condition of only elastic deformation.

Keywords: Elastic-plastic; Minimum energy principle; Torsion; Stress field.

1. Introductions

The elastic-plastic solutions and full plastic solutions from plastic mechanics cannot be confirmed by the torsion tests of metallic materials.[1] Deformation occurred in the member when loaded, which is an objective natural phenomenon; therefore, the deformation distribution must follow certain natural rules. When the deformation regularity is described in stress and strain, it is required that not only the stress solutions corresponding to true deformation meet static equilibrium equations,[1] but also the strain solutions meet deformation

† Corresponding author, email: lizhihui@xaut.edu.cn

compatibility equations. When described in energy principle, it is required that the true deformation meets minimum deformation energy principle. That is, in all the possible deformations, the true deformation should make the deformation energy reserved in member be the minimum. This is the natural rule which the deformation distribution must follow. In the case of elastic-plastic torsional circular shaft, taking into account the assumption that plastic deformation is small and the fact that the deformation energy reserved in circular shaft follows minimum energy principle, it is concluded that the original distribution regularity of elastic deformation doesn't change. To sum up, the small plastic deformation has no influence on the elastic property of materials, and it doesn't change the distribution regularities of elastic deformation in member.

2. The Minimum Deformation Energy Principle

In elastic body, the work done by external forces moving through possible displacement equals to the work done by corresponding stress moving through possible strain, [2] which can be expressed as Eq. (1).

$$\int_v f_i \delta u_i dv + \int_s p_i \delta u_i ds = \int_v \sigma_{ij} \delta \varepsilon_{ij} dv \tag{1}$$

This is the elastic deformation principle of virtual displacement, Where, f_i and p_i refer to body force and surface force respectively, σ_{ij} is stress tensor considering that δu_i and $\delta \varepsilon_{ij}$ are respectively virtual elastic displacement and strain. The stress-strain relation $\sigma_{ij} = \partial A(\varepsilon_{ij}) / \partial \varepsilon_{ij}$ was introduced, and then Eq. (1) can be expressed as:

$$\begin{gathered} \delta\left(\int_v f_i u_i dv\right) + \delta\left(\int_s p_i u_i ds\right) = \delta\left(\int_v A\left(\varepsilon_{ij}\right) dv\right) \\ \delta\pi = 0 \quad \pi = \int_v A\left(\varepsilon_{ij}\right) dv - \int_v f_i u_i dv - \int_s p_i u_i ds \end{gathered} \tag{2}$$

Formula (2) is the principle of elastic deformation minimum potential energy. Where, $A(\varepsilon_{ij})$ refers to the elastic strain energy density. To metal materials, because of the one-to-one correspondence linear relationship between the elastic deformation and external force, the deformation energy U equals to one half of the total product of every force and its corresponding displacement. Therefore, formula (2) can also be expressed as:

$$\pi = -U = -\int_v A(\varepsilon_{ij}) dv \qquad \delta U = 0 \tag{3}$$

Formula (3) is the principle of minimum deformation energy, and the physical significance of which is that, among all the possible deformations meeting deformation compatibility condition and displacement boundary condition, the deformation energy of true deformation equals the minimum. As it were, among all the possible stress meeting equilibrium condition and force boundary condition, the deformation energy of true stress equals the minimum.

3. Stress and Strain Distribution of Torsion of Circular Shaft

A circular shaft is loaded torque moment in the elastic stage (shown in figure 1). In the elastic body: $\tau = G\gamma, A = \tau\gamma/2$

The total deformation energy U equals to:

$$U = l\int_0^R \frac{1}{2}\tau\gamma \cdot 2\pi\rho d\rho$$
$$= \pi l G\int_0^R \gamma^2 \rho d\rho = \frac{\pi l}{G}\int_0^R \tau^2 \rho d\rho \tag{4}$$

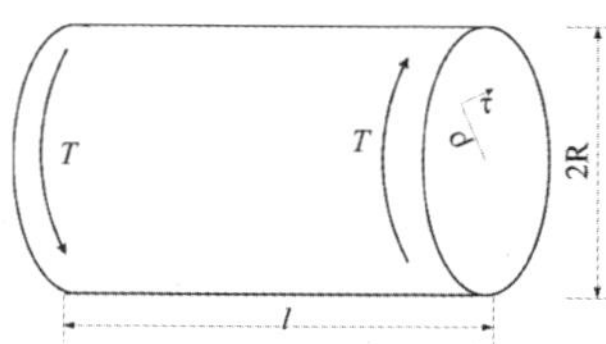

Fig. 1. Free torsion of circular shaft.

Where, G is shear modulus, γ is shear strain. By means of Schwartz's inequality,

$$\left|\int f(x)g(x)dx\right|^2 \le \int (f(x))^2 dx \int (g(x))^2 dx \tag{5}$$

Where the equality holds up if and only if $f(x)/g(x)$ is a constant. Making substitution $x = \rho, f(x) = \rho^{1/2}\gamma, g(x) = \rho^{3/2}$ to formula (5), the formula (5) can be expressed as follows:

$$\left(\int_0^R \rho^{\frac{1}{2}}\gamma\rho^{\frac{3}{2}} d\rho\right)^2 \le \int_0^R \rho\tau^2 d\rho \int_0^R \rho^3 d\rho \quad U \ge \frac{4\pi l G}{R^4}\left(\int_0^R \rho^2 \gamma d\rho\right)^2 \tag{6}$$

The condition of the minimum deformation energy is that, $f(x)/g(x) = \gamma/\rho = \text{constant}$. (7)

Another substitution $x = \rho$, $f(x) = \rho^{1/2}\tau$, $g(x) = \rho^{3/2}$ is made to formula (5), then, formula (5) can also be expressed as follows:

$$U \ge \frac{4\pi l}{GR^4}\left(\int_0^R \rho^2 \tau d\rho\right)^2$$

The condition of the minimum deformation energy is $\tau/\rho = \text{constant}$. (8)

The conclusions of formula (7) and (8) indicate that, in the elastic stage, the shear stress and strain values at any point on the cross-section are in direct proportion to the distance between the point and the axes, which is determined by the principle of minimum energy.

4. The Elastic Strain Distribution in Small Plastic Deformation

In the condition of small plastic deformation, both elastic and plastic deformations occur in the member. The elastic deformation is formed by the change of the relative position of atoms in crystal cells, therefore, the elastic deformation is proportionate to the corresponding force, and the relation between them complies with generalized Hooke's law. The deformation energy stored in materials corresponding to elastic deformation is a kind of potential energy. The elastic strain and stress value are popularly used to measure the change of relative position of atoms in crystal cells. The plastic deformation is formed by intracrystal dislocation motion, and because the plastic deformation bears no more loads, [3] there are no certain correspondence between the plastic deformation and the force. The deformation energy corresponding to plastic deformation is a kind of plastic dissipative energy, which is dissipated (converted into heat) in the process of deformation and not stored in materials. In essence, elastic and plastic deformations are two completely different physical processes. The former corresponds to conservative system and the latter corresponds to nonconservative system.

After plastic deformation, the position and quantity of intracrystal dislocation have been changed largely; this makes metal materials hardening and changes the mechanical property. But the researchers found that the occurrence of plastic deformation changes the plastic behaviours of materials, [4] but not elastic behaviours. That is to say, in the condition of elastic plastic deformation, the relationship between elastic strain and stress still conforms to generalized Hooke's law.

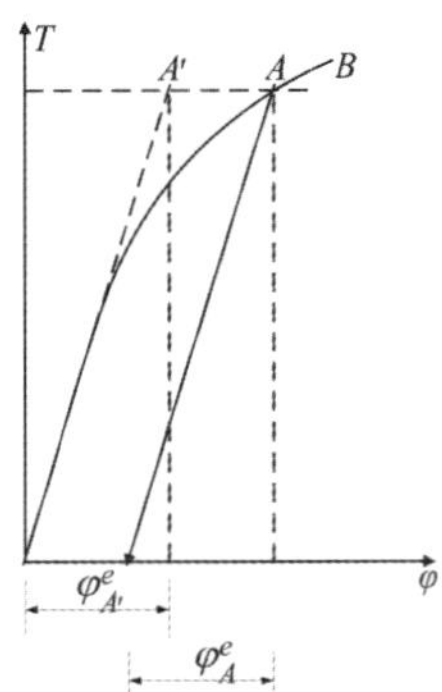

Fig. 2. The torque-twist curve of reinforced materials in torsion test.

When the circular shaft is loaded by torque moment to point A in the stage of elastic plastic deformation, the deformation energy in the member can be calculated (shown in figure 2). At point A, the plastic deformation has occurred, but the corresponding plastic deformation energy has been dissipated in the process of deformation. For this reason, only elastic deformation energy is stored in the member. At point A, the elastic deformation still needs to satisfy deformation compatibility and displacement boundary conditions, and the stress still needs to satisfy equilibrium and force boundary conditions and these two are equivalent to minimum deformation energy stored in the member. According to the derivation mentioned above, the following equations could be obtained:

$$U \geq \frac{4\pi lG}{R^4}\left(\int_0^R \rho^2 \gamma d\rho\right)^2, U \geq \frac{4\pi l}{GR^4}\left(\int_0^R \rho^2 \tau d\rho\right)^2$$

The deformation energy equals the minimum value on the condition that: $\gamma_e/\rho = $ constant and $\tau/\rho = $ constant. Namely, in the condition of small plastic deformation, the distribution law of which is the same as in the elastic stage.

Test results show that, if the specimen is unloaded from the point of the elastic plastic deformation stage, the unload-deformation curve is linear (shown in figure 2). The unloading process is the recovery of elastic strain and deformation energy. For instance, the released energy during the unloading process from point A is $U_A = T_A \varphi_A^e / 2$, which equals to the deformation energy U at point A' corresponding to linear elasticity. So the unloading test results show that, the linear distribution of elastic strain in the small plastic stage is perfectly in agreement with the experiment results.

5. Conclusion

Elastic deformation and plastic deformation are two physical phenomena with different natures, and the occurrence of plastic deformation can't change the property of elastic deformation. Therefore, for a circular shaft loaded torque moment in the elastic and small plastic stage, the shear stress and elastic shear strain value at any point on the cross-section are still in direct proportion to the distance between the point and axes.

Acknowledgments

This project is supported by the National Natural Science Foundation of China (No. 10972180) and Shaanxi Natural Science Foundation (No. SJ08A18).

References

1. Chen Mingxiang: Elastic-Plasic Mechanics [M]. Science Press, 2007, Beijing (in Chinese).
2. Jiu jin-jiu-yi-lang: Variational Method in Elastic and Plastic Mechanics [M]. Science Press, 1984, Beijing (in Chinese).
3. Tang Anmin, Xu Mingda, Zhao lei: New method for elasto-plastic analysis of free-form deformation structure [J]. Journal of Guangxi University, 2009,34(1): 36-39(in Chinese).
4. Chen Jianqiao: Strength of Engineering Materials [M]. HuaZhong University of Science and Technology Press, 2008, Wuhan(in Chinese).

PLASTICITY-DRIVEN DAMAGE DETECTION USING LAMB WAVES FOR STIFFENED PLATE

SHEN YANFENG

Department of Engineering Mechanics, Wuhan University, Wuhan 430072, China

QU WENZHONG † XIAO LI

Department of Engineering Mechanics, Wuhan University, Wuhan 430072, China

The process of implementing a damage detection strategy for aerospace, civil and mechanical engineering infrastructure is referred to as structural health monitoring (SHM). Aerospace structural components such as fuel tanks, wings, etc. are often comprised of substructures that consist of plates with integral stiffeners. The use of guided waves for stiffened plate becomes increasingly difficult as the geometric complexity of the structure increases. In this paper, a transient dynamic finite element simulation of Lamb waves with piezoelectric transducers for damage detection in integrally stiffened plate structures is carried out. Transient dynamic finite element simulations using coupled field elements on the commercial finite element code ANSYS 9.0 platform were used to model both the structure and transducers. The plasticity damage zone is produced by reducing the material stiffness. A time reversal imaging method is adopted to detect and locate the damages in the plate. The results of the numerical simulation demonstrate that the approach can detect the stiffened plate structural damage.

Keywords: Integrally stiffened plate; Damage detection; Finite element analysis; Piezoelectric wafer active sensor; Lamb wave; Time reversal theory.

1. Introduction

Stiffened plates are extensively used in most industries such as the aeronautic and automotive industry, because of their light weight and desirable stiffness in the designed direction. However, cyclic work loads and unpredictable collisions may cause possible plastic zone and other damages which severely decrease the structure's performance. In order to respond to any possible damage leading to failure of the structures, damages should be detected, evaluated, and if possible

† Corresponding author, email: qwz0803@yahoo.com.cn

monitored, even though the structures are in services [1]. A promising damage detection method is the active system based on the propagation of Lamb waves [2]. In recent years, a significant amount of research on the modeling and analysis has been carried out to further understand the properties of Lamb waves excited by piezoelectric transducers [3].

The Lamb-wave time reversal method is a new and tempting baseline-free damage detection technique for SHM. With this method, certain types of damage could be detected immediately without prior baseline data [4]. Time reversibility of Lamb waves and its tuning effect were further studied by Victor Giurgiutiu and his colleagues [5]. Recently, imaging process based on time reversal theory was investigated to detect damages in composite plates by Yuan Shenfang and her colleagues [6].

2. Time Reversal Imaging Method

Time reversal process is achieved by Time Reversal Mirrors (TRM). Lamb waves monitored by the sensors can be mainly separated into two parts in time domain: waves directly come from the source, and those from scatters or boundaries. A typical signal sensed by an active sensor is shown in Fig. 1.

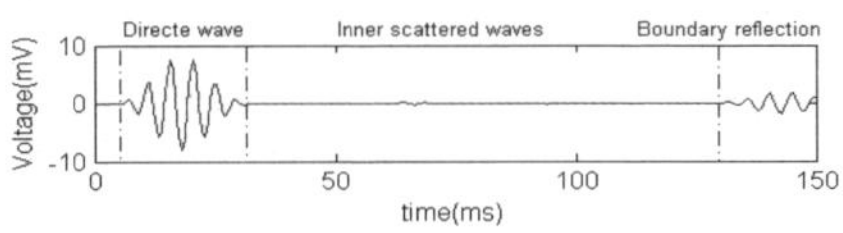

Fig. 1. Typical signal sensed by sensor.

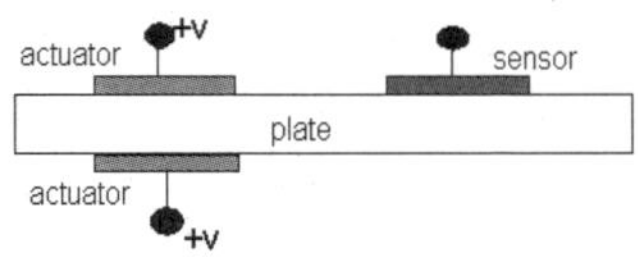

Fig. 2. Actuator/sensor function group.

As the signal coming directly from the original source and those from the scatters often mix up with each other, a sensor is put in the vicinity of an actuator to overcome this problem. The actuator/sensor pair array arrangement in this study is presented in figure 2. This imaging method also requires a single mode inspection wave to achieve good result. Since pure S0 or A0 mode excitation is hard to be achieved by a single actuator, in this particular discussion the double-sided installation was chosen to permit symmetric and antisymmetric excitation. Thus, axial and flexural waves could be selectively generated. In this paper, pure S0 Lamb wave mode is excited by applying in-phase excitation signals onto both actuators.

According to time reversal theory, after the reversed signals are sent back, they will retrace the propagation path and self-focus on the source point and the energy will also converge on the source location. In this damage imaging method, the scatters in the structure, including damage, stiffener and boundaries, are considered as the second sources during Lamb wave propagation. So after being re-emitted, the reversed signal will self-focus on the scatters. The amplitude of waves at the focusing location is much higher than other points where signals randomly overlap. The damage locating and imaging process could be achieved by creating a digital image with the time reversal algorithm below. Consider the distributed PZT array bonded in the monitored area consisting of N pairs of active actuators/sensors. The contrast S at a particular location or imaging pixel (i,j) can be expressed as:

$$s(i,j)=\sum_{m}^{N} A_m f_w\left(t_{mij}\right) f_m\left(t_{mij}\right), \quad m=1,2,\ldots,N$$

$$t_{mij}=\frac{R_m^a+R_m^s}{c_g}, \quad A_m=\frac{10}{\max\limits_{m}\left|f_m\right|} \tag{1}$$

where $f_m(t)$ denotes the received scattered signal by sensor m due to the excitation of actuator m. t_{mij} is the propagation time of the Lamb wave from the actuator to the imaging pixel (i,j) and then to the sensor. Parameters R_m^a and R_m^s stand for the distance from the imaging pixel $S(i,j)$ to senor m, and the one from the imaging pixel $S(i,j)$ to actuator m respectively. c_g represents the group velocity of the propagating wave. A_m refers to the appropriate aperture weight to each function pair, which compensate the performance difference among the various actuator/sensors pairs. Since the excitation is a five-cycle Hanning window modulated sine signal, the imaging result is speckled. An envelope curve method is adopted to process the received signals, which will present the energy distribution character of these signals in time domain.

3. Damage Detection and Localization in Stiffened Plate

In this paper, a transient dynamic finite element simulation of Lamb wave with piezoelectric transducers for damage detection in an integrally stiffened plate is carried out. The stiffened plate is shaped at 1000mm*1000mm*4mm and a stiffener shaped at 1000mm*40mm*1mm is vertically integrated in the middle of the plate. The plate was instrumented with a surface bonded array of eighteen $8\times8\times0.2mm^3$ piezoelectric transducers.

Numerical simulations of the wave propagation process were performed using the commercially available finite element code ANSYS. The stiffened plate specimen was modeled using two-dimensional four-node structural shell element (SHELL63) and the three-dimensional coupled element (SOLID5) was used for modeling the piezoelectric sensors. Voltages DOFs were coupled for the nodes of PZT sensors at top and bottom surface to simulate the electrodes. The plate is free-free. The discretization process of the stiffened plate specimen used 62500 elements. The finite element model of stiffened plate, the location of damage zone and the PZT active sensors are illustrated in Fig. 3 (a) and (b).

The Lamb waves are excited by applying voltage to the top nodes of the actuator. The applied excitation is in the form of a wave packet (5-cycle sine modulated by the Hanning window) with 100 v p-p. The main frequency of this packet is 200 KHz, as shown in Fig. 3 (c).

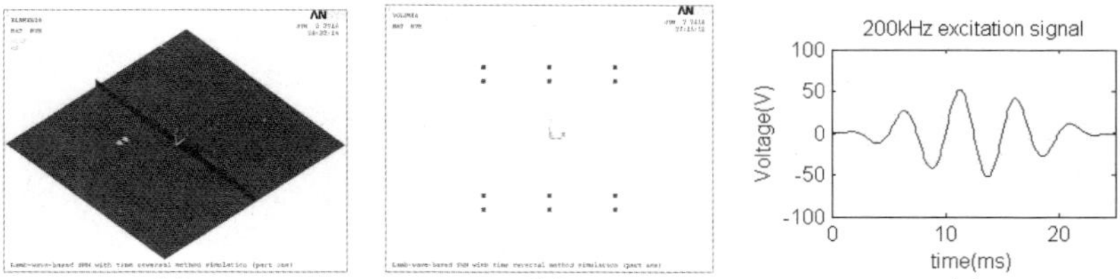

(a) (b) (c)

Fig. 3. Finite element model, actuator/sensor pairs array and the excitation signal.

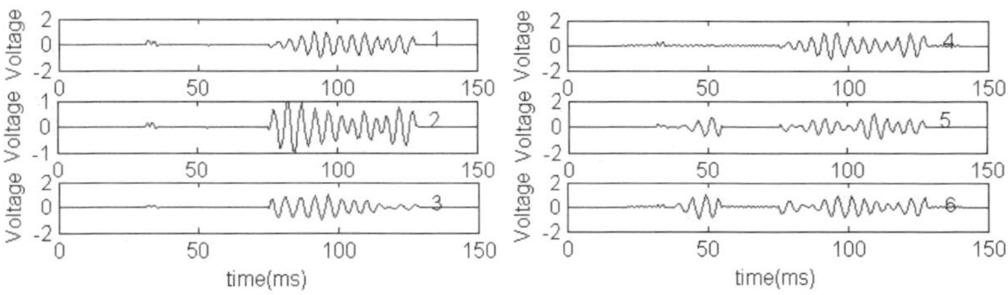

Fig. 4. Normalized signals of actuator/sensor after time window function.

Examples of the corresponding time domain signals of six active PZT sensors are shown in Fig.4. Fig.5 shows the enveloped inner scattered signals after spline interpolation and normalization.

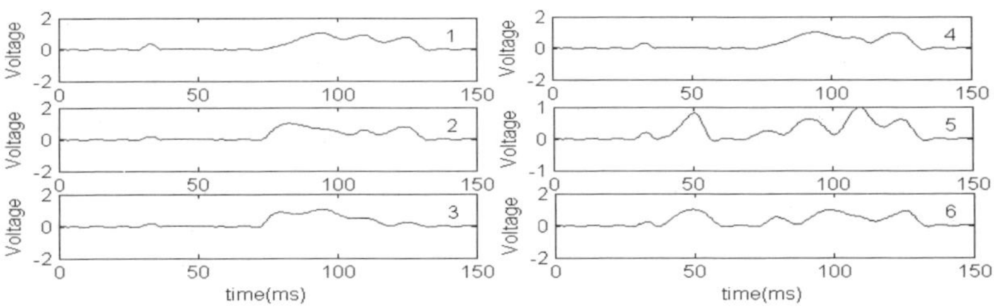

Fig. 5. Envelope curve of the signals.

In the simulation, a damage zone displayed in red in the finite element model is created in the upper right area of the stiffened plate, and the stiffener is in the middle across it. To achieve good imaging result, stiffener scattered waves are removed from the signals. The finite element model and imaging result are shown in Fig.6 (a). A high pixel value area appears in the digital image, which represents the converging and focusing phenomenon of time reversal procedure. Its location agrees well with the damage zone created in the FEM. Without losing generality, a third simulation is carried out when there exist two damages at the same time. The two damage zones displayed in red are created in the upper right and lower left area of the stiffened plate. The FEM and imaging result are shown in Fig.6 (b). The locations of the high pixel value areas accord with the created damage zones. In these two simulations, stiffener scattered waves are all removed, so the stiffener is not revealed.

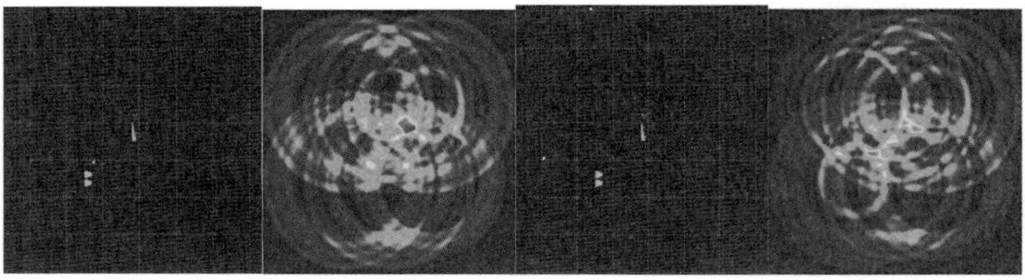

(a) (b)

Fig. 6. Locations of the damage zones and the imaging results.

4. Concluding Remarks

In this research, we developed a full multiphysics finite element simulation using coupled field elements for the analysis of Lamb-wave-based damage detection of stiffened plate structure using piezoelectric transducers. The time reversal imaging technique of Lamb waves is used to locate and reveal damage in the stiffened plate. This method of this study can be easily extended to practical geometries such as stiffened or riveted plate-like structures.

Acknowledgments

This study is funded by China Science Grant (Grant number 51078293).

References

1. Giurgiutiu V, 2000. Active Sensors for Health Monitoring of Aging Aerospace Structures. In: SPIE's 7th International Symposium on Smart Structures and Materials and 5th International Symposium on Nondestructive Evaluation and Health Monitoring of Aging Infrastructure, 5-9 March, Newport Beach, CA, Paper # 3985-103.
2. S. S. Kessler. Piezoelectric-Based In-Situ Damage Detection of Composite Materials for Structural Health Monitoring Systems. PhD thesis, Massachusetts Institute of Technology, 2002.
3. Wang X, Lu Y and Tang J 2007 Damage detection using piezoelectric transducers and the Lamb wave approach: I system analysis Smart Mater. Struct. 17: 025-033.
4. Victor Giurgiutiu, 2008. Structural health monitoring with piezoelectric wafer active sensors. ELSEVIER. 2008.
5. Xu, B. and Giurgiutiu, V. 2007. Single Mode Tuning Effects on Lamb Wave Time Reversal with Piezoelectric Wafer Active Sensors for Structural Health Monitoring. Journal of Nondestructive Evaluation, 26:123-134.
6. Wang Qiang and Yuan Shenfang 2009. Baseline free imaging method based on new PZT sensor arrangement. Journal of intelligent material systems and structures.20:16-63.

STAMPING-FORGING PROCESS OF AN ALUMINUM ALLOY PAN

LUO JIAN-CHENG, WANG XIN-YUN†, GUO MEI-LING, XIA JU-CHEN, LUO YUN-HUA

State Key Laboratory of Material Processing and Die & Mould Technology, Huazhong University of Science and Technology, 1037 Luoyu Road, Wuhan, 430074, China

Stamping-forging process of sheet metal is a newly-developed metal composite processing technology. The drawing process and the forging process are organically combined, which is applied in the overall plastic forming of complex shell parts with large thickness difference. In the paper, the stamping-forging process of an aluminum alloy pan was studied. The pan was formed by one-step stamping-forging process, which can substitute the original two-step forming process.

Keywords: Stamping-forging process; Overall plastic forming; Aluminum alloy; Sheet metal.

1. Introduction

Metal plastic forming is a universal method in parts manufacturing process. As products' properties were required more and more strictly, shapes of parts evolve toward integration and complication, correspondingly, some new technology appeared. Early ^{90th} of last century, Takano Takashi proposed Flow Control of blank sheet Forming (FCF for short) for the first time. C. J. Tan proposed a method to forming parts by two-step stamping-forging process and drawing process: first, the local thickening area was formed by two-step stamping-forging process; second, the parts were formed by drawing process, and the influence of punch shape and clearance between punch and die on blank piece thickening was studied.[1] H. H. Lin introduced metal flow control process in consecutively positive and negative upsetting forming.[2] H-C. Wu studied the flanging and reshaping processing of a clutch hub and proposed an optimized technique to prevent local crack.[3] Ching-Lun Li studied the punch load-time curves, the thickness variation and forming limit in hole flanging process.[4]

For shell structure parts with large surface area and large-thickness-difference, the traditional methods are casting, cutting after hot /warm forging or

† Corresponding author, email: wangxy_hust@163.net

welding with thick and thin parts, as a result, these parts usually have a poor mechanical performance. Furthermore, there are too many working procedures, high energy consumption, low efficiency and high machining cost in these processes. Therefore, the author proposed sheet metal stamping-forging process in which a car fly panel part was formed by warm drawing and forging process;[5] a double-cup shaped part with thickness-difference between inner and outer side wall was formed by strengthen drawing and cold forging process.[6]

In the paper, the stamping-forging process of an aluminum alloy pan was studied (showed in Figure 1) with the help of FEM software. In order to cut down the processing steps and improve the production efficiency, an one-step stamping-forging process is proposed to substitute the original two-step method.

2. Part Size and FEM Model

The aluminum alloy pan is a rotational part. The upper bed diameter is Φ258.8mm, the end lap diameter is Φ216mm, and the thickness of the end lap is 4.5mm, while the thickness varies gradually from 2.6mm to 4.5mm along the sidewall. Two round corners approximate rectangular corner exist at the edge of the upper bed and end lap.

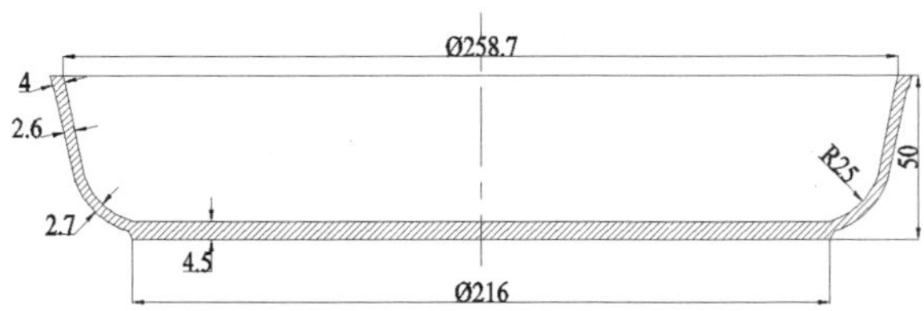

Fig. 1. Part size.

From the part size, we can reversely calculate that the diameter of the plate slab is Φ300mm, and the thickness of the slab is 4.5mm. The side wall is formed by ironing, and round corners at the edge of the upper bed and end lap are formed by forging process. The original process is: first, drawing the side wall by ironing process; second, turning the flange; third, forming the round corners at the edge of the upper bed and end lap by forging process; the last, forming the part by debarring process. In the whole process, two steps of plastic forming process and two steps of machining process are used. In the paper, an one-step method by stamping-forging process is proposed.

Due to the axisymmetric shape, the forming process is simulated by two-dimensional model, four node quadrilateral elements are generated. Considering large plastic strain in the process, mesh adaptation is activated.

The material properties of the slab is showed in table1, and the initial thickness of the slab is 4.5mm, the diameter is Φ300mm, the time of the punch motion is 1s.

Table 1. Material properties.

ρ (g /cm^3)	σ_s (MPa)	σ_b (MPa)	E (GPa)	n	λ
2.7	130	280	70	0.226	0.3

3. Results and Analysis

The final shape of the pan is showed in Figure 2, which is the 1900 step of the whole simulation process. From the result, we can see that the pan is formed well, especially the round corners at the edge of the upper bed and end lap are formed successfully.

Fig. 2. The shape of the part (1900 step).

In order to clearly express the forming process of the pan, six intermediate processes are showed in Figure 3. From the results, we can see that the plastic strain mainly occurs in the side wall, and the end lap deforms little. The whole process can be described by three stages. The first stage, before the 800 step, the side wall is formed by drawing process. The second stage, between the 800 step and 1600 step, the side wall is gradually thinned by drawing and extruding process on large contact area between the surface of punch and die mould. The third stage, after the 1600 step, the bottom surface of the punch and die mould all get into contact status, and the round corners at the edge of the upper bed and end lap are formed successfully.

Before the 1800 step, the strain is mainly caused by drawing process, so the punch load is not very large. The load rises observably after the 1850 step. The load-time curve between the 1800 step and 1900 step is showed in Figure 4. From the curve, we can see that the punch load is less than 5000 KN, while the punch load at the final step is 44700KN.

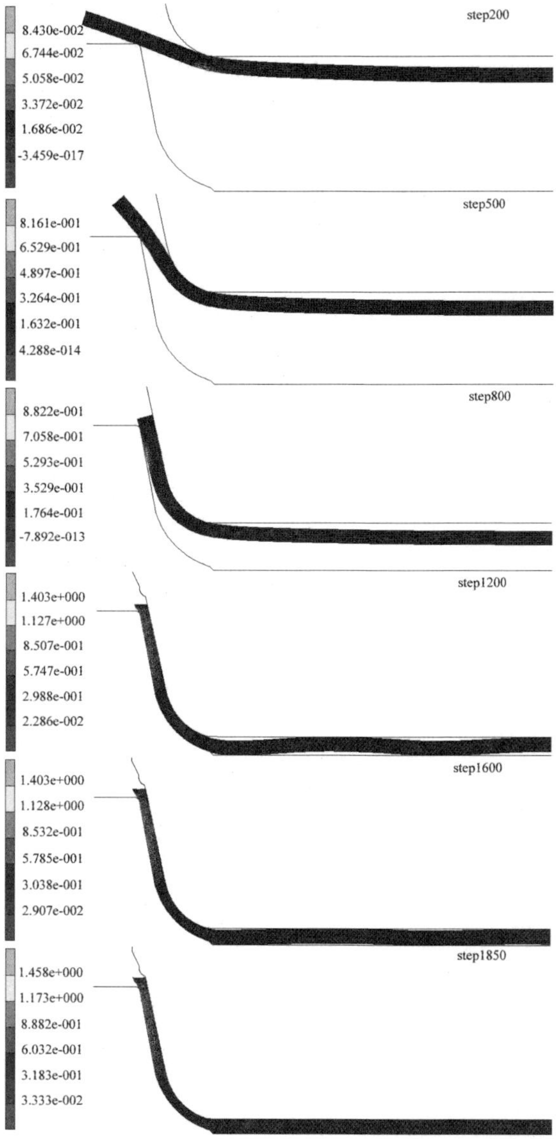

Fig. 3. Intermediate processes.

The reason is that the bottom surface of the punch, die mould and the work piece get into large area contact status after the 1850 step, the work piece is mainly formed by forging process, especially the side wall is ulteriorly thinned by pressing and the metal flows from the side wall to the round corner. The punch load will observably rise.

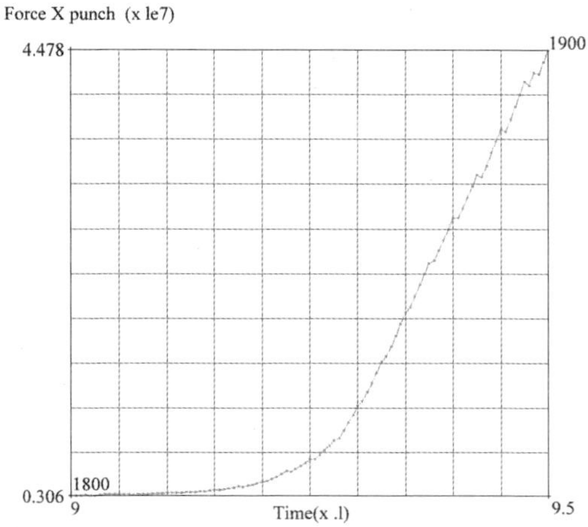

Fig. 4. Punch load-time from 1800~1900 step.

4. Conclusion

In the paper, the stamping-forging process is applied to the forming of an aluminum alloy pan. A one-step method is proposed to substitute the original two-step method, which is proved by FEM simulation. The new method can reduce the production cost, increase production efficiency and improve the quality of products.

Acknowledgments

This work was supported by NSFC (50705034), and State Key Laboratory of Materials Processing and Die & Mould Technology.

References

1. C. J. Tan, K. Mori and Y.Abe: *Journal of Materials Processing Tech.*, 2008, **202**: 443-449.
2. H. H. Lin: *CIRP Annal-Manufacturing Tech.*, Volume**37**, issue1, 1988, page231-234.
3. H-C.Wu and T. Altan: *Jou rnal of Materials Processing Tech.* **146**(2004) 8-19.
4. Ching-Lun Li: *Journal of Materials Processing Tech.* **201** (2008) 256-260.
5. Xin-Yun. Wang: *Journal of Plasticity Engineering*, 2008.**15**:4.
6. Xin-Yun. Wang: *Forging & Stamping Technology*, 2008.**34**:4.

3D SIMULATION OF ELECTROMAGNETIC FORMING FOR TUBE BULGING BASED ON FEM

XIAOHUI CUI†, JIANHUA MO

State Key Laboratory of Material Processing and Die & Mould Technology, Huazhong University of Science and Technology, Wuhan 430073, China

A loose coupling method and a 3D model are used to simulate the electromagnetic tube bulging process. ANSYS/EMAG is used to obtain the nodes force, then, the nodes force is used as the loading condition to model the high-rate deformation of the tube with ANSYS/LS-DYNA. The two models are updated at 17.5μs intervals. The maximum axial force is not in the location 40mm away from the center, but 37mm.Because some edge effects in real geometries could not be taken into consideration due to the nature of 2D simulations. Radial force is much larger than the axial force, so the tube is deformed by radial force. The simulation result of displacement at the center of tube with the time and the final tube shape in this paper is very close to the experiment.

Keywords: Electromagnetic tube bulging process; Loose coupling method; 3D simulation model.

1. Introduction

Aluminum alloy has lower density than steel alloy. So it is a candidate material to be used in lightweight automotive in order to increase fuel efficiency. However, high spring-back is intrinsic to aluminum because it has lower elastic stiffness than steel.[1] EMF is one of the high velocity forming processes which can promote significant increases in strain to failure in low-ductility materials due to high strain rate and inertial effect.[2] Therefore, electromagnetic forming is suitable for high electrical conductivity and hard forming materials, such as aluminum alloy.

The EMF is a complicated task to numerically simulation, which is because the finite element model contains electromagnetic, structure, thermal. Thermal effect has little effect on the result of EMF. So the EMF study mainly consists of electromagnetic field and dynamic deformation analyses.[3] To simulate the EMF, three approaches are developed: No coupling approach, sequential-coupled approach and loose coupling approach.[2, 4-6]

† Corresponding author, email: suiyuan841125@163.com

No coupling approach is the simplest method because the Lorentz force is calculated without considering the deformation of the work-piece during the process. However, if the work-piece has a large deformation in the initial conditions, the method will get a large calculation error.[7] Sequential-coupled approach is calculated by ANSYS-MULTIPHYSICS code. It will get accurate results because of the deformation of the work-piece which can affect the body force calculated during EMF process and is easy to transfer date. However, this method is difficulty in obtaining a converging solution and high computational cost for solving equations. Loose-coupling approach combines the advantages of the two methods above. The finite element code ANSYS/EMAG for electromagnetic field analysis is used to obtain the nodes force that is imparted to the work-piece. The elastic-plastic element code LS-DYNA is used to predicate the deformation process of the work-piece with the nodes force passed by the interface code form the EM code to the structural model. The two models are updated at specific intervals.[2]

Nowadays, 2D finite element models are all used to predict the deformation of axisymmetric work-pieces.[4~7] The calculated error become lager due to ignoring the end effect in 2D finite element models.[8] Therefore, 3D finite element and loose-coupling approach are used to obtain an accurate result in this paper.

2. Model

The geometry and the mesh of the tube expansion system are presented in figure. 1a and 1b.[9] The surrounding air is meshed in order to propagate the magnetic field generated by the coil.

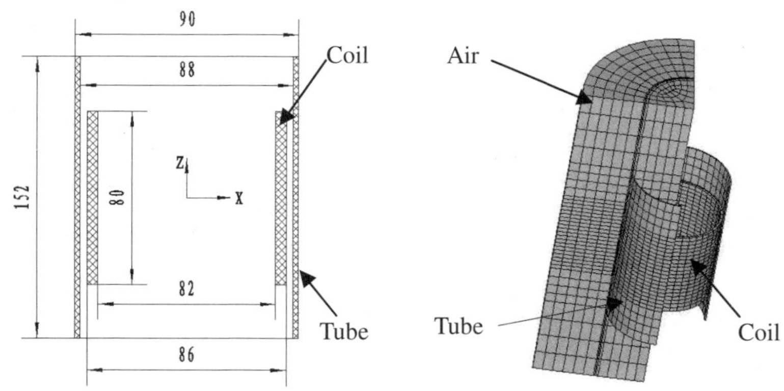

(a) Model for electromagnetic tube bulging (b) 3D mesh

Fig. 1. The model of the tube expansion system.

The material properties of the tube expansion system are given in table 1. The primate current flow through the coil is presented in figure 2.

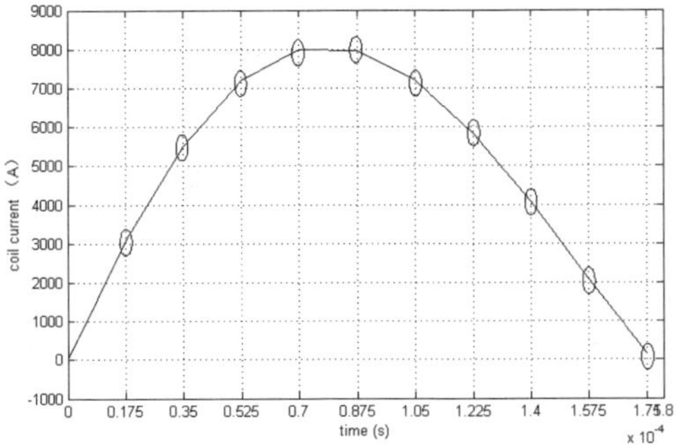

Fig. 2. Discharge current used in analysis.

The current rises sharply to about 8KA and then decays exponentially. During this ramp-up of current, the tube is driven away from the coil, resulting in a decrease in the result of the node forces. So only the first half wave of the current is considered.[3] The code of ANSYS/EMAG is used to obtain the nodes force, and then the codes of LS-DYNA calculates the deformation of the tube and update the EM models. The two models are updated at 17.5μs intervals.

The strain rate effect is neglected in the analysis.[3] Flow stress of the tube material is approximated by

$$\delta = 162 \cdot (\varepsilon + 0.000663)^{0.411} \tag{1}$$

Table 1. Material properties of the tube expansion system.

Air	Relative permeability	1
Coil	Resistivity	5.85e7 Ω m
Tube	Resistivity	2.7e-8 Ω m
	Elastic modulus	71.5Gpa
	Poisson ratio	0.3

3. Result and Discussion

Loose-coupling approach considers the deformation of the work-piece which can affect the body force calculated by EM code. In the post-processing, the result show in the cylindrical coordinate. Figure 3 presents the radial force and axial force of tube outside surface at different time values. Radial force mainly in the center to 40mm away from the center has the same length as tube, and drop sharply from 40mm to 60mm and -60mm to -40mm. Axial force mainly distributed in 15mm to 45mm and -45mm to -15mm. The maximum axial force located between 37mm and -37mm instead of 40mm and -40mm.This is because ignoring the end effect in 2D finite element models. Radial force remains invariant to the axial location from -30mm to 30mm at time from 17.5μs to 70μs and form -15mm to 15mm to the axial location at time form 87.5μs to 122.5μs.

Changes in distribution of radial force happened due to the deformation of tube with the increase of discharge time. Radial force is much larger than the axial force, so the tube is deformed by radial force.

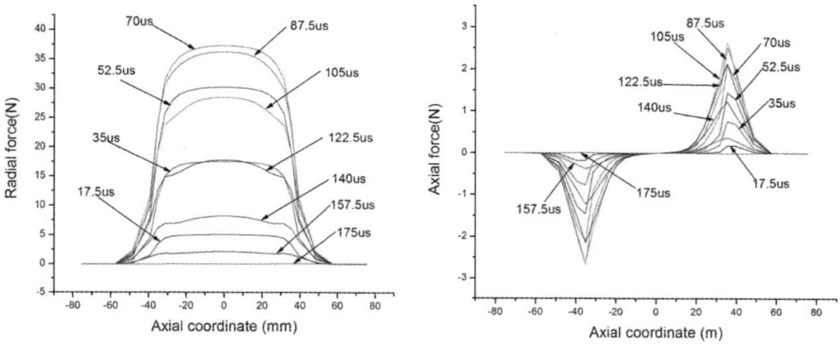

(a) Radial force (b) Axial force

Fig. 3. Lorentz force of tube outside surface at different time values.

Figure 4 compares the simulation results for radial displacement of tube outside surface and the experiment value with the calculated one by Suzuki et al. The present result is very close to the experiment, which indicates that the loose-coupling method is accurate to predict EMF process.

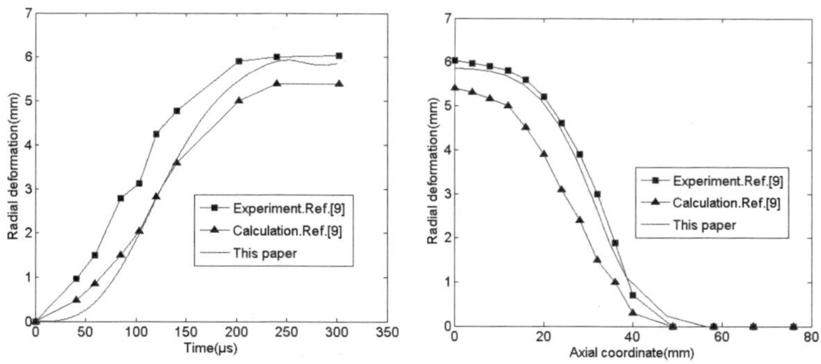

(a) Radial displacement with time (b) Final tube profile

Fig. 4. Comparison of the simulated and experiment results.

4. Conclusion

1. A 3D finite model which considered the end effect in real geometry is used to obtain the nodes force. The distribution of radial force has the same length as tube. The maximum axial forces are not in the end of tube which is different with the result calculated by 2D finite element. Radial force is much larger than the axial force, so the tube is deformed by radial force.

2. Loose coupling method is used in this paper to predict the deformation of the tube. The calculated results are very close to experiment values. So it is feasible to solve the tube bulging with loose coupling method. The code of ANSYS/EMAG is used to obtain the nodes force and then the code of LS-DYNA calculates the deformation of the tube. The two models are updated at 17.5μs intervals.

Acknowledgments

This work was supported by the National Natural Science Foundation of China (No. 50875093).

References

1. G. K. Fenton, G. S. Daehn: *J. Mater. Process. Tech.*, 1998, **75**, 6-16.
2. D. A. Oliveira, M. J. Worswick, M. Finn, D. Newman: *J. Mater. Process. Tech.*, 2005, **170**, 350-362.
3. S. H. Lee, D. N. Lee: *J. Eng. Mater.*, 1994, **116**: 250-254.
4. L. F. Wang, Z. Y. Chen, C. X. Li, S. Y. Huang: *Int. J. Adv. Manuf. Tech.*, 2006, **30**, 395-400.
5. Y. U. Haiping, L. I. Chunfeng: *J. Mater. Process. Tech.*, 2009, **209**, 1053-1059.
6. Y. U. Haiping, L. I. Chunfeng, D. E. N. G. Jianghua: *J. Mater. Process. Tech.*, 2009, **209**, 707-713.
7. Y. U. Haiping, L. I. Chunfeng: Chinese Journal of Mechanical Engineering, 2006, **42**, 231-234. (In Chinese)
8. M. A. Bahmani, K. Niayesh, A. Karimi: *J. Mater. Process. Tech.*, 2009, **209**, 2295-2301.
9. H. Suzuki, H. Negishi, Y. Yokochi: Journal of the Japan Society for Technology of Plasticity, 1986, **27**, 1254-1260.

ANALYSIS OF LIGHT-WEIGHT STRUCTURE FOR VEHICLE USING METAL AND CARBON FIBER

JUNG HO KIM

Dept. of Advanced Parts and Materials Engineering Graduate School, Chosun University, 375 Seosuk-dong, Dong-gu, Gwangju 501-759, Korea

JUNG MIN BANG

Dept. of Advanced Parts and Materials Engineering Graduate School, Chosun University, 375 Seosuk-dong, Dong-gu, Gwangju 501-759, Korea

JI HOON KIM[†]

Dept. of Mechanical Design Engineering, Chosun University, 375 Seosuk-dong, Dong-gu, Gwang-ju, 501-750, Korea

It is developing and making the vehicle, wind power and railroad train structures for light-weight because of exhaust gas emission, fuel economy and energy efficiency. Light-weight materials are normally alloy aluminum, various ceramics and fiber composites. But ceramics and fiber composites have Double Edged word. That is, they have merit and demerit according to mechanical method. This study tested and analyzed the hybrid specimens that compound the carbon fiber composite and the light-weight metal for synergy. It had better strength and stiffness using hybrid specimens than each metal and carbon fiber composite.

Keywords: Vehicle; light-weight; fiber composite; mechanical method; strength; stiffness.

1. Introduction

The integration and advancement of functionalities in automobile and airplane, researches on the structural member of light-weightiness and stability have been actively conducted. Among the possible candidate for aluminum alloy is mainly used. Recent restriction on hazardous gas emission and needs for improved gas mileage, researches and applications of fiber reinforced composite material, which has the numerous benefits over aluminum alloy in weight, stiffness, and strength, have been actively explored.

[†] Corresponding author, email: kjh@chosun.ac.kr

Fiber reinforced composite material can design the optimized performances of structural member to achieve appropriate mechanical properties with cross-sectional shape, fiber direction, stacking sequence, and thickness.

There are extensive database for optimal design of CFRP structural member for compression through preparation of different shape, interface number, thickness, and stacking angle. When pressure is applied to structural member, the corresponding member will show compression, bending, and torsion. For the effective utilization of fiber reinforced composite material as main structural member, optimized design technology should be established to maximize mechanical properties for compression, bending and torsion.

In this paper, CFRP (carbon fiber reinforced plastics) prepreg sheet with different stacking angle is prepared in circular and square shape. According to flexure test result, the corresponding strength and stiffness is compared and analyzed to provide optimal design guideline. In addition, hat-shaped member, which is used for front shock absorption in vehicle, is prepared to compare the corresponding strength and stiffness.

2. Experimental Method

To provide optimal design with high strength and stiffness according to change in cross-section and stacking angle, square and circular tube is prepared via autoclave forming method, and 3 point Flexure test is conducted to compare the corresponding strength and stiffness.

2.1. *Test specimen*

CFRP (carbon fiber reinforced plastics) is stacked in square (30mm of edge) and circular shape (38mm of diameter). Changes in stacking angle are 0°, 15°, 45°, 90°, 0°/90° and 90°/0°. Stacking sequence is 7 interfaces and 8 ply, which is the greatest performance from the previous study on axial compression test. Thickness is 1mm. stacking specimens were made at ±0°, ±15°, ±45°and ±90° for coupling effect, 0°/90° and 90°/0°. Test specimen is expressed as [A°/B°]4. Test specimen is manufactured on square and circle mandrel as shown in Figure 1. CFRP prepreg sheet is HANKUK carbon, CU125NS and Vacuum tape is covered for autoclave forming.

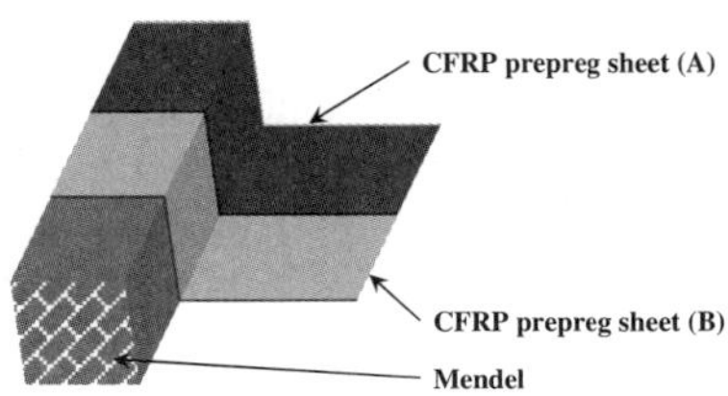

Fig. 1. Schematic view of fabrication.

2.2. *Flexure tester*

Flexure test method can be divided into 3 and 4 point flexure test. 3 point flexure test features easy set-up. Before starting experiment, conventional test set-up for 3 point flexure with round tube will induce stress concentration from point contact. Therefore, separate test equipment for round tube is prepared.

Conventional flexure tester is consisted of supports in 30 mm radius and loading nose in 10 mm radius. Loading nose device and support is installed at load cell and actuator in universal test machine (UTM), respectively. Test specimen is placed on support, and loading nose is moved in 10 mm/min of speed as gravitational direction.

For round tube flexure test set-up, supports with 30 mm of lateral radius and 30 mm of vertical radius is prepared to show constant line load. In addition, loading nose with 10 mm of lateral radius and 10 mm of vertical radius is prepared and installed at loading cell. The rest of test condition is same as conventional test.

3. Results and Discussion

CFRP members show fracture at the maximum load, and significant decrease in load after that. This fracture behavior is considered as layer to layer fracture in mode II shear, and load-displacement behavior up to maximum load is referred as flexure. Among the samples, $[+45^{\circ}/-45^{\circ}]_4$ shows the slow parabolic graph of fracture.

Strength and stiffness of CFRP in square shape over stacking angle is shown in Figure 3. The greatest strength can be found at the maximum load in load-displacement graph, which is $[90^{\circ}/0^{\circ}]_4$, $[0^{\circ}/90^{\circ}]_4$ and $[+45^{\circ}/-45^{\circ}]_4$. Stiffness of CFRP in square shape shows the greatest value at the steep slope (small deformation before fracture) that corresponds to $[90^{\circ}/0^{\circ}]_4$, $[0^{\circ}/90^{\circ}]_4$ and $[+0^{\circ}/-0^{\circ}]_4$. However, $[+0^{\circ}/-0^{\circ}]_4$ displayed 2 or 3 times of fracture before reaching maximum load in the graph, and compression is exhibited rather than flexure. Therefore, $[+0^{\circ}/-0^{\circ}]_4$, which shows high stiffness in value, is excluded from flexure member for actual application.

Figure 4. is shown result of strength and stiffness in CFRP circle shape. The greatest stiffness of CFRP circular shape is exhibited at $[90^{\circ}/0^{\circ}]_4$, $[0^{\circ}/90^{\circ}]_4$ and next is $[+15^{\circ}/-15^{\circ}]_4$.

CFRP square and circular shaped member shows the greatest strength and stiffness at $[90^{\circ}/0^{\circ}]_4$, $[0^{\circ}/90^{\circ}]_4$. Therefore, the application requiring greater strength and stiffness as major structural member is preferred to be $[90^{\circ}/0^{\circ}]_4$, $[0^{\circ}/90^{\circ}]_4$ stacking (fabric).

We need to test more for comparing exactly. CFRP Hat members manufactured and carried out by same method and show figure 5. is comparing the strength and stiffness of square, circle, hat shaped of $[90^{\circ}/0^{\circ}]_4$, $[0^{\circ}/90^{\circ}]_4$ that had good thing. In the result of strength Circle is the highest and in the result of stiffness hat shaped is the highest.

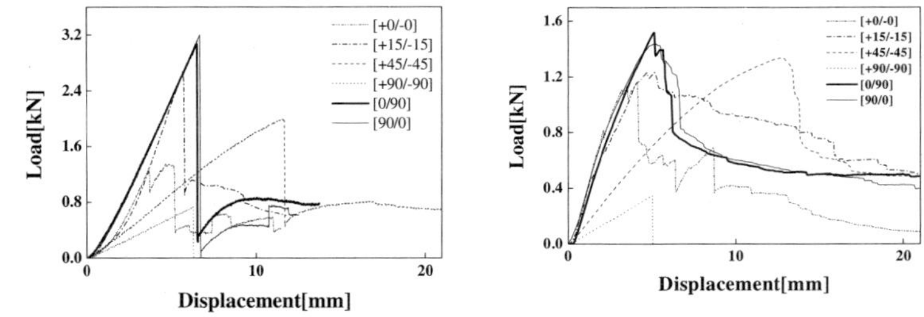

Fig. 2. L-D Graph of CFRP Square(Left Figure) and Circle(Right Figure) after Flexure Test.

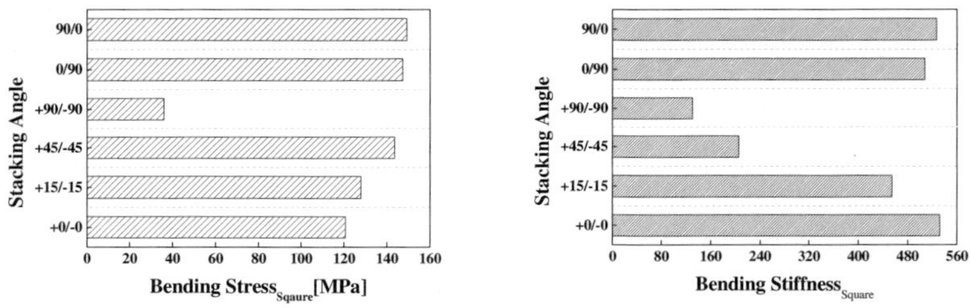

Fig. 3. Flexure Stress(Left Figure) and Stiffness(Right Figure) of CFRP Square.

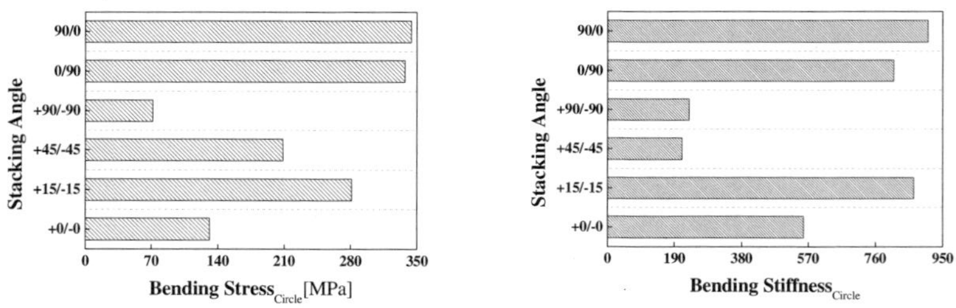

Fig. 4. Flexure Stress(Left Figure) and Stiffness(Right Figure) of CFRP Circle.

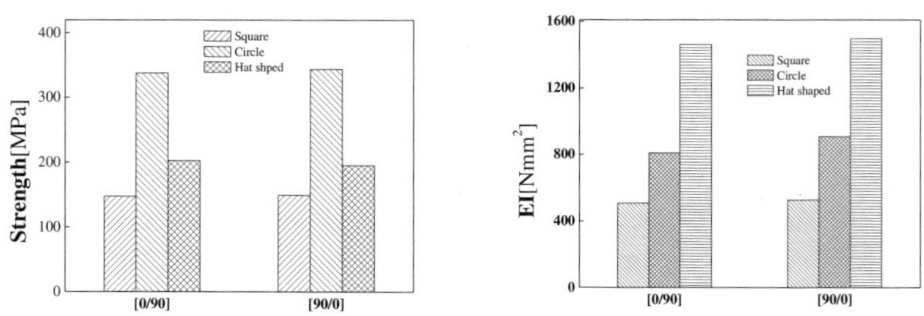

Fig. 5. Flexures Stress(Left Figure) and Stiffness(Right Figure) of CFRP Circle.

4. Conclusion

Strength and stiffness is calculated from flexure test using equation (1) and (2), respectively, and exhibited on graph according to stacking angle.

1) Most of specimen showed abrupt decrease in load in the graph after showing maximum load. However, $[+45^{\circ}/-45^{\circ}]_4$ displayed parabolic fracture pattern.

2) To prepare CFRP square member with greater strength and stiffness, $[90^{\circ}/0^{\circ}]_4$ or $[0^{\circ}/90^{\circ}]_4$ stacking is preferred.

3) For CFRP round member, greater strength and stiffness as major structural member can be achieved at fabric stacking.

4) In the result of strength Circle is the highest and in the result of stiffness hat shaped is the highest.

5) Comparison on strength and stiffness between square and round member, circular shape exhibited 2 times of square. For same stacking condition, round structure is preferred.

Acknowledgment

Following are results of a study on the "Human Resource Development Center for Economic Region Leading Industry" Project, supported by the Ministry of Education, Science & Technology (MEST) and the National Research Foundation of Korea (NRF).

References

1. C. S. Cha, K. S. Lee, S. H. Kim, J. O. Ching, I. Y. Yang, Axial Collapse Characteristics of Aluminum/CFRP Compound circular Tube. *Key Engineering materials*, 2005, Vols. 297-300 166-171.
2. J. H. Kim, I. Y. Yang, J. K. Sim, Evaluation of Fracture Toughness of Dynamic Inter-laminar for CFRP Laminate Plates by Resin Content. *KSMTE*, 2003, Vol. 12, No. 4, 43-49.
3. Y. N. Kim, K. H. Im, J. W. Park, I. Y. Yang, Experimental approach on the collapse mechanism of CFRP composite tube. *Reviews of progress in Q&DE*, 2000, 369-376.
4. A. G. Mamalis, D. E. Manolakos, M. B. Ioannidis, D.P. Papapostolou, Crashworthy characteristics of axially statically compressed thin-walled square CFRP composite tube: experimental, *Composite structure*, 2004, Vol. 63, 347-360.
5. M.J. Robert, Crushing characteristics of comtinuous fiber-reinforced composite tubes, *Journal of composite materials*, 1992, Vol. 26, No. 1.
6. Standard Test Method for Density. ASTM D6272-02.

STUDY ON SPRINGBACK OF HIGH STRENGTH STEEL MPV FRONT SIDE RAIL BASED ON BP NEURAL NETWORK MODEL

WENPING LI [†], YUNWANG MA, MIAO JIN, BAOFENG GUO SHAOMIN NIE

College of Vehicle and Energy, Yanshan Universtiy, 438 Hebei Street Qinhuangdao, Hebei 066004, China

ZHONG WANG, JIYAN LIU

Jilin Province Yuanlongda Mould Manufacture Co. Ltd. Liaoyuan, Jilin 136600, China

This paper deals with the optimization of process parameters in high strength sheet metal forming in order to reduce the springback effects after forming. The optimum parameters including holder force, friction coefficient and die radius which influence springback are studied through numerical simulation and orthogonal experiment. For predicting the springback accurately, the uniform experiment was used and BP neural network model was built up. The results show that springback of the front side rail is seriously influenced by these parameters, and the springback under the case of giving a group of process parameters can be predicted by the BP neural network model.

Keywords: Front side rail; Springback; Neural network; High strength steel.

1. Introduction

In order to meet government regulations of vehicle safety, fuel-efficient and low-emission, the automotive industry has been in a quest to increase the proportion of high strength steels (HSS) in manufacturing car body. However, application of HSS materials in body encounters many issues such as formability, springback and dies life. Especially, parts made of HSS materials demonstrate more springback than parts made of mild steel do. The finite element method (FEM) is an effective method to simulate sheet metal forming process and obtain the optimal parameters to reduce costs and shorten cycle of development of new car. There are many factors affect the process, and the number of combination for these factors is very large, which leads to increase times of simulation analysis and the amount of work for technologists. In the study, the prediction model of springback in a MPV HSS rail (see Figure 1) stamping was developed using BP

[†] Corresponding author, email: liwenping1968@yahoo.com.cn

neural network approach. Several numerical simulations using FEM were performed with the parameter combinations formed by orthogonal design to obtain the learning data of neural network. The neural network model was numerically verified and used to optimist the process parameters.

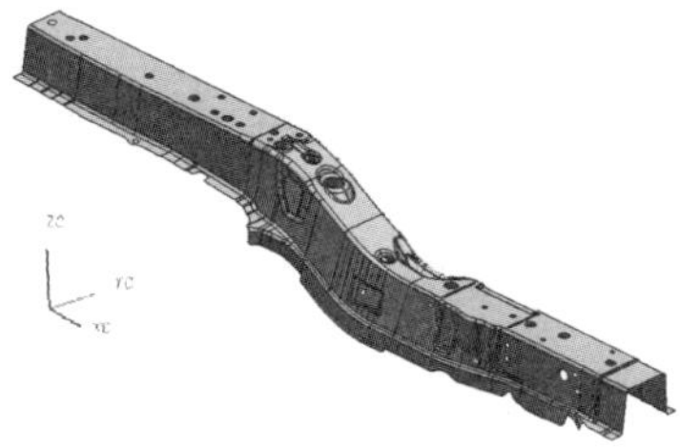

Fig. 1. Front side rail model.

2. Stamping Experiments Based on Uniform Design and Its Springback Simulation

2.1. *Factors Affecting Springback*

The factors which affect springback mainly contain two aspects: the material performance parameters and the tool geometry and production conditions. [1,2] In our study, the sheet material was DP600 and its properties were shown in Table 1. In optimization of stamping progress parameters, the blank holder force, friction coefficient and die radius were chosen as optimization variables which have a great influence on springback and can be adjusted in actual production condition. The maximal springback value in transverse section passing through point (300, 0, 0) was selected as evaluated criteria.

Table 1. Material properties of DP600.

Material	Direction	E /GPa	σ_s /MPa	σ_b /MPa	v	r	n	K /MPa
DP600 (t=1.8 mm)	0°	198.8	420.0	688.8	0.355	0.837	0.152	1082.0
	45°	215.4	423.2	689.6	0.298	0.882	0.150	1063.8
	90°	200.9	424.1	694.3	0.378	0.911	0.151	1092.9
	Avg.	207.6	422.6	690.9	0.33	0.878	0.151	1080.0

2.2. *Uniform Experiment*

In this paper, the ranges of three experimental factors were selected as below: the blank holder force (BHF) was chosen in range of 100–250 kN according to results of tentative calculation.[3] The range of friction coefficient was 0.1–0.2.[4,5] Based on tool adjustment experience in the factory, the range of die radius was

5–17 mm. The uniform design layout $U_{31}^{*}(31^{10})$ was selected to define uniform experimental program.[6]

2.3. *Springback Simulation*

In simulation of the front side rail forming and springback, the stamping speed was 5 ms^{-1}, the initial size of elements was 10 mm, element number on die radius was 10, element type was 16th fully integrated element, and clearance between punch and die was 1.98 mm. The springback simulation results of all experiment groups were shown in Figure 2. There was defection of cracking and wrinkling occurred, so the springback were not calculated under the 14th, 17th, 21st, 24th, 27th and 28th level combination.

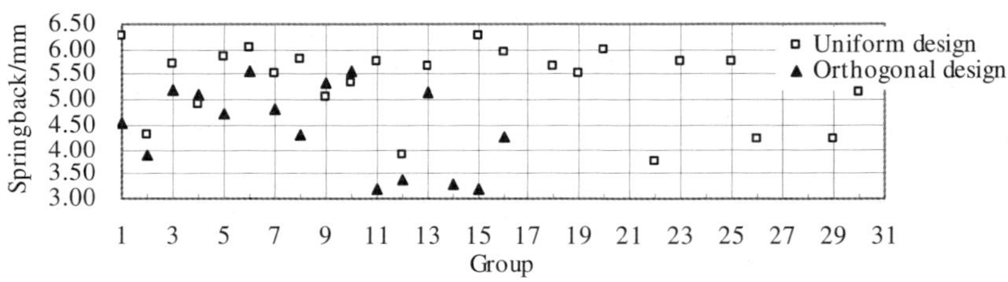

Fig. 2. Simulation results of uniform design groups and results of orthogonal design groups.

3. BP Neural Network Model and Its Results

3.1. *Neural Network Structure*

In the paper, we chose back propagation algorithm in MATLAB software to study relationship of the springback and process parameters. A three-layer network model was created, in which the uniform design combinations factors were taken as input and the maximal springback value along the transverse section as output. After lots of tentative simulations, the network structure of 3-10-1 for the issue was defined and used to train samples.

3.2. *Network model*

To validate the accuracy of the network model, the experimental samples were divided into learning samples and test samples. The 9th and 19th groups in uniform experiment were used as test samples, the others were learning samples.

In the network, the convergence goal was 0.01, the number of training was 5000, and other parameters were the default values.

Based on the above parameters, the target reached after 177 iterations.

3.3. *Verifying the Protection Capability of the Network Model and Optimizing Process Parameters*

The test samples were chosen to verify the established neural network model. The results of neural network were 5.16 mm and 4.98 mm, which had deviation of 1.98% and 9.45 % compared with results of the 9th and 19th groups, respectively (Figure 2). It proves that the network can accurately predict springback value that meets the quality requirement.

The minimal springback value occurs in the 22nd group of the uniform experimental design (Figure 2). Theoretically, there would be better solution near the data of the group. Thus, the layout L_{16} (4^3) was chosen to designed orthogonal experiment. Four levels for each factor are: 203, 204, 206, 207 kN; 0.10, 0.11, 0.13, 0.14; 5.2, 5.3, 5.5, 5.6 mm. The data of every group were input the network to get springback value, shown in Figure 3. The minimum was 3.19 mm in 15th group that reduced 15.3% compared with 3.76 mm in the 22nd group (see Figure 2).

In order to check the inhibitory effect on the springback with the optimized parameters, six sections parallel to *X-Y* plane were chosen on the front side rail, and the maximal springback values on each section were obtained. The actual values of the part were measured (Figure 3), and the simulation results with optimization parameters from the orthogonal design experiments were obtained. The results were shown in Figure 4.

Compared with the actually measured value, the springback value can reduce significantly on the condition of the optimization process parameters, which were obtained from the initial optimization of uniform design and the second optimization of the neural network. The springback value could reduce 23.5% at least.

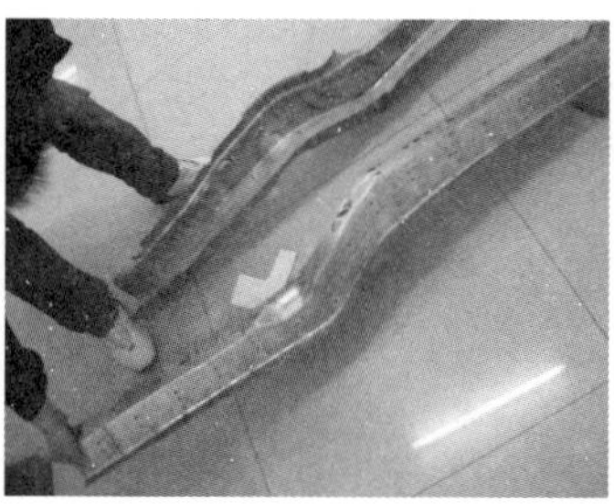

Fig. 3. Stamping parts of the front side rail.

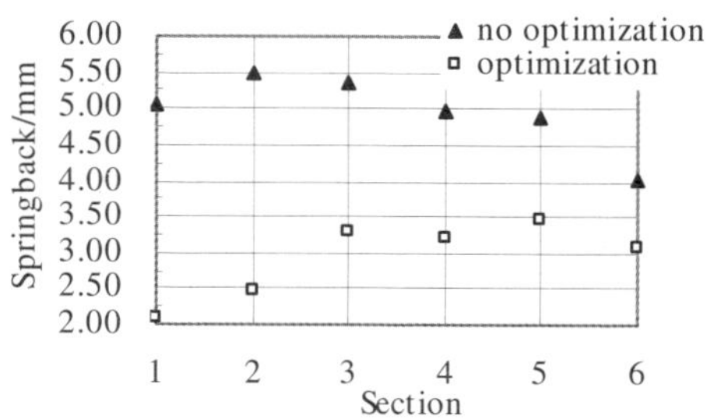

Fig. 4. Comparison of the springback value.

Compared with the results, the neural network model established in the paper could get accurate prediction value of springback for front side rail. Meanwhile, it can be used to optimize process parameters to reduce springback value. So, it proved that the method, combined with the neural network and finite element analysis to optimize the combination of process parameters and control the rebound, is feasible and reliable.

4. Conclusion

A study on the prediction and controlling of springback in high strength sheet metal forming by using neural network is carried out in this paper. From the foregoing discussion, the following conclusion can be made:

1. The neural network model is capable of accurate prediction on the springback of high strength sheet metal forming with less time and better convergence.
2. The neural network combined with orthogonal experimental design and numerical simulation can be used to optimize the forming progress parameters to reduce springback of high strength sheet metal part.

Acknowledgments

This work is supported by the National Natural Science Foundation of China (No.50305031, No.50945031), the State Key Laboratory of Material Processing and Die & Mould Technology of China (No.200404) and Education Department of Hebei Provence of China (No.200504).

References

1. R. Bahloul, S. Ben-Elechi and A. Potiron: Journal of Materials Processing Technology, 2006, 173, (1), 101–110.
2. F. C. Lan, J. Q. Chen and X. Yu: Journal of Plasticity Engineering, 2004, 11, (5), 78–84.
3. Y. X. Luo: in 'Newest manual of stamping technology and die design', 795; 2004, Beijing, Yinsheng Audio-video Publishing House.
4. S. Z. Long, S. G. Hu and X. W. Song: Light Industry Machinery, 2007, 25, (3), 81–84.
5. S. S. Han: Journal of Materials Processing Technology, 1997, 63, 129–133.
6. K. T. Fang: in 'Uniform design and layout', 94, 1994, Beijing, Science Press.

STUDY ON SAFETY EVALUATION METHODS AND FAILURE MECHANISM OF HIGH ROLLER COMPACTED CONCRETE DAM

SU PEI-FANG †

State Key Laboratory of Water Resources and Hydropower Engineering Science, Wuhan University, 430072, Wuhan, P.R.China

WANG WEI-MING

State Key Laboratory of Water Resources and Hydropower Engineering Science, Wuhan University, 430072, Wuhan, P.R.China

LU XING-LI

State Key Laboratory of Geomechanics and Geotechnical Engineering, Institute of Rock and Soil Mechanics, Chinese Academy of Sciences, 430071

In order to explore the integral safety degree, probable failure type and failure mechanism of high dam, the combination of overloading and strength reduction method is adopted to simulate the progressive failure and potential destabilization pattern of dam-foundation system. By using finite element method, the overloading of dam-foundation system is carried out firstly and then keeping the load combination unchanged, by reducing rock material strength parameters step by step. The general safety degree of dam-foundation system is obtained. The method mentioned above has been used in Guangzhao roller compacted concrete (RCC) gravity dam, the finite element calculation results show that integrated safety degree of GuangZhao RCC Gravity Dam is greater than 3.0 and the safety degree of the dam foundation surface which is the weakest position of the dam-foundation system is greater than 2.0. The research results indicate that the safety coefficient can satisfy the engineering requirement and the method is efficient and accurate in high dam safety evaluation.

Keywords: High roller compacted concrete dam; degree of safety; overloading; strength reduction method; failure.

1. Introduction

The safety and stability analysis of RCC dam is an important technique problem in dam design and operation period. Guangzhao roller compacted concrete gravity dam is the highest RCC gravity dam under construction in the world, the

---

† Corresponding author, email: pfsu@163.com

structural particularity and running conditions complexity of the dam determine that the actual safety stability and stress-strain state are related to many factors of the dam and foundation in design, construction and operation, so the integrated carrying capacity and safety degree analysis will be a very complicated and extremely difficult research.

At present, the major methods of integrated safety evaluation are geomechanical model test method[1] and numerical test method. The instability process and failure mode of structure are performed by model test, and the integrated safety degree is concerned. But the model has obvious differences with the on-site engineering in geological conditions and load combination etc[2]. and it also consumes a lot of people and material resources.

With the development of science and technology, the numerical test methods have gained more and more widely engineering application, such as finite element method, discontinuous deformation analysis, numerical manifold method[3], composite element method[4] and so on. Among these, the finite element method is the most popular and intensive method, But the shortages of the research at present are as follows: The simulation in the construction period and the safety degree evaluation in the operation period are not closely associated; The coupling computation of temperature, seepage, stress/strain and safety degree isn't carried out perfectly[5].

In light of this situation, the following researches are carried out: on the basis of simulation in the construction period, gradual destroy progress and potential destabilization pattern of the dam-foundation system are obtained by overloading and strength reduction method and the integrated safety degree of Guangzhao RCC gravity dam is evaluated. The series of simulation analysis adopted in this paper can response the stability and safety of the dam more accurate and has a significant meaning for reasonable evaluating the integrated safety degree and they are of some reference value to similar engineering.

2. Calculation Principle and Analytical Method

2.1. *Overloading Method*

The overloading method studies the distortion and destruction process from part to the whole of the dam-foundation system by gradually increasing the hydrostatic load at upper dam face. The safety degree obtained by this method is called overloading safety degree. In analysis, only the hydrostatic load at upper dam face is overloading, the others are remained unchanged.

2.2. *Strength Reduction Method*

The strength reduction method studies the distortion and destruction process by gradually reducing rock material strength parameters including cohesion c and friction angle φ step by step[7-8]. This method can reflect the strength index gradual weakening of the weak structural plane because of the water soak and seepage after long term operation of the geotechnical engineering.

2.3. *Comprehensive Method*

Comprehensive method combines the overloading method and the strength reduction method, so as to comprehend the advantages of the two methods and make up for the inadequacy of independent overloading method and the strength reduction method[8-9]. This method can reflect the possible flash floods in engineering and the strength index gradual weakening of weak structural plane because of water after long term operation of the geotechnical engineering.

3. Finite Element Model and Calculation Scheme

3.1. *Project Description*

Guangzhao Hydropower Station is an RCC gravity dam, approximately 200.5m high, which is the highest RCC dam under construction in the world. A 3D model including dam and foundation is built, there are 1034634 elements and 1012042 nodes of the dam-foundation system. Thin-layer elements are adopted to simulate the base surface, Figure 1 is the global model of GuangZhao and Figure 2 is the dam model of GuangZhao.

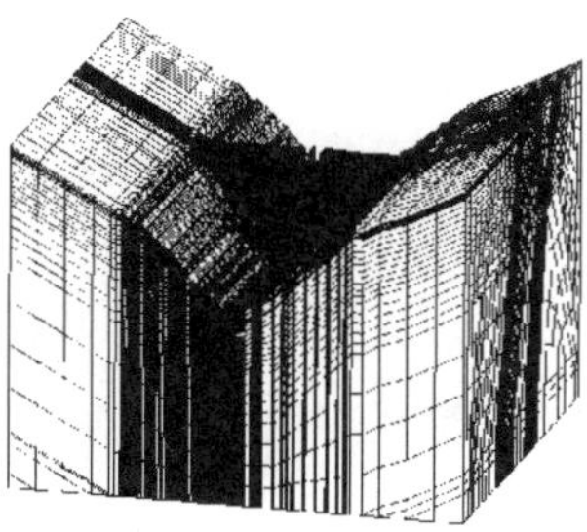

Fig. 1. Global model of GuangZhao.

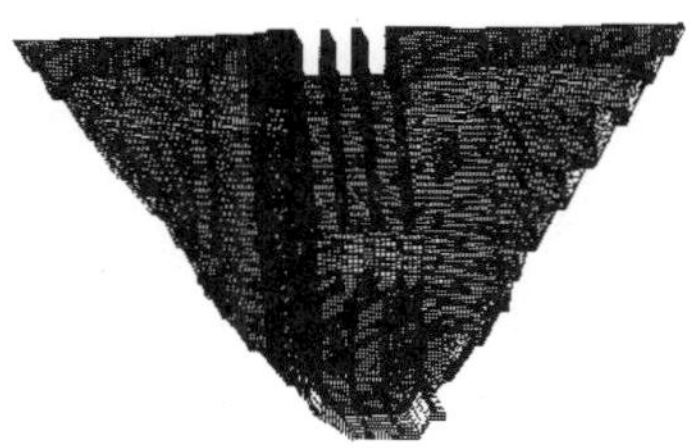

Fig. 2. Dam model of GuangZhao.

3.2. *Material Properties*

The mechanical parameters of materials are summarized in Table 1.

Table 1. Mechanical parameters of materials.

Concrete and stratum	Deformation modulus /MPa	Poisson's ratio	Gravity / KN.m^{-3}	Cohesive strength /MPa	Friction coefficient	Compression strength/MPa	Allowable bearing capacity/MPa	Tensile strength /MPa
C1	31.1	0.167	2.45	2.0	1.0	31.2	/	2.55
R I C25	37.6	0.167	2.46	2.0	1.0	30.5	/	2.32
R II C20	35.5	0.167	2.44	2.0	1.0	23.6	/	2.12
RIIIC15	34.1	0.156	2.44	2.0	1.0	20.2	/	1.96
RIVC25	37.0	0.167	2.46	2.0	1.0	29.6	/	2.36
R V C20	35.9	0.167	2.45	2.0	1.0	23.9	/	2.18
T_1f^{2-2}	3.0	0.33	2.75	0.8	0.5	/	2.5	0.5
T_1f^{2-3}	7.75	0.28	2.75	0.8	0.8	/	3.8	0.7
T_1yn^{1-1}	10.5	0.26	2.75	1.05	1.1	/	5.0	1.3
T_1yn^{1-2}	10.75	0.26	2.75	1.0	1.2	/	5.5	1.1
T_1yn^{1-3}	11.75	0.25	2.75	1.0	1.2	/	6.5	1.2
T_1yn^2	7.0	0.3	2.75	0.6	0.7	/	2.5	0.6

3.3. *Calculating Scheme and Load*

The safety evaluation is carried out on the basis of integrated whole course simulation of Guangzhao RCC gravity dam, and the analysis course is as follows:

1. Temperature calculation, considering the whole process of construction and impoundment, actual simulating all kinds of temperature control methods.
2. Seepage calculation. The seepage control measures such as the drain holes and impervious curtain are simulated accurately, and the seepage calculation is carried out for every impoundment levels.
3. Stress-strain calculation. The loads considering are: (1) self weight; (2) the hydrostatic load at dam face (3) temperature load (4) Seepage load acting on the base of gravity dam.
4. Overloading analysis, using the method increasing water level gradually, the Drucker-Prager criterion is adopted and the overloading coefficients are 1.0, 1.25, 1.5.
5. Strength reduction method. The strength reduction is carried out on the basis of the 1.0, 1.25, 1.5 overloading, and the strength reduction coefficients are 1.0，2.0，3.0，3.2，3.5，3.6，3.8，4.0，4.5，5.0 and so on.

4. Analysis of Results

4.1. *Displacement Curve*

Roller concrete gravity dam is spatial integral structure for it is adopting full-silo casting method. The abnormal of dam operating condition can be reflected by the displacement saltation of typical points. Figure 3 is the along river displacement of dam crest and Figure 4 is the vertical displacement of dam crest. Results show that:in the normal working condition, the along river displacement curve jump at the place that strength reduction coefficient is 3.8. When the overloading coefficient is 1.25, the curve saltate at 3.2. When it is 1.5, the curve jump at 3.0.

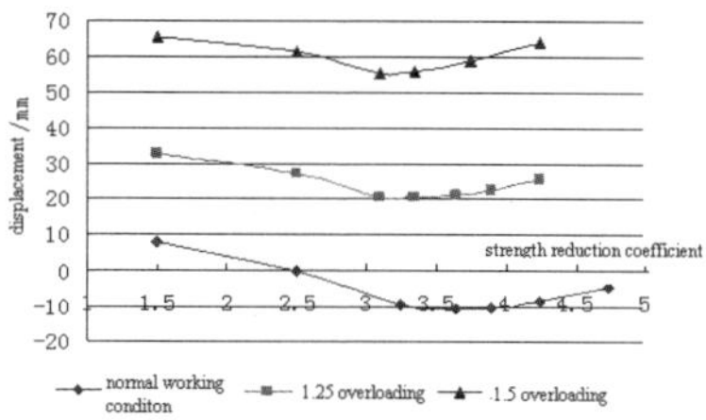

Fig. 3. Along river displacement of dam crest.

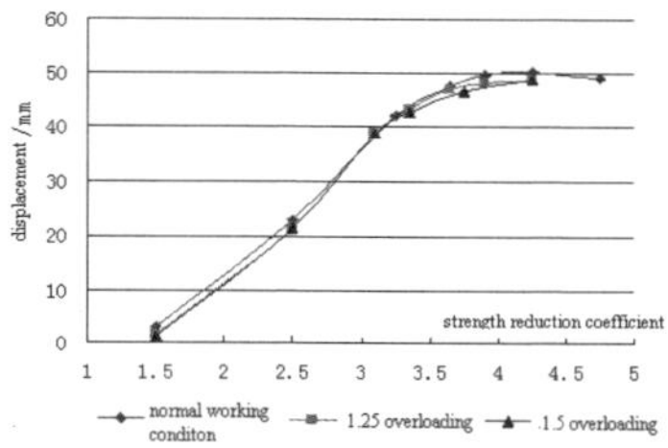

Fig. 4. Vertical displacement of dam crest.

4.2. *Yield Region*

Figure 5 shows the development of connectivity in yield region in the normal working condition, with the corresponding strength reduction coefficient is 1.0, 2.0, 3.0, 3.5, 3.8, 4.0, 4.5, 5.0 respectively. The yield region in the base gradually expands with the increase of strength reduction coefficient. As can be seen, the failure type and failure process of the base along the reducing of strength parameters is that tensile crack takes place at dam heel and compressive yield occurs at dam toe, and then the base surface and faults begin yield, yield area exceeds the curtain center line when the strength reduction coefficient is 3.5, and then expand until the plastic zone become run-through at 3.8. In this process, the yield region expansion is quick in downstream base, and the dam-foundation system loss bearing capacity when the plastic zone become run-through.

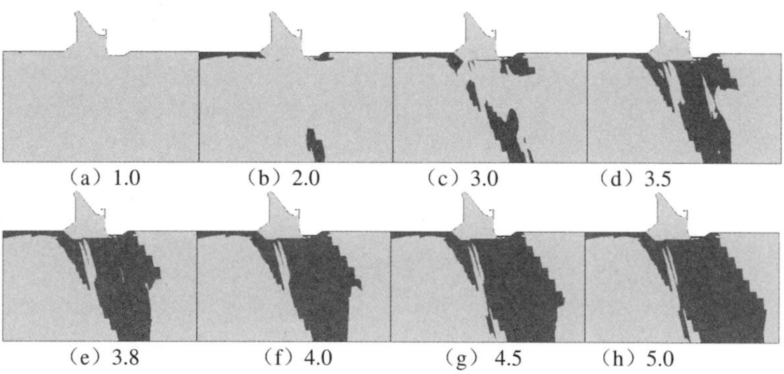

Fig. 5. The development of connectivity in yielded region.

5. Conclusion

According to the displacement jump of typical position and plastic zone pierce method， the integrated safety degree of dam system is gained and slide resistance of the dam foundation surface is evaluated, then gets these conclusions:

1) In the normal working condition, the safety degree is 3.8 according to the displacement jump of typical position. When the overloading coefficient is 1.25, the dam-foundation system become instability at the place that strength reduction coefficient is 3.6. When the overloading coefficient is 1.5, the safety degree is 3.2

2) The plastic zone expands gradually until it becomes run-through with the increase of strength reduction coefficient. The dam-foundation system is in the limit state when the strength reduction coefficient is 3.5, so the integrated safety degree of GuangZhao RCC Gravity Dam is between 3.5 and 3.8, and the safety degree of the surface of the dam foundation which is the weakest position of the dam-foundation system is greater than 2.0. The research results show that the safety coefficient can satisfy the engineering requirement.

3) Since the construction process, construction technology, stress/strain, seepage and temperature in construction period have a great impact on safety degree of RCC gravity dam, the safety evaluation based on the necessity of integrated whole course simulation for RCC gravity dam is significant.

References

1. DONG JH, XIE HP. Chinese Journal of Rock Mechanics and Engineering, 2007, 26, (10), 2027-2033.
2. LI TC, WANG Rk, YOU QS. Journal of Hydraulic Engineering, 2007, 10, (supp), 78-83.
3. Shi GH. 'Numercial Manifold Method and Discontinuous Deformation Analysis', 1997, Beijing, Tsinghua University Press.
4. Chen SH. 'Computational Rock Mechanics and Engineering', 2006, Beijing, China WaterPower Press.
5. Zhu BF. Water Resources and Hydropower Engineering, 2007, (1), 1-6.
6. 6. DUAN QW, GENG KQ, WU YP, JIA ZX. Rock and Soil Mechanics, 2008, 29(supp), 15-20.
7. JIANG Q, FENG XT, XIANG TB. Rock and Soil Mechanics, 2009, 30(supp), 2483-2488.
8. Yu TT,Ren QW. Chinese Journal of Rock Mechanics and Engineering, 2007, 26, (4), 787-794.
9. NING Y,XU Wy, ZHENG WT. Chinese Journal of Rock Mechanics and Engineering, 2008, 27, (9), 1890-1898.

ANALYTICAL STUDY ON THE MECHANICAL BEHAVIOR OF TRUSS CORE STRUCTURES UNDER COMPRESSIVE AND SHEAR LOADS

SANG-WOO KIM†

Materials Deformation Group, Korea Institute of Materials Science, Changwon, Gyeongnam 641-831, South Korea

KWANG-SEOK LEE

Materials Deformation Group, Korea Institute of Materials Science, Changwon, Gyeongnam 641-831, South Korea

BEOM-SOO KANG

Dept of Aerospace Engineering, Pusan National University Busan 609-735, South Korea

The study focused on analytical approach to investigate compression and shear quality of the unit cell of PCM with pyramidal and tetrahedral configuration. With various unit cell models which have the same core weight per unit area but different strut angle, analytical solution for effective stiffness and strength have been predicted and compared with each other. With this approach, the most optimal core configuration was predicted. The result has become the efficient guidelines for the design of PCM core structure.

Keywords: PCM; Sandwich panel; Pyramidal; Tetrahedral; Equivalent properties.

1. Introduction

Metallic sandwich panels based on lattice cell structure have been developed for a wide range of potential applications with their lightweight and multi-functionality [1-3]. Numerous studies to evaluate the performance of PCM have been carried out using analytical and numerical approach [4,5].

† Corresponding author, email: kimsw@kims.re.kr

Analytical investigations on the unit cell of truss cores provide the approximated guidelines for the design of overall core structures for a specific application in a short time. In this study, the effect of geometrical parameters on the mechanical properties of pyramidal and tetrahedral shape of unit cells was investigated with analytical models. Unit cell with strut angle of 45° was considered as a reference model and other models were designed to have the same weight and bottom area but different core strut angle. All core struts were assumed to be connected with pin joint in analytical model. Under the assumptions, effective strength and effective modulus of unit cell for each case under compressive and shear loads were predicted and compared. And finally, optimum core member angle to have maximum mechanical property could be calculated.

2. Analytical Models

Consider a pyramidal and a tetrahedral shape of unit cell under compressive or shear load as shown in Fig. 1. All struts with the length of L are identical and have regular square cross sections of width t. Each strut is inclined from the bottom face with the angle of θ.

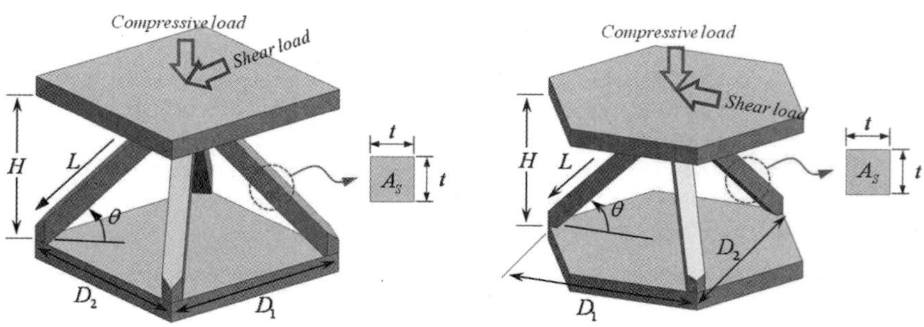

Fig. 1. Schematic illustrations of pyramidal and tetrahedral unit cell.

Relative density ($\bar{\rho}_C$), which is defined as the density of the core divided by that of which it is made, can be expressed as Eq. (1) for each core shape.

$$\bar{\rho}_C = \frac{V_C}{V} = \Phi\left(\frac{t}{L}\right)^2 \frac{1}{\cos^2\theta\sin\theta} \quad \left(\Phi_{pyramidal} = 2, \quad \Phi_{tetrahedral} = \frac{2}{\sqrt{3}}\right) \qquad (1)$$

Where, V and V_C is the volume of unit cell and the core volume within unit cell, and Φ is a geometrical parameter. In order to simplify the problem, it is assumed that truss joints are idealized as pin joints and therefore, rotational resistance at the joint is neglected. From force equilibrium between the struts of unit cell, analytical expressions for effective compressive $(\bar{\sigma})$ and shear stress $(\bar{\tau})$ of unit cell can be represented as follows,

$$\bar{\sigma} = \bar{\rho}_C \sigma_S \sin^2\theta, \quad \bar{\tau} = \Pi \bar{\rho}_C \sigma_S \sin 2\theta \quad \left(\Pi_{pyramidal} = \frac{1}{2\sqrt{2}}, \quad \Pi_{tetrahedral} = \frac{1}{2} \right) \tag{2}$$

Where, σ_S is the applied stress on the strut and Π is another geometrical parameter. Maximum stress occurs when the stress applied on a strut (σ_S) equals to critical buckling stress (σ_{cr}). Thus, replacing σ_S in Eq. (2) with σ_{cr}, analytical expression for peak stresses $(\bar{\sigma}_{peak}, \bar{\tau}_{peak})$ can be obtained as follows,

$$\bar{\sigma}_{peak} = \bar{\rho}_C \sigma_{cr} \sin^2\theta, \quad \bar{\tau}_{peak} = \Pi \bar{\rho}_C \sigma_{cr} \sin 2\theta \tag{3}$$

From force equilibrium and work balance on a unit cell and the above equations, effective Young`s modulus $(\bar{E})$ and shear modulus $(\bar{G})$ of unit cell under compressive and shear load can be represented a function of relative density, strut angle and Young`s modulus of the strut material (E_S) as follows,

$$\bar{E} = \bar{\rho}_C \sin^4\theta E_S, \quad \bar{G} = \frac{1}{8} \bar{\rho}_C \left(\sin^2 2\theta \right) E_S \tag{4}$$

3. Optimization of Truss Core Angle

It is known that strength and stiffness of truss core structures are dependent on relative density, strut angle and material. In this work, in order to find out the optimum strut angle to maximize strength or stiffness, core volume (or weight) per unit bottom area and length D_1 is considered as constraints for optimization. Under the assumption, the effects of strut angle on effective stiffness and strength are investigated. By geometrical relationship between a reference model and other models, $(t/L)^2$ can be represented as constant C and strut angle θ as follows,

$$\left(\frac{t}{L} \right)^2 = C \cos^3\theta \quad \left(C_{pyramidal} = \frac{\sqrt{2} V_{C,Ref}}{2 D_{1,Ref}^3}, \quad C_{tetrahedral} = \frac{9 V_{C,Ref}}{8 D_{1,Ref}^3} \right) \tag{5}$$

Also, substituting Eq.(5) into Eq.(1), the relative density can be expressed as follows,

$$\bar{\rho}_C = \frac{C\Phi}{\tan\theta} \tag{6}$$

Using Eq.(6) and Eq.(3), effective compressive and shear strength are expressed as follows,

$$\bar{\sigma}_{peak} = C\Phi \sin\theta \cos\theta \sigma_{cr}, \quad \bar{\tau}_{peak} = 2C\Phi\Pi \cos^2\theta\sigma_{cr} \tag{7}$$

Assuming that the material is hardened with the rule of $\bar{\sigma} = K\bar{\varepsilon}^n$ during the deformation, critical buckling stress can be expressed as follows,

$$\sigma_{cr} = \frac{\pi^2}{12}\left(\frac{te}{L}\right)^2 E_t = K\left[\frac{n\pi^2 e^2}{12} C\cos^3\theta\right]^n \tag{8}$$

Where, E_t is the tangent modulus at σ_{cr} and e is the end condition of the strut. Finally, effective strength and stiffness under compressive and shear loading can be obtained from Eq.(4),(7) and (8).

$$\bar{\sigma}_{peak} = KC^{n+1}\Phi\left(\frac{n\pi^2 e^2}{12}\right)^n \sin\theta\cos^{3n+1}\theta, \quad \bar{\tau}_{peak} = 2KC^{n+1}\Phi\Pi\left(\frac{n\pi^2 e^2}{12}\right)^n \cos^{3n+2}\theta \tag{9}$$

$$\bar{E} = 2C\cos\theta\sin^3\theta E_S, \quad \bar{G} = C\cos^3\theta\sin\theta E_S \tag{10}$$

From Eq.(9) and (10), It is known that effective strength and stiffness of unit cell of truss core can be represented as only a function of strut angle (θ) for a given material (K, n, E_S) and geometrical constraint (C) of unit cell.

4. Results and Discussion

In order to investigate the effect of strut angle on effective stiffness and strength, mechanical properties obtained from the uniaxial tensile tests of Al 5052 sheet with thickness of 2mm are used. Elastic modulus and yield strength of the material are 70.3GPa and 141.5MPa, and strain hardening exponent (n) and strength coefficient (K) are 0.216 and 400MPa. Geometrical conditions of the reference model are L = 21.21mm, t = 2mm, θ =45°, V_C = 339.4mm^3 for pyramidal core and L = 21.21mm, t = 2.63mm, θ =45°, V_C = 440.2mm^3 for tetrahedral core. The core volume per unit bottom area of both reference models are the same as 0.75mm.

Figure 2 shows the relative density variations for strut angles. The relative density is decreased with increasing strut angle because the height and volume of unit cell are increased with increasing strut angle under the same bottom dimensions. But, it is clear that the core weight per unit cell is the same for all models.

Constant C in Eq. (5) is calculated from geometrical relationship of the reference model and other models with strut angles from 10° to 80°. And the effective stiffness and strength for various strut angles were calculated as shown in Fig. 3 and 4. Figure 3 shows peak stress variations under compressive load. The peak stress is increased with increasing strut angle and reaches the maximum value at θ =37.92° for pyramidal and tetrahedral unit cells. However, peak stress is decreased after maximum point. For all ranges of strut angle, the compressive strength of tetrahedral core unit cell is higher than that of pyramidal core unit cell and maximum difference is 0.43MPa (12.6%) at θ =37.92°.

Figure 4 shows peak stress variations under shear load. Differently from the results of compressive load, the peak stress shows a trend to decrease with increasing strut angle throughout all strut angles. Shear strength of tetrahedral core unit cell is higher than that of pyramidal core and the differences between both core models get enlarged as strut angle is high.

Figure 5 shows the variations of effective compressive and shear stresses at various strut angles. From Eq. (5), effective compressive and shear modulus are dependent on the relative density and the strut angle of unit cell. And the models considered in this work have the same relative density when the strut angle is the same regardless of core geometry as shown in Fig. 2. Therefore, effective

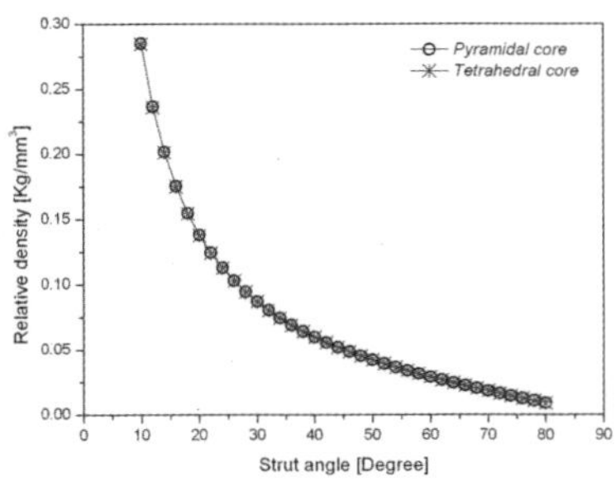

Fig. 2. Relative density variations for strut angles.

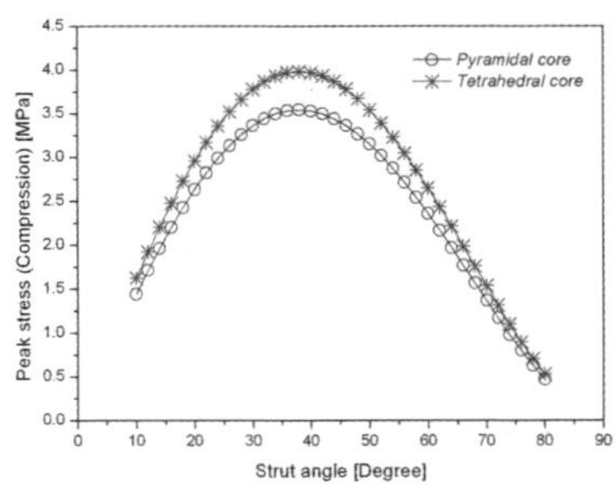

Fig. 3. Effect of strut angle on peak stress under compressive load.

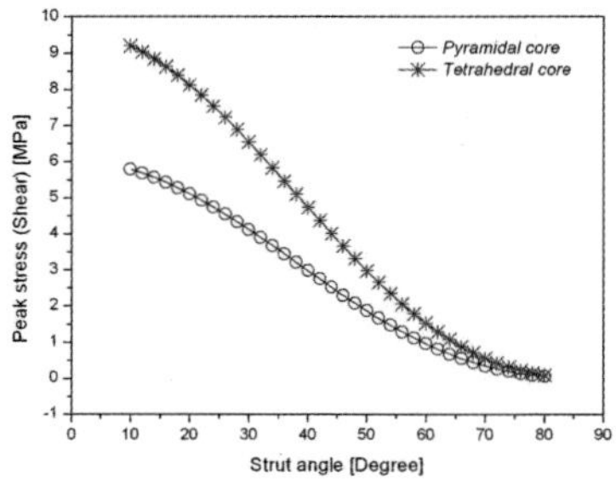

Fig. 4. Effect of strut angle on peak stress under shear load.

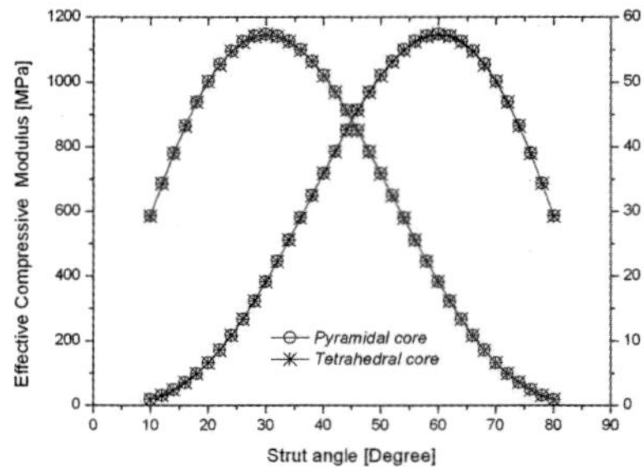

Fig. 5. Effect of strut angle on effective moduli.

stiffness values of pyramidal and tetrahedral core are only dependent on strut angles and have the same value at the same strut angle. And effective compressive modulus shows the maximum value of 1148MPa at θ=60° and effective shear modulus shows a maximum value of 574MPa at θ=30°.

References

1. L.J.Gibson and M.F.Ashby: 'Cellular solids: structure and properties', 2nd edn; 1997, Cambridge UK, Cambridge university press.
2. A.G. Evans, J.W. Hutchinson, M.F. Ashby: *Prog. Mater. Sci.*, 1999, **43**, 171-221
3. A.G. Evans, J.W. Hutchinson, N.A. Fleck, M.F. Ashiby, H.N.G. Wadley: *Prog. Mater. Sci.*, 2001, **46**, 309-327
4. V.S. Despande, N.A. Fleck: *Int. J. Solids Struct.*, 2001, **38**, 6275-6305.

INSTRUMENTATION OF INFRARED THERMOGRAPHY DURING TENSILE TEST OF GLASS FIBER REINFORCED POLYMER

JIN WOO KIM

Center for NDE, Iowa State University,
Ames, IA, 50011, USA

DONG GI LEE, JAE KI SIM AND JAE YEOL KIM

Dept. of Mechatronics Engineering, Chosun Univ.,
Gwang-ju, 501-759, Korea

SEUNG HYUN CHOI†

Dept. of Advanced Parts and Materials Engineering, Chosun Univ. Graduate School,
Gwang-ju, 501-759, Korea

Property of glass fiber reinforced polymer (GFRP) is naturally dependent on the property of matrix, but receives of temperature effects by external force. Therefore, determination of mechanical property by effect of temperature in GFRP is most crucial factor. In this paper, temperature distribution during crack propagation from temperature change under tensile test was proposed through IR thermography camera. Lock-in thermography method, which is one of technique in IR thermography camera to measure minute change in temperature, was utilized to monitor temperature distribution and change during crack propagation. Method to analyze temperature distribution during crack propagation under tensile test of GFRP via IR thermography camera was suggested. Anisotropy in fiber orientation showed longer fracture time and lower maximum temperature.

Keywords: Infrared Thermography (IT); Fiber Orientation Function; Lock-in Thermography; Fracture Velocity; Glass Fiber Reinforced Polymer (GFRP); Tensile test.

1. Introduction

GFRP (glass fiber reinforced plastic composite materials) has superior strength/density and modulus/density ratios to metallic materials. In addition, GFRP has been widely used as aerospace and automobile application due to high shock absorbing property. Among automobile components, structural parts in interior/exterior application are the one of promising applications for GFRP.[1]

† Corresponding author, email: cshddubi@hanmail.net, +82-62-230-7745

Infrared thermography (IT) is known to detect defects in material and structures without contact. IT can quantitatively measure the location and shape of delamination, and is useful for maintenance work of concrete structures, electric facility, leakage, etc. Infrared from object at more than 0 Kelvin is detected by IR sensor in camera from certain distance. Temperature distribution of object is indexed as red for higher temperature and blue for lower temperature. Temperature distribution of object surface can be acquired as single image. In addition, temperature of specific location can be acquired through thermography.[2]

Along with metal, fracture of composite material is started from the defects in material. These defects might be broken fiber, defect in matrix, debonded interface, etc. Once fracture is started, propagation of crack can be described with simple model. For instance, model in crack tip can be described by Fig. 1.

In this paper, temperature distribution during crack propagation from temperature change under tensile test was proposed through IR thermography camera. Lock-in method, which is one of technique in IR thermography camera to measure minute change in temperature, was utilized to monitor temperature distribution and change during crack propagation.

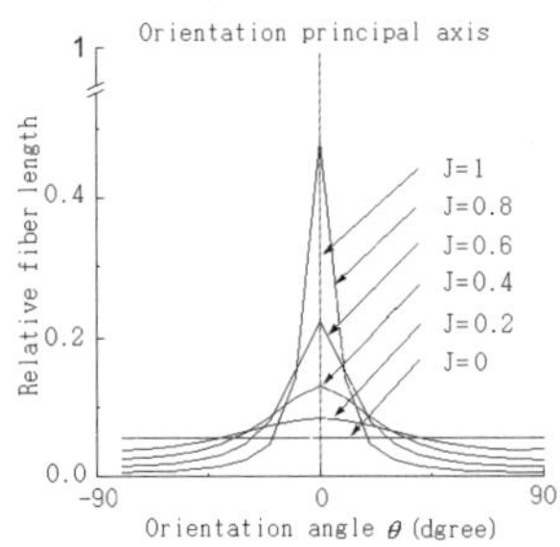

Fig. 1. Distribution of relative fiber length according to the fiber orientation angle.

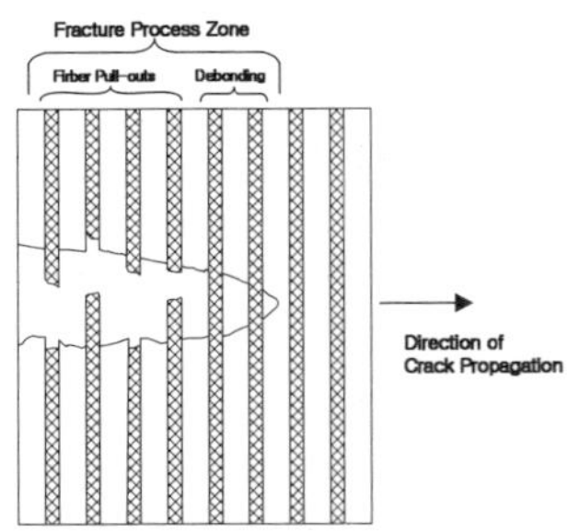

Fig. 2. Model of fracture process.

2. Fiber Orientation Function

Now, compared with width and length, the thickness of molded fiber reinforced polymer composite is very thin so we consider the fiber as a two dimensional set. In case when the fiber is distributed in thin fiber reinforced polymer composite's x-y surface(plane), fiber distribution angle θ with x axis ranges from 0 to π. Accordingly, Fig. 2 shows a relationship of Eq. (1) between variable which shows the fiber orientation distribution state (hereinafter referred to as fiber orientation function J) and fiber orientation angle distribution, $q(\theta)$.[1]

$$q(\theta)=\frac{1}{\pi}\left[\left(\frac{1-J}{1+J}\right)\cos^2\theta+\left(\frac{1+J}{1-J}\right)\sin^2\theta\right]^{-1} \quad (1)$$

where, J is a fiber orientation function which indicates two-dimensional fiber orientation distribution state. Let the x axis, the direction of resin flow, is 0. If the fiber runs parallel with the flow, fiber orientation function J is 1 and if the fiber runs at a right angle to the flow, J is -1. In this way, fiber orientation function J ranged from 0 to 1 and if J approaches to 0, the state of fiber orientation will be randomized.

3. Heat Measurement

The IR electromagnetic spectrum can be divided into bands of wavelength. There is no fundamental difference between bands, and same principle holds. Only difference between bands is wavelength. IR thermography camera utilizes the band with 2 ㎛~13 ㎛. Where, SW (shortwave) utilizes the band of 2 ㎛~5 ㎛, and LW (long wave) region is corresponded to the band of 8 ㎛~12 ㎛. Band of 5 ㎛~8 ㎛ shows the 0% of transmittance in vapor, and IR detection is not available.

Corresponding measurement equation is assigned from included ERIKA software in system controller. Equation for measuring heat can be expressed as Eq. (2) First term indicates the radiation from object, and second term indicates the reflected radiation from background. Third term is radiation from air.[3]

$$I_m = I(T_{obj})\times\tau\times\varepsilon+\tau(1-\varepsilon)\times I(T_{amb})+(1-\tau)\times I(T_{afm}) \quad (2)$$

IR thermography camera is Thermovision 900 system from AGEMA IR SYSTEM AB. For controller, 2 68020 processors, which is based on dual channel microprocessor, are utilized with 32 bit VME bus and two 16-bit IR bus. Cooling method for scanner is thermoelectric type, and it features 2 split, serial scanning, and 2-5.4 micron spectrum response. Temperature range is -10℃~500℃ (up to 2,000℃ with high temperature filter), and sensitivity is 0.1℃ at 30℃. Spatial resolution is 140 element/line (50% modulation), and IR line frequency is 3.5 kHz with simple/line 204.

To measure temperature distribution from thermography information during tensile test, sample was mounted in UTM. IR thermography camera is installed at 0.1m from specimen, and it was installed to synchronization with temperature signal from camera and loading signal from UTM. Frame setting of IR camera was set to 1 frame per second, and ambient temperature was 20℃. For the condition as dark room, black cloth was hanged, and 30 minute of soak time was taken before commencing experiment.

4. Test Results and Considerations

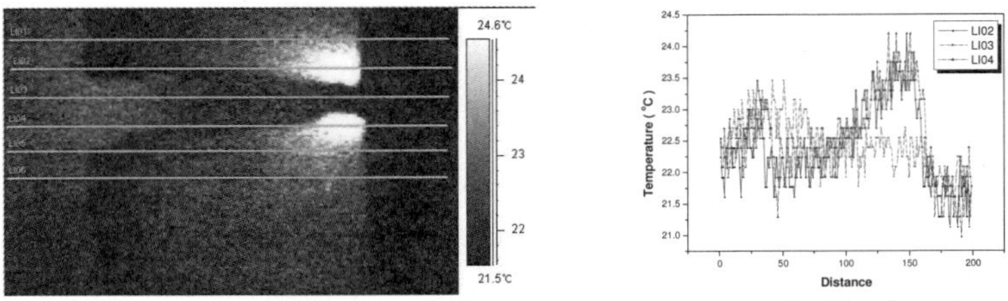

(a) Temperature distribution at 491sec Profile of tread

Fig. 3. Temperature distribution measurement (Polypropylene).

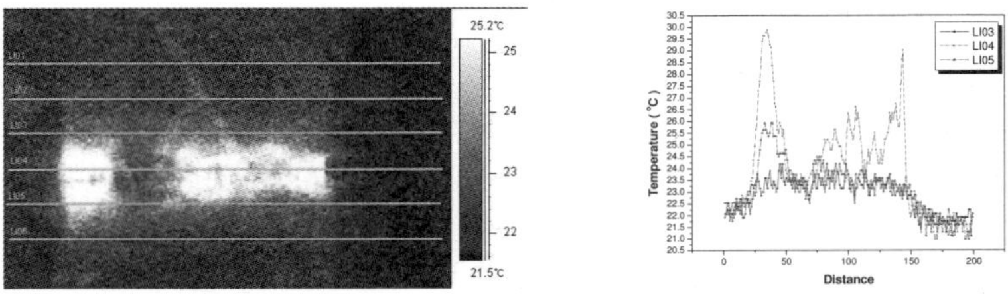

(a) Temperature distribution at 349sec Profile of tread

Fig. 4. Temperature distribution measurement (*J*=0.1, 10wt%).

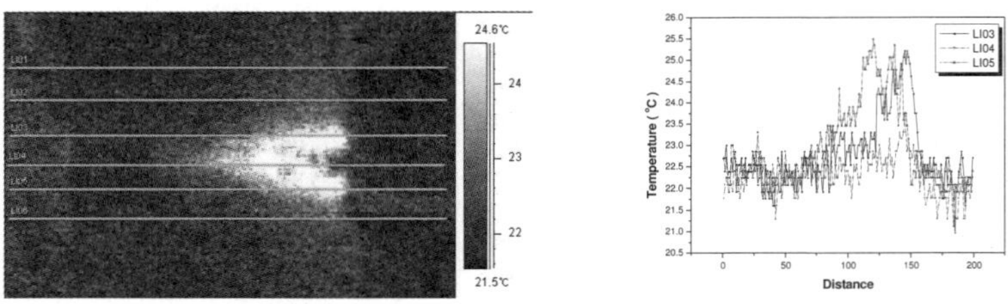

(a) Temperature distribution at 313 sec Profile of tread

Fig. 5. Temperature distribution measurement (*J*=0.1, 20wt%).

To measure temperature distribution at fracture area during tensile test of GFRP, thermography of fractured area is divided into 6 parts. 5 straight lines are drawn around fracture area. Among 5 lines, line with less fractured area confirms its temperature distribution with 3 lines. 6 thermography images on significant temperature distribution for fractured area are taken per specimen. Each time for corresponding point is studied with fracture time.

Fig. 3 shows the thermography image for polypropylene, which showed intermittent crack growth. After 1st fracture, crack growth stopped temporarily. On 150 seconds, 2nd fracture increment occurred. Maximum temperature for fractured area was 24.7°, and fracture area was increased from 2nd fracture. Fig. 4, 5 shows fracture image of GFRP with fiber content (10wt%, 20wt%) and fiber orientation function (*J*=0.1).

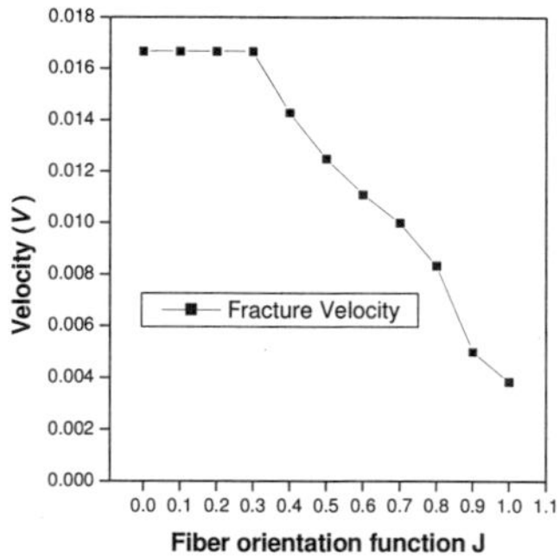

Fig. 6. Relationship between fracture velocity and fiber orientation function *J*.

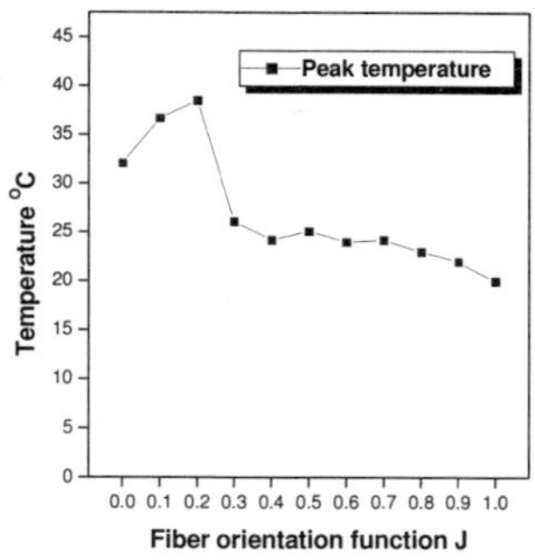

Fig. 7. Relationship between peak temperature and fiber orientation function *J*.

Fig. 6 shows the relationship between fiber orientation function and fracture speed. As fiber orientation function increases (anisotropy), fracture speed decreases. This indicates the slow propagation of fracture when fiber orientation is changed to anisotropy. When fiber orientation function, J is 0, 0.1, 0.2, and 0.3, fracture speed is the same. Fig. 7 shows the relationship between fiber orientation function and maximum temperature is described. This figure has the correlation to Fig. 6. When fiber orientation exhibits anisotropy (J=1), the maximum temperature is lowered. On Fig. 6, fracture speed is the same for J=0-0.3. However, maximum temperature is the highest for J=0.2. After J=0.2, maximum temperature is lowered.

5. Conclusion

Method to analyze temperature distribution during crack propagation under tensile test of GFRP via IR thermography camera was suggested. As fiber orientation function increases (anisotropy), fracture speed decreases. This indicates the slow propagation of fracture when fiber orientation is changed to anisotropy. When fiber orientation exhibits anisotropy (J=1), the maximum temperature is lowered.

References

1. J. W. Kim and D. G. Lee: *J. Mater. Processing Technol.* 2008, **201**, 755.
2. X. Maldague: *Trends in Optical Nondestructive Testing(Invited Chapter)*, 2000, 591–609.
3. M. Rovert and O. Gray: INFRAMATION Vol. 5, 2004.

DEFORMATION CHARACTERISTICS IN COLD ROTARY FORGING OF A 20CrMnTi SPUR BEVEL GEAR

XINGHUI HAN

School of Automotive Engineering, Hubei Key Laboratory of Advanced Technology of Automotive Parts, Wuhan University of Technology, Wuhan, 430070, China

LIN HUA†

School of Automotive Engineering, Hubei Key Laboratory of Advanced Technology of Automotive Parts, Wuhan University of Technology, Wuhan, 430070, China

Cold rotary forging is an advanced metal plastic forming technology with continuous and local deformation. Previous research concentrates mainly on studying the cold rotary forging process of the simple cylindrical and ring workpiece and thus it is essential to explore the deformation characteristics of cold rotary forging of a complicated spur bevel gear. In the current paper, a 3D rigid-plastic FE model of cold rotary forging of a 20CrMnTi spur bevel gear is developed using the DEFORM-3D FE software. Through simulation, the metal flow and distributions of different field-variables such as metal flow velocity, effective strain and force and power parameters are investigated in detail. The results of this research reveal the fundamental deformation laws of cold rotary forging of a spur bevel gear and provide valuable guidelines for better understanding the complex forming process.

Keywords: Cold rotary forging; Spur bevel gear; 20CrMnTi alloy; FE modeling.

1. Introduction

Cold rotary forging is an advanced metal plastic forming technology with continuous and local deformation and it offers the following advantages: lower level of noise and vibration, uniform quality, smooth surface and considerable savings in energy and materials cost. These significant advantages have determined that cold rotary forging has been used to manufacture a wide variety of mechanical components such as disks, rings, flanges and gears in many industry fields such as automobile, machine tool, electrical equipment, cutting tool and hardware.

Because of the above significant advantages, many scholars have carried out the research on the deformation mechanisms of cold rotary forging of the

---

† LIN HUA, email: lhua@whut.edu.cn

cylindrical and ring workpiece by analytical[1], experimental[2-6] and simulation[7-12] methods. Obviously, all of these efforts promote a greater knowledge and understanding of the basic deformation characteristics and mechanisms of the cold rotary forging process. However, it should be noted that the particular attention is paid to simple symmetrical geometries, the boundary conditions of which differ greatly from those of other complicated rotary formed parts such as gears, inevitably suffering from a lack of practical significance.

Due to the complex shape, complicated tooling design and high load requirements, spur bevel gears have raised a formidable challenge to the conventional precision forging technology beyond its current level of technology. For the purpose of breaking these stubborn forming barriers, cold rotary forging technology is more and more used for the fabrication of spur bevel gears owing to its flexibility and low tool load requirement. Unfortunately, relative studies on cold rotary forging of spur bevel gears have been still scant. Since FE simulation is an efficient and reliable approach to explore the dynamics of the system, it can be widely applied in many applications such as structural dynamics[13-14], fluid-structure coupling problems[15-16], as well as many other multi-field problems. In the current paper, the numerical simulation of cold rotary forging of a 20CrMnTi spur bevel gear has been carried out on the basic of the theory of rigid-plastic FE method under the DEFORM-3D platform and the fundamental deformation laws have been revealed.

2. 3D FE Modeling of Cold Rotary Forging

The spur bevel gear served as a component of the automobile differential mechanism is regarded as the subject investigated. The forging drawing is illustrated in Fig. 1.

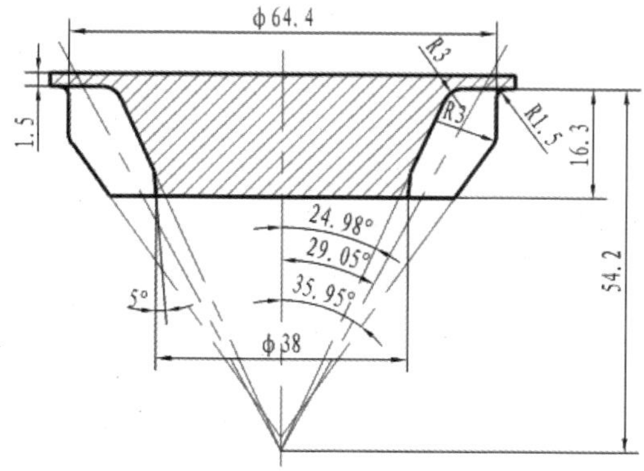

Fig. 1. Forging drawing of spur bevel gear (unit: mm).

The material of the workpiece used in this current study is a low-carbon alloy steel, 20CrMnTi (China specification). Up to date, the mechanical properties of 20CrMnTi alloy at room temperature have not been reported in the previous literature. So it is necessary to conduct the tensile test to obtain its mechanical properties so as to meet the FE modeling requirements. The tensile test is carried out on the Zwick Z100 materials testing machine. Fig.2 shows the representative plot of engineering stress vs. engineering strain of 20CrMnTi alloy.

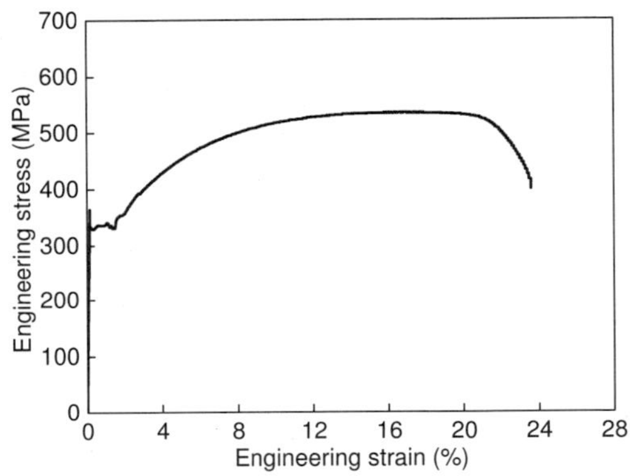

Fig. 2. Engineering stress-engineering strain curves of 20CrMnTi alloy.

Based on the above analysis, a 3D rigid-plastic FE model of cold rotary forging of a 20CrMnTi spur bevel gear is developed using the DEFORM-3D FE software, as shown in Fig.3. The feed amount per revolution is 2 mm r-1 and the inclination angle of the upper die is 2 °. The friction factor of 20CrMnTi alloy under the lubricated (MoS2) condition is equal to 0.14.

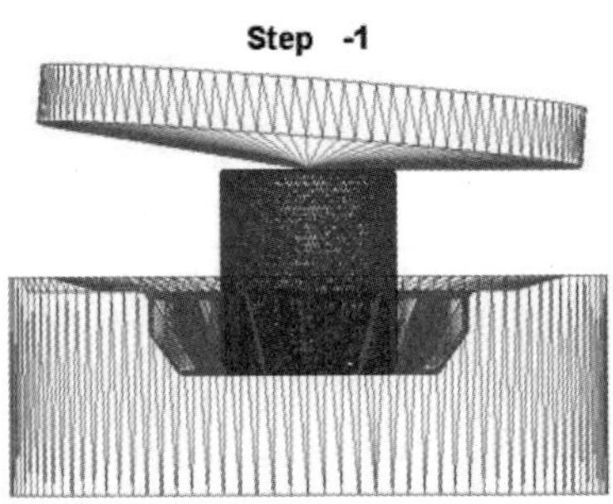

Fig. 3. 3D FE model of cold rotary forging of a spur bevel gear.

3. Results and Discussion

3.1. *Forming process of gear shape*

Fig.4 illustrates a series of typical forming steps at different time, which clearly presents the detailed forming process of gear shape in cold rotary forging of a spur bevel gear. As can be seen clearly that the mesh deformation exhibits a great difference with time and position. At the very beginning of the forging process, the upper die contacts the upper surface, and the meshes near the upper die initiate to deform, thus resulting in the reduction of the workpiece height and the slight increase of the diameter of the upper surface, as shown in Fig.4 (b). Over time, the height of the workpiece decreases by degrees, and the lower die begins to contact the cylindrical surface near the lower surface, which causes the metal in these areas to begin to yield (Fig.4 (c)). As the deformation progresses, the metal near the upper die flows more acutely than that near the lower die, thus leading to the significant increase in diameter of the upper surface, as shown in Fig.4 (d) and (e). This is so-called "mushroom" effect, which is one of main deformation characteristics of cold rotary forging. Meanwhile, it is also clearly shown that the cylindrical surface is progressively in contact with the lower die from the small-end gear shape to the large-end one (Fig.4 (c)-(e)). Then the workpiece enters a steady deformation stage and the profile of gear shape is gradually deformed (Fig.4 (f)). As the forging process continues, the flash is observed initially between gear teeth (Fig.4 (g)). At the final forging stage, the flash grows slowly due to the increasing radial flow resistance and the filling of large-end gear shape precedes that of small-end gear shape, as shown in Fig.4 (h).

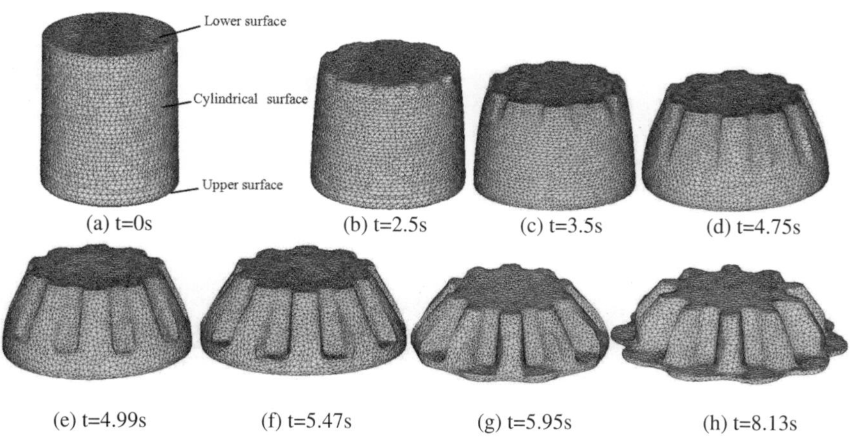

Fig. 4. Forming process of gear shape in cold rotary forging.

Apart from the above-mentioned deformation characteristics, it is also interesting to note that the ultimate flash shape exhibits the "petal-form" effect caused by the difference of radial flow resistance.

3.2. *Distribution of metal flow velocity fields*

In the cold rotary forging process of a spur bevel gear, the workpiece under the local load can be divided into two deformation areas: active deformation area contacting the upper conical die and passive deformation area separating with the upper conical die. For this reason, the metal flow in cold rotary forging is much more complicated as compared to the conventional forging processes and hence the metal flow laws during the cold rotary forging process must be clarified. In DEFORM-3D, metal flow velocity fields are represented by line segments with arrows, in which colors of arrows stand for flow velocities, and directions of arrows represent directions of metal flow.

Fig.5 presents the distribution of flow velocity fields with time in cold rotary forging of a spur bevel gear. The observation from this figure clearly shows that during the whole cold rotary forging process, the flow velocities of metal in the active deformation area are much greater than those in the passive deformation area. This phenomenon can be ascribed to the fact that under the action of axial feed of the lower die and oscillation of the upper die, the active deformation area is under local heavy load, thus leading to severe metal flow in this area; by contrast, the metal flow velocities of the passive deformation are far smaller. Moreover, it can also be seen that the metal flow directions in the active deformation area differ from those in the passive deformation area: before the

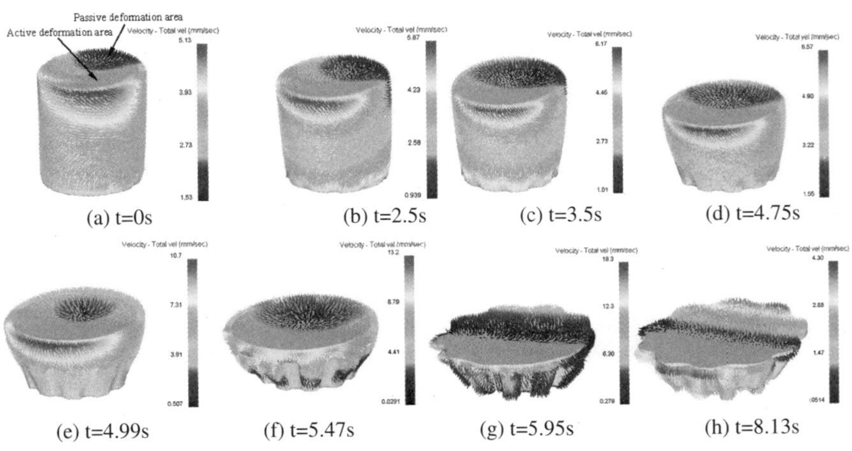

(a) t=0s (b) t=2.5s (c) t=3.5s (d) t=4.75s

(e) t=4.99s (f) t=5.47s (g) t=5.95s (h) t=8.13s

Fig. 5. Flow velocity fields distribution with time in cold rotary forging of a spur bevel gear.

finishing stage, outward radial flow to fill the die cavities is predominant for the metal in the active deformation area, which causes the metal to expand outwards remarkably, as shown in Fig.5 (a)-(g). That is what leads to the "mushroom" effect. At the final finishing stage, the main metal flow directions in the active deformation area change from radial direction to axial direction due to the great radial flow resistance, which helps cause the complete filling of die cavity (Fig.5 (h)). The passive deformation area is under lateral compression from the active deformation area, thus resulting in the metal flow in the opposite direction of deviating from the die cavities.

3.3. *Distribution of effective strain*

The distribution of effective strain (PEEQ) at different time during the cold rotary forging process is illustrated in Fig.6. As described above, the lower die contacts the workpiece on the down-side surface near the lower surface firstly, so the metal in this area is easier to satisfy the yield condition to arrive at the plastic deformation state. Naturally, the plastic deformation zone is firstly formed in this area (Fig.6 (a)). Under the action of axial feed of the lower die and oscillation of the upper die, the plastic deformation zone gradually expands along the circumferential direction and along the gear tooth orientation, as shown in Fig.6 (b) and (c). As the deformation progresses, the plastic deformation zone has penetrated the whole down-side surface of the workpiece (Fig.6 (d)). After the gear shape emerges, the PEEQ accumulates rapidly in large-end dedendum areas owing to the "mushroom shape" effect, as shown in Fig.6 (e)-(g). At the final forging stage, the large-end gear shape achieves filling prior to the small-end

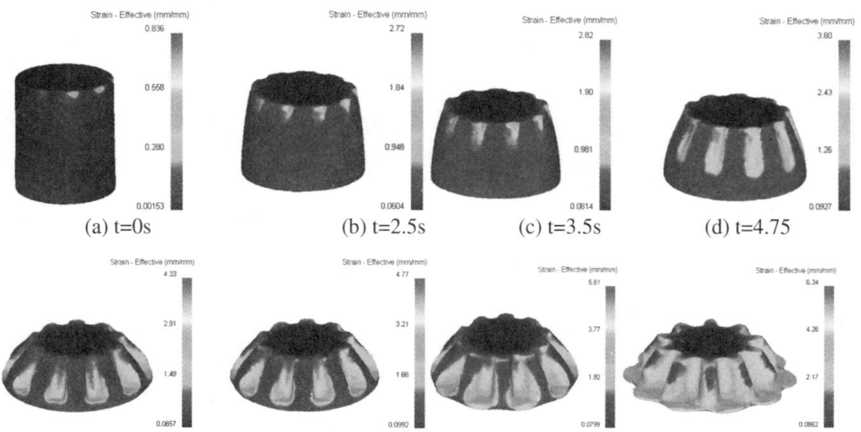

(a) t=0s (b) t=2.5s (c) t=3.5s (d) t=4.75

(e) t=4.99s (f) t=5.47s (g) t=5.95s (h) t=8.13s

Fig. 6. PEEQ distribution with time in cold rotary forging.

gear shape, so the PEEQ value of the former is larger than that of the latter (Fig.6 (h)). Moreover, it can also be seen from Fig.6 (h) that the largest PEEQ value locates in the large-end dedendum areas in the forged gear, which indicates the metal of this area undergoes the highest degree of deformation. By contrast, the PEEQ value is relatively lower in the central areas of the forged gear close to the lower die since the metal in this area is not pressed against the wall of the die cavities.

3.4. *Evolution of forming load*

Forming load requirement is very significant for accurate design of dies, reasonable choice of equipment and proper determination of process specification. Fig.7 illustrates the evolution of the axial forging force obtained numerically in function to forming time. It can be seen from Fig.7 that the whole cold rotary forging process can be divided into four stages as marked in the figure. At the first stage, i.e. in the very beginning of the forging process, the upper die contacts the workpiece, thus resulting in the plastic deformation of the contact area. Consequently, the axial forging force increases rapidly from zero to a certain value. With increasing deformation the workpiece reaches a steady deformation stage and thereby the force increases slowly. At the third stage, as the process proceeds, the axial forging force increases sharply up to the maximum value (about 2475 kN) because of the great metal flow resistance of complete filling of die cavity. At the final stage, i.e. the finishing stage, the lower die stops the axial feed, resulting in the significant decrease in the axial forging force. Apart from these, it is also clear that the variation curve of axial forging force presents a slight oscillation with time, the reason for which is that the cold rotary forging is a complex dynamic contact, highly nonlinear and non-steady-state deformation process.

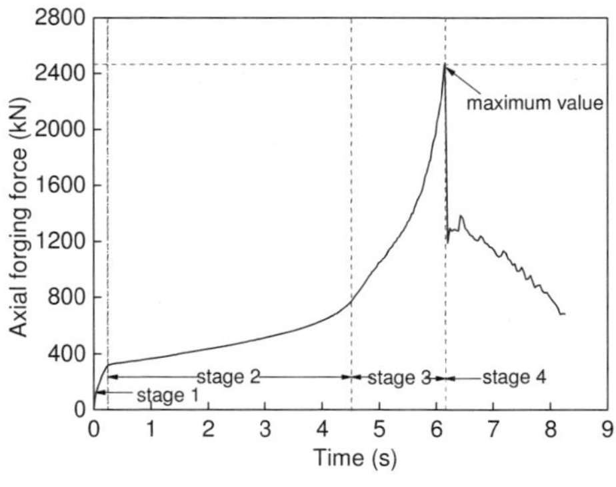

Fig. 7. Variation curve of the axial forging force with time.

4. Conclusion

In the current paper, a 3D rigid-plastic FE model of cold rotary forging of a 20CrMnTi spur bevel gear is developed using the DEFORM-3D FE software. Through simulation, the metal flow and distributions of different field-variables such as metal flow velocity, effective strain and force and power parameters are investigated in detail. The results of this research show the following:

(1) At the beginning of cold rotary forging of a spur bevel gear, the workpiece metal near the upper die flows more acutely than that near the lower die, so the diameter of the upper surface of the workpiece increases significantly and the "mushroom" effect occurs. Meanwhile, the forming sequence of gear shape is from the small-end gear shape to the large-end one. At the final forging stage, the flash is formed.

(2) During the whole cold rotary forging process, the flow velocities of metal in the active deformation area are much greater than those in the passive deformation area. Before the finishing stage, outward radial flow to fill the die cavities is predominant for the metal in the active deformation area. At the final finishing stage, the main metal flow directions in the active deformation area change from radial direction to axial direction due to the great radial flow resistance. The metal flow in the passive deformation area is in the opposite direction of deviating from the die cavities.

(3) The effective strain gradually expands from the small-end gear shape to the large-end one. At the final forging stage, the effective strain value of the large-end gear shape is larger than that of the small-end one. Moreover, the largest effective strain locates in the large-end dedendum areas in the forged gear. By contrast, the effective strain is relatively lower in the central areas of the forged gear close to the lower die.

(4) The whole cold rotary forging process of a 20CrMnTi spur bevel gear can be divided into four stages and the maximum axial forging force is about 2475 kN.

Acknowledgment

The authors would like to thank the Natural Science Foundation of China for Distinguished Young Scholars (No. 50725517) for the support given to this research.

References

1. J. Oudin, Y. Ravalard, G, Verwaerde and J. C. Gelin: *Int. J. Mech. Sci.*, 1985, **27**, 761-780.

2. J. B. Hawkyard, C. K. S. Gurnani and W. Johnson: *J. Mech. Eng. Sci.*, 1977, **19**, 135-142.
3. P. M. Standring, J. R. Moon, E. Appleton: *Met. Technol.*, 1980, **7**, 159-166.
4. X. H. Pei, D. C. Zhou and Z. R. Wang: Proc. 2nd Int. Conf. on 'Rotary Metalworking Processes', UK, October 1982, Paper 81.
5. D. C. Zhou, S. J. Yuan, Z. R. Wang, Z. R. Xiao: *J. Mater. Process. Technol.*, 1992, **32**, 471-479.
6. G. C. Wang, J. Guan, G. Q. Zhao: *J. Mater. Process. Technol.*, 2005, **169**, 108-114.
7. G. C. Wang, K. M. Xue and Y. Lu: *J. Mater. Process. Technol.*, 1997, **71**, 299-304.
8. S. J. Yuan, X. H Wang, G. Liu and D. C. Zhou: *J. Mater. Process. Technol.*, 1998, **86**, 252-256.
9. G. Liu, S. J. Yuan, Z. R. Wang and D. C. Zhou: *J. Mater. Process. Technol.*, 2004, **151**, 178-182.
10. L. Hua and X. H. Han: *Mater Des.*, 2009, **30**, 2133-2142.
11. X. H. Han and L. Hua: *Mater Des.*, 2009, **30**, 2802-2812.
12. X. H. Han and L. Hua: *J. Mater. Process. Technol.*, 2009, **209**, 5353-5362.
13. T. H. Yan, X. D. Chen, R. M. Lin: *Mechatronics*, 2008, **18**, 466-474.
14. T. H. Yan, X. D. Chen, W. F. Dou, R. M. Lin: *Mech. Based. Des. Struc.*, 2008, **36**, 283-305.
15. X. D. Chen, X. M. He: *Tribol. Int.*, 2006, **39**, 1336-1341.
16. X. M. He, X. D. Chen: *Int. J. Adv. Manuf. Tech.*, 2007, **32**, 978-984.

NUMERICAL SIMULATIONS AND OPTIMIZATION OF B-PILLAR FORMING BASED ON DIFFERENT BINDERS

JUN LIU, HUA-JIE [†] MAO, RONG-CHUANG CHEN

School of Material Science and Engineering, Wuhan University of Technology, Wuhan, 430070, China

WEI CAO

Zhejiang Geely Automobile Research Institute Co., Ltd., Zhejiang, 317000, China

Weld line motion is one of the difficulties in High Strength Steel Tailor Welded Blanks (HSSTWBs) forming. Studies show that an appropriate process design can reduce weld line movement and improve the HSSTWB formability. In this paper, three process schemes based on different binders were studied using Dynaform software in order to produce a qualified B-pillar. The first method employed a common binder with a constant blank holder force. The second method used a segmented binder with different constant blank holder forces applied at each part. The third method applied a stepped binder with a constant gap between die and binder. Different binders lead to different stress-strain state and metal flow. Simulations demonstrated that the last two methods can effectively improve weld-line movement, and a stepped binder is better at reducing material thinning and springback. Finally, the third process scheme was carried out to obtain a good quality B-pillar.

Keywords: High Strength Steel Tailor Welded Blank (HSSTWB); Weld line movement; Material thinning; Springback; Stepped binder; B-pillar.

1. Introduction

In recent years, the automotive sector has paid a great deal of attention to weight reduction and fuel economy. As an important development tendency of vehicle weight reduction technology, High Strength Steel Tailor Welded Blanks (HSSTWBs) are commonly used in automotive panels manufacturing. A HSSTWB usually consists of several flat sheets with different materials, thickness, and coatings welded together before forming. The main advantages of using a HSSTWB are as follows: (1) weight reductions, (2) reduction of the fuel consumption, (3) improved safety performance, (4) improvement of dimensional accuracy, and (5) reduced manufacturing costs. However, the different deformation-resistant of dissimilar materials may result in uneven deformation,

[†] Corresponding author, email: maohj@whut.edu.cn

which can lead to tearing/wrinkling failures in the thinner/weaker material, [1] thus reduce the overall HSSTWB formability. And the uneven deformation can be reflected by the lateral movement of weld line during stamping process.

Hence, it is feasible to improve the formability of HSSTWB by minimizing the weld line motion. The control method of weld line movements in HSSTWBs has been studied by many researchers, and various process schemes have been proposed to solve this difficulty. Youngmoo Heo *et al.* [2] indicated that the drawbead adds additional restraining forces to the blank, so that the movement of the weld line is reduced. CHEN Wei *et al.*[3] found that the VBHF control is an effective method to improve the formability of TWB. Brad Kinsey *et al.* proposed a clamping method along the weld line of the tailor-welded blank during forming to reduce the maximum strains. [4] However, these three methods are practically hard to be implemented in real productions because (1) spatial allocations of drawbead/BHF along the TWB flanges are challenging for complex parts; (2) general press failed to provide regularly changed BHF during stamping and (3) clamping mechanism is too complex, hence increase the manufacturing cost.

In practice, binder design is a simple and effective way to reduce weld line movement for HSSTWB with similar material but dissimilar thickness (the emphasis of this research). In this paper, three process schemes have been simulated by finite element simulation software Dynaform in order to produce qualified B-pillars. The three process schemes are as follows： (1) common binder: a flat binder used for comparison, (2) segmented binder: blank holder forces were imposed on both sides of the binder, ensuring the continuity of pressure distribution on both sides of the weld seam, and (3) stepped binder: a binder designed to accommodate the thickness differential in order to press the HSSTWB seamlessly between the binder and die. These three schemes are respectively called case A, case B, and case C. Weld line motion, thickness reduction and springback under the three processes were compared. Finally, the best of the three process schemes was conducted to obtain a qualified part.

2. Numerical Simulation

2.1. *Finite element modeling*

During this research, the weld seam was replaced by a line containing 56 spots. The blank meshing was mainly done with quadrilateral shell elements of the Belytschko–Tsay formulation, with five through-thickness integration points. And the elements number of blank is 79577 after adaptive remeshing. The die, punch and binder elements were defined as rigid shell elements, and the maximum size was set to 20mm.

Simulation modeling including die, binder, punch and blank is show in Figure 1(a). The equivalent drawbeads were adopted in this simulation. And the layout of drawbeads in this simulation is shown in Figure 1(b). The number of drawbeads was 12.

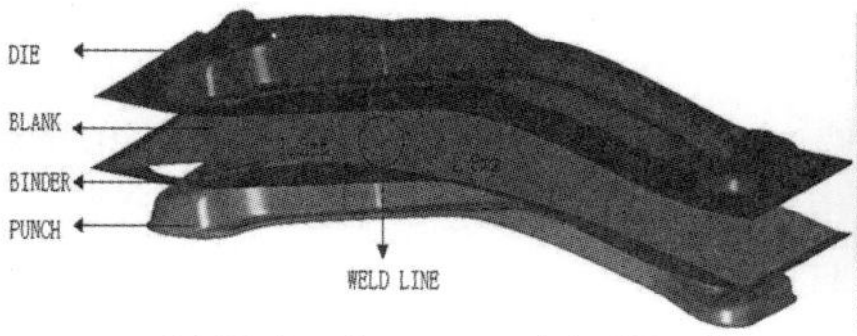

(a) Finite element model of B-pillar.

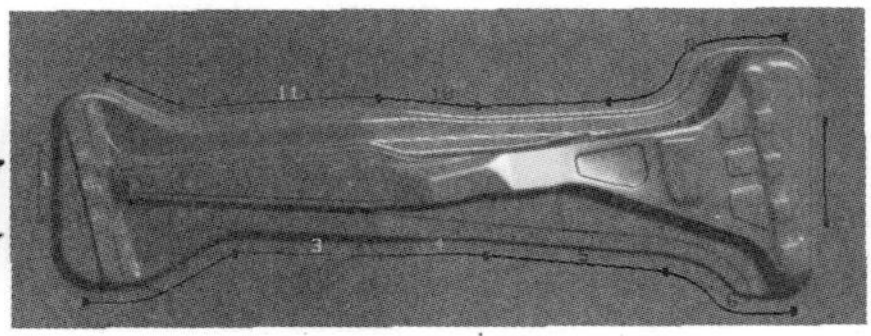

(b) Layout of drawbeads.

Fig. 1. Finite element model and drawbeads.

2.2. *Simulation program and material property*

Simulation program totally contained three process schemes. The first scheme adopted a common binder, which applied constant blank holder force of 1000kN. The second scheme used a segmented binder, which applied constant blank holder force of 400, 600kN, respectively for two material thicknesses of 1.2, 1.8mm. The third scheme applied a stepped binder, which maintained a fixed gap of a material thickness between the binder and die.

To meet the performance requirements, High Strength Steel B340/590DP was selected to make vehicle B-pillar. The material properties are as follows: $\sigma_s = 456.6\text{Mpa}$, $\bar{r} = 0.88$, $n = 0.21$.

3. Results and Discussion

3.1. *Weld line movement*

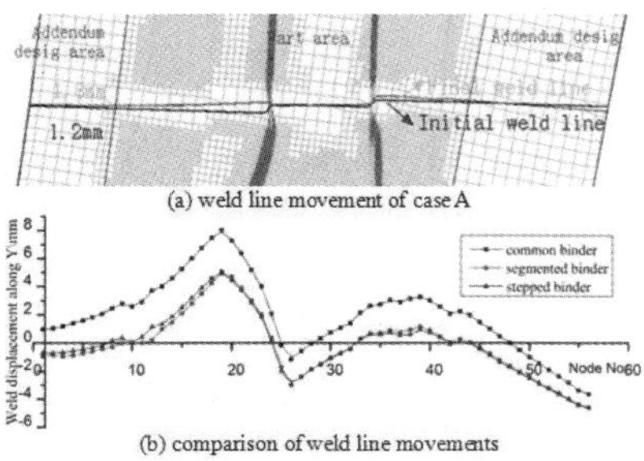

Fig. 2. Weld line movements comparison.

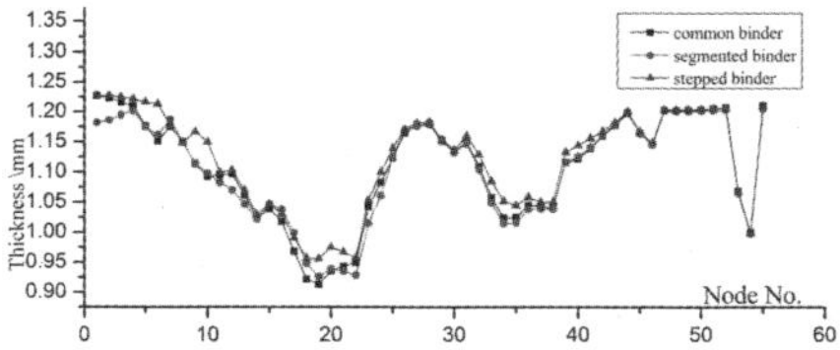

Fig. 3. Thickness distributions of three cases.

Figure 2 illustrates the weld line movements along Y direction under the three process schemes. The displacements are defined to be positive when materials flow from the thin blank toward the thick part. The weld line movements predicted by simulation results of the three process schemes show almost the same tendency: the weld line moves toward the thicker side at most of part area and moves toward the thinner side in addendum design area, refer to Figure 2. The maximum positive weld line movement of the three cases ranks as case A > case B ⩾ case C, while the maximum negative weld line movement occurs in the reverse order. The maximum weld line movement occurs at the end of left flange.

As show in Figure 2, case B and case C are better than case A in the controlling of weld line movement. And this illustrates that case B and case C are good at reducing uneven deformation of HSSTWBs.

3.2. *Thickness reduction*

The thickness distributions of the drawn B-pillars are measured along weld line. Figure 3 compares the thickness distributions along weld line of case A, case B, and case C. By comparison of the three curves, a similar tendency is observed. The minimum thickness is observed at the corner of the left punch shoulder. This might be due to the large biaxial tension stress state here. Relatively, the thickness distributions in case C are better than that in case A and case B. And this means that a stepped binder can effectively reduce material thinning.

3.3. *Springback*

The springback results are listed in Table 1. The maximum springback of the whole part of case C is the smallest of the three cases, and this is more important in practice. It is obvious that a stepped binder can reduce the springback of HSSTWBs.

Table 1. The maximum springback of parts (mm).

Schemes	Left part (1.2mm)	Right part (1.8mm)
Case A	6.64	4.42
Case B	6.32	3.96
Case C	4.83	5.21

In summary, the use of stepped binder and segmented binder helps reduce weld line motion, but stepped binder is better at reducing material thinning and springback. Hence, the best quality of B-pillar forming can be achieved by using a stepped binder.

4. Experiment

According to the stepped binder process scheme, the drawing mold of B-pillar was designed and manufactured. The experiment was conducted in a single-action press with a capacity of 630 t. The result is shown in Figure 4(a). The forming status of this part is found to have good agreement with the simulation result, see Figure 4(b). There is no crack in part area, most wrinkles are located in binder surface, and the maximum weld line movement is less than 6 mm. It is obvious that this part is qualified. From the experimental research, it can be seen that the stepped binder process scheme is indeed an effective way to improve the formability behavior of HSSTWB, and the FE simulation results are dependable.

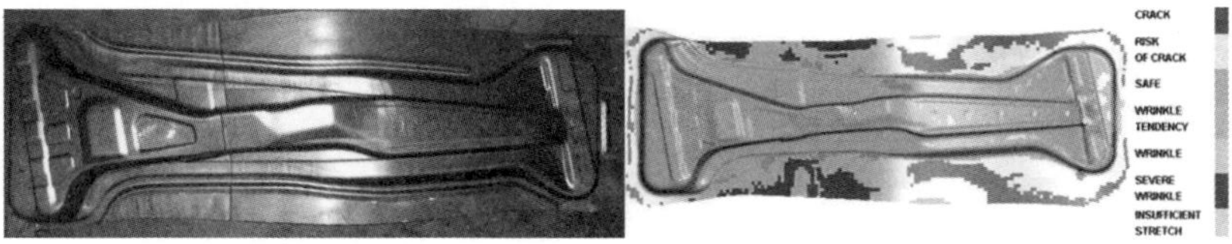

(a) Experimental result. (b) Simulation result.

Fig. 4. Simulation result and experimental result.

5. Conclusion

- FE simulations are able to predict the forming behaviors of HSSTWB with acceptable accuracy;
- Compared to common binder, segmented binder and stepped binder can effectively reduce weld line movement;
- Though both of segmented binder and stepped binder are good at controlling weld line motion, stepped binder are better at reducing material thinning and springback, offering the best HSSTWB forming quality.

Acknowledgments

The authors of this paper are grateful to the support from International Scientific Cooperation and Exchange Special Project: the development and application of vehicle weight reduction core technology (2008DFB50020).

References

1. Brad Kinsey, Zhihong Liu, Jian Cao: *J. MATER. PROCESS. TECH.*, 2000, **99**, (1–3), 145–153.
2. Youngmoo Heo, Youho Choi, Heon Young Kim, Daegyo Seo: *J. MATER. PROCESS. TECH.*, 2001,**111**, (1-3), 164-169.
3. Chen Wei, Lv Dun, Lin Guo-Song, Hou Bo: 'International Technology and Innovation Conference 2006', Hangzhou, China, November 2006, Chinese Mechanical Engineering Society, 1215–1219.
4. W. Chen, G.S. Lin, S.J. Hu: *J. MATER. PROCESS. TECH.*, 2008, **207**, (1–3), 204–210.

HIGH TEMPERATURE COMPRESSION DEFORMATION BEHAVIOR OF 7050 ALUMINUM ALLOY

HUIE HU [†]

Department of Chemistry and Materials, Naval University of Engneering, Wuhan, 430033, China

XIAODONG KONG

Department of Chemistry and Materials, Naval University of Engneering, Wuhan, 430033, China

DEBIN CHEN

Department of Chemistry and Materials, Naval University of Engneering, Wuhan, 430033, China

Compression tests of 7050 aluminum alloy have been conducted at different temperatures (340, 380 and 420°C) and different strain rates of 0.1, 1 and $10s^{-1}$. The microstructure characteristics of the alloy after deformation are investigated using EBSD technique. The results show that the stress level of the alloy can be presented by a Zener-Holloman parameter in a hyperbolic sine-type equation with the hot activation energy of 137.9 kJ/mol. The dynamic restoration mechanism of the 7050 aluminum alloy is dynamic recovery mechanism, and the main dislocation movement mechanism is cross-slipping of crew dislocation.

Keywords: Aluminum alloy; Dynamic recovery; Compression; Microstructure.

1. Introduction

Aluminum alloys with high specific strength and excellent mechanical properties have been always the primary structural materials for aircrafts in the space and vehicles at ground,[1]. Dynamic recovery and dynamic recrystallization often occur in aluminum alloys with high stacking fault energy during high temperature deformation,[2]. Several dynamic recrystallization mechanisms of aluminum alloys at different deformation conditions had been proposed, and continuous dynamic recrystallization is more prone to occur,[3-4]. Many reports,[5-6] about high temperature deformation behavior and dynamic soften mechanism of aluminum alloy have been published.

[†] Corresponding author, email: huhuie@163.com

In this paper, the deformation behavior and dynamic restoration process of the 7050 aluminum alloy during high temperature compression were studied.

2. Experimental Method

The 7050 aluminum alloy used in the present work was provided by Beijing Institute of Aeronautical Materials with as-rolled plate with 80 mm thickness. Compression specimens with a diameter of 8mm and a height of 12mm were machined with the compression axis parallel to the normal direction. Before compression test, the compression specimens were solid solution treated at 477 °C for 1 h and then quenched into water. Compression tests were conducted to a predetermined strain of 0.69 on a thermal mechanical simulator (Gleeble-1500) at 340, 380 and 420 °C and strain rates of 0.1, 1.0, and 10 s^{-1}, respectively. The specimens were resistance heated to compression temperature at a heating rate of 10 °C/s by thermocoupled feedback-controlled AC current and homogenized for 5min before comepression was initialed. Samples for analyses were sectioned parallel to the compression axis, and then mounted. Samples for EBSD measurement were mechanically polished followed by electro polishing using 10 vol. % $HClO_4$ acids in alcohol. The samples were examined and analyzed using HKL Channel 5 software in a JEOL 733 electron probe with an accelerating voltage of 20 kV. The demarcate ratios of EBSD in the present research are all higher than 94%.

3. Results and Discussion

The true stress-strain curves of the 7050 aluminum alloy deformed at different deformation conditions are presented in Fig.1, which shows the effects of deformation temperature and strain rate on flow behavior of the experimental alloy. It is shown that all the curves exhibit a peak stress, and the peak stress increases with increasing strain rate and decreasing deformation temperature. Furthermore, deformation temperature and strain rate have great effect on the rates of strain hardening and strain softening. When the 7050 aluminum alloy was deformed at 0.1 s^{-1}, the true stress-strain curve with any strain rate is flat, and large plastic deformation was achieved at almost constant stress, suggesting that the 7050 aluminum alloy has good uniform deformation ability at 0.1 s^{-1}.

Following constitutive equations have commonly been applied in high temperature deformation of crystal materials:

$$\dot{\varepsilon} = A'\sigma^{n'} \exp(-\frac{Q}{RT}) \tag{1}$$

$$\dot{\varepsilon} = A'' \exp(\beta\sigma) \exp(-\frac{Q}{RT}) \tag{2}$$

$$\dot{\varepsilon} = A(\sinh \alpha\sigma)^n \exp(-\frac{Q}{RT}) \tag{3}$$

$$Z = \dot{\varepsilon} \exp(\frac{Q}{RT}) \tag{4}$$

Where A, A', A'', n, n', α and $\beta(=\alpha n)$ are constants, Q is an activation energy for deformation, R is the gas constant and Z the Zener-Holloman parameter.

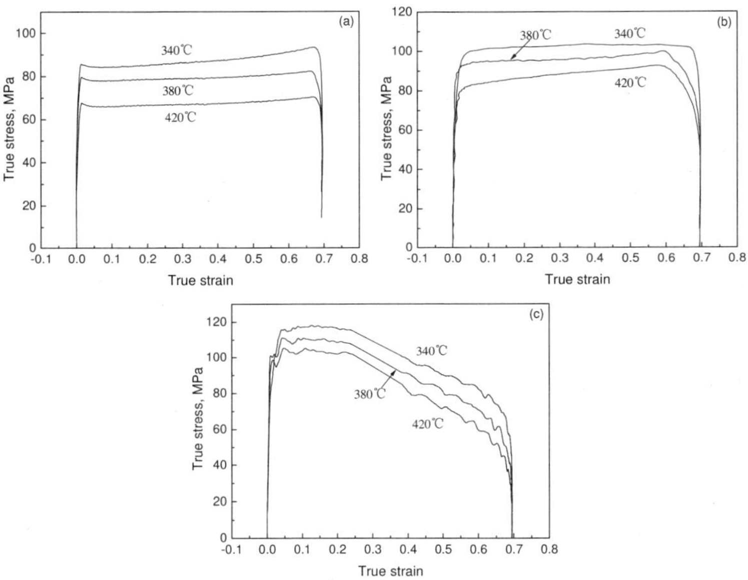

Fig. 1. True stress-strain curves of the alloy deformed at different deformation temperatures (a) 0.1 s^{-1}; (b) 1 s^{-1}; (c) 10 s^{-1}.

The power law, equation (1), and the exponential law, equation (2), are suitable for a low stress and a high stress, respectively. The hyperbolic sine law, equation (3), is a more general form suitable for stresses over a wide range.

As shown in Fig.2, the peak stresses obtained form the tensile tests can be approximated by a group of parallel and straight lines in the high temperature deformation conditions, demonstrating that the peak stress for the 7050 aluminum alloy can be represented by a Zener-Holloman parameter in the hyperbolic sine-type equation. The high temperature deformation activation energy derived from Arrhenius plots of versus $1/T$ shown in Fig.2(b) is 137.9 kJ/mol for the 7050 aluminum alloy, which is similar to self-diffusion activation energy of pure aluminum (142 kJ/mol).

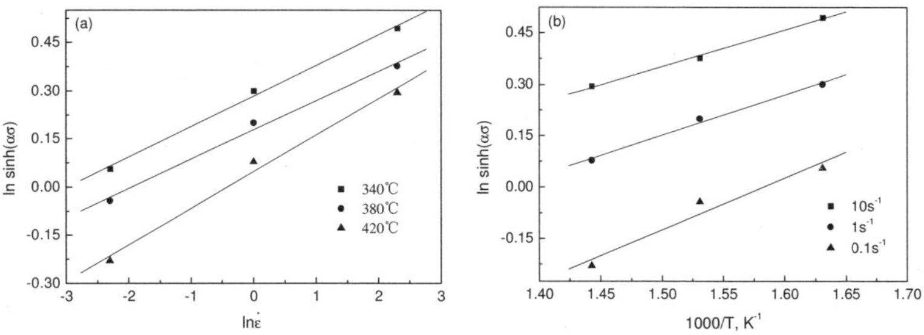

Fig. 2. Relationships between peak stress with: (a) strain rate and (b) deformation temperature.

Fig. 3 is boundary misorientation angle distributions of the 7050 aluminum alloy deformed at different deformation temperatures (340, 380 and 420 ºC) and the strain rate of 0.1 s^{-1}. The EBSD results show that all the recrystallized grain volume fractions are less than 15% when the deformation temperatures are 340, 380 and 420 °C. So the dynamic restoration mechanism of the 7050 aluminum alloy at high temperature compression is dynamic recovery.

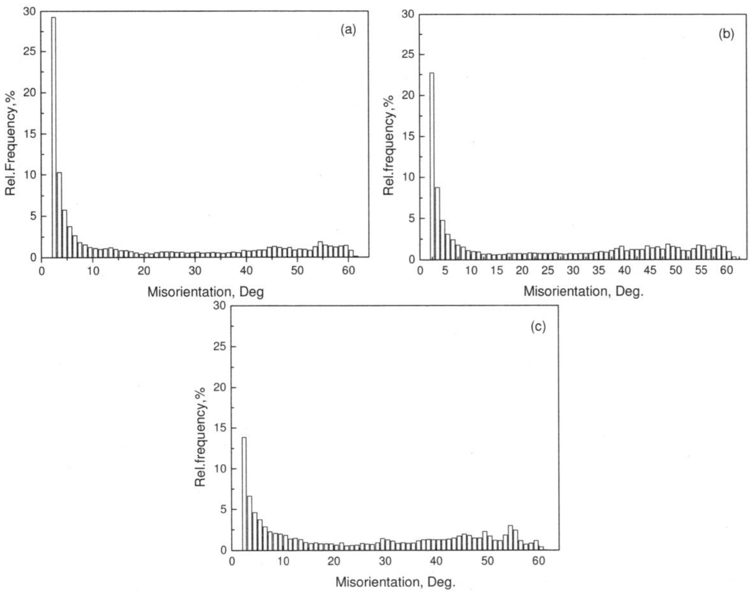

Fig. 3. Histograms showing boundary misorientation angle distributions of the 7050 aluminum alloy compressed to the strain of 0.69 at different compression temperatures with the strain rate of 0.1 s^{-1} (a) 340 °C ; (b) 380 °C; (c) 420 °C.

Table 1 and Table 2 are thermo activation parameters of pure aluminum and the 7050 aluminum alloy. Table 2 show that apparent activation volumes of the 7050 aluminum alloy compressed at 340, 380 and 420 °C are 51.8 b^3, 59.9 b^3

and 54.0 b^3, respectively. So the main dislocation movement of the 7050 aluminum is cross-slipping of crew dislocation. The higher deformation activation energy of the 7050 aluminum alloy deformed at high temperature comes from the accession of Mg in 7050 aluminum alloy and dynamic precipitation during high temperature compression of the 7050 aluminum alloy.

Table 1. Thermo activation parameter of pure aluminum for dislocation mechanism.

Dynamic recovery mechanism	Activation energy /kJ.mol^{-1}	Activation volume /2.863×10^{-30}m^{-3}
Cross-slipping	117	10-100

Table 2. Activation energy for deformation and activation volume of 7050 aluminum alloy during high temperature compression.

Deformation temperature	Activation energy /kJ.mol^{-1}	Activation volume /2.863×10^{-30}m^{-3}
340 °C	116.90	51.8
380 °C	130.27	59.9
420 °C	166.55	54.0

4. Conclusion

(1) The stress level of 7050 aluminum alloy, which can be presented by a Zener-Holloman parameter in a hyperbolic sine-type equation with the hot activation energy of 137.9 kJ/mol.
(2) The dynamic restoration mechanism of the 7050 aluminum alloy is dynamic recovery mechanism, and the main dislocation movement mechanism is cross-slipping of crew dislocation.

Acknowledgments

This research is supported by Science College Fund of Naval University of Engineering.

References

1. J. Wloka, T. Hack and S. Virtanen: Corros. Sci., 2007, **49(3)**, 1437-1449.
2. S. Gourdet and F. Montheillet: Mater. Sci. Eng. A, 2000, **283**, 274-288.
3. H. J. McQueen: Mater. Sci. Eng. A, 2004, **387-389**, 203-208.
4. M. E. Kassner and S. R. Barrabes: Mater. Sci. Eng. A, 2005, **410-411**, 152-155.
5. H. Yamagata: Scripta Metal. Mater., 1992, **27(9)**, 1157-1160.
6. J. P. Lin, Y. L. Wang and Z. Lin: J. Univer. Sci. Technol. Beijing, 2003, **10(2)**, 45-48.

CFRP/FOAM CIRCULAR MEMBERS FOR ENERGY ABSORPTION CHARACTERISTICS UNDER AXIAL COMPRESSION

JU-HO CHOI

Department of Advanced Parts and Materials Engineering, Graduate School, Chosun University, 375 Seosuk-dong, Dong-gu, Gwang-ju 501-750, Korea

JAE-KI SIM

Department of Mechatronics Engineering, Chosun University, 375 Seosuk-dong, Dong-gu, Gwang-ju 501-750, Korea

GIL-YOUNG HAN

Department of Mechanical Engineering, Chosun University, 375 Seosuk-dong, Dong-gu, Gwang-ju 501-750, Korea

HYUN KOOK

Department of Mechanical Design Engineering, Chosun University, 375 Seosuk-dong, Dong-gu, Gwang-ju 501-750, Korea

IN-YOUNG YANG[†]

Department of Mechanical Design Engineering, Chosun University, 375 Seosuk-dong, Dong-gu, Gwang-ju 501-750, Korea

In this study, one type of circular shaped composite tube was used, combined with reinforcing foam and without foam. Furthermore, CFRP (Carbon Fiber Reinforced Plastic) circular member manufactured from CFRP prepreg sheet for lightweight design. CFRP is an anisotropic material which is the most widely adapted lightweight structural member because of their inherent design flexibility and high specific strength, elastic modulus, lower thermal transformation. The aim was to examine which structure was the most serviceable in absorbing the crushing energy in a safe way, as well as to find out the most financially efficient one.

Keywords: Energy absorption; CFRP (Carbon Fiber Reinforced Plastic); Foam.

[†] Corresponding author, email: iyyang@chosun.ac.kr

1. Introduction

Major targets for recent automobile design trend are environment-friendliness and shock safety performance. Therefore, lightweight vehicle, which will be inevitable for carbon dioxide reduction and fuel efficiency, is essential part of development. Therefore, automobile design should be made for reduced weight once the safety of vehicle is maintained. If the shock energy during collision is applied to passenger without any retarding measure, the safety of passenger will be greatly endangered. Especially, lighter material based on aluminum alloy exhibited significant improvement over time, and highly functional composite material have been researched consistently. [1~4]

In this study, polyurethane foam and carbon fiber reinforced plastic (CFRP), which is a well-known lightweight structural material, are used for fabrication of test specimen, and crushing property during static and collapse load application is quantitatively evaluated. CFRP is anisotropic material that changes mechanical performances according to lamination stack-up. Therefore, changes in interface layer and foam density are selected to compare crushing mode and energy absorption properties.

2. Specimen

Test specimen used in this experiment was made of CFRP prepreg sheet using autoclave, and polyurethane foam was filled into circular CFRP member. Filling foam for CFRP member's density is 40Kg m^{-3}. Shape for test specimen is shown in Fig.1. Energy absorption of CFRP circular member is affected by lamination condition. To understand the impact of change in interface number on energy absorption, number of interface was changed as 2, 3, 4, 6 and 7. To investigate energy absorption according to density of CFRP member and filled, density was set to 40Kg m^{-3}.[5~6] Forming of test specimen was made through autoclave as shown Fig. 2, and curing temperature from circumference heater

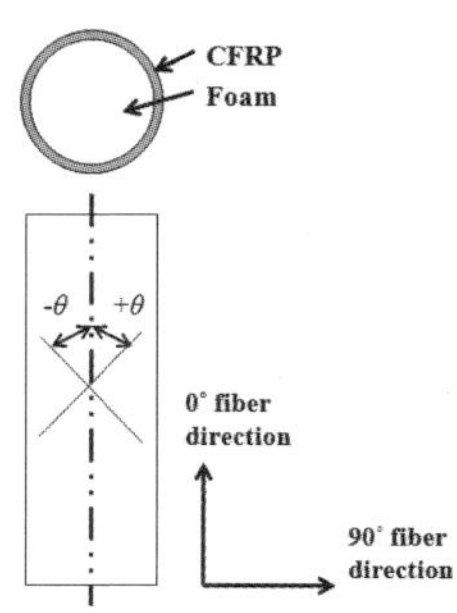

Fig. 1. Configuration of specimen.

on chamber was set to 130°C, and curing time was 90minutes. Chamber was vacuumed up to 10^{-1}Pa by compressor. It was pressurized to $3x10^5$Pa from outside. In addition, specimen was periodically cut for 120mm in length using diamond cutter for providing repetitive crushing if the specimen was not buckled without cutting of the end. To prevent the low absorption energy after initial crushing load. It was triggered as 45° at the one end of specimen to induce sequential crushing and higher energy absorption. [7]

Fig. 2. Autoclave.

3. Crushing Test

Static crushing test was made with universal test machine (UTM), and strain rate was controlled as 10mm min^{-1}. To provide homogeneous compressive load, crushing test in axial direction was made. Displacement was controlled up to 58.3% (70mm) for total length of 120mm, and sequential crushing test was conducted. Calculation of energy absorption amount on specimen was considered as the area from load-displacement diagram as shown in Fig. 3. By integrating Eq. (1) of load-displacement curve, it is used for evaluating absorbed energy in member.

$$E_a = \int_0^\delta PdS \quad (1)$$

Where, E_a is absorbed energy, P is crushing load, and S is deformation of test specimen during crushing.

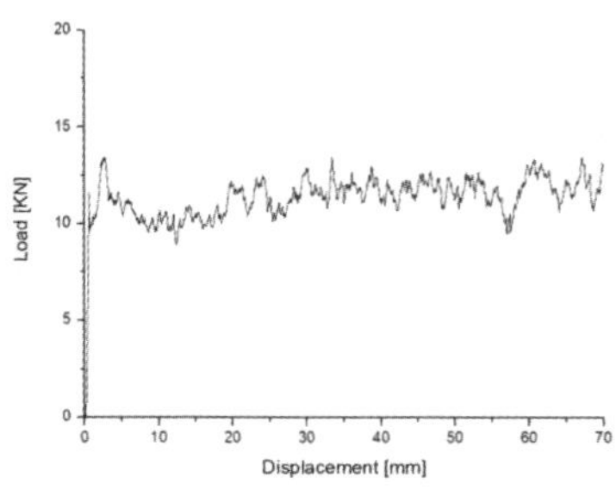

Fig. 3. Load-Displacement curve of $[0/90]_{2S}$.

4. Results and Discussion

In this study, absorption energy (E_a), average crushing load (P_{avg}) were calculated from load-displacement curve, which was acquired from static crushing test on the number of interface and density on circular member of CFRP/foam. Average crushing load was calculated by dividing absorbed energy with strain. After conducting static crushing test, initial deformation was started from trigger in specimen.

4.1. *Collapse mode*

Test specimen with 0° of the outer layer regardless of foam filling as shown Fig. 4 displayed spline mode. Inside of test specimen revealed brittle fracture mode with bending and breakdown, which was observed in 90°of the outer layer. For 90° of the external layer, breakdown mode from fracture of external ply fiber and brittle fracture mode from fiber bending was detected at the same time. If foam is not filled, fracture residue from the inside of specimen was impregnated into empty space, and external fiber was agglomerated.

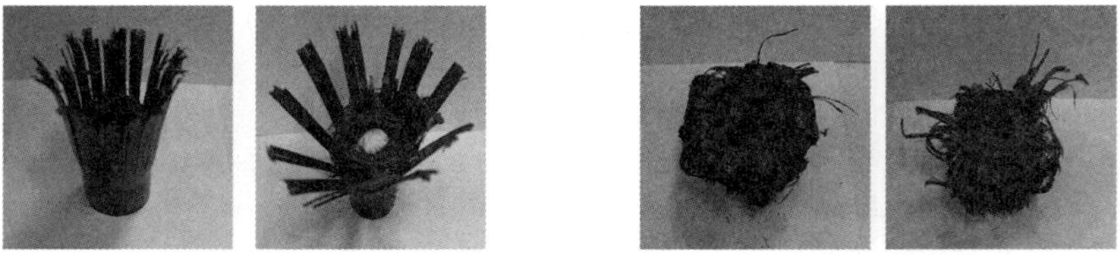

(1) Outer angle 0° CFRP&CFRP/Foam (2) Outer angle 90° CFRP&CFRP/Foam

Fig. 4. The collapsed shape of CFRP and CFRP/Foam (40) according to outer angle.

4.2. *Energy absorption*

Total absorption energy and average load were highly affected by the external layer. As shown in Fig. 5, where external layer was 0°, total energy absorption was increased along with the number of interface. In case of 90° of external layer, absorption energy was decreased with interface number, and it was increase later. 2 interface with 90° of external layer lamination showed higher total absorption energy for non-filled specimen. In case that external layer was laminated as 0°, higher absorption energy was achieved for larger number of interface and with foam filling. Similar results from the previous cases were acquired. According to lamination sequence of specimen, 0° of external later showed spline mode during crushing, and energy absorption was reduced due to distribution of force. However, 90° of external layer can absorb more energy by limiting the extension of 90° external fiber even though of elongation of 0° fiber.

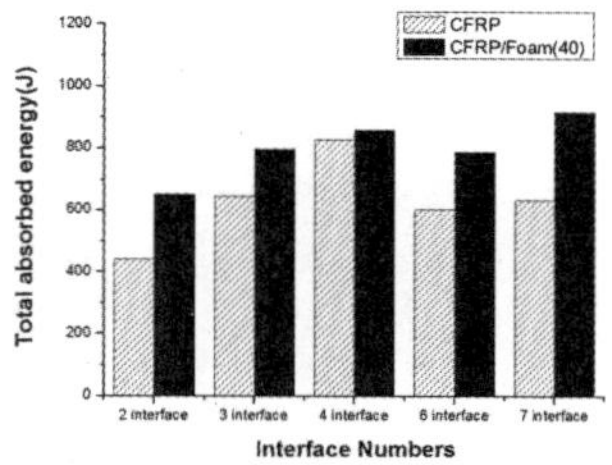

(1) Outer angle 0°

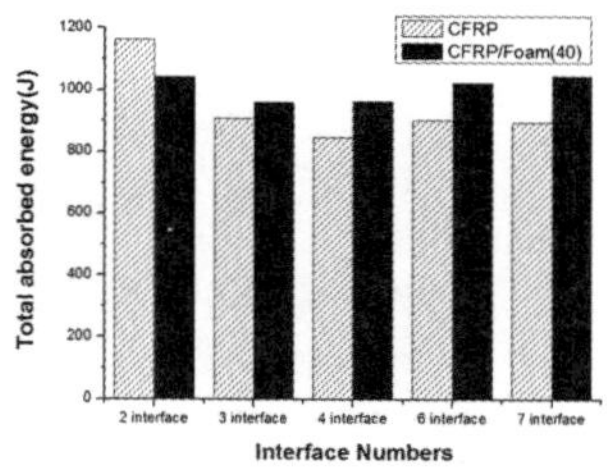

(2) Outer angle 90°

Fig. 5. Relationship between total absorbed energy and variation interface number of foam density according to outer angle.

5. Conclusion

In this study, static axial crushing test was conducted to evaluate CFRP circular member with foam filling as lighter structural body in vehicle.

1) Test specimen showed the crushing on triggered direction, and external layer was affected significantly.
2) For 0° of external layer, absorption energy was increased with number of interfaces, but 90° displayed the reduction in absorption energy and increase in afterward.
3) By filling of empty space in CFRP circular member with foam, total absorption energy and average load was increased for 30~40% and 20~25%.

Acknowledgments

Following are results of a study on the "Human Resource Development Center for Economic Region Leading Industry" Project, supported by the Ministry of Education, Science & Technology(MEST) and the National Research Foundation of Korea(NRF).

References

1. Abramowicz, W: *Engineering. Transaction.*, 1981, **29**, 5–26.
2. Abramowicz, W., Wierzbicki, T: *J.Appl. Mech.*, 1986, **50**, 157–174.
3. Donald F. Adams: *MTechnom. Pub. Company.*, 1990.
4. Farley, G. L., Jone, R. M.: *Jour. Compo. Mater.*, 1991, **26**, 388–404.
5. Lee K. S., Cha C. S. And Yang I. Y: *Transaction of KSAE.*, 2005, **13**, 50–56.
6. Yang I. Y, Lee K. S, Kim Y. N, Chung J. O and Cha C. S: *Key Engineering materials.*, 2006, 306–308.
7. Wang, Q, Fan, Z and Gui,L: *Int. J. Mech. Sci.*, **49**, 515–522.

THE COMPARATIVE STUDY OF INTENSITY AND RIGIDITY OF THE HYBRID MEMBER AND CARBON FIBER

JI HOON KIM[†]

Department of Mechanical Design Engineering, Chosun University, 375 Seosuk-dong, Dong-gu, Gwang-ju, 501-750, Korea

JUNG MIN BANG, JUNG HO KIM, HYUN GOO KANG

Department of Advanced parts and Materials Engineering, Mold-design Laboratory, 375 Seosuk-dong, Dong-gu, Gwang-ju, 501-750, Korea

In the recent car industry, lightweight of vehicles is on issue considering our environment. But it's getting weight in accordance with passenger's safety, convenience, high quality and various functions. For these reasons, a design for a car should be made lightly by considering safety. So materials for lightweight's demands are getting higher instead of Aluminum. CFRP, one of the materials, has high specific intensity and specific rigidity. So it can make the weight 20~50% less. Also it has elastic modulus and low thermal deformation ratio. CFRP is used widely in car, sport, leisure and astronautic industry. But it has a characteristic which is changed according to the orientation angle and it is weak from exterior weight because of the fragility. A great effort for this section is carried out many fields. On this study, laminated CFRP to aluminum tube and made it like hat shaped and conducted three point bending and Axial collapse tests and then compared intensity and rigidity depending on bending and axial collapse.

Keywords: CFRP; Collapse; Bending; Specific intensity; Specific rigidity.

1. Introduction

Today in the car industry, the most important goal is lightweight of vehicles such as fuel expense improvement, limitation of waste gas.[1,6,7] On the other hand, it's getting weight depending on passenger's safety and convenience, vehicle's technical capacity and high quality. So the design of the car is made by safety and lightweight the vehicle's whole weight. In these reasons, intensity and light material's demands are getting higher instead of aluminum.

Side member is an important part in passenger's safety. The side member is typically hat shaped section. There are elasticity of bending and collapse when the load works from the exterior. The study of bending and collapse is in progress so well for appraisal of side member's safety.

[†] Corresponding author, email: kjh@chosun.ac.kr

CFRP (Carbon Fiber Reinforced Plastics), one of the composite materials, has high specific intensity and rigidity.[2] So it is well known for the material of lightweight. CFRP can be applied structural member and lightweight easily as it maintains intensity and rigidity. Also, it is used in sport, leisure, astronautic industry due to high elastic modulus and low thermal deformation ratio. But, collapse and bending characteristics of CFRP using one-sided prepreg sheet are presented differently depending on orientation angle and CFRP has fragility that is so weak about load of the external.[3,4] A great effort for this section is carried out many fields.

In this study, made hat shaped CFRP laminating [15°, 45°, 90°, 0°/90°, 90°/0°] on the aluminum tube that can be applied in structural member. Searched three point bending and collapse for comparing hardness by adding universal testing machine(UTM). Also we made comparison with intensity and rigidity about bending and an axial collapse.

2. Method of Test

A standard used in bending and axial collapse is made by KS. In KS standard of this test, support span is fixed by specimen's thickness.

In ASTM, specimen's length and thickness are fixed by 1:16. The width should be used over 1/4 of length.

2.1. *Specimen*

The specimen already manufactured laminated length direction as based on 0°. Everything in this test is made by angle of lamination using CFRP prepreg sheet(HANKUK carbon, CU125NS, Carbon Fiber uni-direction 125g/m^2). They are all 1mm in thickness. The size of specimen in axial collapse test is 30mm× 30mm and length is 120mm. The size of bending test's specimen is same with axial collapse. The length is 495mm on ASTM standard. This specimen is made by autoclave on the ground of compression formed under vacuum in 130°C and cut it by using diamond cutter for removing residual stress.

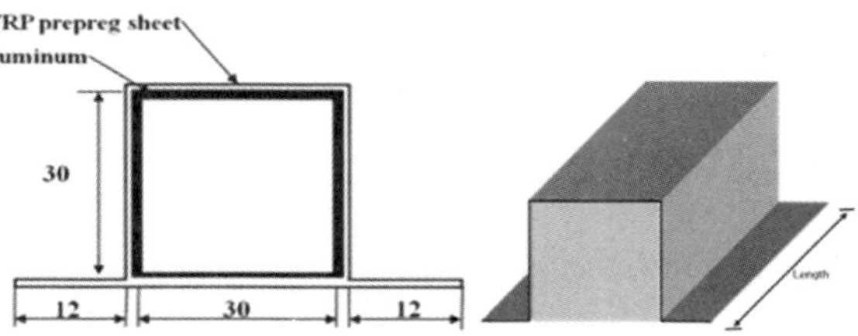

Fig. 1. Shape of Specimen.

2.2. *Method of the Test*

2.2.1. *Three point Bending Test*

Generally, there's three and four point bending tests. Three point bending test is used much more due to much better way. In this study, equipment of three point bending test is fixed by ASTM D-6272 like Pho 2.[5]

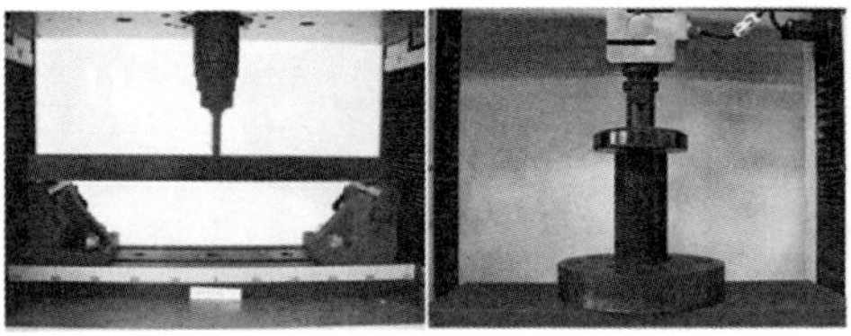

Fig. 2. Three point Bending Test and Axial collapse Test Device.

Next, we measured Load-Displacement. Bending intensity is calculated by formula(1). Bending rigidity is calculated by formula(2).

$$\sigma = \frac{M}{Z} \tag{1}$$

$$EI = \frac{PL^3}{48\delta} \tag{2}$$

2.2.2. *Axial collapse Test*

In this study, controlled displacement toward an 50%(60mm) in axis using UTM and made an static experiment in strain speed of 10mm/min.

After axial collapse test, measured Load-Displacement, got axis intensity applying formula(3) and axis rigidity in formula(4).

$$\sigma = \frac{P}{A} \tag{3}$$

$$AE = \frac{PL}{\delta} \tag{4}$$

3. The Result and Discussion

3.1. *Discussion of Load-Displacement*

Compared max Load measuring Load-Displacement of hat shaped member CFRP and hat shaped AL-CFRP hybrid member after three point bending and Axial collapse tests in $[0°/90°]_4$ and $[90°/0°]_4$.

Fig. 1 is Load-Displacement of three point bending and Axial collapse tests in hat shaped CFRP laminated in $[0°/90°]_4$ and $[90°/0°]_4$. After the test, the specimen laminated $[0°/90°]_4$ and $[90°/0°]_4$ is similar with Load-Displacement. Max load's hybrid member is high in both tests.

As a result of this, hat shaped AL-CFRP hybrid member is better than CFRP.

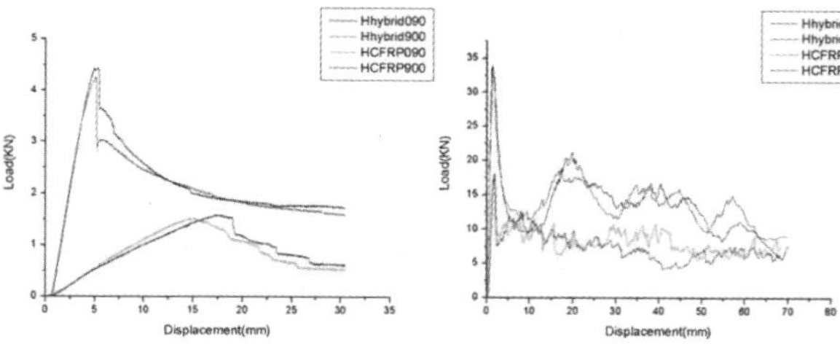

Fig. 1. Load-Displacement of Three point test and Axial collapse test.

3.2. *Intensity and Rigidity in Laminated Angle*

Examined Bending and Rigidity in accordance with orientation angle of hat shaped AL-CFRP hybrid member. Angle is laminated by $[+15°/-15°]_4$, $[+45°/-45°]_4$, $[+90°/-90°]_4$, $[0°/90°]_4$, $[90°/0°]_4$.

Fig. 2 (a) presents bending and axial intensity of hat shaped AL-CFRP hybrid member according to orientation angle. $[+15°/-15°]_4$ was the highest one in bending.$[+45°/-45°]_4$ and $[0°/90°]_4$, $[90°/0°]_4$ are almost same. The lowest one was $[+90°/-90°]_4$. In the test of axial intensity, $[+15°/-15°]_4$ was the highest one. It was similar both $[0°/90°]_4$ and $[90°/0°]_4$. The low axial intensity rank was $[+45°/-45°]_4$ and $[+90°/-90°]_4$ in order. When make hat shaped AL-CFRP hybrid member which has high intensity, the laminated member with $[+15°/-15°]_4$ is the best one.

Fig. 2 (b) presents bending rigidity and axial rigidity of hat shaped AL-CFRP hybrid member on orientation angle. $[+15°/-15°]_4$ was the highest one in bending rigidity. As we saw in the Load-Displacement, the member laminated $[0°/90°]_4$, $[90°/0°]_4$ was similar in bending rigidity. The next ones were $[+90°/-90°]_4$ and $[+45°/-45°]_4$ in order. In the axial rigidity, $[0°/90°]_4$, $[90°/0°]_4$ and $[+15°/-15°]_4$ are highly presented in order. The reason there are some differences in axial rigidity is that there are differences in the orientation angle of outer.[5]

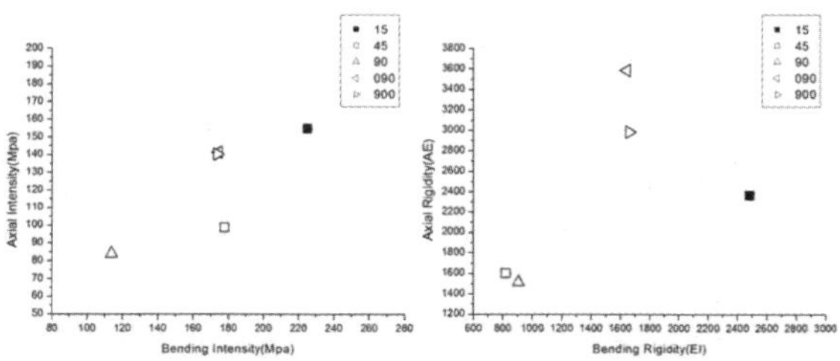

(a) (b)

Fig. 2. Relationship between Axial and Bending on orientation angle : (a) Intensity, and (b) Rigidity.

4. Conclusion

1) As compared Load-Displacement after tests, Max-load is highly more presented in AL-CFRP than in CFRP. So AL-CFRP is better than CFRP.

2) Upon the consideration and comparison of bending and axial strength/rigidity, $[+15°/-15°]_4$ was the most superior one. For this reason, $[+15°/-15°]_4$ of composite member is suggested as the best candidate for high strength and rigidity application.

3) $[0°/90°]_4$ and $[90°/0°]_4$ of hat-like Al-CFRP composite member showed similar bending and axial strength. However, bending and axial rigidity displayed the discrepancy. It is considered to be happened because of lamination angle for the external layer.

Acknowledgment

Following are results of a study on the "Human Resource Development Center for Economic Region Leading Industry" Project, supported by the Ministry of Education, Science & Technology(MEST) and the National Research Foundation of Korea(NRF).

References

1. Kim, J.H, Park, S.G, Kim, J.H, Sim, J.K, Yang, I.Y., "*The axial compression and bending characteristics of CFRP hat shaped section members according to orientation angle.*" KSMTE, pp.426-429, 2006.
2. Kim, J.H, Kim, J.H, Cha, C.S and Yang, I.Y., "*Bending Characteristic of CFRP & Hybrid Shaped Hat Structure Member According to Stacking Orientation Angle.*" KSMTE, Vol. 17, No.3, pp.34~39, 2008.
3. Yang, I.Y, Kim, J.H, Sim, J.K and Kim, J.H., "*Evaluation of Characteristics of CFRP Structural Member's Bending Strength and Rigidity.*" KSME, pp.556~560, 2008.
4. Yang, I.Y, Kim, J.H, and Kim, J.H., "*The Characteristics of Flexure Strength and Rigidity in Light-weight CFRP Members.*" KSPE, Vol. 25, No.1, pp.1~7, 2008.
5. Standard Test Method for Density, ASTM D6272-02.
6. Hwang, W.C, Seo, H.K, Lee, K.S, Kim, J.H and Yang, I.Y., "*The Development of light-weight thin-walled member which has collapse characteristics of optimum.*" KSAE spring conference proceedings, Vol.3, pp.1502~1506, 2007.
7. Lee, K.S, Cha, C.S, Pyeon, S.B, Yang, I.Y, and Sim, J.K. "*Axial Collapse Characteristics of Aluminum CFRP Compound Square Members for Vehicle Structural Members.*" KSME, Vol.29, No.10, pp.1329~1335, 2005.
8. Seo, H.K, Hwang, W.C, Min, H.K, Im, K.H and Yang, I.Y., "*Energy Absorption Capability of Aluminum CFRP Compound Tubes According to Change of Interlaminar Numbers.*" KSAE spring conference proceedings, Vol.3, pp.1496~1501, 2007.
9. KS B 0804 : 2001.

EFFECT OF INITIAL STATE ON GRAIN REFINEMENT STRENGTHENING OF AZ31 MAGNESIUM ALLOY

JIANMIN YU†

College of Materials Science and Engineering, North University of China
Taiyuan Shanxi /030051, China

ZHIMIN ZHANG

Engineering Research Center of Magnesium-base Material Processing Technology,
Ministry of Education, China
Taiyuan Shanxi /030051, China

The effects of different initial condition (as-cast, homogenized) on grain refinement strengthening of AZ31B magnesium alloy were investigated by optical microscopy and tensile testing. The results show that the inhomogeneous second-phase γ-Mg17Al12 along the grain boundary in as-cast was still distributed unevenly after extrusion. Compared with as-cast, tensile strength of homogenization was increased greatly after extrusion. At 250℃ and extrusion ratio of 50, tensile strength reached to 391Mpa and elongation of 15.2%.The grain size after extrusion is bigger than that of as-cast after extrusion, so the Tensile strength is lower than that of as-cast. Besides, non-equilibrium phase γ-Mg17Al12 has been dissolution after homogenized and grain distributed uniformly. Therefore, plasticity was improved greatly at the same condition as upon.

Keywords: Grain refinement; Mechanical properties; AZ31B Magnesium alloy.

1. Introduction

The microstructure of as-cast AZ31B magnesium alloy consists of α-Mg and $Mg_{17}Al_{12}$, in which $Mg_{17}Al_{12}$ mainly distributes in grain boundary with the shape of coarse "bone" but only a little amount of $Mg_{17}Al_{12}$ distributes in the grain [1]. Due to segregation and the second phase $Mg_{17}Al_{12}$ existing interdendritic and in grain boundary, the thermoplasticity of the magnesium alloy is reduced and processability becomes worse. In the process of plastic deformation, the coarse $Mg_{17}Al_{12}$ in grain boundary is easily fractured to form crack source [2, 3]. The size of magnesium alloy grain has more influence on yield strength than that of iron alloy and aluminum alloy [4-7]. It is well known that grain refinement can significantly improve the mechanical properties and plastic deformation capacity

† Corresponding author, email: minyu889@163.com

of magnesium alloy, especially for AZ31 magnesium alloy [8]. Recent research demonstrates that the grain of AZ31 magnesium alloy ingot can be obtained the size of 0.5-3μm after extruded 4 times under the condition of 160-220℃ by equal channel angular pressing(ECAE)[9]. But it is not widely application because extrusion device still at the stage of laboratory research. AZ31 magnesium alloy ingot extruded with extrusion ration 100/1 at 350℃and its size of the grain will decrease from 15μm to 5μm [10]. However high-ratio extrusion process has the higher requirement for equipment so it is difficult to achieve mass production. In the process of hot extrusion, grain is elongated and cracked to small particles because of high extrusion ratio; friction between grains accelerated the crushing process. This paper will research the effect of different initial state on grain refinement strengthening.

2. Experiment

Experimental materials used in the experiments were as-cast AZ31B magnesium alloy ingots with dimensions φ67×60mm、φ89×60mm and φ106×60mm. The chemical composition of the AZ31B Mg alloy is presented in Table 1.

Table 1. The chemical composition of AZ31B magnesium alloy ingots.

Al	Zn	Mn	Si	Fe	Cu	Ni	Be
2.95	0.90	0.30	0.0209	0.0036	0.0018	0.0007	0.0003

Samples were randomly divided into two groups: group A and group B. Cases in group A were extruded directly, while cases in group B were extruded after homogenization. Homogenization annealing process was identified as 400℃×16h, and then was taken out air-cooled to room temperature. Extrusion experiments were performed on the 6300KN hydraulic press. Extrusion temperatures were 250 °C, 300 °C, 350 °C, 400 °C and 450 °C, and extrusion ratios were 20, 35 and 50 respectively. The tensile tests (Figure 1) were performed according to the standard of GB6397-86, by WDW-E100D electronic universal testing machine. Microstructure observation was to be performed on digital metallographic microscope ZEISS.

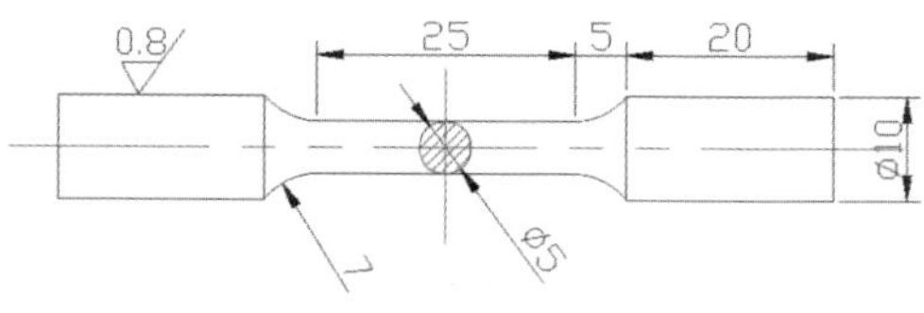

Fig. 1. Schematic diagram of tensile sample (mm).

3. Results and Discussion

3.1. *Mechanical properties*

Figure 2 is tensile strength-temperature curve of AZ31B magnesium alloy of different state. At a certain extrusion ratio, the variation range of the strength of as-cast is greater than that of homogenization as the temperature increasing. When the extrusion ratio is 50 and temperature is 250℃, the tensile strength of cast structure after extrusion is maximum, 391MPa; when the temperature increase to 450℃, strength is 342MPa and the difference value is 49MPa. While the strength of homogenization structure is relatively stable as the temperature increasing. In different initial states, both cast extruded directly and extrusion after homogenization, the strength increases with the increase of the extrusion ratio at the same temperature. Figure 3 is elongation temperature curve of AZ31B magnesium alloy of under different temperatures and extrusion ratios. It is shown that when the cast alloy is extruded directly, the curve appears as wavy which means that with the increase of the temperature, the elongation variation has no regulation. While the elongation temperature curves of homogenization extrusion appears the trend of decreasing with the increase of the temperature. Compared Figure 2 with Figure 3, Homogenization can improve formability and mechanical properties of AZ31B magnesium alloy effectively.

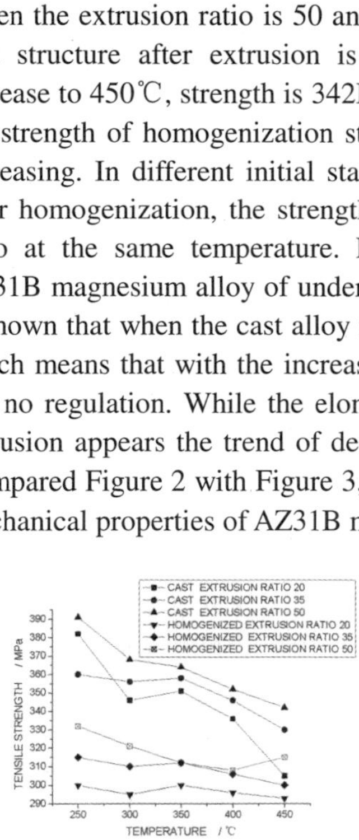

Fig. 2. Tensile strength-temperature curve.

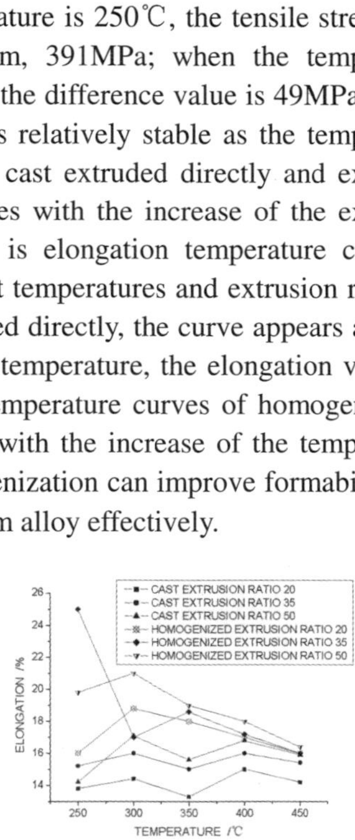

Fig. 3. Elongation-temperature curve.

3.2. *Microstructure*

Figure 4(a) is the cast microstructure of AZ31B magnesium alloy. The average size of grain is about 50μm. There are mass coarse alloy phases which distribute along the grain and a little amount of second phase $Mg_{17}Al_{12}$ distributes in grain boundary and interdendritic. Dendrite segregation makes the structure appear non-equilibrium brittle-phase and result in the plasticity of the alloy decreases when continuous coarse network shell brittle compounds are produced in the edge of the dendrite bronchus[1]. Figure 4 (b), (c) and (d) show that the grain of the as-cast samples extruded directly is fine and distributed with much second phase nonuniformly. When the extrusion ratio is 20 and the size of the grain is about 5.5μm, the deformation microstructure is nonuniform; with the extrusion ratio increases, recrystallized grain is deformed sufficiently. When the extrusion ratio is 50, the average size of the grain is 4.5μm, the second phase is relative well-distributed.

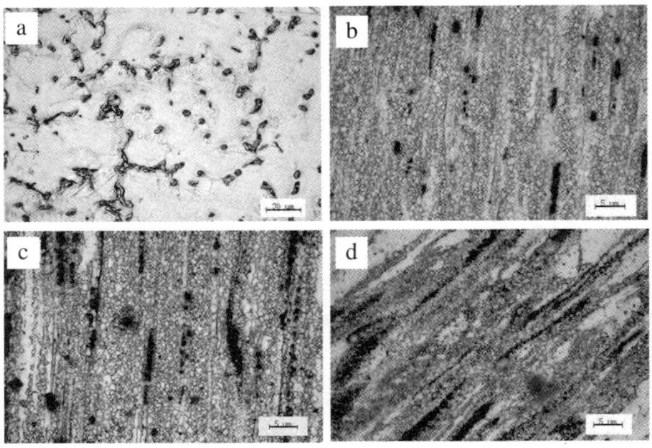

(a) initial cast state; (b) λ=20, T=350℃; (c) λ=35, T=350℃; (d) λ=50, T=350℃.
Fig. 4. Microstructure of as-cast AZ31B magnesium alloy after extrusion directly.

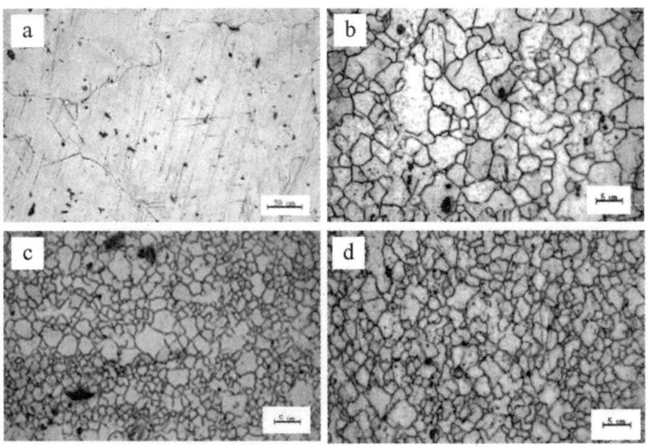

(a) homogenization; (b) λ=20, t=350°C; (c) λ=35, T=350°C; (d) λ=50, T=350°C.
Fig. 5. Microstructure of homogenized AZ31B magnesium alloy after extrusion.

Figure 5 (a) is the microstructure of homogenized AZ31B magnesium alloy before extrusion. Compared with the initial cast microstructure, the grain has grown. The number of dendritic grain in the matrix decreases greatly. $Mg_{17}Al_{12}$ has become fine particle and uniformly distributed in α -Mg phase matrix. Figure

5(b), (c) and (d) are the microstructures of the homogenized AZ31B magnesium alloy after different extrusion ratio at 350℃. With the extrusion ratio increases, metal crystal structure is distorted seriously because of mass plastic deformation. The crystalline grain is refined obviously. The average grain size is 3μm and well-distributed. Therefore, plasticity and strength will increase with the increase of the extrusion ratio. Compared Figure 4 with Figure 5, The bigger initial grain size is, the coarser new grain size will be, and vice versa.

4. Conclusion

1. Dynamic recrystallization of AZ31B magnesium alloy has close connection with the initial grain size. The bigger initial grain size is; the coarser new grain size will be.
2. Both as-cast alloys extruded directly and extrusion after homogenization, tensile strength increases with the increase of extrusion ratio. While the strength of homogenized alloy is relatively stable with the increasing of the temperature.
3. When the cast alloys is extruded directly, the curve appears as wavy with the increase of the temperature, while the elongation curves of homogenized alloy after extrusion appears the trend of decreasing.

Acknowledgments

This work was supported by National Natural Science Foundation of China under grant No.50735005, No.50605059, Youth Foundation of Shanxi Province No.2008021030 and 2010021024-2.

References

1. HUANG Guangsheng,WANG Lingyun,PAN Fusheng: Journal of Chongqing University (Natural Science Edition), 2004, 27(11), 18-21.
2. CHEN Zhenhua, XU Fangyan:Chemical Industry and Engineering Progress, 2006, 25(2),140-146.
3. LIU Chuming:The Chinese Journal of Nonferrous Metals,2006,16(1),1-12.
4. Kojima Y:Materials Science Forum, 2000, 350-351,3-18.
5. Mabuch M, Nakamura M: Materials Science Forum, 1999, 304-306, 67-72.
6. Watanabe H, Mukai T: Materials Science Forum, 1999, 304-306, 303-308.
7. Watanabe H,Tsutsui,et al:Materials Science Forum, 2000,350- 351,171-176.
8. WANG Lingyun,HUANG Guangsheng:The Chinese Journal of Nonferrous Metals,2003, 13 (3),594-598.
9. Zhang Shichang,Duan Hanqiao,Cai Qizhou,: Special Casting & Nonferrous Alloys,2000,(6): 51-54.
10. Mukai T, Watanabe H: Materials Science Forum, 2000, 350-351, 159-170.

SUPERPLASTIC FORMING OF AZ80A MAGNESIUM ALLOY AUTOMOBILE WHEEL*

YANG YONGSHUN†

School of Materials Science and Engineering, Henan University of Science and Technology
Luoyang, 471003, P.R.China

YANG XI

School of Materials Science and Engineering, Northwestern Polytechnical University
Xi'an, 710072, P.R.China

This paper proposes a forming technology of AZ80A magnesium alloy automotive wheel, which collects the superplastic compound extrusion and the bulging technology of rigid die. By using dynamic recrystallization superplasticity, the new method completes the forming for complex auto wheel successfully. It has the advantages of simple process, high forming efficiency and compact structure.

Keywords: Magnesium alloy; Superplasticity; Dynamic recrystallization; Forming.

1. Introduction

As a basic load bearing component of car movement, the performance of the automobile wheel is playing a direct impact on safety traffic, energy saving and appearance quality of automobile. Comparing with aluminum alloy, magnesium alloy has lower density, higher specific strength and specific rigidity. Moreover, magnesium alloy has better impact resistance[1], just 100 times more than aluminum alloy[2,3]. So it is very suitable for the automobile wheel. Considering the advantages above, magnesium alloy wheel could play a great role for the lightening weight of automobile, improving fuel efficiency and reducing the discharge of CO_2[4,5]. So it will be prevalent in the future.

As an advanced forming method, superplastic precision forming technique can form complex and difficult machining products with precision forming by one step, utilizing low flow resistance and high plastic deformation of material

* Foundation item: Science & Technology development program of Henan province (Grant No. 072102230007) and the fund of the State Key Laboratory of Solidification Processing in NWPU (Grant No. SKLSP201008)

† Corresponding author, email: yys@mail.haust.edu.cn

presented at the superplastic state. So the products can be cheaper and higher quality.

2. The Superplasticity of AZ80A Magnesium Alloy

2.1. *The superplastic method*

This paper proposed a new technics which combining the grain refinement with superplastic extrusion forming process employing the grain refinement mechanism of dynamic recrystallization. It will simplify the forming process and reduce the production cost, significantly. Firstly, the material is required to meet the dynamic recrystallization conditions by quickly isothermal compression deformation technics. Thus the material will have an enough driving force of recrystallization. Then the complex product can be obtained by reducing deforming rate under the superplastic forming strain rate where the material has come into the superplastic state by dynamic recrystallization.

2.2. *The superplasticity of AZ80A*

Experiment sample is as-cast AZ80A bar with size Φ10×15mm. The superplastic experiment of AZ80A is completed on the gleeble-1500 thermal testing machine. Heat the samples to the different required temperature (320℃ to 410℃) and keep warm. Then compress the samples with different strain rates ($1\times10^{-2}s^{-1}$ to $1\times10^{1}s^{-1}$). The tests stop when total strain capacity equals ε=0.8.

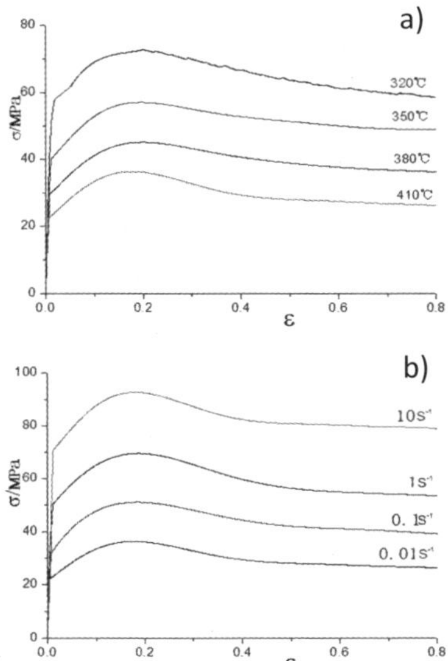

Fig. 1. The strain-stress curve
a) at different temperatures
b) at different strain rates.

From the strain-stress curves at different temperatures and strain rates shown in Fig.1, we can see that at the first stage of deformation, the flow stress goes up with the increase of true strain. This indicates that work hardening is dominant, and dynamic recovery or dynamic recrystallization occurs little. With the increase of deformation, the dislocation energy goes up continuously. Then the drive of dynamic recrystallization and softening increased. So work hardening is countervailed by the softening of dynamic recovery gradually and the curve slope decreases, simultaneously. Work hardening and softening of dynamic recrystallization strike a balance when the flow stress comes to the peak. The effect of dynamic recrystallization becomes more obviously with the deformation increasing. It shows that dynamic recrystallization is the typical mechanism during the hot deformation[6]. The flow stress of AZ80A magnesium alloy decreases with the deforming temperature increasing and the stress rate decreasing. The plasticity of as-cast AZ80A comes to optimum value. The maximum deformation resistance of compression is about 35MPa under the conditions of 410℃ and $\dot{\varepsilon} = 1\times10^{-2}s^{-1}$.

3. Forming Scheme of Auto Wheel

Fig.2 shows the sketch of automobile wheel, as an asymmetric double-cup. Due to the large diameter of products and the split cavity die for demoulding of ring groove, the size of die is large and the structure is more complex, causing the difficulty in mould machining, preheating, fitting die space of extrusion equipment and forming process. So it leads to poor manufacturability for the one-forming method. Therefore, this paper presents a two-step forming technology. Firstly, the circle blank is made into double-cup structure by combined extrusion. Secondly, the near-net-forming of wheel is completed by bulging.

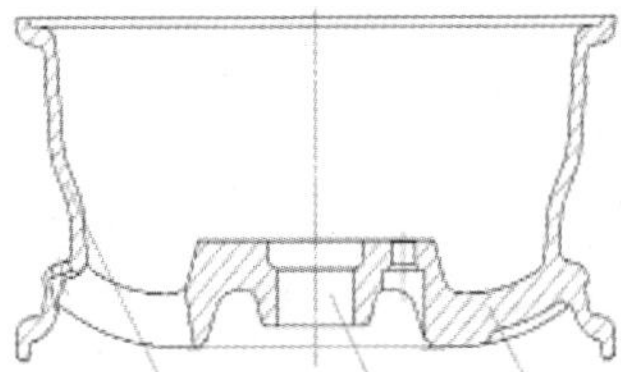

Fig. 2. Wheel structure.

3.1. *Combined extrusion*

Using superplastic extrusion of recrystallization technics, the diameter of blank would be reduced properly and the blank could come to critical reduction during upsetting deformation, to enable the produce of recrystallized structure. Thus it will increase plasticity and reduce deformation resistance before the blank going into the rim of mould.

The combined extrusion die of magnesium alloy automotive wheel is shown in fig.3. The forming temperature is the key to ensure forming in time of superplastic extrusion. Both dies and blanks should keep at superplastic temperature to implement superplastic isothermal extrusion

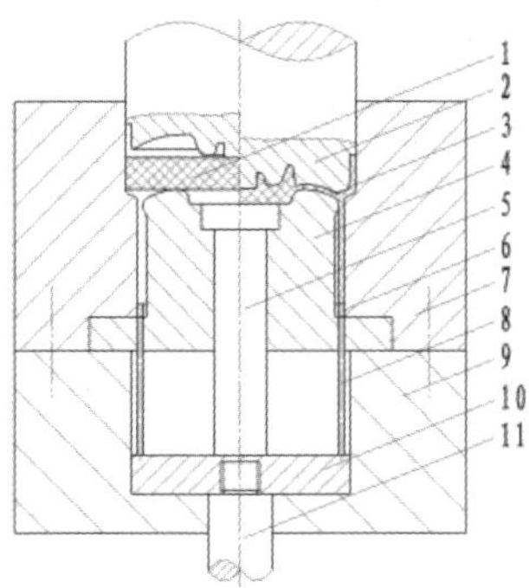

Fig. 3. Combined extrusion mould
1-blank 2-upper punch 3-extrusion 4-lower punch 5-ejector 6-ejector ring 7- die 8-small ejector pin 9- lower die holder 10-ejector plate 11- ejector pin.

successfully. So besides heating the blanks, it also should set heating equipment on the related site of male and female die. Before extrusion, the die is heated to the superplastic forming temperature. The blanks coated with water-based graphite are also putted into preheating furnace to heat and keep warm. When up to the required temperature, the blank is moved from the heating furnace into the mould quickly. Then pressure head puts down and the blank produce superplastic flow until the die cavity is filled with the blank. Finally, the workpiece demoulds with the pressure head returning.

3.2. *Bulging technology*

This paper adopts combined bulging of rigid die technics. That is, the punch with the consonant shape of the rim is pressed into middle-blank bidirectionally, to make the inner wall with the straight-barrel shape flaring into the rim. It should adopt the die constraints to prevent tube blank collapsed during bulging process.

Fig.4 shows the die of magnesium alloy automotive wheel. The side parting of split cavity die is achieved by the lateral hydraulic cylinder. The rim is too thin to maintain the temperature, which resulted in the workpiece losing plasticity and bursting. The solution is using an effective heating and thermal insulation measures.

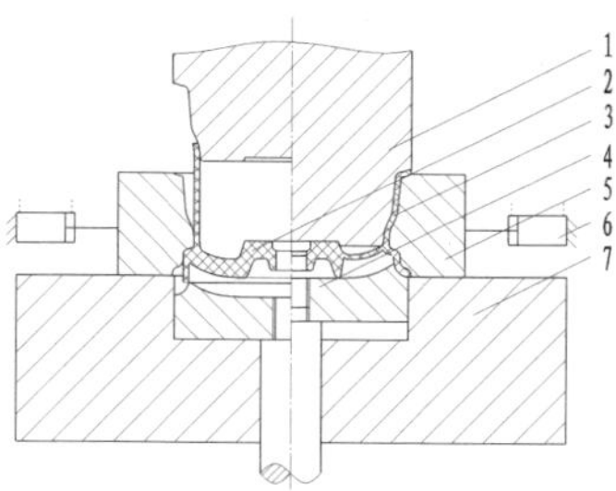

Fig. 4. Bulging mould
1-upper punch 2-extrusion part 3-bulging part 4-lower punch 5-combined die 6-lateral hydraulic cylinder 7-lower die holder.

Both the bulging deformation and the force of wheel are small. So a lower bulging temperature can be used. First, put the heated extrusion blank into the preheating die in the die-opening state and the lateral hydraulic cylinder drives the die closed. Then, with the ejector cylinder working, the lower punch goes up by ejector pin, and the lower rim is bulged. With the movable beam going down, the upper punch presses into the upper rim and makes it bulged. At last, the product demoulds after pressure head returned and split cavity die opened.

4. Results and Discussion

During the superplastic compound extrusion, the maximum forming force is 3200KN, while the unit extrusion pressure is 128MPa. The whole process is stable, lasting 4 minutes. And the metal fills well. The bulging pressure is less than 100KN and the bulging time is about 1 minute. As shown in Fig.5, after

Fig. 5. Product after extrusion and bulging.

extrusion and bulging, the part surface is smooth and the sizes meet to design requirements. Through inspection, the product has good quality without pore.

After sampling from the magnesium alloy wheel, the tensile test was conducted on the AG-I250KN precision universal testing machine. It can be obtained that the tensile strength is 335MPa, while the elongation is 12.8%. Compared with as-cast specimen, all of these are enhanced greatly. Furthermore, the tensile strength is also higher than that of the low-pressure casting magnesium alloy wheel. Thus, the wheel can meet the requirement of mechanical performance for auto wheel completely.

5. Conclusion

By using dynamic recrystallization superplasticity, the superplasticity of as-cast AZ80A can be achieved, with the best plasticity at the temperature of 410℃ and $\varepsilon=1\times10^{-2}s^{-1}$, while the resistance of compression deformation is 35MPa.The magnesium alloy auto wheel can be obtained by the two-step forming technology, which includes the superplastic compound extrusion and the bulging technology of rigid die. The new technology has advantages of simple process and high production efficiency. Moreover, the product has compact structure without pore and shrinkage, and the mechanical properties could satisfy the use requirement as well. In short, the two-step forming technology proposed in this paper shows very important reference value.

References

1. B. T. Zi, H. Wang: Chinese Journal of Rare Metals, 2004, 2, 229-232.
2. S. S. Li, B. Tang, D. B. Zheng, S. Lv, H. Q. Wu: *Proceedings of the Fourth International Conference on Non-ferrous & Special Casting Processes*, 2005, 143-149.
3. X. Y. Jin, S. S. Li, X. G. Liu, D. B. Zeng, X. Z. Xu, J. Y. Lai, C. M. Liu: *Ordnance Material Science and Engineering*, 2005, 28, (2), 30-34.
4. A. William, R. Anna, L. Melissa, J. Loscher, A. Alpas: *International Journal of Impact Engineering*, 2004, 30, 117-142.
5. D. S. Mehta, S. H. Masood, W. Q. Song: *Journal of Materials Processing Technology*, 2004, 155-156, 1526-1531.
6. Q. Guo, H. G. Yan, Z. H. Chen, H. Zhang: *The Chinese Journal of Nonferrous Metals*, 2005, 6, 900-906.

THE EFFECT OF DIE CURVATURE AND MATERIAL ANISOTROPY IN U-BENDING OF AZ31 MAGNESIUM ALLOY SHEET ON SPRINGBACK

H.H. KIM

Department of Mechanical and Precision Engineering, Pusan National University, Busan, 609-735, Republic of Korea

C.G. KANG[†]

School of Mechanical Engineering, Pusan National University, Busan 609-735, Republic of Korea

A springback is become a significant factor inhibition in precision part forming of magnesium alloy sheet material. There are many factors that affect springback such as blank holding force, punch speed, material thickness, material strength, die roughness and clearance during the hot forming of magnesium alloy sheet. The factors which effect on springback have been not systematized until now. Almost researcher and company have been depended on trial and error or experimental until now. Therefore, investigation of factors that affect springback is needs to optimize and systemize. In this study, we conduct a study on the effect of punch radius, die radius and material anisotropy on the springback for AZ31 Mg alloy sheet.

Keywords: AZ31 Mg alloy; Springback; Punch and die radius; Anisotropy.

1. Introduction

Magnesium, which is the lightest amongst metals, has excellent material properties, such as specific strength and stiffness, ability to be cast, ability to be machined, vibration and shock absorption ability, and fatigue resistance. Therefore, it is widely used in transportation vehicles, electronic appliances, portable electronics, sports and leisure supplies, military supplies, and medical equipment.

However, magnesium is poorly formable at room temperature due to its crystalline structure, which is hexagonal close-packed (HCP). It is difficult to produce engineering components by means of a pressing and press-forging

[†] Corresponding author, email: cgkang@pusan.ac.kr

process due to the forming difficulty and the low heat maintenance of magnesium. In addition, spring back is main problem in sheet forming process of Mg alloy because electronic parts require high precision, dimensional stability, and fine surface quality. Therefore, there has been as increasing focus on formability in magnesium alloy such as AZ31. Cho et al.[1] predicted the springback by finite element method (FEM). Carden et al.[2] studied the springback effect of High strength low alloy steel (HSLA) and drawing-quality silicon-killed steel (DQST). Geng et al. studied the springback of Al6022 aluminum alloy by FEM. F.K Chen et al.[3] investigated that various punch radius and temperature has an effect on springback in V-bending.

With many study of springback in Mg alloy sheet, springback database is still not systemized by various shape function such as punch radius and anisotropy property in AZ31 Mg alloy. Therefore, this study investigates the springback effect with various die radius and anisotropy in U-bending warm forming.

2. Experimental Equipment

In warm forming, it is very important to precisely control for temperature of AZ31 Mg sheet. Therefore, the material is heated up to same die temperature before bending test. Figure 1 (a) shows the dimension of blank and attached k-type thermocouples to measure the material temperature. Thicknesses of blanks are 0.5mm and 1.0mm. The U-bending experiments used hydraulic presses equipment which possible to control the blank holding force and punch velocity as shown Fig. 1 (b). The heat-cartridges are inserted to heater holes so that heating the die and punch. Blank temperature is increased by heat transfer from die.

The experimental conditions are listed at Table 1. The radius of the die and punch are main variable parameters. In addition, blank temperature, rolling direction and etc. are also selected as minor variable parameters.

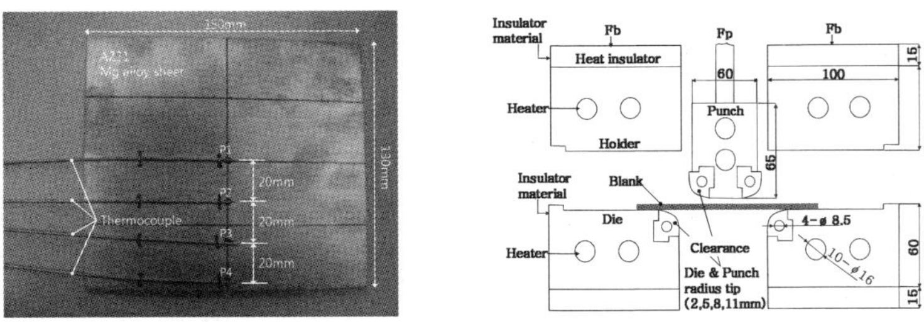

(a)Blank dimension and thermocouples

(b) dimensions for U-bending test

Fig. 1. Shape and dimensions of AZ31 Mg blank to measure temperature and dimension of bending equipment.

Table 1. Process parameters of simple U-bending test.

Process parameters	Unit	Experiential condition
Temperature	℃	RT, 100, 150, 200, 250, 300
Velocity	mm/s	2.0
Radius of the die & punch	mm	2, 5, 8, 11
Rolling direction	°	0, 90
Sheet thickness	mm	0.5, 1.0

3. Results and Discussion

Deformed profile of AZ31 magnesium sheet according to punch corner radius are shown as Figure 2. In constant clearance and thickness of ratio, the springback with 0.5mm blank thickness, is more than 1.0mm thickness.

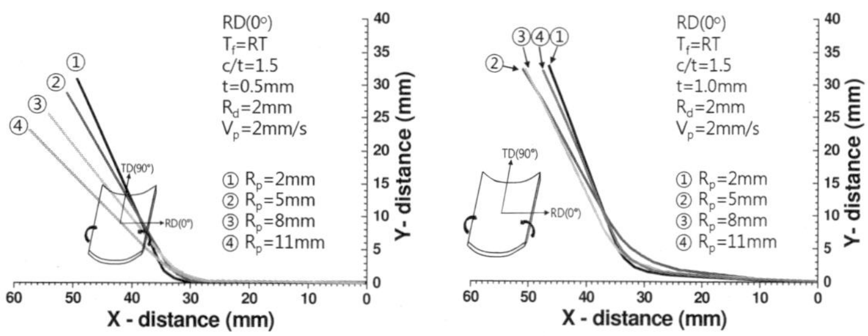

(a)0.5mm thickness (b)1.0mm thickness

Fig. 2. Deformed profile of AZ31 magnesium sheet according to punch corner radius.

Table 2. Spring-back of AZ31 magnesium sheet according to punch corner radius.

Thickness (mm)	Die corner radius (mm)	Punch corner radius (mm)	Spring-back	
			Left	Right
0.5	2	2,5,8,11	27,34,42,48	32,39,46,51
1.0	2	2,5,8,11	23,26,29,33	24,27,31,35

Figure 3 (a) and (b) show the comparing of springback profile when, by changing the radius of the shoulder radius. At sheet thickness is a 1 mm, the spring back phenomenon, regardless of the die radius and springback was found to be around 24~25 percentage. However, if the thickness is 0.5, springback depends on the die radius and the springback was found to be around 30~35 percentage.

Figure 4 (a) and (b) show the deformed profile of AZ31 magnesium sheet according to rolling direction material of the bending direction. The springback is decreased by increasing material temperature. An elastic recovery of the material is decreased with increasing temperature and decreasing springback is natural. Figure 5 shows the effect of rolling direction on springback with respect to die temperature. The amount of springback did not change according to rolling direction in room temperature.

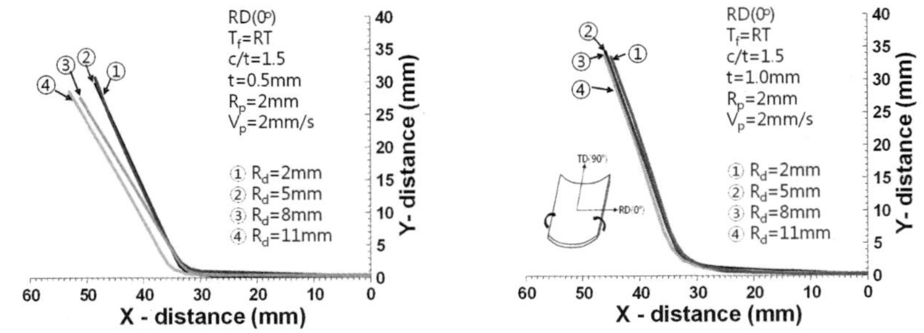

(a)0.5mm thickness (b)1.0mm thickness

Fig. 3. Deformed profile of AZ31 magnesium sheet according to die corner radius.

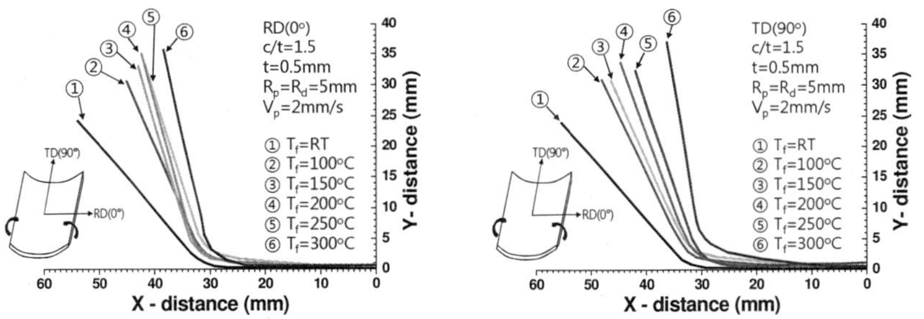

(a)Rolling direction (0°) (b)Transverse direction (90°)

Fig. 4. Deformed profile of AZ31 magnesium sheet according to rolling direction (a) rolling direction(0°), (b) transverse direction(90°).

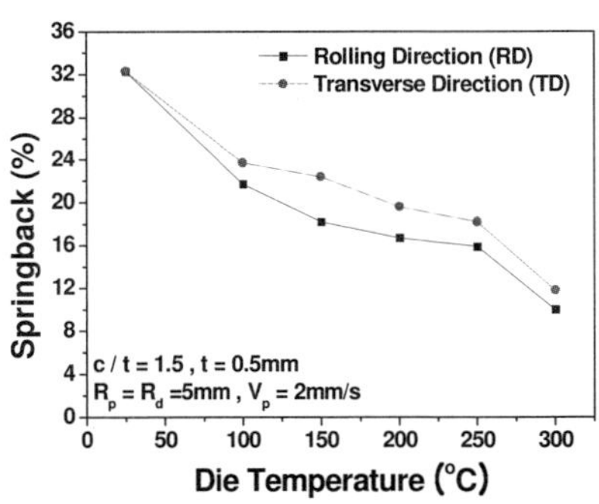

Fig. 5. Effect of rolling direction on spring-back according to variable die temperature.

However, amount of springback is changed by increasing die temperature and rolling direction in warm forming. The rolling direction (0 °) of the material can be seen less springback than perpendicular to the rolling direction (90 °). The difference of springback has the largest at 150 °C die temperature. After 150 °C die temperature, differences of springback between two materials are reduced according to increasing die temperature

4. Conclusion

In this study, the springback changes, with respect to die radius and material anisotropy, are investigated by simple U-warm bending experiments for AZ31 Mg alloy. The results were as follows:

(1) At room temperature, the springback was increased with increasing punch shoulder radius (bend radius). In 0.5mm thin material, the slight increase in spring back as increase the radius of the die shoulder, but according to the radius of the die shoulder is almost no change in springback.
(2) At room temperature, the springback are not changed according to the rolling direction. Meanwhile, the rolling direction (0 °) of the material can be seen less springback than perpendicular to the rolling direction (90 °) in warm forming.
(3) In room and warm temperature, more springback was appeared as material thickness decreasing because more wide plastic deformation is occurred as increasing the material thickness.

Acknowledgments

This work was financially supported by the Ministry of Knowledge Economy(MKE) and Korea Institute for Advancement in Technology(KIAT) through the Workforce Development Program in Strategy Technology and was supported by the National Research Foundation of Korea(NRF) grant funded by the Korea government (MEST)(No.2009-0081077) and has been supported by the National Core Research Center(NCRC). The authors would like to express our deep gratitude to the Ministry of Science & Technology(MOST) for its financial support

References

1. J.R. Cho, S.J. Moon, Y.H. Moon, S.S. Kang, *Journal of Materials Process Technology*, 2003, **141**, 109-116.
2. W.D. Carden, L.M. Geng, D.K. Matlock, R.H. Wagoner, *International Journal Mechanical Sciences,* 2002, **44**, 79-101.
3. F.K. Chen, T.B. Huang, *Journal of Materials Processing Technology*, 2003, **14**, 643-647.

INFLUENCE OF PREFORMED BILLET ON THE MULTI-DIRECTIONAL FORGING OF THE BLADE WITH A DAMPER PLATFORM

Y.L. Liu [†], Y. Wang, H. Yang

State Key Laboratory of Solidification Processing, Northwestern Polytechnical University, P.O.Box 542, Xi'an, Shaanxi, 710072, PR China

Q. Guo

Xi'an Aerospacemotor Machine Factory, Xi'an, Shaanxi, 710072, PR China

The shape and size of preformed billet have a significant influence on die cavity filling, property of the blade and forging load in the multi-directional forging process of the blade with a damper platform. An appropriate preformed billet can ensure good die cavity filling and uniform plastic deformation. Therefore, based on specifying the loading method, die parting lines and loading sequence, a 3D thermo-mechanical finite-element model for multi-directional forging process of the blade with a damper platform was built using DEFORM-3D. Moreover, three different preformed billets were obtained based on the principle of designing the billet of blade. Then, the influence laws of different preformed billets on material flow, die cavity filling, distribution of stress and strain were investigated. The results obtained show that the preformed billet with elliptical damper platform enables completely die cavity filling, no fold defects, regular flow lines and uniform distribution of tress and strain

Keywords: Blade with a damper platform; Multi-directional forging; Preformed billet; Fold defect; Numerical simulation.

1. Introduction

The blade with a damper platform is an important component in the aero-engine. The blade is complicated in shape, such as variable cross-section, great twist angle, thin body and an irregular damper platform with concave area[1], as shown in Fig. 1. Due to the concave area in the damper platform, the blade can not be separated from the die as the traditional forging method is used. Therefore it is necessary to adopt multi-directional forging technology in order to demould successfully. Moreover, the shape and size of the preformed billet significantly affect die cavity filling, plastic deformation and forging load in the multi-

[†]Corresponding author. Tel.: (029)88460212-803;E-mail address: lyl@nwpu.edu.cn (Y.L. Liu)

directional forging process of blade with a damper platform. Using an appropriate preformed billet, fold defect can be avoided, and good die cavity filling as well as uniform plastic deformation can be achieved[2].

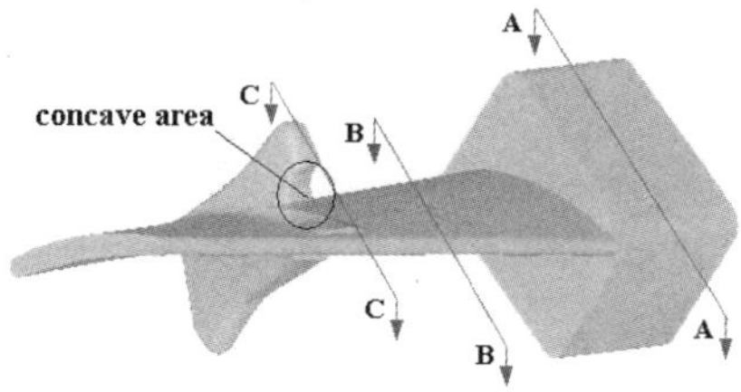

Fig. 1. Blade with a damper platform.

Many studies have been done to design the preformed billet in forging of the blade. Kang *et al.*[3] attempted to study the preformed design in forging of an airfoil section blade as a two-dimensional plane-strain problem using backward tracing simulations. H.Ou *et al.*[4] analyzed three kinds of preformed billets of turbine blade in forging process. Liu *et al.*[5] acquired a reasonable preformed billet by analyzing forming laws of different preformed billets in forging blade with a single tenon. Gao *et al.*[6] applied backward tracing scheme method to search the reasonable preformed billet in precision forging blade with a single tenon.

However, the shape of blades involved in references[4–6] is simple. The shape of blade with a damper platform is very complex and the billet is difficult to design. Therefore, in this paper, three different preformed billets were designed. The loading method, die parting face as well as loading sequence were also determined. Then, a 3D coupled thermo-mechanical finite-element model for this multi-directional forging process of the blade with a damper platform was built using DEFORM-3D. Furthermore, influence laws of preformed billets on die cavity filling, fold defect, distribution of effective stress and strain were investigated based on the model.

2. Design of Preformed Billet

The size of the billet was achieved mainly according to the design principle of axial forgings, the steps were shown below. Firstly, the billet was determined to be axisymmetrical; then, the area and diameter of axial typical cross-sections, tenon section A-A, body section B-B and damper platform section C-C shown in Fig.1, could be calculated; finally, perform billet Ⅰ was obtained, as shown in Fig.2(a). modifying the shape of damper platform section C-C, preformed billets

II and III with irregular and elliptical geometry were achieved respectively based on preformed billet I, as shown in Fig.2(b)-(c).

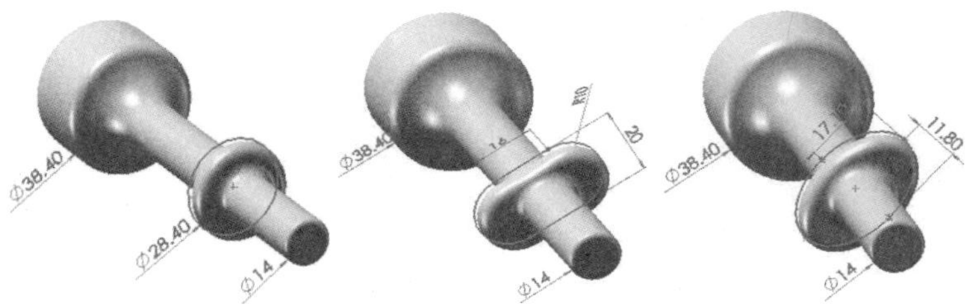

(a) Preformed billet I (b) Preformed billet II (c) Preformed billet III

Fig. 2. Geometry models of preformed billets.

3. Establishment of 3D-FE Model

The parting lines of upper dies and lower dies were selected in the place with maximum projected area of the blade at horizontal plane, and the parting lines of left dies and right dies were selected at the middle of the damper platform. Then, the left upper die, right upper die, left lower die and right lower die were obtained. Furthermore, the 3D thermo-mechanical finite-element model for multi-directional forging process of the blade with a damper platform was built, as shown in Fig.3.

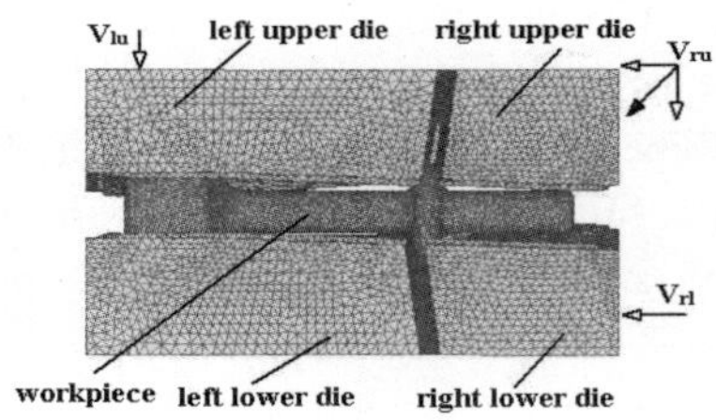

Fig. 3. 3D FE model of multi-directional forging blade with a damper platform.

4. Conditions of Numerical Simulation

The materials of the workpiece and dies are Ti6Al4V and 4Cr5MoSiV1 respectively. The material property is defined by DEFORM-3D itself. The initial temperature of 940℃ is assumed for the workpiece, whilst the initial temperature of dies is 300℃. The heat transfer coefficient is 4KW $(m^2)^{-1}$ K^{-1} and convection coefficient is 0.00295KW $(m^2)^{-1}$ K^{-1}. The friction factor is prescribed as 0.3[7].

Dies are loaded simultaneously, as shown in Fig.3. The right upper die moves in slope direction, whilst the left upper die and right lower die move in vertical direction and horizontal direction respectively, and the left lower die maintains static. The velocity of the right upper die(V_{ru}) is 20.0mm s^{-1}, so in order to keep loading consistent, the velocity of left upper die(V_{lu}) and right lower die(V_{rl}) is 18.8mm s^{-1} and 7.1mm s^{-1} respectively.

5. Results and Discussion

5.1. *Influence of preformed billets on die cavity filling and fold defect*

The die cavity filling at section A-A and section B-B is well under different preformed billets, whereas that at damper platform section C-C varies notablely. It can be obtained that die cavities at section C-C are not filled sufficiently using billet Ⅰ, moreover, fold defect appears at both sides of blade body near damper platform. The die cavities are filled sufficiently, and there is no fold defect at both sides of blade body using billet Ⅱ and Ⅲ. These show that the shape and size of billet Ⅱ and Ⅲ are more reasonable.

5.2. *Influence of preformed billets on effective strain*

The standard deviation of effective plastic strain(SDP) is used to quantify the deformation uniformity. With the decrease of SDP value, the uniformity of plastic deformation increases and the possibility of generating defects decreases.

The SDP value under different preformed billets were obtained. The SDP value of preformed billet I is maximum which indicates the deformation is most inhomogeneous and the possibility of generating defects is the highest. the SDP value of preformed billet Ⅲ is minimum which means the deformation uniformity are the best. Thus the uniformly distributed deformation can be improved by selecting preformed billet Ⅲ.

In order to investigate the deformation behavior of damper platform further, the preformed billets Ⅱ and Ⅲ with smaller SDP value are selected. A piece of deformation body with the thickness of 1mm at C-C section is selected, and the deformation body is divided to be flow nets with $25\times25\times1$ elements, as shown in Figs.4(a) and 4(c). The flow lines at the center region of both preformed billets are distinct and continuous, while deformation of flow lines at region A and B are different. disturbance of flow lines of preformed billet Ⅱ occurs at regions A and B. Moreover, the flow lines at region B are cut off, as shown in

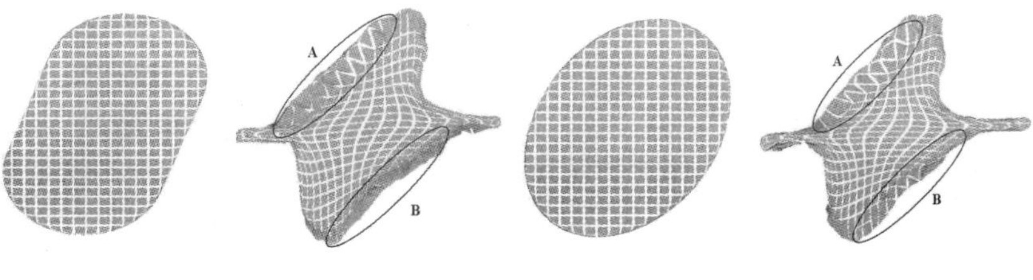

(a) Initial nets of Ⅱ (b) Eventual nets of Ⅱ (c) Initial nets of Ⅲ (d) Eventual nets of Ⅲ

Fig. 4. Flow nets of damper platform under different preformed billets.

Fig.4(b). The flow lines at regions A and B of preformed billet III are regular and continuous(see Fig.4(d)). Therefore, distribution of flow lines at damper platform can be improved and defects caused by material flow can be eliminated by using preformed billet III.

5.3. *Influence of preformed billets on effective stress*

The effective stress values under different billets at the end of forging process were obtained. The average and maximum effective stress of billet III are much less than that of the other two billets, which indicates that forming force of billet III is the lowest and the life of dies can be improved. The high effective stress may cause early failure of dies and component-form errors. The standard deviation of effective plastic stress(S.D.) of billet III is also the minimum which represents effective stress distribution of billet III is the most uniform.

Further studies on distribution of effective stress at damper platform are carried out by selecting preformed billets II and III with smaller S.D. values. Fig.5 shows the distribution of effective stress of the blade concave and convex.

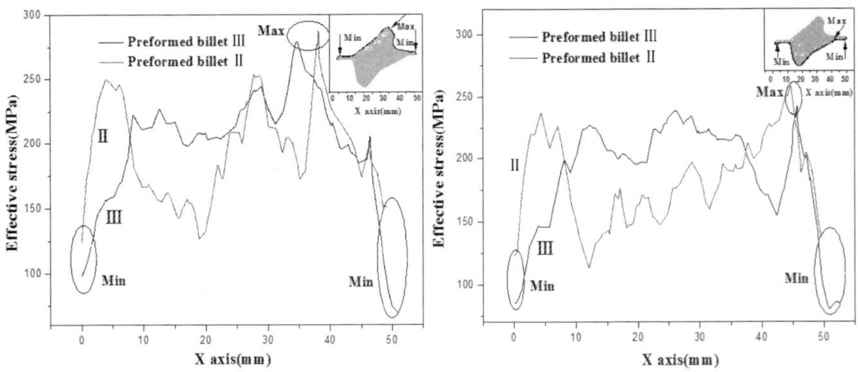

(a) Blade convex (b) Blade concave

Fig. 5. Distribution of effective stress of blade concave and blade convex.

The minimum stress of both regions occur at the flash land. The stress peak of the blade concave occurs at the top of damper platform, while that of the blade convex occurs at the transition region between damper platform and flash land. The maximum and minimum effective stress of billet II are higher than that of billet III, moreover, the effective stress values of billet II fluctuate greatly while stress fluctuation of billet III is relatively stable. Consequently, The billet III would lead to lower effective stress and smaller stress fluctuation.

6. Conclusion

(1) With consideration of the shape of damper platform, three different preformed billets are designed. Using the 3D coupled thermo-mechanical FEM, process simulation for multi-directional forging of a blade with a damper platform has been realized.

(2) Die cavities of billet Ⅰ can not be filled sufficiently, and fold defect appears at both sides of blade body near damper platform. Using billets Ⅱ and Ⅲ, the die cavity can be filled completely, fold defects can be avoided.

(3) Disturbance of flow lines of billet Ⅱ occurs at damper platform and the flow lines are cut off. While disturbance of flow lines at damper platform can be eliminated using billet Ⅲ.

(4) The distribution of effective stress at damper platform of billet Ⅲ is homogeneous in addition that forming force is the lowest. The minimum and maximum stress at damper platform of billet Ⅱ are higher than that of billet Ⅲ, and the stress fluctuation of Ⅲ is stable.

Acknowledgments

This project has been supported by Project of Science and Technology of Xi'an Shaanxi Province (GG06046) and the 111 Project (B08040).

References

1. Z.W. Zhang: 'Blade Forging', 76; 1987, Xi'an, Xi'an Jiao Tong University Press.
2. X.H. Zhao, G.Q. Zhao and G.C: *J. Plasticity Eng.*, 2000, **7**, (3), 1–6.
3. B.S. Kang, N. S. King and S. Kobayashi: *Int. J. Mach. Tools Manuf.*, 1990, **30**, (1),43–52.
4. H. Ou, R. Balendra: *J. Mater. Proc. Technol.*, 1998, **80–81**, 144–148.
5. Y.L Liu: '3-D FEM Analysis of Forming Laws in Precision Forging Process of Blade', Ph.D Dissertation, 73–78, 2001, Northwest Polytechnical University.
6. T. Gao, H. Yang and Y.L Liu: *Trans. Nonferrous Met. Soc. China*, 2006, **16**, 639–644.
7. Z.M. Hu, J.W. Brooks and T.A. Dean, *J. Mater. Proc. Technol.* 1999, **88**, 251–265.

EMBEDDED DELAMINATION GROWTH IN INTERLAMINAR SHORT-FIBER REINFORCED COMPOSITE PANELS UNDER COMPRESSIVE LOAD

ZHANG FENGPENG†

College of Resources and Civil Engineering, Northeastern University, Shenyang, Liaoning Province, China

QIU ZHAOGUO

School of Science, Northeastern University, Shenyang, Liaoning Province, China

LIU DDCHENG

School of Science, Northeastern University, Shenyang, Liaoning Province, China

Crack growth analyses on interlaminar short-fibers (ISFs) reinforced composite panels containing embedded delamination has been performed using a geometrical nonlinear FEM code. Strain energy release rate (ERR) criterion is used to rule the propagation of the cracks. The interface fracture elements are placed among brick elements lying along the delamination front. The code has been improved with an effective virtual crack closure technique to evaluate ERR and with penalty method to evaluate contact forces. The crack propagation has been simulated by node coupling/partition and the technology of element's living or death. The influence of ISFs on ERR and delamination front shapes have been analyzed. Compared with the composite panels without ISFs, the results showed that ISFs can effectively reduce the delamination propagating and enhance the interlaminar toughness.

Keywords: Composite laminates; delamination; damage tolerance; energy release rate; interlaminar short-fiber; bridge effect.

1. Introduction

The poor resistance to delamination due to the brittle nature of the thermoset matrix is a major weakness and potential danger of laminar composite. There are different methods available for improving the delamination toughness of carbon-fiber/epoxy composites, such as interleaving, stitching and z-pinning et al[1-3]. Using chopped Kevlar fibers randomly distributed in interfaces, a simple and low-cost interlaminar toughening method was developed by Sohn and Hu[4]. The

† Zhang Fengpeng, email: zhangfengpeng@126.com

effectiveness of the ISFs for interlaminar toughening have been proved by the experiments[4] and the micro-modelling[5]. However, current understanding in this field of research is still in a primitive stage.

The aim of the present work is to study the influence of ISFs on the strain ERR distribution along the embedded delamination's front and its growth.

2. Finite Element Approach

In order to perform the numerical analyses to predict delamination growth, the nonlinear finite element code features have been enhanced by means of:
1. Introduction of a contact element;
2. Introduction of an interface fracture element.

2.1. *Contact element*

In order to prevent inter-penetrations between the two contacting surfaces, contact forces should be taken into account during analyses, so a 3D node-to-node contact element has been introduced in our FEM code. It should be pointed out that the node-to-node contact approach is applicable wherever the slip between contacting surfaces is not large. This element is based on most effective methods for contact problems in engineering, because of its minimum requirements of computational cost.

2.2. *Fracture element*

From fracture mechanics of composites, it is well known that strain ERR rules propagation of cracks.

Three basic fracture modes in laminates are depicted. For each fracture mode a strain ERR can be defined: G_{I} for mode I, G_{II} for mode II and G_{III} for mode III.

For general load conditions, it is possible to predict crack propagations in laminates by mean of suitable criteria. We have adopted the power-law criterion. It states that when propagation happens:

$$\left(\frac{G_{\mathrm{I}}}{G_{\mathrm{IC}}}\right)^2+\left(\frac{G_{\mathrm{II}}}{G_{\mathrm{IIC}}}\right)^2+\left(\frac{G_{\mathrm{III}}}{G_{\mathrm{IIIC}}}\right)^2=E_{\mathrm{d}}\geq 1 \tag{1}$$

The critical ERR values (G_{IC}, G_{IIC} and G_{IIIC}) are evaluated from classified experimental procedures.

The delaminations' growth model was implemented in our finite element code by means of a new interface fracture element based on the modified virtual crack closure technique [6] to evaluate the strain ERR.

3. Embedded Delamination Analysis

3.1. *Laminates and ISFs parameters*

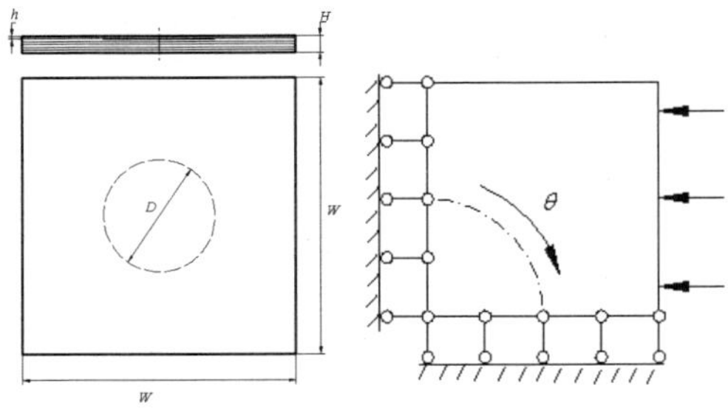

(a) Full model (b) 1/4 model

geometrical parameters		geometrical parameters	
h (mm)	0.375	D (mm)	80
H (mm)	6	W (mm)	150

Fig. 1. Model of composite laminates.

Laminates, two cases with different stacking sequence: $[0^{o}]_{16}$, and $[0^{o}/90^{o}]_{4s}$ containing embedded circular delaminations, have been selected to perform delamination grown behavior. The geometrical characteristics of the analyzed specimens are shown in Figure 1.

A compressive displacement filed has been assumed to act on the specimen. Due to the symmetry only a quarter of the structure has been modeled. Twenty nodes, brick elements, fracture interface elements and contact elements have been used to discretize the problem.

3.2. *Distribution of ERR along the delamination front*

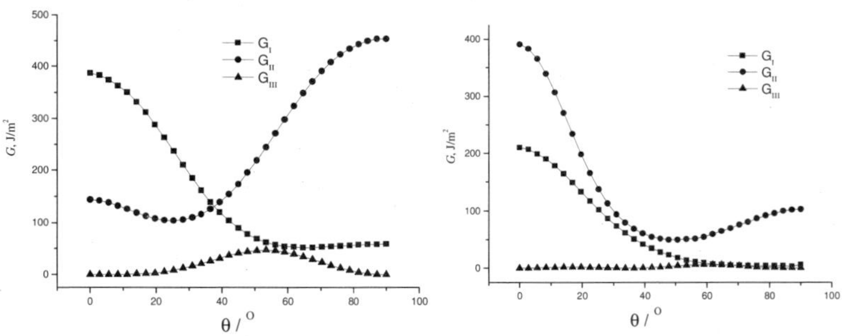

(a) $[0^{o}]_{16}$ (b) $[0^{o}/90^{o}]_{4s}$

Fig. 2. ERR distributions along the delamination front as $\varepsilon = 0.248\%$.

The distribution of ERR along the front of delamination as $\varepsilon = 0.248\%$ in case of stacking sequence $[0^{\circ}]_{16}$ are plotted in the Figure 2(a). It is shown that G_{I} is the largest at θ=0°, G_{II} is the largest at θ=90°, and G_{III} is smallest always.

In Figure 2(b), it is shown that G_{I} is the largest at $\theta=0^{\circ}$, and G_{II} is the largest at $\theta=0^{\circ}$ also in case of stacking sequence $[0^{\circ}/90^{\circ}]_{4s}$, and G_{III} is great smaller than G_{I} or G_{II} as same as the case of $[0^{\circ}]_{16}$.

Because of the ERR for mode III is great smaller always, so G_{III} is ignored in this paper, and the Eq. 1 is modified as:

$$\left(\frac{G_{\mathrm{I}}}{G_{\mathrm{IC}}}\right)^2 + \left(\frac{G_{\mathrm{II}}}{G_{\mathrm{IIC}}}\right)^2 = E_{\mathrm{d}} \geq 1 \tag{2}$$

If relation (2) is satisfied, the node will be released leading to a modification of the delamination front.

3.3. *Propagation of delamination*

3.3.1. *The critical ERR increment values of laminates with ISFs*

The critical ERR values of laminates with ISFs are modified as:

$$G_{\mathrm{IC}}^{\mathrm{br}} = G_{\mathrm{IC}} + \Delta G_{\mathrm{IC}}^{\mathrm{br}}\ ,\ G_{\mathrm{IIC}}^{\mathrm{br}} = G_{\mathrm{IIC}} + \Delta G_{\mathrm{IIC}}^{\mathrm{br}} \tag{3}$$

The critical ERRs increment values are computed as:

$$\Delta G_{\mathrm{IC}}^{\mathrm{br}} = \int_0^{w_{\max}} \sigma_{\mathrm{br}} dw\ ,\ \Delta G_{\mathrm{IIC}}^{\mathrm{br}} = \int_0^{u_{\max}} \tau_{\mathrm{br}} du \tag{4}$$

The bridge stresses in Ref. 5 are used to compute the increments.

3.3.2. *Propagation of delamination of laminates*

If relation (2) is satisfied, the propagation occurs, and the delamination front shapes will be modified.

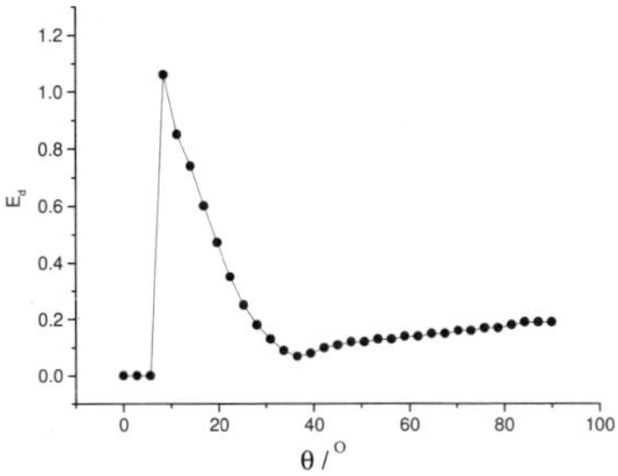

(a) with ISFs

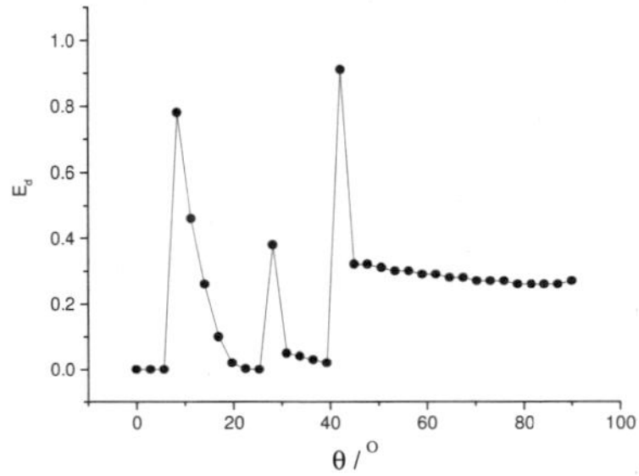

(b) without ISFs

Fig. 3. The variation of E_{d} along delamination front as ε =0.304%.

In laminates with ISFs, there is only one delamination growth point, and the growth zone is among 0° to 10° (in Figure 3(a)). But in laminates without ISFs, there are three delamination growth points, and the zone of growth is among 0° to 43° (in Figure 3(b)). It is to say that the delamination grown zone of laminates with ISFs is much smaller than that of laminates without ISFs, and ISFs can effectively reduce the delamination propagating and enhance the interlaminar toughness.

4. Conclusion

A numerical investigation on delaminated composite panels with and without ISFs has been presented. The results show that:

The propagations in all specimens (with different stacking sequence, with ISFs or without ISFs) are all in the zone of θ=0°, but the propagation zones are different. The zone of $[0^{\circ}]_{16}$ is larger than $[0^{\circ}/90^{\circ}]_{4s}$, and zone of laminates with ISFs is much smaller than that of laminates without ISFs at the same compressive load. The ISFs can effectively enhance the interlaminar toughness and reduce the delamination propagating.

Acknowledgments

This work was financially supported by the National High-Tech Research and Development Program ("863" program) of China (No.2007AA06Z107), the Natrual Science Foundation of China (No.50974031), the National Key Technology R&D Program of China (No.2008BAB34B02), the Program for New Century Excellent Talents in Chinese universities (No.NCET-07-0163), and the Fundamental Research Funds for the Central Universities (No.90401006, No. 90405008).

References

1. K. Dransfield, C. Baillie, Y. M. Mai. *Compos. Sci. Technol.*, 1994, 50, 305-317.
2. S. F. Chen, B. Z. Jang. *Compos. Sci. Technol.*, 1991, 41, 77–97.
3. M. Grassi, X. Zhang. *Compos. Sci. Technol.*, 2003, 63, 1815–1832.
4. M. S. Sohn, X. Z. Hu. Compos., 1995, 26, 849–858.
5. B. Z. Huang, X. Z. Hu, J. Liu. *Compos. Sci. Technol.*, 2004, 64, 2165–2175.
6. E. F. Rybicki, M. F. Kanninen. *Eng. Fract. Mech.*, 1977, 9, 931–938.

ANALYSIS OF THE FORMING DEFECTS AND ITS INFLUENCING FACTORS OF THE POLY-WEDGE PULLEY SPINNING

XIUQUAN CHENG †

Guangzhou Civil Aviation College, Guangzhou, 510403, China

JIAZI WANG, QINXIANG XIA

School of Mechanical & Automotive Engineering, South China University of Technology, Guangzhou, 510640, China

The forming defects occurred during each forming processes of poly-wedge pulley spinning, such as the drumming failure, flanged opening-end, folded side-wall, insufficient bottom size, flashed opening-end, cutting-off bottom, are introduced, and the factors influencing the defects are analyzed. The corresponding preventive measures are put forward, which establishes a reliable base for practical manufacture and theoretical research of poly-wedge pulley spinning.

Keywords: Spinning; Poly-wedge pulley; Defect; Influencing factor.

1. Introduction

The poly-wedge pulley is an important part of the belt driving system. The strength, rigidity the service life of the belt, and the driving efficiency and precision of the system are influenced directly by the forming quality of the belt[1]. Spun pulley made by metal sheet as a new pulley structure is widely applied in various industrial fields with its high accuracy, energy saving, material saving, good dynamic equilibrium and non-pollution, etc.[2]. For poly-wedge pulley spinning, a thicker blank should be adopted to meet the requirement of tooth forming, resulting in the increasing of the material and consequential enhancement of manufacture cost. This shortage can be overcome well by thickening process[3].

The forming defects during spinning of poly-wedge pulley based on thickening process are analyzed, and the corresponding preventive measures are put forward to prevent the defects.

† Xiuquan Cheng, email: chengxiuquan@yahoo.com.cn

2. Processing Analysis of Poly-wedge Pulley deformed by Spinning

The material used for the poly-wedge pulley is 08Al. The bottom thickness of the pulley is 2.5mm (as shown in Fig. 1h). To form the tooth, the thickness of the side wall should be 2.8mm. Thickening process is necessary before tooth forming. The whole manufacturing procedures include blanking, drawing and spinning[4], and the spinning procedure includes five processes, performing, drumming, thickening, toothing and finishing (as shown in Fig. 1c ~ Fig. 1g).

The deformation of the tooth portion of pulley by spinning belongs to the near-net shape forming. The demanded tooth shape and size can be formed directly by spinning without machining. Only slight machining is carried out at the outlines marked by A and B (as shown in Fig. 1).

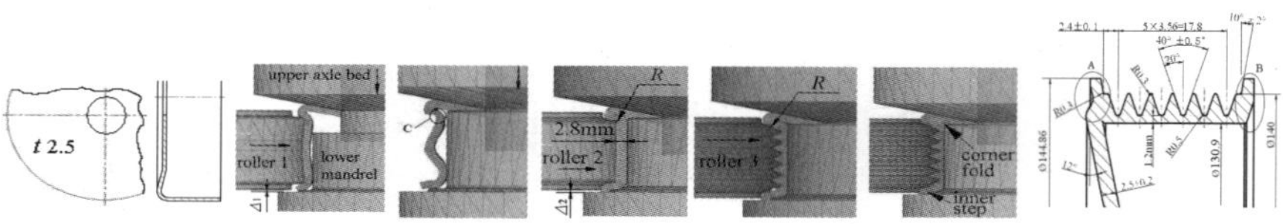

a) blanking b) drawing c) preforming d) drumming e) thickening f) toothing g) finishing h) spun poly-wedge pulley

Fig. 1. Manufacturing procedures of the poly-wedge pulley.

3. Defects Analysis of Poly-wedge Pulley during Deforming Step by Spinning

3.1. *Defects during drumming*

The defects occurred during drumming process are mainly diameter expanding at bottom portion, flanging at opening end of workpiece and unsymmetrical drumming (as shown in Fig.2).

The defect of diameter expanding at bottom during drumming is shown in Fig. 2a). The reason for the defect is that the radial feed of roller 1 is too small (as shown in fig. 1c), causing a flat curvature at the side wall of the cup-shaped blank after preforming. Therefore, diameter expanding occurs at the bottom of workpiece due to buckling deformation on condition that the supporting rigidity of the side wall is strong enough to prevent the drumming deformation.

The defect of flanging at opening end during drumming is shown in Fig. 2b). The reason for the defect is the unreasonable setting height of roller 1. The gap Δ_1 between the roller 1 and the surface of the lower mandrel (as shown in fig. 1c) is greater than the designed one. Meanwhile, the radial feed of roller 1 is a little large. The diameter of the opening end of workpiece doesn't decrease and the diameter near the opening end decreases greatly by necking. It results in an unstable deformation of opening end, and flanging when moving downward the upper mandrel.

The defect of unsymmetrical drumming is shown in Fig. 2c). The reason for the defect is that the distribution of the side wall thickness of the cup-shaped workpiece obtained by deep drawing is uneven. The thickness reduces gradually from the opening end to the bottom along the side wall of the cup-shaped workpiece, hence the drum tends to occur at the bottom during drumming. It may result in the other defects in the subsequent thickening process.

a) diameter expanding b) flanging c) unsymmetrical drumming

Fig. 2. Defects during drumming process.

The main measures to prevent the above defects are as follows: selecting a reasonable setting height of roller 1, decreasing the gap Δ_1 between roller 1 and lower mandrel as small as possible on condition that the roller 1 doesn't contact with the lower mandrel; controlling the thickness deviation of the side wall of the cup-shaped workpiece obtained by deep drawing to less than 7% strictly.

3.2. *Defects during thickening*

The defects occurred during thickening process are mainly side wall folding, insufficient bottom size and flash at opening end (as shown in Fig. 3).

The defect of side wall folding is shown in Fig. 3a), a swallowtail annular trough occurs on the side wall of workpiece during thickening. The reason for the defect is the unsymmetrical drum during drumming process (as shown in Fig.2c). Because the drum tends to occur at the workpiece bottom and the wall thickness near the bottom is thinner than another portion, the material flows more easily to the bottom than to the opening end of workpiece. Meanwhile, a large resisting force exists at the bottom arc portion, under the action of the pressing forces from both the bottom and opening end directions, the local accumulated material is extruded, which results in the swallowtail annular folding.

The defects of insufficient bottom size and flash at opening end are shown in Fig. 3b). The reason for the defects is that the setting height of thickening roller 2 is unreasonable. The gap Δ_2 between the roller 2 and the lower mandrel is greater than the designed one (as shown in Fig. 1e).

The radial extrusion occurs on the corner portion *A* of workpiece (see Fig.1h) under the pressing of upper surface of roller 2, which results in the material flows towards the workpiece center and results in the insufficient bottom diameter. On the other hand, the material at the opening end flows easily towards the workpiece edge under the action of lower surface of roller 2, which results in the flash at the opening end of workpiece.

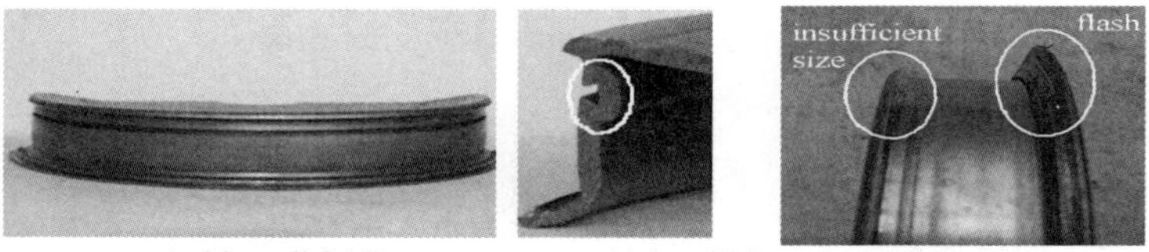

a) side wall folding b) insufficient size and flash

Fig. 3. Defects during thickening.

To prevent the above defects, a suitable setting height of roller 2 should be guaranteed. The gap Δ_2 between the roller 2 and the lower mandrel should be as smaller as possible under the condition that the roller 2 doesn't contact with the surface of lower mandrel.

3.3. *Defects of bottom cutting-off*

The defects of bottom cutting-off may occur during both thickening and toothing processes (as shown in Fig.4). The reason for the defect is the roundness radius R of roller 2 or roller 3 is too small (see Fig. 1e and Fig. 1f). Because the thickness reduction at the position C (see Fig. 1d) is large during preforming, the bottom cutting-off occurs easily during thickening or toothing. The problem could be solved well by increasing the roundness radius R from 1mm to 2mm. Experimental results showed that roundness radius R should be greater than 0.8 times of the side wall thickness of cup-shaped drawn workpiece.

a) bottom cutting-off during thickening b) bottom cutting-off during toothing

Fig. 4. Bottom cutting-off.

4. Trial Manufacturing of the Poly-wedge Pulley

The experiment was carried out in the HGQX-LS45-CNC vertical spinning machine[5]. The material of blank sheet is 08Al, the gap $\Delta_1=\Delta_2=0.2$mm, the radial feed of the preforming roller 1 is 4.8mm.

Fig. 5 shows the qualified spun poly-wedge pulley. Fig. 5a) shows that the spun pulley has the regular contour, distinct outline and good surface quality.

Fig. 5b) shows that the tooth profile on the side wall is plump and uniform. The inner surface of spun workpiece shown in Fig. 5c) is smooth.

a) complete outline b) amplified tooth c) inner surface

Fig. 5. Qualified spun pulley.

5. Conclusion

1. The defects occurred during drumming are mainly diameter expanding at bottom, flanging at opening end, and unsymmetrical drumming. The defects can be prevented well by a small thickness deviation of the side wall of the cup-shaped drawn workpiece.
2. The defects occurred during thickening are mainly side wall folding, insufficient bottom size and flash at opening end. The defects can be prevented well by controlling the thickness deviation of the side wall of the cup-shaped drawn workpiece and the reasonable gap between the roller 2 and the lower mandrel.
3. The defect of bottom cutting-off may occur during both thickening and toothing. The defect can be prevented well by increasing the roundness radius of thickening and toothing rollers properly.

Acknowledgments

The authors gratefully acknowledge the funding obtained from Production, Teaching & Research Collaborative Project between Guangdong Province and Ministry of Education of China (Project No. 2006D90304021).

References

1. Pan Jie, Xu Wei-cheng: *Rural Science & Technology*, 2007, **7**, 78.
2. Wang Zhong-qing, Yang Dong-fa: *Metalforming Machinery*, 1999, **34**, (1), 7~9.
3. XIA Qin-xiang, WANG Jia-zi, WANG Ying-pin, LIN Ye-hai, MA Zhe-ming: *Forging &Stamping Technology*, 2009, **34**, (6), 101~106.
4. XIA Qin-xiang, WANG Ying-pin, WANG Jia-zi, LIN Ye-hai,WU Zhen-bang: *Die & Mould Industry*, 2010, **36**, (1), 26~30.
5. Qinxiang XIA, Baixiang LIANG, Saijun ZHANG, etc.: *Journal of Materials Science & Technology*, 2006, **22**, (2), 261~268.

THE BUCKLING OF THE STRETCHABLE ELECTRONIC STRUCTURE

ZHANG XIAOQING[†], OU ZHICHENG, YAO XIAOHU, WANG WEITONG
School of Civil Engineering and Transportation, South China University of Technology, Guangzhou, Guangdong 510640, China

The structure of stiff thin film on a compliant substrate has important applications in international advanced stretchable electronics. Based on the Von Karman plate theory, we model a stiff thin film on a compliant substrate to deduce the buckling governing equations through energy method. The relations between characteristic parameters of buckling, material properties and geometric parameters are obtained. Then the numerical simulation are achieved by ABAQUS, and compared with the theoretical solutions. It is shown that, the equilibrium wavelength and critical strain are dependent on the plane strain elasticity modulus of the film and the substrate and the film thickness

Keywords: Stretchable electronics; Compliant substrate; Stiff thin film; Buckling.

1. Introduction

The structure of stiff thin film on a compliant substrate, specially, the buckling of such structure is applied in international advanced stretchable electronics[1].

The researchers have built the model of stiff thin film on a compliant substrate to study the buckling behavior in many cases, such as the isotropic, orthotropic elasticity thin film on an elasticity, viscous, viscoelastic substrate[1-5]. In these researches, the substrate is described by small deformation theory, while the film is described by large deformation plate theory. Based on the previous researches, we model a stiff thin film on a compliant substrate and the buckling governing equations are obtained. And the characteristic parameters are discussed. In addition, we apply ABAQUS to make the numerical simulation.

2. Theory

2.1. *Model of stiff thin film on a compliant substrate*

A stiff elastic film is bonded to a compliant elastic substrate without slipping, which in turn is bonded to a rigid support, as Figure 1. The system buckles under the axial pre-strain.

[†] Corresponding author, email: tcqzhang@scut.edu.cn

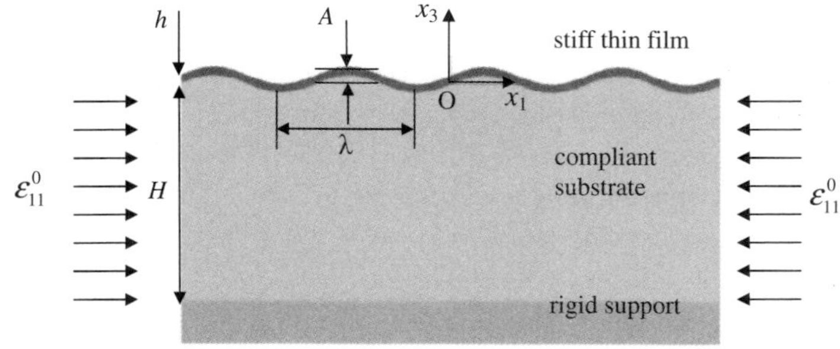

Fig. 1. Model of stiff thin film on a compliant substrate.

2.2. *Energy of the film*

Let u_1 and u_2 be the in-plane displacement of the film. And u_3 is the deflection. So the nonlinear strains are

$$\varepsilon_{\alpha\beta} = \varepsilon^0_{\alpha\beta} + \frac{1}{2}(u_{\alpha,\beta} + u_{\beta,\alpha}) + \frac{1}{2}u_{3,\alpha}u_{3,\beta} \tag{1}$$

By the Hooke's law, we know the membrane forces are

$$N_{\alpha\beta} = h\sigma_{\alpha\beta} = h\bar{E}_{\mathrm{f}}[(1-\nu_{\mathrm{f}})\varepsilon_{\alpha\beta} + \nu_{\mathrm{f}}\varepsilon_{\alpha\beta}\delta_{\alpha\beta}] \tag{2}$$

where E_{f} and ν_{f} are Young's modulus and Poisson's ratio of the film. And the plane strain effective modulus is $\bar{E}_{\mathrm{f}} = E_{\mathrm{f}} / (1-\nu_{\mathrm{f}}^2)$.

Because the buckling mode is similar to cosine curve, deflection of the midplane in the film and the in-plane displacement are assumed as

$$u_3 = A\cos(kx_1)\ ,\ u_1 = \frac{1}{8}kA^2\sin(2kx_1)\ ,\ u_2 = 0 \tag{3}$$

The energy of the film contains bending strain energy and compression strain energy. We calculate the elastic energy per unit area in one period

$$U_b = \frac{h^3\bar{E}_{\mathrm{f}}}{24}(\nabla^2 u_3)^2 = \frac{h^3\bar{E}_{\mathrm{f}}}{24}\frac{k}{2\pi}\int_0^{2\pi/k} u_{3,11}^2\,\mathrm{d}x_1 = \frac{h^3\bar{E}_{\mathrm{f}}}{48}k^4A^2 \tag{4}$$

And the compression strain energy is

$$U_m = \frac{1}{2}N_{11}\varepsilon_{11} = U_m^0 + \frac{1}{4}N_{11}^0k^2A^2 + \frac{1}{32}h\bar{E}_{\mathrm{f}}k^4A^4 \tag{5}$$

where $U_m^0 = \frac{h\bar{E}_{\mathrm{f}}}{2}(\varepsilon_{11}^0)^2$ is constant, the initial membrane forces $N_{11}^0 = h\bar{E}_{\mathrm{f}}(\varepsilon_{11}^0 + \nu_{\mathrm{f}}\varepsilon_{22}^0)$.

2.3. *Energy of the substrate*

We introduce a stress function $\phi(x_1, x_3)$ as $\sigma_{11} = \phi_{,33}$, $\sigma_{33} = \phi_{,11}$, $\sigma_{13} = \sigma_{31} = -\phi_{,13}$. According to the stress compatibility equation, the general solution is

$$\phi(x_1, x_3) = [C_1 \cosh(kx_3) + C_2 \sinh(kx_3) + C_3 x_3 \cosh(kx_3) + C_4 x_3 \sinh(kx_3)]\cos(kx_1) \quad (6)$$

By the small deformation theory, the strains are

$$\varepsilon_{\alpha\beta} = \frac{1}{2}(u_{\alpha,\beta} + u_{\beta,\alpha}) \quad (7)$$

The physical equations of plane strain are

$$\varepsilon_{11} = \frac{1}{\bar{E}_s}(\sigma_{11} - \bar{v}_s \sigma_{33}),\ \varepsilon_{33} = \frac{1}{\bar{E}_s}(\sigma_{33} - \bar{v}_s \sigma_{11}),\ \varepsilon_{13} = \frac{2(1+\bar{v}_s)}{\bar{E}_s}\sigma_{13} \quad (8)$$

where the plane strain effective modulus and Poisson's ratio are $\bar{E}_s = E_s/(1-v_s^2)$ and $\bar{v}_s = v_s/(1-v_s)$.

The boundary conditions are

$$\sigma_{13}(x_1, 0) = 0,\ u_3(x_1, 0) = A\cos(kx_1),\ u_1(x_1, -H) = u_3(x_1, -H) = 0 \quad (9)$$

According to (6)-(9), the constants C_i can be solved. So we get the stress on the upper surface of the substrate

$$\sigma_{33}^* = \sigma_{33}(x_1, 0) = -Ak\bar{E}_s \cos(kx_1)\frac{2H^2k^2 - 4v_s[\cosh(2kH)+3] + 3\cosh(2kH) + 8v_s^2 + 5}{(6-8v_s)\sinh(2kH) - 4Hk}$$

For the thick substrate, $kH \to \infty$, it can be simplified

$$\sigma_{33}^*(x_1) = \frac{1}{2}Ak\bar{E}_s \cos(kx_1) \quad (10)$$

The energy of the substrate comes from the interaction stress. Considering the periodicity similarly, we have the energy density

$$U_s = \frac{k}{2\pi}\int_0^{2\pi/k} \frac{1}{2}\sigma_{33}^* u_3 \mathrm{d}x_1 = \frac{1}{8}k\bar{E}_s A^2 \quad (11)$$

2.4. *Buckling analysis*

The total energy of the film/substrate system

$$U(A,k) = U_m + U_b + U_s = U_m^0 + \frac{1}{4}(N_{11}^0 + h\bar{E}_f\zeta)k^2A^2 + \frac{h\bar{E}_f}{32}k^4A^4 \quad (12)$$

Where $\zeta(k) = \frac{(hk)^2}{12} + \frac{\bar{E}_s}{2(hk)\bar{E}_f}$.

The total energy is a fourth order polynomial of A. When $N_{11}^0 + h\bar{E}_f\zeta \geq 0$, the energy minimizes at the state A=0, and the film is flat. When $N_{11}^0 + h\bar{E}_f\zeta \geq 0$, the energy minimizes when the film buckles. By solving $\partial_A U(A,k) = 0$, we have the amplitude A. The critical condition of buckling is A=0. We know the minimum of $\xi(k)$ and the critical prestrain

$$\zeta_{\min} = \frac{1}{4}\left(\frac{3\bar{E}_s}{\bar{E}_f}\right)^{2/3} = -\varepsilon_{cr} \quad (13)$$

To obtain the equilibrium wave number k_{eq}, we set $\partial_k U(k) = 0$ when $|N_{11}^0| > h\bar{E}_f \zeta$. The equilibrium wavelength is

$$\lambda_{eq} = \frac{2\pi}{k_{eq}} = 2\pi h \left(\frac{\bar{E}_f}{3\bar{E}_s} \right)^{1/3} \tag{14}$$

We know that the critical prestrain and equilibrium wavelength are dependent on the plane strain elasticity modulus of the film and substrate. The equilibrium wavelength is proportional to the film thickness.

3. Finite Element Analysis

The film/substrate structure is simplified as a plane plain problem. It consists of a 0.1 mm-thick substrate and a 100 nm-thick thin film. The length of the system is 0.1 mm.The substrate is modeled by the 6-node and 8-node plane-strain element. The film is modeled by the beam element with the smallest element size 0.3 μm. The upper surface of the substrate is attached to the thin film by sharing the same nodes at the interface. Here we give an example of silicon film and PDMS (Polydimethylsiloxane) substrate. The material parameters are E_f $=1.30\times10^{11}$Pa, v_f=0.27, $E_s = 1.8\times10^6$Pa and v_s=0.48. We applied the prestrain to the substrate through temperature field. The buckle analysis is performed to measure the equilibrium wavelength and get the critical prestrain.

4. Solutions and Conclusion

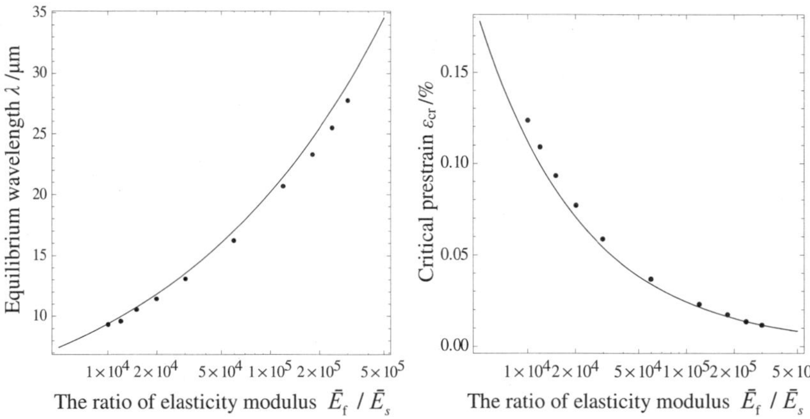

Fig. 2. The equilibrium wavelength and the critical prestrain versus the ratio of elasticity modulus.

Based on the theoretical and numerical analysis, we change the material parameters and the film thickness to study the equilibrium wavelength and the critical prestrain, as shown in Figure 2 and 3. Obviously, the solutions of

FEA coincide with the theoretical solutions. It is shown that, with the increment of the modulus ratio, the equilibrium wavelength increase, while the critical prestrain decrease. And the equilibrium wavelength is proportional to the film thickness.

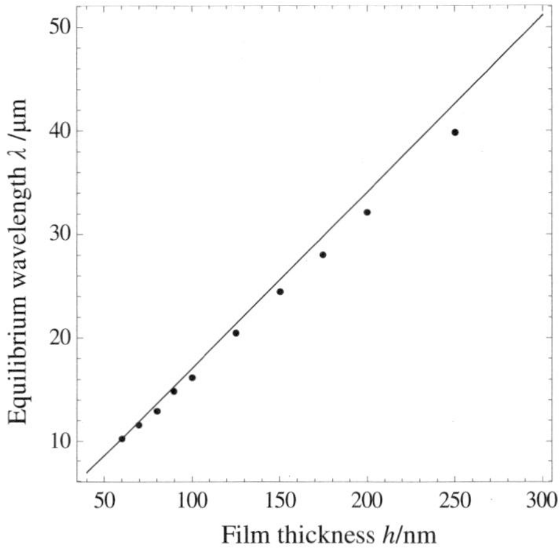

Fig. 3. The relation between the equilibrium wavelength and film thickness.

To summarize, the buckling of a stiff thin film on a compliant substrate are analyzed through the theory and FEA . And some important results are given.

Acknowledgments

This work was supported by the National Natural Science Foundation of China (10902040/A020602), the Specialized Research Fund for the Doctoral Program of Higher Education(New Teachers), the Natural Science Foundation of Guangdong Province (8451064101000229) and the Fundamental Research Funds for the Central Universities, SCUT(2009ZM0238，2009ZM0280).

References

1. J. Song, H. Jiang, Z. J. Liu, D. Y. Khang, Y. Huang, J. A. Rogers, C. Lu, and C. G. Koh: *Int. J. Solids Struct.*, 2008, **45**(10): 3107-3121.
2. Z. Y. Huang, W. Hong, Z. Suo: *J. Mech. Phys. Solids*, 2005, **53**: 2101-2118.
3. R. Huang, Z. Suo:*Int. J. Solids Struct.*, 2002, **39**(7): 1791-1802.
4. R. Huang, Z. Suo: *J. Appl. Phys.*, 2002, **91**(3): 1135-1142.
5. R. Huang, *J. Mech. Phys. Solids*, 2005. **53**(1): 63-89.

CLOSED FORM STRESS DISTRIBUTION IN 2D ELASTICITY FOR ALL BOUNDARY CONDITIONS

YOUZHEN YANG†

Institute of Solid Mechanics of Ningxia University, Yinchuan 750021, China

This paper applies a Hamiltonian method to study analytically the stress distributions of anisotropic two-dimensional elasticity for arbitrary boundary conditions without beam assumptions. The homogenous solutions consist of the eigensolutions of the derogatory zero eigenvalues zero eigensolutions and that of the well-behaved nonzero eigenvalues. The Jordan chains at zero eigenvalues give the classical Saint-Venant solutions. On the other hand, the nonzero eigen-solutions describe the exponentially decaying localized solutions usually ignored by Saint-Venant's principle. Completed numerical examples are newly given to compare with established results.

Keywords: Anisotropic; Eigenfunctions; Symplectic expansion.

1. Introduction

Stress singularity analysis is always a difficult point in plane elastic research. References[1~10] had investigated the stress singularities with different kinds of methods. Although lots of related papers have been published, there are only a few of them using rigorous two-dimensional elasticity theory to give complete stress distribution solutions. As pointed out by Spence[11], it is difficult to guarantee convergence of the expansion of biharmonic stress functions. So a symplectic approach based on the two-dimensional elasticity is introduced here to investigate the stress distributions at the clamped end. Convergence is guaranteed automatically by the completeness of symplectic expansion as proved by Yao, eta1[12]. Leung[13] used this method to analyze the end effect for orthotropic beams. On this base, Stress distribution diagrams are given for the first time for anisotropic beams in this paper.

2. Equations of Motion

Consider xz-plane domain and length is L, the thickness is 2h. The cross section in the yz-plane is rectangular with unit width. The constitutive relation is

† YOUZHEN YANG, email: hemai@163.com

$$\begin{Bmatrix} \sigma_x \\ \sigma_z \\ \tau_{xz} \end{Bmatrix} = \begin{bmatrix} c_{11} & c_{13} & c_{15} \\ c_{13} & c_{33} & c_{35} \\ c_{15} & c_{35} & c_{55} \end{bmatrix} \begin{Bmatrix} \varepsilon_x \\ \varepsilon_z \\ \gamma_{xz} \end{Bmatrix} \tag{1}$$

Where c_{ij} are elastic constants. The variation principle is[27]

$$\delta E_p = \delta \left\{ \int_0^L \int_{-h}^h [\boldsymbol{L}(u, w, \dot{u}, \dot{w}) - F_x u - F_z w] dxdz + v_e^1 + v_e^2 \right\} = 0 \tag{2}$$

Where $(\dot{\ })$ is the partial derivative with respect to x.

$$\sigma = \frac{\partial L}{\partial \dot{u}} = c_{11}\dot{u} + c_{13}\frac{\partial w}{\partial z} + c_{15}(\frac{\partial u}{\partial z} + \dot{w}), \tau = \frac{\partial L}{\partial \dot{w}} = c_{15}\dot{u} + c_{35}\frac{\partial w}{\partial z} + c_{55}(\frac{\partial u}{\partial z} + \dot{w}) \tag{3}$$

Then, the mixed energy variational principle is deduced as

$$\delta E_p = \delta \left\{ \int_0^L \int_{-h}^h [\sigma\dot{u} + \tau\dot{w} - H(u, w, \sigma, \tau)] dxdz + v_e^1 + v_e^2 \right\} = 0 \tag{4}$$

The Hamiltonian equations are constructed as

$$\dot{\boldsymbol{v}} = \boldsymbol{H}\boldsymbol{v} + \boldsymbol{Q} \tag{5}$$

Where σ_z can be expressed by the dual variables,

$$\begin{aligned} \sigma_z = &-((c_{13}^2 c_{55} + c_{11}c_{35}^2 - 2c_{13}c_{15}c_{35})/c_{11}c_{55} - c_{15}^2 - c_{33})\partial w/\partial z + \\ &(c_{15}c_{35} - c_{13}c_{35})/(c_{11}c_{55} - c_{15}^2)\sigma + (c_{13}c_{15} - c_{11}c_{35})/(c_{11}c_{55} - c_{15}^2)\tau \end{aligned} \tag{6}$$

The solution are of the form

$$\boldsymbol{v} = e^{\mu x}\boldsymbol{\psi}(z) \tag{7}$$

Where μ is the eigenvalue, $\psi(z)$ is the eigenvector

$$\boldsymbol{H}\psi(z) = \mu\psi(z) \tag{8}$$

2.1. *Eigensolutions of zero eigenvalue*

Solving Eq.(8) yields the basic solutions of zero eigenvalue

$$\boldsymbol{\psi}_f^{(0)} = (1 \quad 0 \quad 0 \quad 0)^T, \boldsymbol{\psi}_s^{(0)} = (0 \quad 1 \quad 0 \quad 0)^T \tag{9}$$

The eigensolutions in Eqs.(9) are also the solutions of the original problem, i.e.

$$\boldsymbol{v}_f^{(0)} = \boldsymbol{\psi}_f^{(0)}, \ \boldsymbol{v}_s^{(0)} = \boldsymbol{\psi}_s^{(0)} \tag{10}$$

The first-grade solutions can be solved from the above two Jordan chains.

$$\boldsymbol{H}\psi^{(1)} = \psi^{(0)} \tag{11}$$

Eigensolutions of the original eigen-problem are in fact

$$\boldsymbol{v}_f^{(1)} = \boldsymbol{\psi}_f^{(1)} + x\boldsymbol{\psi}_f^{(0)}, \boldsymbol{v}_s^{(1)} = \boldsymbol{\psi}_s^{(1)} + x\boldsymbol{\psi}_s^{(0)} \tag{12}$$

Similarly, the second-order vector $\psi_s^{(2)}$ can be obtained

$$\boldsymbol{H}\boldsymbol{\psi}_s^{(2)} = \boldsymbol{\psi}_s^{(1)} \tag{13}$$

The solutions of the original problem is formed as

$$\boldsymbol{v}_s^{(2)} = \boldsymbol{\psi}_s^{(2)} + x\boldsymbol{\psi}_s^{(1)} + 0.5x^2\boldsymbol{\psi}_s^{(0)} \tag{14}$$

Therefore, the third-order vector exist

$$\boldsymbol{H}\boldsymbol{\psi}_s^{(3)} = \boldsymbol{\psi}_s^{(2)} \tag{15}$$

The eigensolutions of original problem are constructed from

$$\boldsymbol{v}_s^{(3)} = \boldsymbol{\psi}_s^{(3)} + x\boldsymbol{\psi}_s^{(2)} + 1/2\,x^2\boldsymbol{\psi}_s^{(1)} + 1/6\,x^3\boldsymbol{\psi}_s^{(0)} \tag{16}$$

2.2. *Eigensolutions of nonzero eigenvalue*

When the eigenvalue μ in (10) is not equal to zero. Solving the above characteristic polynomial gives four roots

$$\lambda_i = \lambda_i \mu \qquad (i = 1 \sim 4) \tag{17}$$

The eigensolutions can be assumed to take the form of

$$(u, v, \sigma, \tau) = \sum_{i=1}^{4} (A_i, B_i, C_i, D_i)\exp(\lambda_i \mu x) \tag{18}$$

2.3. *Integral path along the boundaries*

The general solution is constructed to be

$$\boldsymbol{v} = \tilde{\boldsymbol{v}} + \hat{\boldsymbol{v}}, \qquad \hat{\boldsymbol{v}} = \sum_{i=1}^{N} A_i \boldsymbol{v}^{(i)} \tag{19}$$

The Hamiltonian mixed energy variation principle is written as

$$\begin{aligned}\delta\Big\{&\int_0^l \int_{-h}^h \left[\mathbf{p}^T\dot{\mathbf{q}} - \mathrm{H}(\dot{\mathbf{q}},\mathbf{p})\right] dxdz - \int_{-h}^h \left[\mathbf{p}^T(\mathbf{q}-\bar{\mathbf{q}}_l)\right]_{\Gamma_{ul}} dz \\ &+ \int_{-h}^h \left[\mathbf{p}_l^T\mathbf{q}\right]_{\Gamma_{\sigma l}} dz + \int_{-h}^h \left[\mathbf{p}^T(\mathbf{q}-\bar{\mathbf{q}}_0)\right]_{\Gamma_{u0}} dz + \int_{-h}^h \left[\mathbf{p}_0^T\mathbf{q}\right]_{\Gamma_{\sigma 0}} dz\Big\} = 0\end{aligned} \tag{20}$$

3. Numerical Examples

Consider a thick composite cantilever beam with a uniformly distributed transverse load q applied at the top and bottom surfaces as shown in Figure1. The geometric parameters and elastic constants used in this example are same as those given by ANSYS. A thick anisotropic beam is of length-to-height ratio $L/(2h)=4$. The elastic constants are listed at Table.1. Figure2 and Figure3 shows the distributions of stresses at the clamped end. The solid line is the results from the symplectic approach while the dotted line is the results using ANSYS. Concentrations of the stresses are observed obviously. The profiles of both curves are similar with each others except that the shear stress calculated from ANSYS does not approach zero at the free surfaces. Figure2 and Figure3 reveals that the peak of normal stress appears near solid line is due to finite expansions of eigensolutions during calculation.

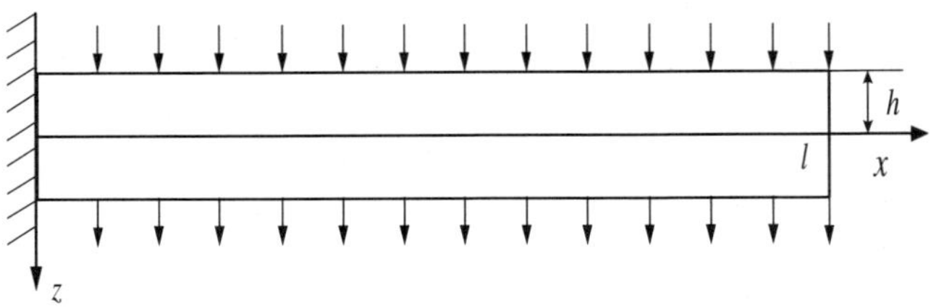

Fig. 1. A cantilever with uniformly transverse loads applied at upper and lower surfaces.

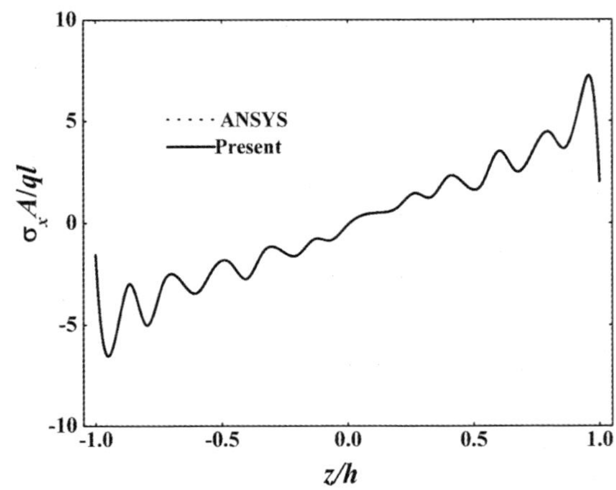

Fig. 2. The Variation of fixed end stress.

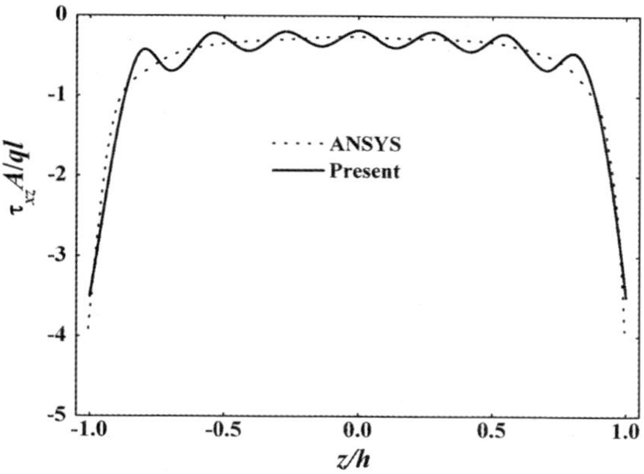

Fig. 3. The Variation of fixed end stress.

Table 1. The material properties (unit: GPa).

s_{11}	s_{13}	s_{15}	s_{33}	s_{35}	s_{55}
103.401	40.867	-8.28046	100.679	-8.70555	30.6174

References

1. M.L.Williams: *J.Appl.Mech.*, 1952, 19(4), 526-528.
2. S.P.Timoshenko:'Theory of Elasticity', 1970, New York, McGraw-Hill.
3. R.D.Gregory: *J.Elast.*,1980, 10(3), 295-327.
4. R.D.Gregory and I.Gladwell: *J.Elast.*, 1982, 12(4), 317-343.
5. R.D.Gregory and F.Y.M.Wan:. *J.Elast.*, 1984, 14(1), 27-64.
6. C.O. Horgan:*Int.J Solids.Struct.*, 1991, 27(15),1895-1914.
7. I. Choi, and C.O. Horgan:*J.Appl.Mech.*, 1977, 44(3),424-430.
8. Y.H.Lin and F.Y.M.Wan:*Comput.Struct.*, 1990, 35(4), 349-359.
9. M.Savoia and N. Tullini: [J]. *Int.J Solids.Struct.*, 1996, 33(17), 2459-2484.
10. A.Y.T.Leung: *J.sound.Vib.*, 1990, 142(3),527-528.
11. D.A. Spence: *J.Appl.Math.,*1983, 30(2), 107-139.
12. W.A.Yao: 'Symplectic elasticity',2002,Beijing, Higher Education Press.
13. A.Y.T.Leung and J.J. Zheng: *Appl.Math.Mech.*, 2007, 28(12), 1629-1642.

THE COMPOUND MATERIALS DEFECT TEST THAT USED ULTRASOUND THERMOGRAPHY

YONG HA YANG, SEUNG HYUM CHOI AND SUNG HYUN KIM

Dept. of Advanced Parts and Materials Engineering, Chosun University, Gwang-ju, KOREA, 501-759

JAE YEOL KIM†

Dept. of Mechatronics Engineering, Chosun University, Gwang-ju, KOREA, 501-759

Ultrasound thermography detects defects by radiating 20-30 kHz ultrasound waves to the samples and capturing the heat generated from the defects with the use of an infrared thermographic camera. This technology is being spotlighted as a next-generation NDT technique for the automobile and aerospace industries because it can test large areas and can detect defects such as cracks and exfoliations in real time. The heating mechanism of the ultrasound vibration has not been accurately determined, but the thermomechanical coupling effect and the surface or internal friction are estimated to be the main causes. When this heat is captured by an infrared thermographic camera, the defects inside or on the surface of objects can be quickly detected. Although this technology can construct a testing device relatively simply and can detect defects within a short time, there are no reliable data about the factors related to its detection ability. In this study, the ultrasound thermography technique was used to manufacture gasoline engine piston specimens, and nondestructive reliability tests to verify the applicability and validity of the ultrasound thermography technique.

Keywords: Infrared Thermography (IT); Ultrasound Thermography; Compound Material; Gasoline piston.

1. Introduction

In the ultrasonic infrared imaging (UII) technology, the ultrasonic energy and infrared image are combined to detect defects in objects. Cracks or delaminations produce heat by applying ultrasonic energy, and their locations and directions are found and analyzed via infrared images. Cutting or radiography is conventionally used to inspect car parts. Cutting inspection is advantageous in that it can identify the internal quality and defects, but the subject parts are damaged, and only the cross-sections can be inspected.

† Corresponding author, email: jykim@chosun.ac.kr

Radiography is advantageous in that it is a non-destructive test, but the precise defects and their locations are not easy to identify. To address these problems, company H is studying computed tomography (CT). The scanner that uses hundreds of kilovolts can efficiently analyze the inside and outside of aluminum parts, but its coverage does not fully include steel or polymer materials due to their permeability and resolution problems. Such materials require a large scanner and a long scanning time (1 min for one CT image of an aluminum part). In the scanning of the entire product, large parts such as engine blocks require about five hours, and small parts require one to two hours. CT is suitable for sampling or inspecting the internal quality of products in small-scale production, but not for mass production lines, where the internal quality must be immediately inspected.

In this study, ultrasonic thermography was used to perform non-destructive reliability assessment of gasoline piston cracks.[1]

2. Ultrasound Thermography

When the ultrasonic wave passes through the defects in objects, the two defective surfaces that face each other do not uniformly vibrate. The friction between the defective surfaces converts part of the elastic wave into heat. If this exothermic phenomenon is observed using an infrared thermographic camera, the defects, including the cracks in structures, can be detected. A high-energy ultrasonic pulse is applied to the specimen to generate local heat from the defects. An infrared thermographic camera is used to observe the exothermic phenomenon in the defects according to the time lapse after the application of the ultrasonic pulse. The ultrasonic wave of the frequency has a tens-of-centimeters wavelength. Even when it progresses over the distance much farther than the wavelength, the ultrasonic wave has sufficient amplitude energy. The ultrasonic wave can progress over a distance several times the wavelength without attenuation if there was no material loss. The speed of the sound wave in a solid material is generally several kilometers per second. At the moment of ultrasonic-wave excitation, the sound field completely passes through the entire inspection area of a structure with a size of 1 m2 or less. An ultrahigh-speed infrared thermographic camera with an imaging time of several kilohertz can detect defects within tens of microseconds. To efficiently excite the heat in the defective areas in structures, the 15-40 kHz excitation frequency that partly overlaps the audible frequency band is used, and 20 kHz band is widely used. Figure 1 shows the defect detection mechanism of ultrasonic thermography [2~3]

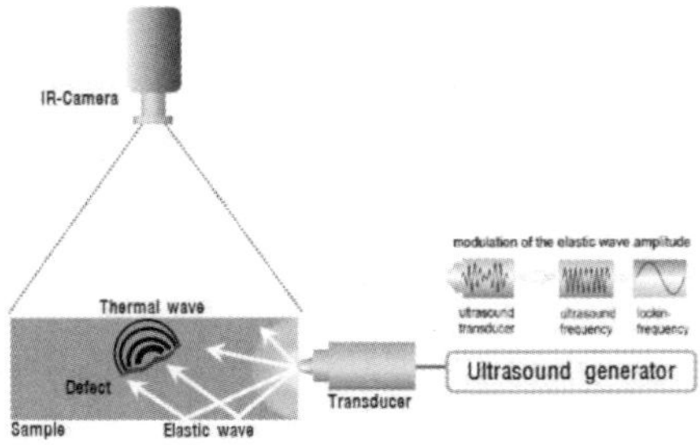

Fig. 1. Diagram of Heating System using Ultrasonic Excitation.

3. The Composition of Experiment Equipment and the Method of Experiment

3.1. *The composition of experiment equipment*

AGEMA's Thermovision 900 system was used as the infrared thermographic camera for this study. The system features the thermal electric (TE) scanner cooling method, two-split detection, serial scanning, 2-5.4 micron spectrum response, a temperature range of -10-500℃ (up to 2000℃ when a high-temperature filter is used), 0.1℃ sensitivity at 30℃, 104 pixel/line spatial resolution (50% modulation), and 3.5 kHz IR line frequency.

The DYW-430H handgun-type booster-integrated vibrator of Daeyoung Ultrasonic was used as the ultrasonic vibrator. Its specifications include 180-245 V input power, 0.1-5s continuous vibration, 30 kHz frequency control range, and 400 W power.

The handgun-type vibrator was modified into a button-type device. The experiment equipment was fixed with a jig, as shown in Figure 2, using bolts, to minimize the movement during vibrating operation.

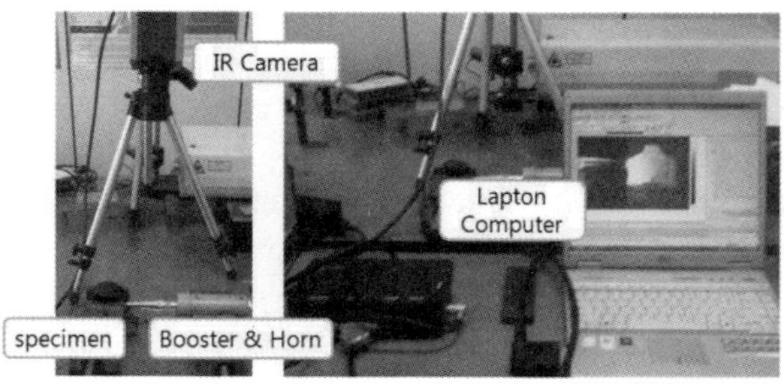

Fig. 2. The composition of experiment equipment.

3.2. *Design and manufacture of the ultrasonic horn*

For the application of the optimal ultrasonic pulse to objects, a cone-type horn was designed and manufactured with duralumin, based on relevant papers and data. Figure 3 shows the 2D design drawing and image of the cone-type horn.

3.3. *Manufacture of the specimen*

The specimen was the three-cylinder aluminum alloy gasoline piston of Company K. An arbitrary crack was made in the specimen by applying an impact. Black-matt paint was applied because infrared thermography is greatly influenced by the radiation rate.

3.4. *Experimental method*

The specimens and vibrator horn were fixed with a jig and were placed as close to each other as possible to allow the maximum incidence of ultrasonic pulses. They were vibrated with 30 kHz ultrasonic pulses for 60s (12-time 5s applications), and images were taken at a rate of 1 frame/s. Figure 3 shows the test specimen and horn fixed at the jig.

Fig. 3. Fixed to press into specimen and horns.

4. Result

To obtain precise temperature data in the crack area, three spots were selected, as shown in Figure 4. Spot 1 was the crack area, and spots 2 and 3 were the areas with no crack. Figure 5 shows the temperature data of each spot in graph.

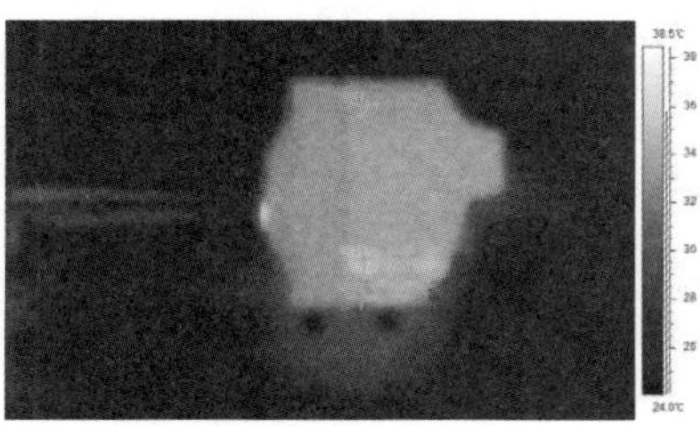

Fig. 4. IR image of piston(10sec).

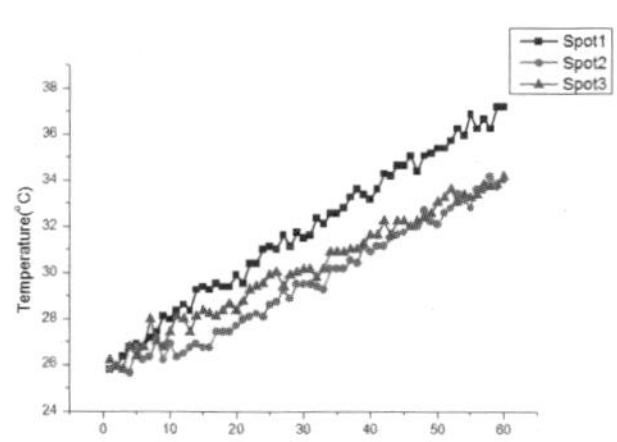

Fig. 5. Spot temperature graph of piston.

Spots 2 and 3 had similar temperature deviations, but the temperature of spot 1 in the crack area continued to increase.

5. Conclusion

In this study, an ultrasonic horn was designed and manufactured based on the relevant data in and outside South Korea, for the application of ultrasonic thermography. It was used to perform a non-destructive reliability test on a gasoline piston, and to detect the piston cracks. The following conclusions were arrived at based on the results of the experiment:

1. Hot spots appeared in the crack area of the specimen due to the application of an ultrasonic pulse.
2. Ultrasonic thermography detected cracks faster than did the conventional NDT technique, which finds cracks by continuously applying heat, because it uses the heat from the vibration due to the ultrasonic pulse.
3. Hot spots appeared in the crack area right after the ultrasonic-wave application, and the temperature difference from the other areas was 2℃.

Acknowledgments

This work was supported by the Korea Research Foundation(KRF) grant funded by the Korea government(MEST)(No. 2009-0075231)

References

1. T. Zweschper, G. Riegert, A. Dillenz and G. Busse, "Ultrasound Burst Phase Thermography (UBP) for Applications in the Automotive Industry", Review of Progress in Quantitative Nondestructive, Vol. 657, 2003, pp. 531～536
2. A. Gleiter, G. Riegert, T. Zweschper, G. Busse, "Ultrasound lock-in thermography for advanced depth resolved defect selective imaging", Insight - Non-Destructive Testing and Condition Monitoring, Vol. 49, 2007, pp. 272～274
3. T. Zweschper, A. Dillenz, G. Riegert, D. Scjerling and G. Busse, "Ultrasound excited thermography using frequency modulated elastic waves", Insight - Non-Destructive Testing and Condition Monitoring, Vol. 45, 2003, pp. 178～182
4. J. W. Cho, Y. S. Choi, S. H. Jung and H. K. Jung, "A Brazing Defect Detection Using an Ultrasonic Infrared Imaging Inspection", Journal of the Korean Society for Nondestructive Testing, Vol. 27, No. 5, 2007, pp. 426～431

5. J. W. Cho, Y. C. Seo, S. H. Jung, S. H. Kim and H. Y. Jung, "Defect Detection of Ceramic Heating Plate Using Ultrasound Pulse Thermography", Journal of the Korean Ceramic Society, Vol. 43, No. 4, 2006, pp. 259～263

THE 20MN ENERGY SAVINGS FAST FORGING PRESS HYDRAULIC SYSTEM DESIGN

DAI MIN-QIANG†
School of Mechanical Engineering, Xi'an Jiao tong University, Xi'an, 710049, China

ZHAO SHENG-DUN, XU FAN, YUAN XIAO-MEI
School of Mechanical Engineering, Xi'an Jiao tong University, Xi'an, 710049, China

In the fast forging press hydraulic system accumulators are often used to absorb pressure pulsation, eliminate pressure pulsation and reduce noise. This paper analyzes the energy saving possibilities with pressure accumulators based on the introduction to the principles of energy saving methods of fast forging press. Two hydraulic system of 20MN fast forging press were designed, one of which was applied pressure accumulators while the other was not. Their simulation models based on the AMESim soft were made and the simulation results were obtained. After analyzing the calculation results and the simulation results of 20MN fast forging press hydraulic system with pressure accumulators, the fast forging press for the energy saving method is made. At last, all the conclusion show that the energy saving methods applying accumulators can reduce the power losses remarkable in the fast forging press hydraulic system.

Keywords: Accumulator; Fast forging press; Hydraulic system; Simulation.

1. Introduction

With the rapid development of modern industries, the concept "Energy saving, green manufacturing"[1] has been actively spread and applied in China manufacturing industry. The accumulator, as the energy storage devices, is widely used in the hydraulic systems of much industrial equipment. Because of the high frequency of fast forging machine, high cylinder speed, the large impact pressure of the system, and the big flow, the required power source system pumps is big. The work cycle of fast forging machine is divided into three working process ----'fast-forward - Forging - fast-back', and each has different working conditions, the power required also varies. In the mechanical system, similar to the working conditions, the system is designed for energy-saving system. In the traditional fast forging press hydraulic system, accumulators are often used to absorb pressure pulsation, eliminate pressure pulsation and reduce

† Dai Min-qiang, email: dmq0739@163.com

noise.[2] Energy-saving measures are often using liquid-filled tank to compensate for high flow that required in fast-forward, but the pressure is so high that liquid-filled tank can not meet .There are high speed, large instantaneous flow and high pressure during the process of quick return. Using the quickly back and forth way with fast energy storage can effectively solve the problem of energy-saving to the process.

2. Design of the Fast Forging Press Hydraulic System

Two return cylinders are used in the 20MN Fast Forging Press system, and the diameter of the piston in return cylinder is 330mm, the diameter of the piston rod is 200mm. When the normal forging stroke is 60-100mm, frequency is 0.3-0.6 Hz, and when the rapid forge stroke is not more than 25mm, frequency is 2Hz. The return speed ratio is 3:1, the maximum forging velocity is 0.09m s^{-1}, and the maximum return velocity is 0.27m s^{-1}. The pressure in the filling tank is 0.28~0.6Mpa, the maximum pressure in the return system is 18.7MPa and the minimum pressure is 13.1MPa. The principle of the hydraulic system is showed in the figure 1 (a).[3]

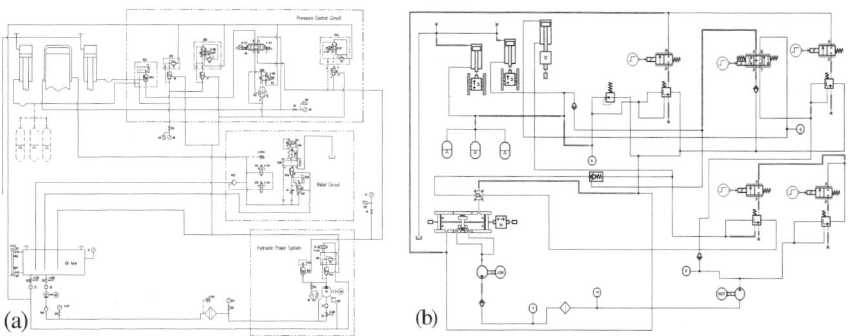

Fig. 1. The hydraulic system of the 20MN Fast Forging Press, (a) is the principle of the hydraulic system and (b) is the simulation model.

The main cylinder is filled by the tank for the rapid drop process of the traditional rapid forging machine. The oil for the process of the forging and the fast return is supplied by the oil source system. Therefore, the oil source power of the hydraulic system is determined by the condition of the forging and the return process.

2.1. *Analysis and Calculation for the System without Accumulator*

The output oil volume of the pump is decided by the condition of the return process.

Supposing the total efficiency of the pump "η" is 0.9, as a result the driving power of the system is

$$N = \frac{PQ}{60\eta} = 320.5kW \tag{1}$$

According to the return speed ratio is 3:1, this power only works in the quarter period of the forging cycle. Therefore, the loss of the system power is very huge.

2.2. *Analysis and Calculation for the System with Accumulator*

The time of the working cycle during the normal forging process is 1.5s, the demanded oil volume for the return process is $V_0=A_2\times L=10.8$L. Comparatively, the time of the working cycle during the rapid forging process is 0.5s, the demanded oil volume for the return process is $V_0= A_2\times L=2.8$L.

As the accumulator, the oil volume of the pump which supplies oil for the return process can be calculated by the formulation as follows

$$Q_b \geq K\frac{\sum V}{T} \tag{2}$$

'K' is losing coefficient, usually be selected as 1.1~1.2.

'$\sum V$' is the maximum oil volume which needed for the circulation cylinder.

'T' is the time of the period of the working cycle.[4]

Supposing the total efficiency of the pump is 0.9, the driving power of the system is $N = \frac{PQ_b}{60\eta} = 192kW$ as a result.

2.3. *Analysis and Conclusion*

As the calculate conclusion, when applied accumulator in the rapid return loops, it may save drive power by 40%. From what have been discussed above, by applying accumulator in the rapid return system, the energy saving effects is very considerable.

2.4. *Design of Accumulator*

The system needs maximum oil volume V_0=10.8L for return when it runs constant forging, and V_0=2.8L when high speed forging. So accumulator is designed according to constant forging, while checked according to high speed forging.

Based on working conditions, gasbag accumulator mostly applied in engineering is collected. After accumulator is full of oil, $V_0=V_{gas}+V_{oil}$, and elastic modulus of oil is larger than them of gas, so only compressibility of gas is taken into consideration during oil-filled and released process of accumulator. By the consideration of the pressure loss, the maximum working pressure for accumulator designed is P_{max}=20MPa, minimum is P_{min}=15MPa. According to empirical formula, advanced gas-filled pressure for accumulator is P_0= (0.8-0.9) P_{min}, and P_0>0.25P_{max}. Suppose $P_0=0.8P_{min}$, and ($P_0=0.8P_{min}$ =12MPa)> (0.25 P_{max}=5MPa), which meets the requirements.

According to law of thermodynamics, the relationship between volume V and pressure P for gas cavity of accumulator is abide by gas state equation $PV^n = K$, where n stands for state index. In fact, state index n changes in accordance with pressure.[5] In this system, average working pressure of system is $P_x = 17.5MPa$, n=1.9. So rated volume of accumulator needed is:

$$V = (V_0(P_{min} / P_0)^{\frac{1}{n}}) / (1-(P_{min} / P_{max})^{\frac{1}{n}}) = 86.68L \qquad (3)$$

According to designed calculation, this system could select 3 accumulators with rated volume 40L. Reckoned by formula, effective volume of 120L accumulator is 9L, which meets the requirements of high speed and constant forging. Loop of return cylinder adopted accumulators is shown in Figure 1 (a) by a dotted line.

3. Simulation

From the hydraulic system of 20MN that energy saving fast forging press applying pressure accumulators to, we establish the Figure 1 (b) which shows the simulation model based on AMESim to test the function of fast forging and energy saving of accumulator in the hydraulic system. Figures 2 respectively show pump output flow curve of the source system that fast press forging hydraulic system with accumulators under the rated forging speed or not.

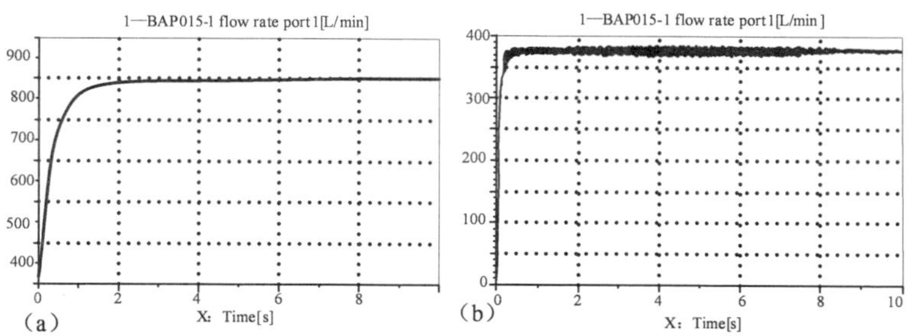

Fig. 2. Simulation curves of the Fast Forging Press hydraulic system, (a) is the flow rate diagram of pump without accumulator and (b) is the flow rate diagram of pump with accumulators.

From the simulation curve that Figure 2 shows, when the system desires to reach the rated speed forging, the pump source output flow that hydraulic system without accumulators required reaches 850L min^{-1}, while it with accumulators only required reaches 385L min^{-1}. We can see from the simulation results that the output power of 20MN fast forging hydraulic system with accumulators reduces by about half. That simulation results are similar to the analysis of results.

4. Conclusion

The article respectively analyzed and calculated the programs when it used the accumulator return and not through the research on hydraulic system of 20MN fast forging machine. The simulation models based on the AMESim soft were made and the simulation results were obtained. Calculation and simulation results show that its oil source drive power decrease by 40% when it uses hydraulic system of fast forging press with the accumulator return. As a kind of hydraulic system of auxiliary devices, accumulator brings obvious effect on the economy, energy saving, safety, reliability, environmental protection to the system. It's especially worth popularizing the system in fast forging machine hydraulic system.

However, there are two problems. The effect that property of accumulator brings to the hydraulic system and how to configure the accumulator so that it serves not only as return-back oil source, and but also can slow down the impact while increasing its frequency response are worth further studying.

References

1. Guoqun Zhao, Xiping Li, and Yanjin Guan: Multi-Objective Optimization of the Heating Rods, Layout for Rapid Electrical Heating Cycle Injection Mold. J. Mech. Des. 132, 061001 (2010).
2. Someday Hibachi, Yamaguchi Heritage: Study on an active accumulator (active control of high-frequency pulsation of flow rate in hydraulic systems [J]). JSME International Journal, 1996 Series B, 39(1):685-690.
3. Kong Xiang-dong, Linen, Yak Jing, Dou Due-chuan Zhao Yuen: Research on Fast Forging Control Characteristic of 22MN Fast Forging Press, Chinese Hydraulics & Pneumatics.
4. Yan Judo: The Application and Research on Energy Accumulators in Hydraulic Pressure Circuits. Journal of Jinzhu university of Guiyang [J]. 2005.03.
5. Bin Xu: Research on Inverter Controlled Hydraulic Elevator Energy-saving System Applying Pressure Accumulators. PhD thesis, Zhejiang University. 2001.05.

EXPERIMENTAL RESEARCH ON SHEAR FATIGUE PERFORMANCE OF CFRP-STRENGTHENED HIGH-STRENGTH CONCRETE STRUCTURE

HONG HAI †

Department of Civil Engineering, Shenyang Jianzhu University, Shenyang, 110168, P.R.China

YINGHUA ZHAO

Road &Bridge Research Institute, Dalian Maritime University, Dalian, 116026, P.R.China

GUANGSHAN ZHU

Department of Civil Engineering, Shenyang Jianzhu University, Shenyang, 110168, P.R.China

The shear resistance at the interface between FRP sheet and concrete is a key problem for the application of fiber-reinforced plastic (FRP)plates, which has emerged as a popular method for the strengthening of reinforced concrete structures. The objective of this study is to discuss the interfacial shear fatigue behavior of a high-strength concrete structure strengthened by carbon fiber-reinforced polymer composites under repeated loads. Considering the different strength of concretes, double-shear fatigue tests were conducted. Based on the test results, the effects of the concrete strength, the failure process, the failure mode and the failure mechanism are investigated. Test results indicate that fatigue failure mode is brittle failure in the concrete layer and fatigue life increases as the strength of concrete decreases.

Keywords: CFRP; High-strength concrete; Interfacial shear performance; Fatigue.

1. Introduction

Fiber Reinforced Polymer (FRP) are being considered for use in the field of civil infrastructure as strengthening and rehabilitation materials because of the characteristics of high strength, high stiffness, light-weight, excellent corrosion resistance, ease and speed of installation, etc. In this strengthening method, the bonding performance of the FRP-to-concrete interface in providing an effective stress transfer is of crucial importance. Much progress in this field includes the

† Corresponding author, email: cehhai@sjzu.edu.cn

results and conclusions on mechanical behavior of interface between concrete and FRP has been made in past decade. [1-4]

The high-strength concrete has been widely used and made significant technical and economic benefits because of its high strength, good durability, small deformation, etc., it can meet the requirement of large span, over loading and high-rise development of modern engineering structures.

Fatigue damage becomes an emerging problem in lots of concrete structures such as concrete pavements, bridges and pressure vessels etc., which will subject to cyclic loadings caused by vehicles, winds, waves and earthquake action etc. during their working life. With the application of high-strength concrete, it is unfavourable factors to structures because of its brittle failure. Under the action of fatigue load, the bonding performance between the surfaces of CFRP and concrete directly decides whether CFRP can fully exert its high-strength characteristics and directly influences the safety of the structure, so the research on its fatigue presents more important. Currently, there is some research on the reinforced concrete of beam, column, [5-9] etc. and the fatigue performance of interface.[10] but, the research result is mainly based on normal concrete, and there is few research on whether the bonding strength relationship, failure mechanism and force failure mode can be directly used in the field of high-strength concrete.

This paper presents the experimental results on the shear performance of the interface between CFRP and high-strength concrete under the fatigue loading with effect of the concrete strength. Based on the experimental results, the effects of the concrete strength, the failure process, the failure mode and failure mechanism are investigated. Compared with the static ultimate strength, fatigue strength reduction. Therefore, A safety factor of the fatigue strength at the interface of CFRP and concrete should be applied in design.

2. Fatigue Test Program

According to the Ref.10, according to elasticity theory, it is found that the formula to calculate the bonding stress at flexion state is identical to that at shearing state. Therefore, the torsional low cycle fatigue experiment is more efficient and precise than the flexural fatigue experiment to study the fatigue properties of the bonding. In the paper, the double-shear experiment is designed for measuring the influence on interfacial properties from different concrete strength.

2.1. *Specimens*

In this test, A total of 10 cube test specimens, each measured 100mm×100mm×100mm, the bonding surface was overall bonded with one-strip

CFPR (namely bp/bc=1) as follows: Batch 1: five specimens for castng with C60 commercial concrete; Batch 2: five specimens for castng with C80 commercial concrete. Details are shown in Table 1. CFRP material adopts UDO CFPR produced by Germany EPO Co., Ltd.. The physical properties of CFRP are listed in Table 2. The adhesive used was a two-part epoxy resin named JGN-C.

Table 1. Test specimens.

CFRP bonding method	concrete	Qty
1- CFPR package	C60	5
	C80	5

Table 2. Physical properties of CFRP.

CFRP	Weight (g/m^2)	Thickness (mm)	Elastic modulus in fiber direction (Mpa)
C300/300	300	0.167	2.5×10^5

2.2. *Experimental setup*

Strain gauges were attached to the top surface of the CFRP/epoxy matrix reinforcement to record the strain at this point (see Fig1). A MTS-318 testing machine was used. The specimens were gripped by steel girder at 5Hz frequency with load control until failure (see Fig2). During the test the electrical resistance strain gauges and the load cell were connected to a data acquisition system thus recording the load-strain relationship for each specimen.

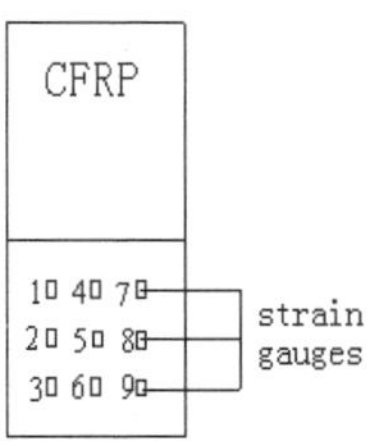

Fig. 1. Layout sketch for strain gauges.

Fig. 2. Specimen and setup.

2.3. *Loading mode*

The loading mode of gradual increase of constant-amplitude load is adopted for exerting fatigue load and R=0.1. Before formal loading, load is exerted in advance in order to make CFPR in tightening state, and the static ultimate load of each test piece was found out. When formal loading in each loading process, the fatigue load is exerted according to sine curve form. In the first step, the maximum of sine wave, namely the maximum stress, starts from 30% of static ultimate load, in the follow steps, the maximum stress is increased by degrees of 10%, in each step, after 5000-time loading, the load level is upgraded until failure.

2.4. *Test phenomenon and failure mode*

When the maximum stress is from 50% to 60% of static ultimate load, obvious crack appears on the test specimen with loud noise, when the maximum stress is from 80% to 90% of static ultimate load, the interface of the test specimen is damaged, carbon fiber sheet tears the concrete on the loading end, the shape of most torn concrete is reverse triangle, the length is from 50 to 70mm, the concrete is cracked into shivers, and the diameter is 20mm. The failure happened in the concrete near the load and is sudden without premonition. The type failure mode is shown in Fig 3 and Fig 4.

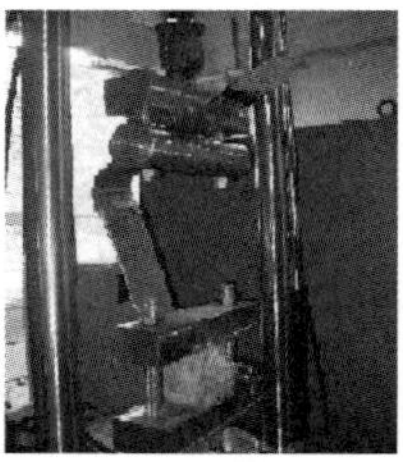

Fig. 3. The type damage mode of the test piece.

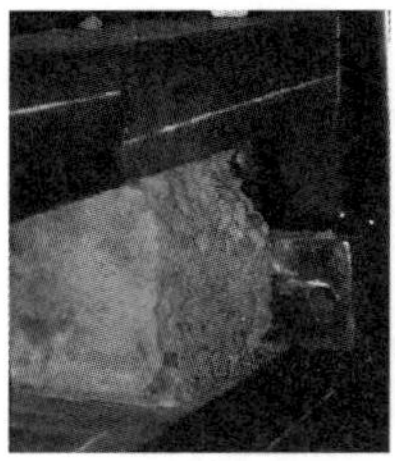

Fig. 4. The type damage mode of the concrete.

2.5. *The influence on interface intensity from concrete strength*

The average life of the test pieces is given in Table 3, the concrete strength has obvious influence on the fatigue life. The fatigue life of C80 is about 10% lower than that of C60. The reason is that the higher concrete intensity is, the more serious the fatigue failure is under heavy loading level; the higher ratio that brittle failure happens, the shorter the fatigue life is. The failure mode is more obvious for the non-uniformity material like concrete.

Table 3. Average life of different concrete strength.

Concrete	CFPR Qty	Average life（time）
C60	1	30000
C80	1	27000

3. Conclusion

By the discussion of the experimental results, the conclusions are as following:

1. Under the action of shear fatigue load, the failure mode of CFRP reinforcement high-strength concrete is the typical brittle failure.

2. Under the action of shear fatigue load, the fatigue life of C80 is about 10% lower than that of C60. Fatigue life increases as the concrete strength decreases.

3. Under the action of fatigue failure, the bonding performance of the interface is obvious reduced and has close relationship with the bonding construction quality of CFPR. If the bonding quality is poor, the damage occurs in adhesive layer, and the carrying capacity is very low. Therefore, A safety factor of the fatigue strength at the interface of CFRP and concrete should be applied in design.

Acknowledgments

The authors would like to acknowledge the financial support of Research Open Project Fund of Construction Ministry with Grant No.2008-k2-19 and Normal Project Fund Education Department of Liaoning Province with Grant No.2008590.

References

1. X. Z. Lu : 'Research on FRP-concrete interface behavior ', Beijing, CHN, 2004, University of Tsinghua, Doctor's Degree Paper.
2. S. S. Wang, Y. C. Chen : *Journal of Jilin University (Engineering Edition)*, 2006, **36**, (1), 53–55.
3. Y. X. Yang, Q. R. Yue : *Journal of Construction Structure*, 2001, **6**, (3), 36–42.
4. J. W. Zhang, Z.T. Lu, J. G. Teng and S.T. Smith : *Industrial Construction*, 2001, **6**, 1–4.
5. D. Yang, H. Toutanji : *Sample* , 2004, 2837–2851.
6. Heffernan P J, Erki M A : *Journal of Composites for Construction*, 2004, **8**, (2), 132–140.
7. K. Lu, H.X. Qiu, L. F. Pang : *Industrial Construction*, 2005, **35**(S1), 962–972.
8. G. W. Yao : *Mechanical Strength* , 2004, **26**(S), 25–27.
9. M. Y. Liu, K. B. Li : *Civil Engineering Journal*, 2005, **38**(9), 32–36.
10. Y. W. Yao, C. L. Li : *Engineering Mechanics*, 2008, **25**(8), 200–204.

INFLUENCE OF PROCESS CONDITIONS ON PROCESSING SURFACE IN SHAVING

MITSUHIRO ASAO[†], HIDENOBU MANO, YOSHINORI YOSHIDA, ZHIGANG WANG

Faculty of Mechanical and Systems Engineering, Gifu University, Yanagido 1-1 Gifu, Gifu 501-1193, Japan

YUKI SATO, HIRONOBU SUZUKI

Research & Development Division, Chuo Spring Co Ltd., 1071-1 Mukaihora, Fukami-cho Toyota, Aichi 470-0441, Japan

Shaving is a process to shave off an edge face obtained in shearing. However, in the shaving process, it is generally difficult to obtain the burnished surface over all aspects on the processing surface, because a little fractured surface often remains. Until now, there are a lot of experimental reports for this process, but analytical researches using finite element method etc. are still few. In this study, analytical investigation of an effect of punch tip shape (edge radius and clearance angle of shaving punch) on stress state and strain distribution on processing surface was performed. Furthermore processing conditions of shaving to make burnished surface all over the shaved surface were examined. As the results, it is observed the fracture morphology of shaving is ductile fracture. The shaving load decreases and the burnished surface area increases with increase of tool tip radius

Keywords: Shearing Process; Shaving; Ductile fracture; Finite Element analysis; Stress triaxiality.

1. Introduction

Shearing is used popularly at various forming today, such as blanking and punching. Because shearing separates some blanks from a work piece using a crack initiation and propagation between punch and die tip, the shearing products have a shear droop, some burnished surfaces, fractured surfaces and a burr. However, shear droop, fractured surface and burr are defects in shearing products. To get more precision products with smooth edge surface, shaving is widely used as next process of shearing.

[†] Corresponding author, email: o3122001@edu.gifu-u.ac.jp

Shaving is a process that chips slightly the sheared surface and can realize a smooth and accurate surface compared with normal sheared surface. However fractured surface often appears and remains, it is generally difficult to get burnished surface over the all processed surface. The study of shaving has many experimental researches[1], but analytical researches using finite element analysis method (FEM) is still few. In the previous analytical studies, deformation analysis of shaving process is performed by means of FEM[2]. In addition, until now influences of removal length and clearance on processing surface[3,4] and mechanism of crack initiation, growth and coalescence[5] has been studied.

In this study, analytically the influence of punch tip shape (punch tip radius and angle) on stress condition and strain distribution of processing surface were considered, and to obtain the burnished surface over all aspects on the processed surface.

2. Experimental Method

An experiment was performed to use the influence of punch tip shape on processing surface. This experiment changed a punch tip radius and angle and was performed. The shaved surface of products was observed. By this observance, the influences of change at punch tip shape on shaved surface and the suitable punch shape on shaving were considered. Figure 1 and Table 1 show the process condition. Changing punch tip radius R as 0.01 mm, 0.05 mm and 0.1 mm, the influences of the punch tip radius R on the shaved surface of products were investigated. In addition, the influences of punch tip angle θ on aspects of processed surface were investigated.

Observation of fracture surface on shaved surface was applied with scanning element microscope (SEM). Occupation ratio of burnished surface and fractured surface area on processed surface area was estimated and compared with FEM analysis result. The fractured surface is predictable becoming elongated dimple pattern. Therefore observation of fractured surface by SEM and elongated dimple pattern was confirmed.

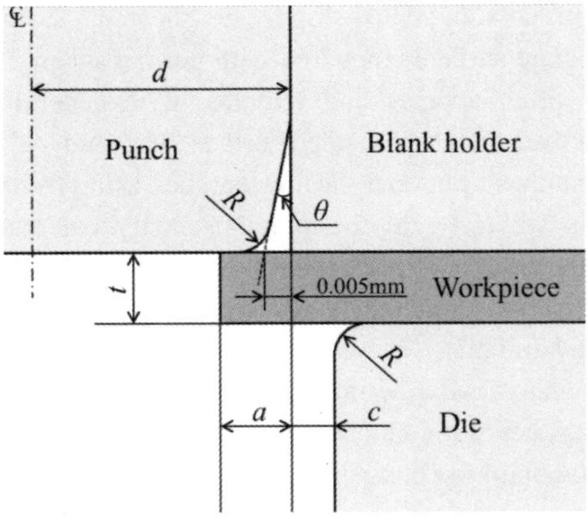

Fig. 1. Process conditions of shaving.

Table 1. Shaving punch geometry.

	R/mm	θ/deg	d/mm	c/mm	t/mm	a/mm
Case 1	0.01	10	5.0	0.03	3.0	0.1
Case 2	0.05	10				
Case 3	0.1	10				
Case 4	0.01	0.0				

R: punch tip radius, θ: clearance angle, d: radius of punch, c: clearance, t: plate thickness, a: cutting length

3. Headings, Text and Equations

The punch tip shape was changed as well as experiment and the process was analyzed using FEM. Distribution of stress and strain within workpiece around the punch tip were calculated. Furthermore the influences of change in punch tip shape on the stress and strain distribution was investigated and a suitable punch shape in shaving was considered. In this study, behavior of stress and strain distribution until just before crack initiation was analyzed. However prediction of fracture including crack initiation and growth by ductile fracture was not analyzed.

DEFORM2D™ was used as a FE code for shaving process. The FE analysis was performed as a 2-dimensional axisymmetric problem assuming that the workpiece is a rigid-plastic material and the tools are rigid body. The material of the workpiece is S45C and that was also used in the experiment. A flow stress curve was obtained from a tensile test of S45C and they were implemented in the FEM analyses. Frictions coefficients of Coulomb's friction model on the boundary between workpiece-workpiece, workpiece-punch were applied and its value was μ=0.2. Element size around punch tip must adequately be tiny because elements in workpiece must not edge into punch and die. To sub-divide elements over all workpiece area take an immense amount of time in calculation and that is not reasonable. Specified area setting of element size (mesh density window) is conducted. If punch displacement per a step is large compare with element size, calculation couldn't be performed normally because elements crashed. Therefore punch displacement per a step let 0.2 μm, and element size around punch tip let approximately 3 μm less or equal of a side length. Crashed elements around punch tip were remedied, to regenerate elements (remesh) per 2 steps.

In stress analysis within the workpiece, equivalent stress and mean normal stress around punch tip were focused. To analyze mean normal stress, fractured surface and crack initiation could be considered. From the results of equivalent stress, stature of work-hardening and influence of crack initiation on next processing were considered.

4. Experimental Result

From observation of processing surface in shaving, elongated dimple pattern was found at fractured surface. Therefore it was confirmed that fracture formation in shaving was involved with initiation, growth of microvoids. Table 2 shows percentage of burnished surface area to that of all processing surface. Comparing with Case 1 and 2, punch tip radius of Case 2 is larger than that of Case 1, burnished surface also is larger. Comparing with Case 1 and 4, Case 1 has punch tip angle and burnished surface area also is larger than that of Case 4.

Table 2. Comparison of sheared surface ratio.

Condition	Sheared surface ratio / %*t*
Case 1	77.6
Case 2	79.2
Case 4	73.9

5. Analytical Result

5.1. *Influence of punch tip radius and punch tip angle on punch load*

5.1.1. *Influence of punch tip radius on shaving load*

Figure 2(a) shows changes in shaving punch load along with punch stroke. The load was the highest in Case 1 and Case 2 showed the lowest load. The maximum punch load of Case 1, 2 and 3 appeared at 82.3, 68.7 and 78.3 %t of each punch stroke, respectively.

Punch load decreased with increase of punch tip radius. Punch load conversely increased in the Case 1 with the smallest punch tip radius. It is thought that material flow is restricted and concentrated at sharp punch tip and work-hardening grew by the strain concentration.

5.1.2. *Influence of punch tip angle on punch load*

Figure 2(b) shows change in punch load to punch stroke. Punch load of Case 4 was higher than that of Case 1. The maximum punch load of Case 1 and Case 4 were obtained 82.3 and 78.4%t of each punch stroke. Punch load is decreased to set up angle at punch tip.

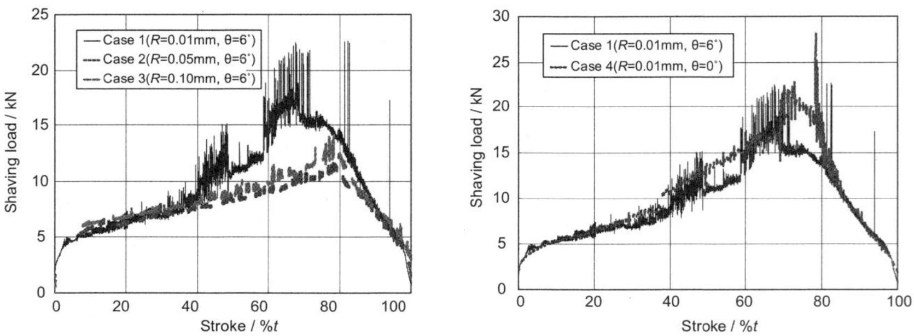

(a) influence of punch tip radius on punch load. (b) influence of punch tip angle on punch load.

Fig. 2. Change in shaving punch load by punch stroke.

5.2. *Influence of punch tip radius and punch tip angle on mean stress distribution*

5.2.1. *Influence of punch tip radius on mean stress*

Figure 3(a) shows relationship between maximum mean stress at punch tip and punch stroke. Mean stress around punch tip was compression stress in the early in a process. This tendency appears regardless of punch shape. Mean stress fields of Case 1, 2 and 3 came to be tensile from compression state when each punch stroke was 80.3, 88.4 and 91.1 %*t*.

Change in equivalent stress around punch tip also was investigated. Equivalent stress was the highest in Case 1 and was the lowest in Case 3. Equivalent stress in Case 1, 2 and 3 were stabilized at 950, 920 and 900 MPa after 20 %*t* of punch stroke regardless of processing conditions.

5.2.2. *Influence of punch tip angle on mean stress*

Figure 3(b) shows change in mean stress around punch tip along with shaving punch stroke. Mean stress field in Case 1 and 4 came to tensile from compression state when each punch stroke exceeded 80.3 and 711 %*t*. The compression area of mean stress was increased to set up punch tip angle. Therefore it is considered that burnished surface area increased.

Figure 10 shows mean stress distribution of Case 1 around punch tip as one of analysis example. Compression mean normal stress was happened in area of contacting punch tip angle. Therefore, it is considered that crack initiation can protect to set up angle at punch tip. It is predictable that crack initiate from punch side, because mean normal stress at punch side become tensile stress as punch stroke progresses.

Figure 11 shows transformation of equivalent stress around punch tip to punch stroke. Equivalent stress of Case 1 was higher than that of Case 4. Equivalent stress of Case 1 and Case 4 were stabilized at 950 and 900 MPa after 20 %*t* of punch stroke.

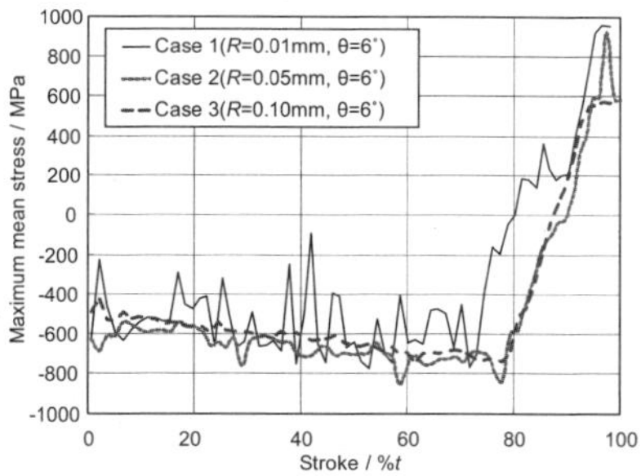

(a) influence of punch tip radius on mean stress.

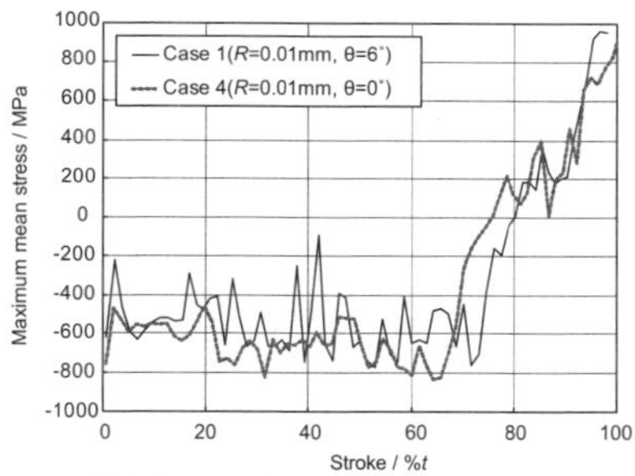

(b) influence of punch tip angle on mean stress.

Fig. 3. Change in maximum mean stress at punch tip.

5.3. *Influence of punch tip angle on punch load, stress and strain*

5.3.1. *Influence of punch tip radius on equivalent strain*

Figure 4(a) shows change in equivalent strain around punch tip to punch stroke. The highest equivalent strain is measured in the Case 1. In Case 3 the equivalent strain was the lowest. Equivalent strain of Case 1, 2 and 3 were stabilized at 7.00, 5.00 and 4.50 after 40 %*t* of punch stroke regardless of punch tip radius.

Equivalent strain increased in accordance with decrease of punch tip radius. The cause of this was concentration of deformation at punch tip, because of restrictions of material flow at punch tip.

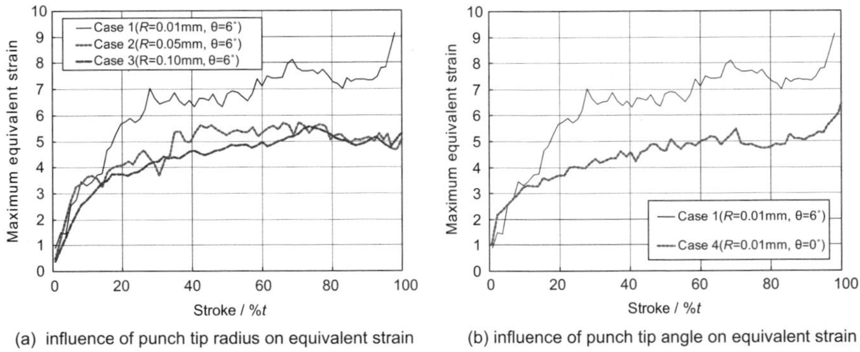

Fig. 4. Relationship between equivalent strain at punch tip and shaving stroke.

5.3.2. *Influence of punch tip angle on equivalent strain*

Figure 4(b) shows influence of punch tip angle on change in equivalent strain around punch tip. Equivalent strain of Case 1 was higher than that of Case 4. Equivalent strain of Case 1 and 4 were stabilized at 7.00 and 5.00 after 40 %t of punch stroke.

6. Conclusion

In this study, the influences of punch tip radius and punch tip angle on stress and strain distribution at punch edge were investigated by implementing experiment and FEM analysis. As the results, following knowledge were obtained,

1. It was confirmed that fracture type in shaving is ductile fracture involved with initiation and growth of microvoids by SEM observation of fractured surface.
2. Burnished surface area increased and punch load decreased with increase of punch tip radius.
3. Burnished surface area increased and punch load decrease to setting up punch tip angle.
4. Equivalent strain and stress value at shaving punch tip increased with sharp punch tip.

References

1. Masuda, M., Jimma, T., Yamauchi, S., *Journal of the Japan Society for Technology of Plasticity*, 1965, **31**-225, 855-863.
2. Sasahara, H., Tabuchi, H., Jin, M., Murakawa, M., *Transactions of the Japan Society of Mechanical Engineers*, 2001, **67**-654, 541-546.

3. Yukawa, N., Inukai, N., Yoshida, Y., Ishikawa, K., Jimma, T., *Journal of the Japan Society for Technology of Plasticity*, 1998, **39**-454, 35-39.
4. Murakawa, M., Kami, M., *Journal of the Japan Society for Technology of Plasticity*, 2003, **44**-513, 53-57.
5. Yoshida, Y., Yukawa, N., Ishikawa, K., Hosono, S., Murase, M., *Journal of the Japan Society for Technology of Plasticity*, 2003, **44**-510, 63-67.

A MODEL OF PLASTIC DEFORMATION AND SURFACE ROUGHENING OF POLYCRYSTALLINE METAL BASED ON R-VALUE OF GRAINS

TAKEJI ABE†

Tsuyama College of Technology
1-8-64-6 Tsushima-Fukui, Kita-ku, Okayama 700-0080, Japan

The r–value (Lankford value) is well known as a measure of formability in the industrial sheet metal forming process. The r-value is defined as the ratio of the width strain to the thickness strain under the uniaxial tensile test of the sheet metal. It is possible to define the r-value for each grain of polycrystalline metal during plastic deformation. It is also known that the free surface of polycrystalline metal roughens during plastic deformation owing to irregular deformation of grains in it. In the present paper, a simple model of plastic deformation of polycrystalline metal is proposed. Various features of the surface roughening are well explained with the present model.

Keywords: Plastic deformation; Polycrystalline metal; Surface roughening; r-value.

1. Introduction

The r-value[1] (Lankford value) is well known as a parameter representing the formability of the polycrystalline sheet metal. When uniaxial stress is applied in x-direction and the surface normal is in z-direction, the r-value is defined as the ratio of the width strain to the thickness strain[1].

$$r = \varepsilon_y / \varepsilon_z \qquad (0 \leq r \leq 1) \tag{1}$$

The relation between slip system and r-value is schematically shown in Fig. 1.

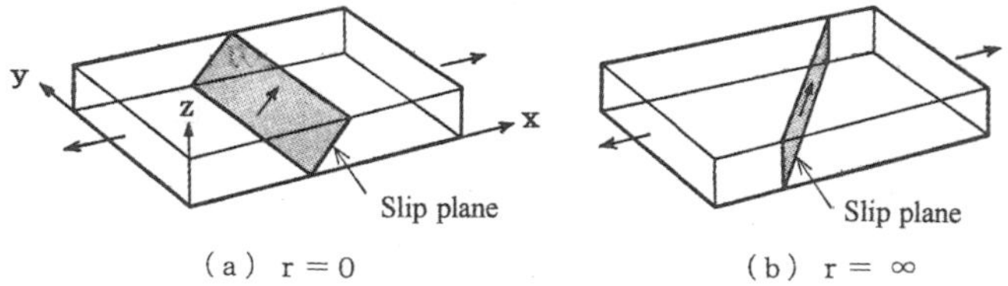

Fig. 1. Relation between slip system and r-value of sheet specimen under uniaxial stress in x-direction.

† email: t-abe@po4.oninet.ne.jp

Song and Abe[2] pointed out that r-value can be defined for respective grains in polycrystalline metal and that r-value for a grain is obtained experimentally by measuring the area strain of the surface grain.

2. Experimental Background

A flat surface of polycrystalline metal is roughened after plastic deformation. Various experimental investigations have been done for the surface roughening of polycrystalline metal after plastic deformation. The characteristic features of the surface roughening are summarized as follows.

(1) The surface roughness increases with plastic strain ε.

(2) The surface roughness increases with the grain size d. Namely, the following relation was experimentally obtained[3].

$$Ra = c_1 \, d \, \varepsilon \qquad (2)$$

Ra is the averaged surface roughness and c_1 is a parameter.

(3) The surface roughness during compressive deformation is larger than that during tensile deformation[3].

(4) The surface roughness slightly increases after reversed tensile and compressive deformation[4].

(5) The wave length of the surface roughening curve is about ten times of the averaged grain diameter[3].

3. Model of Polycrystalline Metal

We consider a simple model (Fig.2) of plastic deformation of polycrystalline metal with different r-values for respective grains and surface roughening after

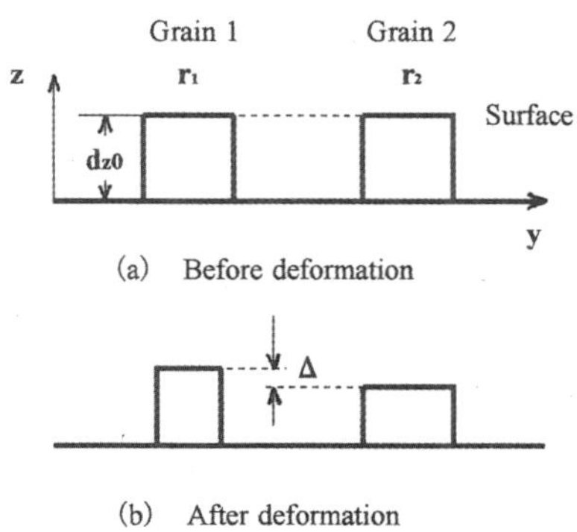

Fig. 2. Model of plastic deformation of polycrystalline metal having different r-values for respective grains and surface roughening after deformation (Uniaxial stress is applied in x-direction).

plastic deformation. For the sake of simplicity, we assume the following for the model.

(1) The strain in the loading direction (x-direction) is equal for all grains.

(2) The grains lie on a certain base plane which is parallel to the surface, as shown in Fig. 2.

In general, the volume constancy is maintained for plastic deformation.

$$\varepsilon_x + \varepsilon_y + \varepsilon_z = 0 \tag{3}$$

From Eqs.(2) and (3), we have

$$\varepsilon_z = -\frac{1}{1+r}\varepsilon_x \tag{4}$$

When the r-values of Grain 1 and Grain 2 are denoted as r_1 and r_2, respectively, the absolute value of the surface roughness Δ of the model shown in Fig.2 is given by

$$\Delta = \left|(\varepsilon_z d_z)_1 - (\varepsilon_z d_z)_2\right| = \left|\frac{r_1 - r_2}{(1+r_1)(1+r_2)}\varepsilon_x\right| d_z \tag{5}$$

Next, we consider the change in grain size in z-direction after plastic deformation. Namely, the grain size after plastic deformation is approximately given by

$$d_z = d_{z0}(1+\varepsilon_z)\ , \tag{6}$$

where d_{z0} is the initial grain size in z-direction (Fig. 2). Then, we have from Eq.(5)

$$\Delta = \left|(\varepsilon_z d_z)_1 - (\varepsilon_z d_z)_2\right|$$

$$= \left|\frac{r_1 - r_2}{(1+r_1)(1+r_2)}\varepsilon_x + \frac{(r_2 - r_1)(2+r_1+r_2)}{(1+r_1)^2(1+r_2)^2}\varepsilon_x^2\right| d_{z0}\ . \tag{7}$$

It is considered that the mean average roughness Ra [Eq. (2)] obtained by experiments is a certain average of the roughness Δ given by Eq.(7) for various values of r_1 and r_2, that is, for various grain orientations. Numerical simulation, however, might be necessary to discuss the relation between Ra and Δ in detail.

We assume here, without loss of generality, that $r_1 \geq r_2$. We consider Δ^+ and Δ^- which denote the absolute value of the roughness under the applied tensile and compressive strains ε_{x0} and $-\varepsilon_{x0}\,(\varepsilon_{x0} \geq 0)$, respectively. The strain

ε_{x0} is usually small compared with 1, and then we have from Eq. (7)

$$\Delta^{+} = \left\{ \frac{r_1 - r_2}{(1+r_1)(1+r_2)} \varepsilon_{x0} - \frac{(r_1 - r_2)(2+r_1+r_2)}{(1+r_1)^2(1+r_2)^2} \varepsilon_{x0}^2 \right\} d_{z0} \quad (8)$$

$$\Delta^{-} = \left\{ \frac{r_1 - r_2}{(1+r_1)(1+r_2)} \varepsilon_{x0} + \frac{(r_1 - r_2)(2+r_1+r_2)}{(1+r_1)^2(1+r_2)^2} \varepsilon_{x0}^2 \right\} d_{z0}. \quad (9)$$

Now, we consider the mean value for the first and the second terms of Eqs. (8) and (9), and denote them with A and B, respectively.

$$A = \frac{1}{2}(\Delta^{+} + \Delta^{-}) = \frac{r_1 - r_2}{(1+r_1)(1+r_2)} \varepsilon_{x0} d_{z0} \quad (10)$$

$$B = \frac{1}{2}(\Delta^{-} - \Delta^{+}) = \frac{(r_1 - r_2)(2+r_1+r_2)}{(1+r_1)^2(1+r_2)^2} \varepsilon_{x0}^2 d_{z0} \geq 0 \quad (11)$$

Eq. (11) shows that the surface roughness under compression is larger than that under tension, which is in consistent with the experimental result[3]. Furthermore, the surface roughness is expected to increase after the reverse tensile and compressive deformation[4].

The ratio of the deviation B to the mean roughness A is given by

$$B/A = \left\{ \frac{1}{1+r_1} + \frac{1}{1+r_2} \right\} \varepsilon_{x0} \quad (12)$$

If we assume, as a representative case, that r_2 $(r_1 \geq r_2)$ is given by

$$r_2 = 1/r_1 \quad (r_1 \geq 1), \quad (13)$$

then we have from Eqs. (10) and (11)

$$B/A = \varepsilon_{x0}, \qquad A = \frac{r_1 - 1}{1+r_1} \varepsilon_{x0} d_{z0}. \quad (14)$$

It should be noted that the ratio B/A is equal to the applied strain ε_{x0}. The relations given by Eq. (14) are shown in Figs. 3 and 4.

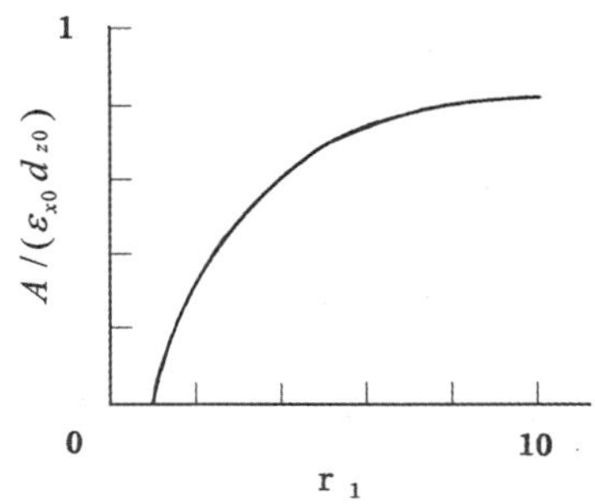

Fig. 3. Relation between r_1 and A.

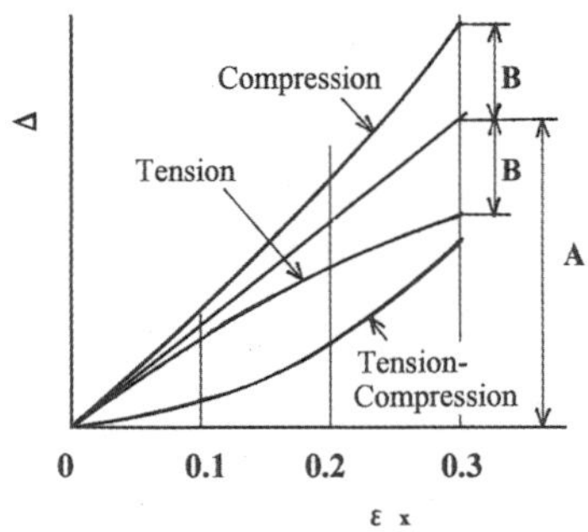

Fig. 4. Surface roughening versus applied strain ε_x.

4. Discussion

4.1. *Wavelength of Surface Roughness Curve*

Osakada and Oyane[3] pointed out that the wavelength of the roughness curve is about ten times of the grain diameter and discussed it in relation with the random work problem in statistics. In the present model, it is explained that the physical meaning of the randomness is resulted from the randomness of r-value of grains.

4.2. *Ridging*

The ridging is a kind of irregular surface and is known as one of defects in the industrial sheet metal forming of a certain kind of metal, such as stainless steel. The present model may be also useful for the understanding of the ridging problem, considering some group of grains having similar orientation, instead of considering respective grains.

5. Conclusion

It was shown that, by adopting the relatively simple model of plastic deformation of polycrystalline metal based on the r-value of grains, the characteristic features of the surface roughening during plastic deformation of polycrystalline metal obtained by various experiments are well explained.

References

1. W. T. Lankford, S. C. Snyder and J. A. Bauscher, ASTM STP 753, 1982, 3.
2. H. Song and T. Abe, Key Eng. Materials, 2007, 340-341, 803.
3. K. Osakada and M. Oyane, Bulletin of JSME, 1971, 14, 171.
4. I. Shimizu, T. Nosho, T. Abe and H. Terai, Proc. AEPA'96, 1997, 277.

STUDY ON THE CASTING-FORGING PROCESS FOR LOW PLASTICITY ALUMINUM ALLOY CONNECTING RODS

WANG YIHUA[†], HU CHENGLIANG

Department of Plasticity Technology, Shanghai Jiaotong University, Shanghai, 200030, China

ZHANG HAIYING, LV JINGLIN

Shanghai Baojie Automobile Components and Parts Forging Co., Ltd., Shanghai, 201818, China

To obtain the low plasticity aluminum alloy connecting rods with good quality, a casting-forging process was put forward in this study. Firstly, the process feasibility of Al-Si-Cu alloy was analyzed by its chemical composition and plasticity chart, and according to the analysis results the casting-forging process was given. Secondly, to solve the crack defect of forging with initial shape of casting billet, an improved shape of casting billet was determined based on divided-flow method. Finally, the wear resistance, mechanical properties and metallographic microstructures of the connecting rods were investigated in details by various tests. Compared with casting, the casting-forging process could improve the wear resistance. The wear resistance of Al-Si-Cu alloy casting-forging connecting rod was better than the other two aluminum alloy forging connecting rods. The tensile strength and elongation of casting-forging process were larger than casting process. The microstructures well explained the performance improvement of the casting-forging connecting rod.

Keywords: Al-Si-Cu alloy; casting-forging process; connecting rod; forgings properties.

1. Introduction

With the light weight demand of auto industry, many kinds of aluminum alloys are used to manufacture structure parts in automobiles because of the high specific strength of aluminum alloys. The connecting rod used in air compressors and internal combustion engines is subject to large inertia and high pressure in the high-speed operation so that it should be as light as possible and able to withstand alternating load. As a typical auto part, aluminum alloy connecting rod can be used to replace the steel one. The casting process[1,2] and forging process[3,4] are two widely used fabrication processes. Lower strength, poor toughness and with void are the main defects of parts made by casting process.

[†] Corresponding author, email: yhw006@sohu.com

The wear-resisting property of parts made by forging process is not good, and the price of rod workpiece for forging is relatively higher. To overcome the shortcomings mentioned above, a casting-forging process for low plasticity aluminum alloy connecting rod is put forward in this study, and some details were discussed.

Forgeability analysis

The low plasticity aluminum alloy in this study is a casting alloy and its chemical composition is shown in Table 1. Si is an important element of this Al-Si-Cu alloy, which makes the aluminum alloy to have better casting flowability and lower shrinkage[5,6]. Generally, the forgeability of aluminum alloy is good when the content of Si is 0.5~1.2%, and the plasticity of aluminum alloy decreased obviously with the content of Si. The content of Cu is 3.0~4.5%, Cu can improve the hot plasticity of aluminum alloy.

Table 1. Chemical composition of the low plasticity casting Al-Si-Cu Alloys.

Element	Si	Cu	Fe	Mn	Ni	Mg	Zn	Ti	Other impurity	Al
Content/%	8.0~14	3.0~4.5	<1.0	<0.5	<0.5	<0.5	<1.0	<0.25	<0.5	balance

To investigate the forgeability of the Al-Si-Cu alloy, plasticity chart shown in Fig. 1 has been built by upsetting tests. The diameter of specimen is 30mm and the length is 35mm. When the temperature is higher than 500°C, the specimen will be overburning. When the temperature is lower than 400°C, the specimen can not recrystallize. Therefore, the testing temperature range is 400°C~500°C. During the test, the deformation degree is 30%, 45% and 60% respectively, and the upsetting force under different temperature with different deformation degree is recorded. Fig. 2 shows the relationship between upsetting force and temperature with 45% deformation degree.

Based on the tests, the alloy can be forged when the deformation degree is less than 45%. If the deformation degree is over 45%, the specimens are easy to crack. It is shown in Fig. 1 that the allowed deformation degree without cracks becomes smaller when the temperature is over 450°C. As shown in Fig. 2, the upsetting force reaches minimum at 400°C. Considering the 20°C temperature drop in practical, the heating temperature of billet before forging should be 420°C.

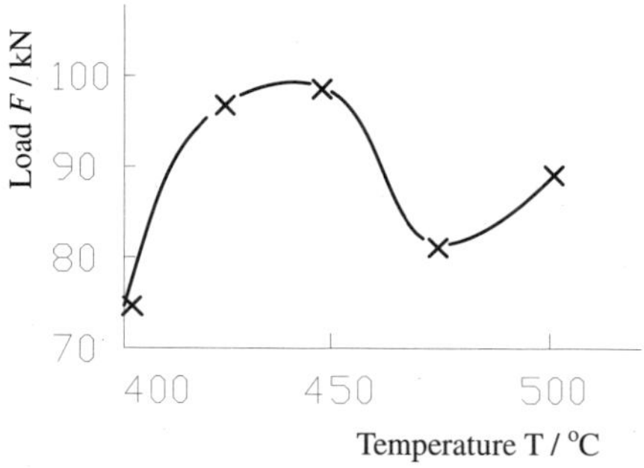

Fig. 1. Plasticity chart of the Al-Si-Cu Alloy.

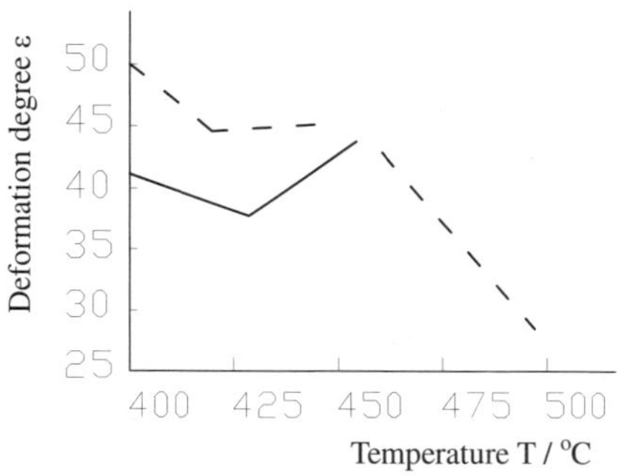

Fig. 2. Upsetting force under different temperature with 45% deformation degree.

According the above analysis, the Al-Si-Cu alloy has poor forgeability and good casting flowability. To form good connecting rods, the casting-forging process is put forward. That is, a billet similar to connecting rod is obtained by casting process, and then the connecting rod is forged with the casting.

2. Determination of the Shape of Casting Billet

To implement the casting-forging process, reasonable shape of casting billet is very important except strictly controlled forging temperature. Fig. 3 shows the initial shape of casting billet. Forging with this billet, there is always crack between the pole and big end of the connecting rod (area J), as shown in Fig. 4. The height of pole and big end of the initial casting billet is just the same. Extra metal of area J almost can not flow in the W and E direction during the forging process. The extra metal can only flow in the N and S direction, and the large tensile stress caused by severe metal flow in the N and S direction is main reason of the crack phenomenon.

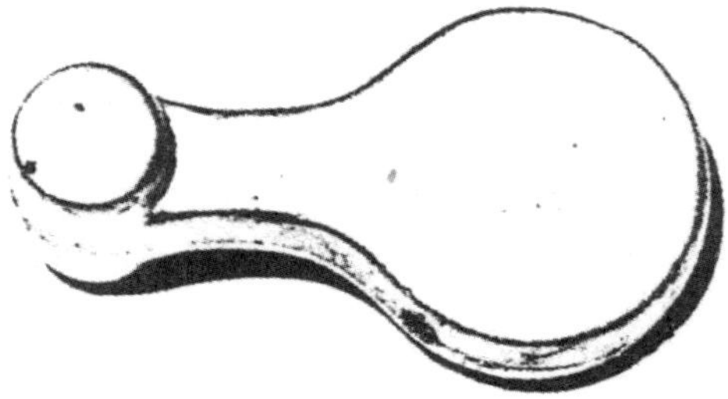

Fig. 3. Initial shape of casting billet.

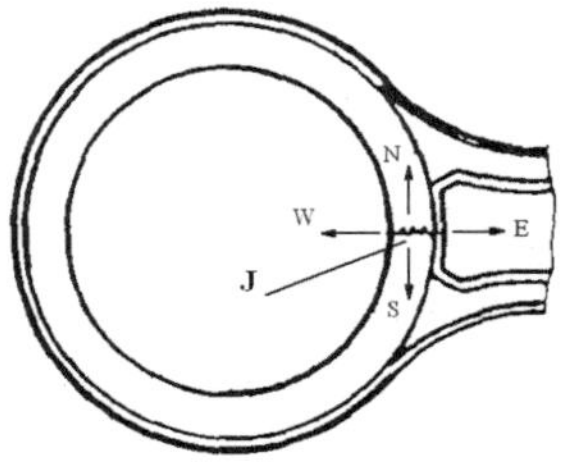

Fig. 4. Crack of final forging part.

To solve the crack defect of connecting rod, an improved shape of casting billet show in Fig. 5 is given by introduce the divided-flow forging theory. The height of pole and big end of the casting billet is different and there is a hole in the big end. During the divided-flow forging process, large part of extra metal flows into the center hole, and then the tensile stress caused by metal flow in the N and S direction is sharply decreased. There is no crack on the forging part with the improved casting billet.

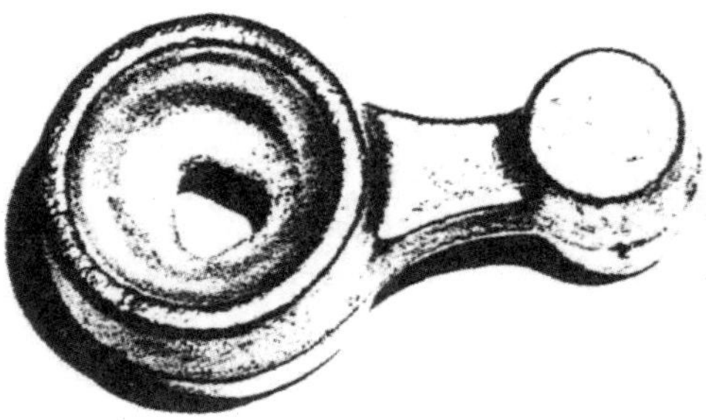

Fig. 5. Improved shape of casting billet.

3. Performance Analysis

3.1. *Wear resistance*

At first, the wear resistance of the connecting rods made by different forming process was evaluated. The wear tests were performed on an Amsler wear tester with pressure 2.24MPa. The upper specimen was taken from connecting rod and was treated by solid solution plus artificial aging. The lower specimen was made of T8 with 62HRC hardness and its rotational speed was 185 rpm. The lubrication medium was 20# oil and testing time was 96 h. From the weight loss data listed in Table 2, the wear resistance of connecting rods was improved by the casting-forging process.

Table 2. Weight loss of different forming process (g).

Forming process	Upper specimen	Lower specimen
Permanent mold casting	0.0072	0.0014
Sand casting	0.0084	0.0020
Casting-forging	0.0065	0.0006

To further study, the wear resistance of connecting rods made by forging process with 2A12 (hard aluminum alloy), forging process with 2A14 (forging aluminum alloy) and casting-forging process with Al-Si-Cu alloy (low plasticity aluminum alloy) were also tested. MM200 wear tester was used. The lower specimen was made of GCr15 with 60HRC hardness and its rotational speed was 30 rpm. Testing load was 200N and each test ran 100,000 runs. The weight loss was measured every 20,000 runs and corresponding data were recorded in Table 3. The wear resistance of Al-Si-Cu alloy casting-forging connecting rod was better than the other two aluminum alloy forging connecting rods.

Table 3. Weight loss of the upper specimens with different materials (g).

Material	20,000r	40,000r	60,000r	80,000r	100,000r	Total
2A12	0.0360	0.0277	0.0291	0.0235	0.0217	0.1379
2A14	0.0167	0.1099	0.0136	0.0164	0.0146	0.0792
Al-Si-Cu	0.0159	0.0145	0.0048	0.0023	0.0079	0.0453

3.2. *Mechanical properties*

Compared with the permanent mold casting connecting rod, the tensile strength of casting-forging connecting rod increased by 50% and the elongation increased about 3 times, as shown in Table 4.

Table 4. Mechanical properties of Al-Si-Cu Alloy connecting rods.

Forming process	No.	Tensile strength σ_b/MPa	Elongation δ/%
permanent mold casting	1	173.0	0.80
	2	196.6	0.13
	3	220.2	0.53
	4	291.2	0.13
Casting-forging	1	357.7	2.70
	2	370.5	2.40

3.3. *Microstructure*

Metallographic samples were taken from the big end of connecting rods, and microstructures of permanent mold casting and casting-forging were observed. As shown in Fig. 6 a), α-Al was primary dendrite morphology and ternary eutectic α-Si-Al2Cu daisy-like scattered between the dendritic branches. As seen in Fig. 6b), most of the dendrite disappeared because of forging, fine eutectic

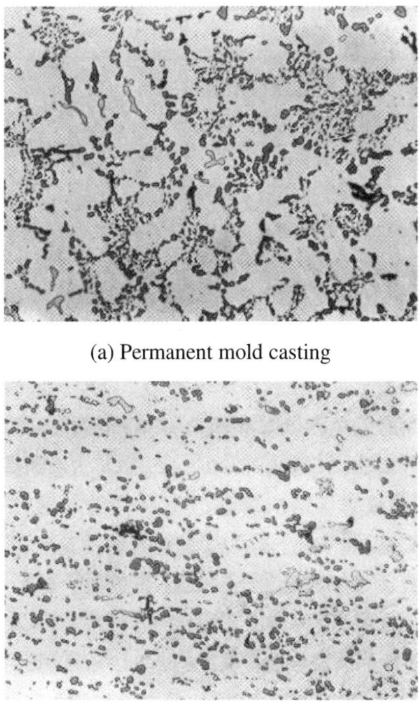

(a) Permanent mold casting

(b) Casting-forging

Fig. 6. Microstructures of the samples with different forming process (320×).

structure was obtained and scattered throughout the a-Al matrix. The fine microstructure of casting-forging could greatly enhance the mechanical properties and uniform scattered hard eutectic phase could improve the wear resistance.

4. Conclusion

(1) According the poor forgeability and good casting flowability of Al-Si-Cu alloy and the geometric feature of connecting rods, a casting-forging process was given. By introducing the divided-flow method, an improved shape of casting billet was determined and the connecting rod without crack was obtained in the end.

(2) The performance of connecting rod including wear resistance, mechanical properties and metallographic microstructures were investigated. Comprehensive performance of Al-Si-Cu alloy casting-forging connecting rods was better than Al-Si-Cu alloy casting connecting rods, also better than hard or forging aluminum alloy forging connecting rods.

(3) The advantages of casting and forging process were fully adopted by the casting-forging process. The casting-billet similar to the part was obtained by casting process, and then the part was formed by forging process. The complex geometry was formed by casting process and the performance was improved by forging process. In addition to the low plasticity

Acknowledgments

The authors would like to acknowledge the support of the National Science & Technology Major Project of China. (Grant No. 2009ZX04013-32)

References

1. P. Zhang, H. Li, Y. F. Chen: *Found.*, 2003, **52**, (1), 70–72. (in Chinese)
2. J. X. Luo, J. H. Hu, L. H. Liu, X. D. Yuan and Y. X. Zheng: *Mech. Sci. Technol.*, 2003, **22**, (2), 275–277. (in Chinese)
3. V. Vazquez, T. Altan: *Mater. Proc. Technol.*, 2000, **98**, (1), 81–89.

A RESHAPING ROLL FORMING TECHNIQUE FOR SQUARE/RECTANGULAR STEEL TUBE MANUFACTURING: NUMERICAL STUDY AND PROCESS DESIGN

DAYONG LI[†1,2]

[1] *State Key Laboratory of Mechanical System and Vibration*
Shanghai, 200240, P. R. China
[2] *School of Mechanical Engineering, Shanghai Jiao Tong University,*
800 Dongchuan Road
Shanghai, 200240, P. R. China

LIMING LI[1], DING TANG[1], YINGHONG PENG1[,2], JILONG YIN[3], FEIZHOU WANG[3]

[3] *Nakata MFG Co. Ltd., 3-7-6, Tagawa, Yodogawa-ku*
Osaka, 532-0027, Japan

Square/rectangular steel tubes are widely used in various industry fields and their quality is related greatly to manufacturing technologies. In this study, a new roll forming technique to manufacture square/rectangular tubes from round pipes is modeled and studied with the finite element method. In the proposed forming process, the horizontal and vertical rolls are installed alternately at the forming stands along the longitudinal direction and the corresponding pass profiles are no longer closed as in conventional reshaping process. The interference between driving horizontal and vertical rolls can be prevented, and more rolls can be used as driving rolls to provide larger driving force. Furthermore, the forming rolls become reusable and can be commonly used in manufacturing the tubes with different sizes. The forming characteristics of the new reshaping process are analyzed and process is designed through comparison among different forming schemes.

Keywords: Reshaping; Roll forming; Square/rectangular tube; Finite element method; Forming characteristics.

1. Introduction

Square/rectangular tubes are widely used as structural components in many important industry fields. The reshaping roll forming process, in which a round tube is progressively formed into a square/rectangular tube through a series of rolls installed at the tandems along the longitudinal direction, is one of the

[†] Corresponding author. Tel.: +86-21-34206313; fax: +86-21-34206313.
E-mail Address: dyli@sjtu.edu.cn.

preferred forming methods in the manufacturing of square/rectangular tubes. The roll forming technique can realize continuous manufacturing of square/rectangular tubes with high efficiency and high throughput.

For the conventional round-to-square roll forming technique, the tube is subjected to severe squeezing deformation, and the forming resistant force is very large. In addition, there is interference problem among the driving shafts; as a result it's hard to manufacture large sizes and high thickness products with this method. Moreover, in the conventional reshaping process, almost all the forming rolls have to be changed when the tubes with different sizes are manufactured, which leads to long downtime and high roll cost.

Nowadays, small-scale and variety are the trends in the manufacturing industry. How to improve capability of the reshaping process and make it more flexible is a topic of interest for tube producers. Due to the complexity of the strip deformation in the roll forming process, the process design of roll forming is still more an art than science in practice. Since Kiuchi[1] pioneered the numerical analysis of the roll forming process, many efforts have been made on the analysis of the roll forming processes by analytical and numerical simulation methods[2-7]. Those numerical methods can be used to predict the deformation and get design guidelines, hence, reduce the design time and cost.

In the present study, the dynamic explicit elasto-plastic finite element method is used for the analysis of the roll forming process of square tubes. A novel reshaping roll forming technique for manufacturing of square steel tubes is modeled and the forming process is numerically studied by simulation. The forming characteristics of the new forming technique are illustrated, through analyzing tube thickness variation, circumferential deformation, and equivalent stress. Then a feasible forming scheme is obtained through comparison among different designs.

2. The New Forming Technique

Based on the idea of flexible manufacturing, a new reshaping roll forming technique is proposed. As shown in Figure 1, the new reshaping process mainly consists of entry stand SH1, forming stands RB1~RB6, shaping stand SH2, and final Turks head stand TH. At the six forming stands, the horizontal and vertical rolls are installed alternately in the longitudinal direction; as a result there is little interference problem and most of the rolls can be designed as driving rolls. Therefore forming of large size or thick tubes is easier. In addition, these forming rolls can be used as the common rolls to produce the tubes with different sizes and realize a kind of flexible manufacturing. At the same time, the alternately positioned forming stands make the squeezing deformation of the tubes is not as severe as in the conventional forming process, which is very helpful to improve the tubes' qualities.

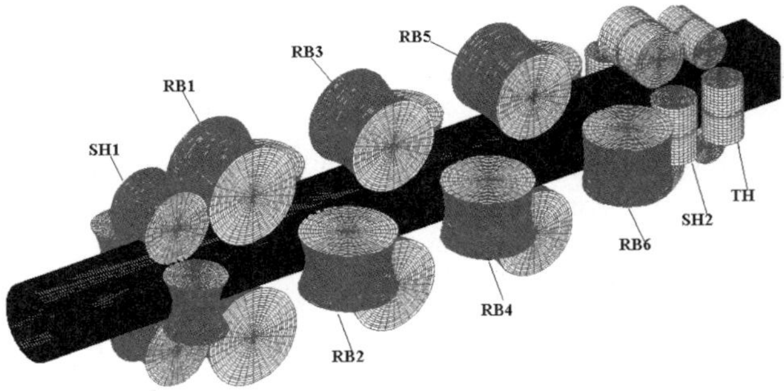

Fig. 1. The new round-to-square roll forming process.

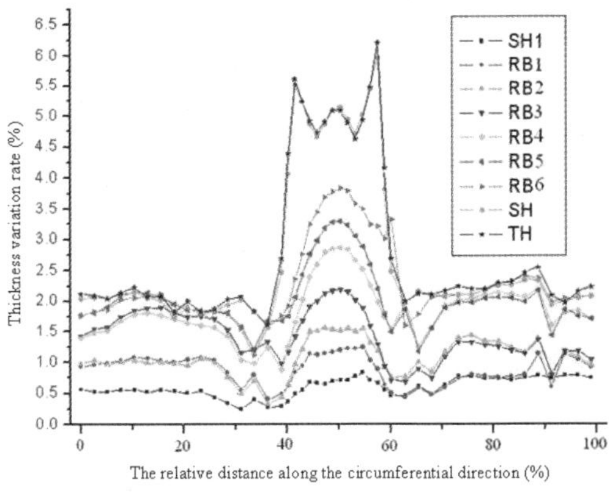

Fig. 2. Thickness variation rate at different stands.

An updated version of the in-house explicit finite element program[2], which is developed for roll forming process simulation and design, is used to simulate the new forming process. In this study, the thickness variation is evaluated to analyze the characteristics of the new forming technique.

The uniformity of thickness is a key factor to post-manufacturing application of tubes. Figure 2. shows the distribution of thickness variation ratio along the circumference direction at different stands. From the entry stand SH1 to exit stand SH2, the thickening rate increases gradually. At the forming stands from RB1 to RB6 only one peak value of thickening rate appears along the circumferential direction. At SH2 and TH stand, two horizontal and two vertical rolls have to be installed together to form the final square section shape, therefore two peak values of thickening rate come into being at these two stands.

3. Process Design of the New Reshaping Technique

Based on the product specification and some experience from Nakata MFG., the following 5 cases are studied and compared to get a feasible process design:

Case 1. Six horizontal and five vertical rolls are equipped. Only the horizontal rolls are designed to be driving rolls.

Case 2. Five horizontal and five vertical rolls are equipped, and all the forming rolls are designed to be driving rolls.

Case 3. Same with Case 2, except that the deformation extent at all the horizontal and vertical rolls is uniformly assigned.

Case 4. Four horizontal and four vertical rolls are equipped. In this case, the rolls at 1/3/4 stands are designed to form small size tubes, while the rolls at 2/3/4 stands to form moderate or large size tubes.

Case 5. Same with Case 4, except that some auxiliary rolls between main forming rolls are additionally used to keep the tube shape and share some deformation.

In case 1~3, the rolls at 1/2/3/4/5 stands are designed to form small size tubes, while the rolls at 2/3/4/5 to form moderate size tubes, and the rolls at 3/4/5 stands to form large size tubes.

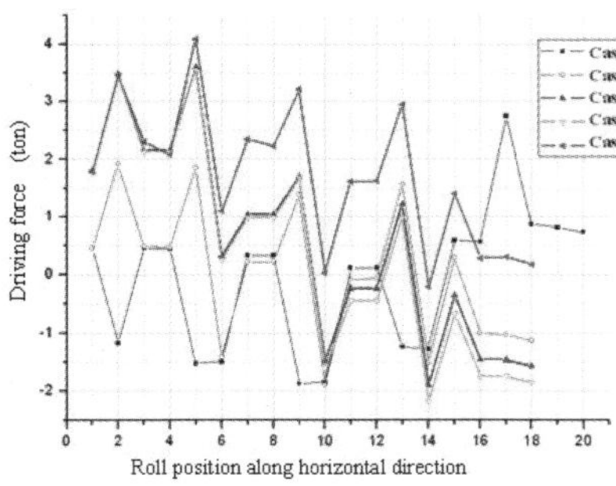

Fig. 3. The driving forces in various cases.

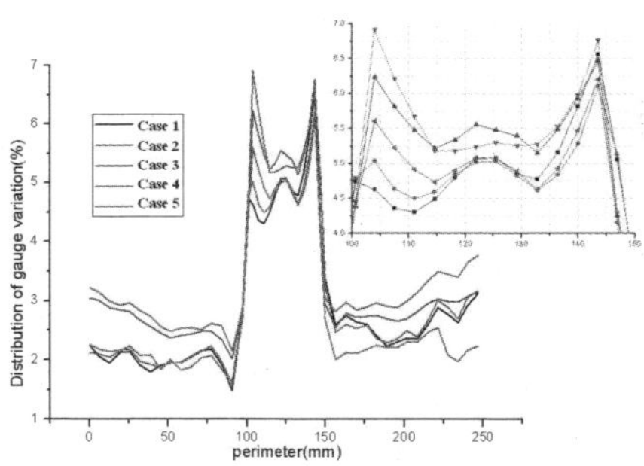

Fig. 4. Thickness variation in various cases.

Figure 3. shows the calculated driving forces in various cases, which is necessary to the design of stands and driving motors. In case 1 and 2, the driving force in the biting stage is not enough. In case 3, the driving force is improved after adjusting deformation extent assignment of all the horizontal and vertical rolls. The design of case 5 apparently can provide the best driving forces during the whole forming process. Figure 4. depicts the thickness variation in various cases. The smallest thickness change is obtained in case 1 because the maximum number of stands leads to a smallest deformation extent at each stand. In case 4, the thickness change is obviously increased because the decrease the number of forming rolls. Through reasonable positioning the auxiliary rolls as in case 5, excessive deformation can be prevented and the thickness variation can be reduced correspondingly. Based on above comparison, case 5 is the best solution, which is used as the guideline of the reshaping machine.

4. Conclusion

In this paper, a new reshaping roll forming technique is modeled and numerically studied. In the new forming process, the forming rolls are distributed alternately along the longitudinal direction. The horizontal and vertical rolls are not located at one section and form a closed pass shape, then the interference problem can be solved and more rolls can be designed as the driving rolls to provide larger driving force. One group of forming rolls can be reused to manufacture the tubes with different sizes, through which the so called “flexible forming” can be realized. Based on the design by the finite element analysis, the new reshaping machine is developed successfully.

Acknowledgments

The authors acknowledge Nakata-SJTU joint research program. The authors from Shanghai Jiaotong University also would like to acknowledge the support of the National Natural Science Foundation of China (Nos. 50634010, 50821003), projects from Shanghai Science & Technology Committee (Grant Nos. 09JC1407000, 10QH1401400) and Program for New Century Excellent Talents in University (NCET-07-0545).

References

1. M. Kiuchi: 'Analytical study on cold roll forming process', Report 23, Report of the Inst. of Ind. Sci., University of Tokyo, Japan, 1973.
2. Proc. 6th Int. Conf. & Wkshp. on 'Numerical Simulation of 3D Sheet Forming Processes', Besancon, France, 1999, Paper 363.
3. A. Alsamhan, P. Hartely and I. Pillinger: *J. Mater. Process. Technol.*, 2003, **142**, 102–111.
4. R. Boman, L. Papeleux, Q. V. Bui and J. P. Ponthot: *J. Mater. Process. Technol.*, 2006, **177**, 621–625.
5. Z. W. Han, C. Liu, W. P. Lu and L.Q. Ren: *J. Mater. Process. Technol.*, 2002, 127, 382-387.
6. M. Lindgren: *J. Mater. Process. Technol.*, 2007, **186**, 77–81.
7. M. A. Sheikh and R.R. Palavilayil: *J. Mater. Process. Technol.*, 2006, **180**, 221–232.

THE SSV CONSTRAINT FOR 0/1 SOLUTIONS IN DENSITY-BASED TOPOLOGICAL DESIGN

XIANGYANG ZHOU †, LIANGZHI FAN
The School of Electrical and Mechanical Engineering, Wuhan Textile University, Wuhan, Hubei, 430073, P.R. China

In density-based topological design, 0/1 solutions are often sought. The SIMP (Solid Isotropic Microstructure with Penalization) method is widely employed to achieve binary values. The solution is not the same when the penalization exponent is different in SIMP. As a possible alternative to SIMP, a new method is introduced. The design variables are still the densities of the finite elements but Young's modulus is a linear function of these densities. In order to drive the solution to a 0/1 layout a new constraint, labeled the sum of squares of the variables (SSV), is added in the topology optimization model. The SSV constraint is convex, which will work effectively with the method of moving asymptotes (MMA) belonging to the convex programming schemes in classical structural topological design. Some examples are given to show the effectiveness of the proposed method. It is turned out that the SSV method improves the total stiffness and can thoroughly remove the grey areas.

Keywords: Topology optimization; SIMP method; SSV constraint; Convex programming.

1. Introduction

In density-based topological design, the solid isotropic microstructure with penalization (SIMP)[1] has got a general acceptance in recent years due to its computational efficiency and conceptual simplicity. However, SIMP does not directly resolve the non-existence of solutions[2] and thus numerical instabilities may occur. Consequently, many numerical techniques are incorporated to make the optimization problem well-posed[2].

Fuchs[3] proposes a possible alternative to SIMP for generating 0/1 structures. The design variables are still the densities of the finite elements but Young's modulus is a linear function of these densities. To drive the solution to a 0/1 layout, the reciprocal variables (SRV) is introduced into the optimization problem. However, The convergence of the SRV algorithm is not smoothly.

In this work, we propose a simple alternative to SRV in the form of the sum of squares of the variables (SSV) constraint. The SSV constraint is convex,

† Corresponding author, email: zxy_paper@126.com

which will work better with MMA than the SRV constraint. The SSV method do indeed produce crisp 0/1 solutions and results in stiffer structures.

2. The SSV Constraint

We begin with the simplest condition, where the design region has been meshed with regular elements with the same size. That means $v_i = 1$.

We introduce the SSV which is the sum of squares of the variables

$$SSV = \sum_i x_i^2 ; (x_{\min} \le x_i \le 1; i = 1 \cdots \cdots n) \tag{1}$$

For every binary instance of the n-vector X in SIMP we have m components with value 1 which represents full-material elements and $n-m$ components with value $x_{\min}$ which represents void elements. Then, the normalized volume of material constraint becomes

$$\sum_i x_i = m + (n-m)x_{\min} \tag{2}$$

Consequently

$$SSV_{discrete} = m \times 1^2 + (n-m)x_{\min}^2 = m + (n-m)x_{\min}^2 \tag{3}$$

Now, we can prove that for any vector X that is not discrete

$$SSV(X) < SSV_{discrete} \tag{4}$$

Now, the optimum design vector is poised to be discrete, $SSV_{opt} = SSV_{discrete}$, therefore we also have $SSV(X) < SSV_{opt}$. In other words, for the solution to be discrete, a necessary condition is that SSV be a local maximum. Since for any design vector $\sum_i x_i^2 \le m + (n-m)x_{\min}^2$, in order to ensure the binary solution we can forcibly

$$\sum_i x_i^2 \ge m + (n-m)x_{\min}^2 \text{ (the SSV constraint)} \tag{5}$$

which is the SSV constraint. Clearly, since the maximum value of the sum of squares of the variables is $m + (n-m)x_{\min}^2$ the only way to comply with the SSV constraint is to satisfy the equality, that is, to be at a discrete design. For similar numerical reasons we have usually relaxed the SSV constraint to the form

$$\sum_i x_i^2 \ge \eta(m + (n-m)x_{\min}^2) \text{ (the SSV constraint)} \tag{6}$$

where the coefficient $\eta \le 1$ and it is typically assigned the value $\eta \ge 0.90$.

The SSV model of the topological 0/1 optimization problem can be expressed as the following formulation

$$\begin{cases} \underset{X=(x_1,x_2,\ldots,x_n)^T}{\text{Minimize}} : f(X) \\ \text{Subject to}: \begin{cases} \sum_i x_i = m+(n-m)x_{\min}; n>m, \\ 0 < x_{\min} \le x_i \le 1 \ (i=1\cdots n), \\ \sum_i x_i^2 \ge \eta(m+(n-m)x_{\min}^2) \ \text{(the SSV constraint)} \\ \text{equilibrium equations.} \end{cases} \end{cases} \tag{7}$$

In the above model, the design region has been meshed with regular elements with the same size. But in SIMP model, the design region can be meshed with free elements with different size. Now, we extend the method to the problems with free mesh and different element sizes. The volume constraint is

$$V^T X = \sum_{i=1}^{n} v_i x_i = \sum_p v_p + x_{\min} \sum_q v_q \quad (p \in m \ ; q \in (n-m)) \tag{8}$$

And the SSV is

$$SSV = \sum_i v_i x_i^2 \quad (i = 1\cdots n) \tag{9}$$

Here, v_i is the volume of the element i. Accordingly, the SSV constraint is

$$\sum_i v_i x_i^2 \ge \eta(\sum_p v_p + x_{\min}^2 \sum_q v_q) \quad (p \in m \ ; q \in (n-m)) \tag{10}$$

This is all the SSV method, in which the penalty exponent p is set to 1. The design variables are still the densities of the finite elements but Young's modulus is a linear function of these densities (in some sense, a SIMP material without penalty). In order to drive the solution to a 0/1 layout a new constraint, labeled the sum of squares of the variables (SSV), is introduced.

3. Numerical Implementation

A Matlab implementation of MMA was kindly provided by Svanberg and all the examples are run on that platform. With regard to MMA we have used the default values given by the author or appearing in Svanberg[4].

To make the convergence of the SSV algorithm fast, we take such tactics: using the optimal solution of SIMP as the starting point of optimization search process of SSV method. That is: firstly, the standard SIMP is employed to generate an intermediate optimal design solution X^{int}; secondly, SSV approach is adopted to produce the final solution X^*, initializing the design variables X with X^{int}. We call this method as SIMP-SSV.

The first example is the cantilever under a tip load from Sigmund[5], in which a group of η are tested and its influence on the results is analyzed. The optimal topologies of cantilever beam under a down-load at the tip with different approaches are shown in Fig.1. The values of the objective function and the iterations can be seen in Table 1.

Table 1. Different compliance values and iterations of cantilever.

	Compliance				
	η =0.90	η = 0.95	η = 0.97	η = 0.99	η = 1.0
SIMP-SSV	50.7246	51.3138	51.6326	52.6355	53.3208
(Percent)	88.47%	89.50%	90.06%	91.81%	93.00%
SIMP	57.3337(100%)				

An important observation is that the grey elements of SIMP-SSV are much less than those of SIMP. With the increase of η, the grey elements can be removed gradually. The topological design must be in clear 0/1 patterns by setting η to 1. As seen in Fig.1 (e), there is no any grey element; this shows that the quantity of grey elements is sensitive to η.

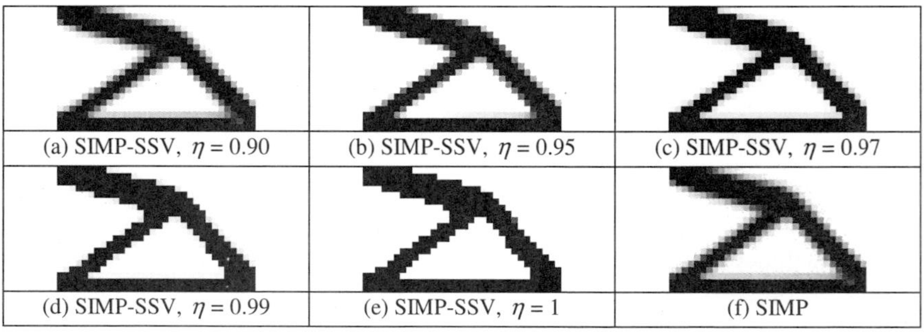

Fig. 1. Optimal topology of cantilever beam under a down-load at the tip: a) b) c) d) e) SIMP-SSV design with different η; f) SIMP design.

In Table 1, it is obvious that the compliance of SIMP is the highest. The objective of the two-pass method can decrease 11.53%. However, in the case of SIMP-SSV, the minimum compliance increases as η. The experiment shows that: the convergence of the SIMP-SSV algorithm are much better than SSV. However, the convergence of the SSV algorithm needs further studies.

The second example is a 3D cantilever under a load as showned in in Fig.2. The domain is meshed with $24 \times 6 \times 15$ cubes. The volume fraction is V=0.35 and η=1. The other relevant parameters in this case are the same as the first example. The optimal topological design is shown in Fig.3.

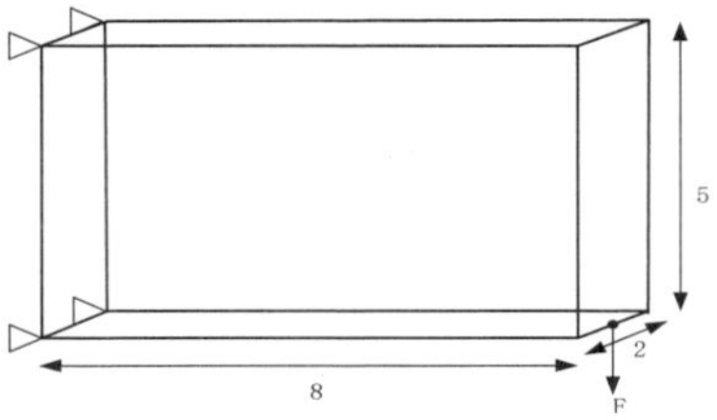

Fig. 2. Design domain.

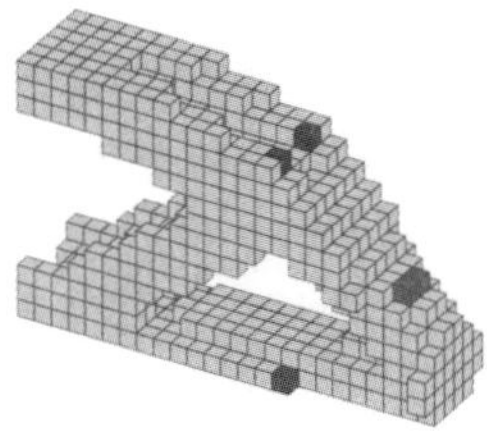

Fig. 3. 3D topology optimized cantilever.

4. Summary and Conclusion

It is an important topic to develop some powerful topology optimization methods though it is quite mature. In density-based topology optimization, 0/1 solutions are the objective sought by researchers. The object of this paper is to present another technique that is also simple: the SSV method. Owed to the strong constraint conditions, the convergence of the SSV algorithm is some slow. But, once the authors combine SIMP approach and SSV constraint for topology optimization: the SIMP-SSV method, the convergence of the SSV algorithm is smooth and speedy. In SIMP-SSV: the first was done by SIMP; in the second, the solution of the previous stage is used for this stage's starting solution and then SSV accomplishes the following optimizing process to obtain the final optimal solution with the help of MMA. It can be turned out SIMP-SSV is more stable than SRV in the implementation of program.

The examples have shown the effectiveness of the proposed combined method SIMP-SSV. SSV tenders the results in the second stage, which is close to the binary set and even is a real binary set. Moreover, compared with the results of SIMP, the total compliance of SIMP-SSV decreases, which is of more important meaning. In a word, the SIMP-SSV method has some obvious advantages and deserves a close attention.

Acknowledgments

This work is supported by the National Natural Science Foundation of China (50805109) and HuBei Provincial Natural Science Foundation(2007ABC003, 2008BDC257).

References

1. M.P. Bendsøe, O. Sigmund: *Archive of Applied Mechanics*, 1999, 69, 635-654.
2. M.P. Bendsøe, O. Sigmund: 'Topology optimization: Theory, Methods, and Applications', 24-28; 2003, New York, Springer, Berlin Heidelberg.

3. M.B. Fuchs, S. Jiny, N. Peleg: *Structural and Multidisciplinary Optimization*, 2005, 30, 320-326.
4. K. Svanberg: *Int J Numer Method Eng*, 1987, 24, 359 – 373.
5. O. Sigmund: *Structural and Multidisciplinary Optimization*, 2001, 31, (6), 419-429.

GALLING BEHAVIOR IN SQUARE CUP DRAWING OF HIGH TENSILE STRENGTH STEEL SHEET

KEISUKE TANABE†, MINGXU YANG, YASUHARU YOSHIKAWA, ZHIGANG WANG

Department of Mechanical Systems Engineering, Gifu University, 1-1 Yanagido Gifu, 501-1193, Japan

A series of experiments have been carried out to investigate the galling behavior in macro-scale in the square cup drawing of a high tensile strength steel sheet. On the die surface, galling occurs at the bottom of the boundary between the straight and corner, and grows upward with the number of drawing cycles. Galling on the drawn cup starts at the top of the same boundary and grows downward with the number of drawing cycles. With decreasing the blank holding pressure or increasing the sliding distance, galling becomes remarked. Galling behavior has been discussed based on the simulation result of FEM.

Keywords: Deep drawing; Tribology; Galling; High tensile strength steel sheet.

1. Introduction

In terms of the galling behavior in forming process, many studies have been carried out, and a considerable number of galling mechanisms have been developed up to now [1]-[3]. However, the galling behavior cannot be predicted in the forming process yet. The object of this study is to clarify the galling mechanism and provide the prediction method of galling behavior in forming processes.

In the previous paper [4], the galling behavior in macro-scale was investigated in square cup drawing of high tensile strength steel sheet by using several die materials. When using the non-coated dies in semi-dry, the macro-scale galling can clearly be observed on die and drawn cup surface after a few drawing cycles. The galling on the die surface occurs at the bottom point of the boundary between the straight and corner edge, and grows upward with the drawing cycles. On the drawn cup, the galling starts at the top of the same boundary and grows downward. When using the coated dies, no remarkable galling is observed on die surface after N = 120 drawing cycles, but the adhesion can be observed on TiCN(PVD) and TiCN(CVD) at the same boundary as the

† Corresponding author, email: o3122019@edu.gifu-u.ac.jp

non-coated die. No conspicuous adhesion can be observed on TiC(CVD) and DLC-Si(DC-PACVD) ever after N = 1000, but micro-scale adhesion can be recognized on TiC(CVD). Any adhesion cannot be found on DLC-Si(DC-PACVD).

In this paper, we will show how the drawing conditions affect the galling behavior. The galling expansion process in the square cup drawing is observed and the frictional condition is analyzed by FEM simulation.

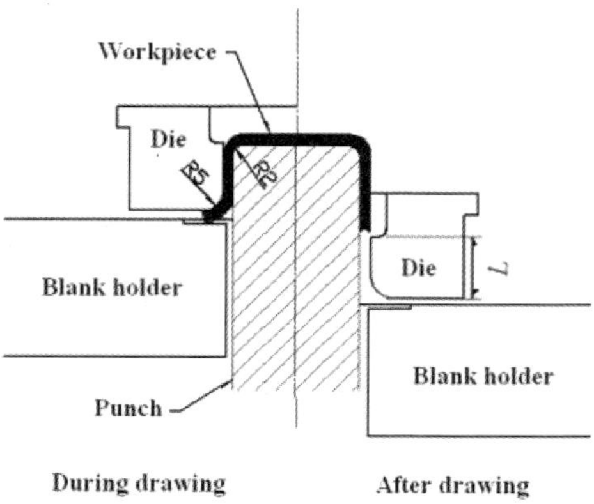

Fig. 1. Principle of square cup drawing test.

2. Experimental Method

Since the presently used tribometer has been reported in the previous paper[4], only the principle of the tribometer is shown in Fig.1. We selected the square cup drawing test because the frictional condition varies in the section position and the peculiarity of galling starting point maybe find out. In this paper, we change the blank holding force and the sliding distance to observe the effects of the drawing conditions on galling behavior. The sliding distance L is defined as the height between bottom and forming edge on die (Fig. 2) and set to 10, 15, 20(mm). The blank holding pressure P is set to 0 and 6MPa. Table 1 shows the experimental conditions. The workpiece material is high tensile strength steel sheet, SPFC590 by JIS, with 2.6 thicknesses. In order to accelerate the galling, the drawing with nominal ironing ratio of 4 % was used with a 2.5 [mm] die-workpiece clearance. The Die was made of D2 die steel (Rockwell Hardness HRC =61), with a 0.04μmRa polished surface finish. In the present experiment, a paraffinic mineral oil P460 was used.

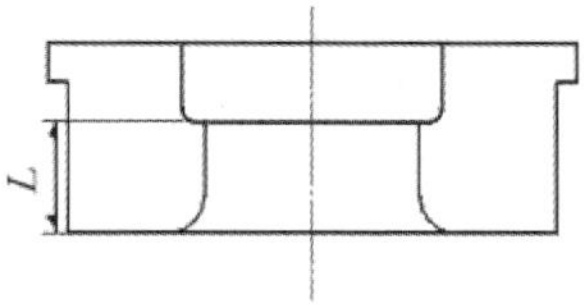

Fig. 2. The definition of L.

Table 1. Experimental conditions.

Workpiece	SPFC590 (Tensile strength 625MPa)
Punch	SKD11
Die	SKD11
Clearance, C	2.5mm
Sliding distance, L	10, 15, 20mm
Blank holding pressure, P	0, 6MPa
Lubricant	P460 (1500×10-6 m 2/s at 25 °C)
Temperature	Room Temperature

3. FEM Simulation Condition

The simulation was carried out using a commercial code for forging DEFORM3D. is used for FEM analysis. For shortening the calculation time, 1/4 model was used as shown in Fig.3. The workpiece is elasto-plastic. The die is elastic and its Young's modulus is 200GPa. The others are rigid. The punch speed is 40 mm/s, the frictional coefficientμis set to 0.06 for all frictional interface. The workpiece' stress-strain is expressed in formula (1) and shown in Fig. 4

$$\sigma=1000\varepsilon^{0.25} \cdot \cdot \cdot (1)$$

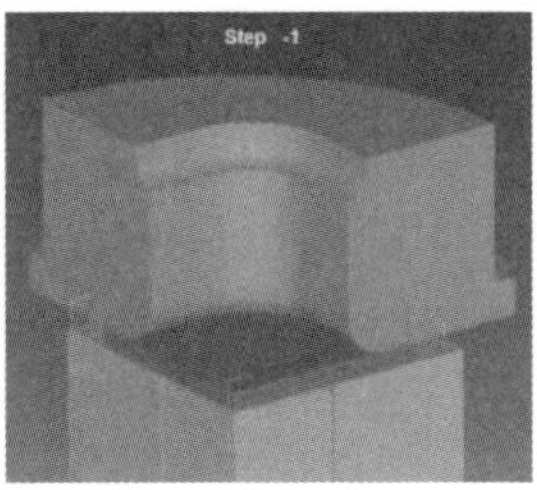

Fig. 3. FEM analysis model.

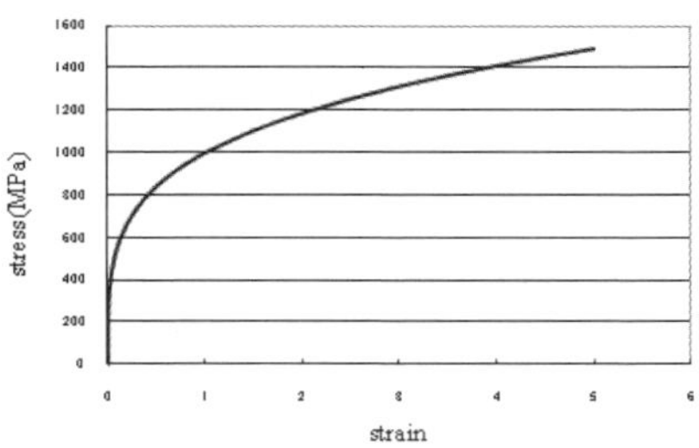

Fig. 4. Stress-strain diagram.

4. Experimental Results

In the textbook and most papers, galling on die occurs around the line. This means that galling will occur at the die shoulder. However, just as shown in the previous paper, this is not true.

The surfaces of die and square cup after drawing are shown in Fig. 5. A few drawing cycles cause galling obviously. According to observation of galling on the die, galling occurs at the point far away from the line.

On the drawn cup, galling occurs from straight bottom edge (Fig. 6) and grows downwards with drawing cycles.

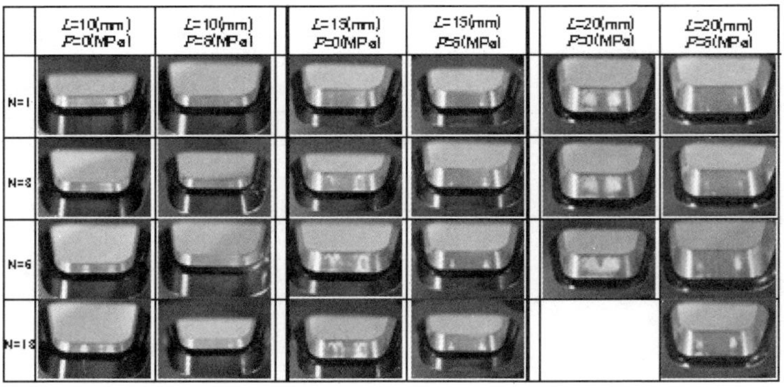

Fig. 5. Surface appearance of die on each condition.

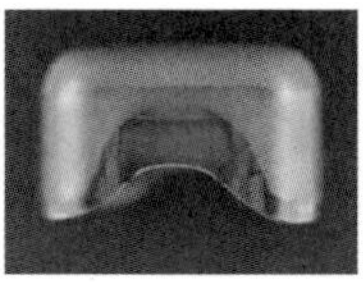

Fig. 6. Galling on the drawn cup.

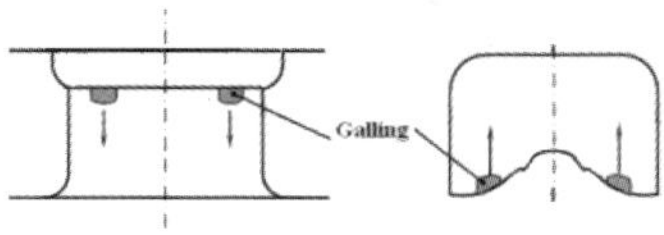

Fig. 7. Schematic image of galling position.

Schematic image of galling position is shown in Fig. 7. On the die surface, galling occurs at the bottom of the boundary between the straight and corner, and grows upward with the number of drawing cycles. Galling on the drawn cup starts at the top of the same boundary and grows downward with the number of drawing cycles.

Fig.8 shows the relationship between the galling starting drawing cycles and sliding distance. Galling occurs after a few drawing cycles for all experimental conditions. Galling is reduced by increasing blank holding pressure and sliding distance. However, the galling starting position and its expansion path are not affected by the drawing conditions such as blank holding pressure and sliding distance.

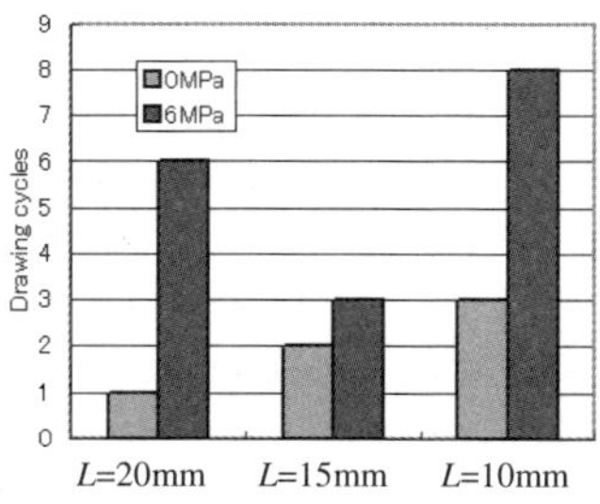

Fig. 8. Effect of blank holder pressure on galling.

5. FEM Simulation Result

The blank thickness change and wrinkle generation can be observed by both FEM simulation and experiment. Fig.9 and 10 show the variation in the sheet thickness during drawing.

Observation points are 2 points: Straight center edge and galling point. Although this was not obvious until after the event, galling point is just the wrinkle position. The sheet thickness increase is remarked at galling point compared to the straight center edge. High BHP reduces the sheet thickness increase. Galling may be caused by hard ironing because of the large increase in the sheet thickness at the wrinkle position.

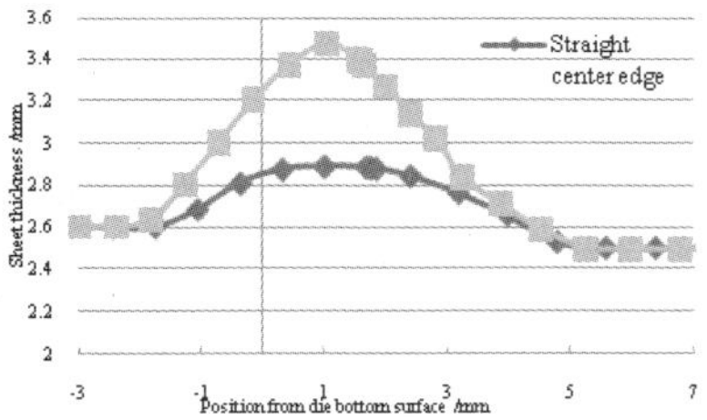

Fig. 9. Variation in sheet thickness during drawing (BHP = 0MPa).

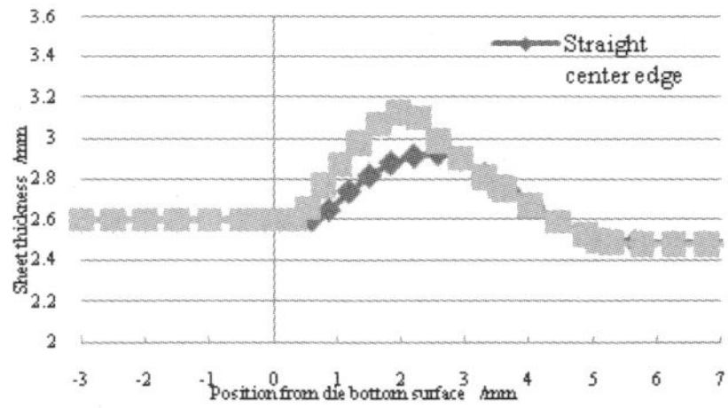

Fig. 10. Variation in sheet thickness during drawing (BHP = 6MPa).

6. Conclusion

The main conclusions can be summarized as follows:

(1) In square cup drawing of high tensile strength steel sheet, the galling on the die surface occurs at the bottom point of the boundary between the straight and corner edge, and grows upward with the drawing cycles. On the other hand, the galling on the drawn cup starts at the top of the same boundary as die surface and grows downward with the drawing cycles.

(2) Larger sliding distance makes galling expansion rapidly but higher blank holding pressure decreases galling.

(3) Galling occurs at the wrinkle position at which the sheet thickness gives its maximum value.

References

1. N. Nakamura: Journal of the JSTP, Vol. 24-265 (1983), p.100
2. N. Nakamura, K. Kato and N. Matsui: Transactions of the JSME Ser. C, Vol.52-473 (1986), p.163
3. K. Kubota, S. Komatsubara, T. Ogihara, M. Narumi and M. Yamaoka: Hitachi Metals Technical Review. Vol. 21 (2005), p.45. Information on http//www.hitachi-metals.co.jp/rad/rad02.html

DEVELOPMENT OF BOTTOM COMPRESSION DRAWING PROCESS FOR SOLID BOSS FORMING

ZHIGANG WANG, TORU ANDO †

Faculty of Mechanical and Systems Engineering, Gifu University, Yanagi-do 1-1 Gifu, 501-1193, Japan

Backward extrusion is the most common method of forming bosses within product cases. However, backward extrusion requires a very large forming load, and has the disadvantage of producing dimple defects when forming thin sheets. To overcome this disadvantage, we investigated a method of bottom compression drawing to replace backward extrusion. Bottom compression drawing is a technique in which, during normal drawing, back pressure from the counter punch that opposes the drawing punch applies a load on the blank, so that compressive force acts on the bottom face of the blank. Bosses are formed at the same time as the case is formed as a result of this drawing. Backward extrusion and bottom compression drawing were compared using both FEM analysis and experiment in the formation of solid bosses. The results showed that bottom compression drawing could form a piece with about 60% of the forming load used with backward extrusion; and that dimple defects were less likely to occur with bottom compression drawing than with backward extrusion.

Keywords: Plastic deformation; Boss; Cold forging; Sheet forging.

1. Introduction

Electric and electronic devises, mobile devises, and similar products are becoming lighter, smaller, and less expensive in recent years. In such devises the case or exterior accounts for a large proportion of the overall mass, and making cases lighter is a major issue. Studies are currently being conducted to form cases and internal supports in a single piece. Because the use of supports inside a case makes positioning and fastening easier and leads to a reduction in the number of parts, it should be possible with such support to make the case more lightweight. Processes used in forming internal supports include welding, casting, machining, powder metallurgy, and backward extrusion. Of these processes, backward extrusion is the most commonly used process based on considerations of accuracy and cost. However, backward extrusion requires a very large processing load. To overcome this disadvantage, we investigated a method of bottom compression drawing to replace backward extrusion.

† Corresponding author, email: o3122003@edu.gifu-u.ac.jp

Bottom compression drawing is a technique in which, during normal drawing, back pressure from the counter punch that opposes the punch applies a load on the blank, and a compressive force acts on the bottom face of the blank. This drawing can be used to form supports at the same time the case is formed.

Tension produced by the drawing load acts on the bottom face of the blank in this processing method, and so it is anticipated that yield pressure will be reduced, the influence of friction will be diminished, and the forming load will become smaller. Another feature is that thinner material can be used because of improved rigidity of the bottom material. Further advantages of this method are that tool life can be extended because of the reduced forming load, and a press with lower power can be used.

The aim of this study is to demonstrate the features of bottom compression drawing while comparing it with backward extrusion and simple compression by using both FEM simulation and experiment.

2. Comparison of Bottom Compression Drawing of Backward Extrusion

The comparison of both processing methods is shown in Fig. 1. In the bottom compression drawing process, a blank is set on the stationary counter punch and compressed by the punch with a given load and then drawn by moving the die toward the punch. In the backward extrusion process, the material is compressed by the punch and flows into the boss portion and cup wall. In the bottom compression drawing process, a tension generated by drawing acts on the compressed material and a tensile stress works radially at the punch shoulder and the bottom material yields easily. In the backward extrusion process, a radial compression stress acts at the punch shoulder and the yield pressure becomes higher than that at the bottom compression drawing process.

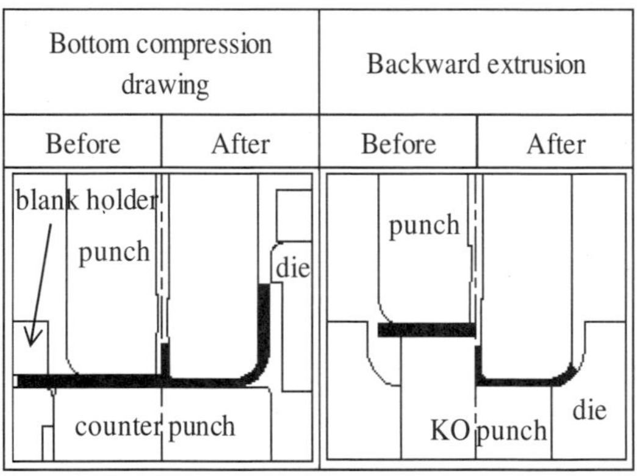

Fig. 1. Comparison of bottom compression drawing and backward extrusion.

In the present study, the comparison of the bottom compression drawing process and the backward extrusion is carried out by using FEM analysis, and the validity of FEM results is confirmed by the experiment. A simply compression process is also done for the comparison.

3. FEM Simulation Result

The relation between the height of the boss and the punch load by FEM simulation is shown in Fig. 2.

Two kinds of friction condition were simulated. Table 2 gives the lubricated area. In the type A, lubricant was applied only for the upper surface of blank and the friction coefficient on an upper surface of the blank was set to 0.03 and the friction coefficient on the lower surface of the blank was set to 0.3 where no lubricant was applied. In the type B, lubricant was applied for both the upper and lower surface of blank and the friction coefficient for all contact area was set to 0.03. Although the boss can be formed at a lower punch load for the type B, the boss height is small because of the rapidly thinning of the bottom thickness. For the type A, the boss height increases satisfactorily with increasing punch load. The forming load required for the same boss height becomes larger in the order of the bottom compression drawing, simple compression and backward extrusion. Only 50 percent of the backward extrusion is required for the bottom compression drawing.

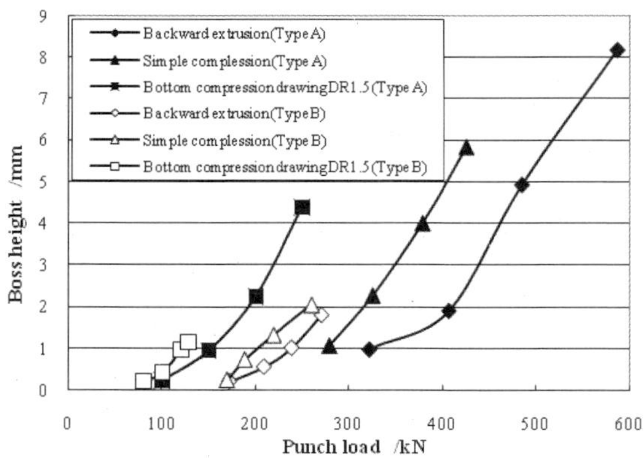

Fig. 2. Relationship between boss height and load by FEM.

4. Experimental Conditions and Results

A series of experiment were carried out by using 1100kN servo press. The blank material was A1050-O. Blank dimensions and the tool shape are shown in Table 1. The punch load was measured with a load cell installed in the servo press. Simple compression was done by using the punch and counter punch for the bottom compression drawing. The punch was common for the three processes. The counter punch in the bottom compression drawing and the KO punch in the backward extrusion were polished to 0.05 μmRa with the diamond paste after every formation test to keep the frictional condition same. A paraffinic mineral oil P460 (460cSt at 40℃) was used and applied according to the table 2.

Table 1. Blank and tool shape.

Blank and tool shape	Bottom compression drawing	Simple compression	Backward extrusion
Blank diameter /mm	60	56	40
Blank thickness /mm	2		
Punch diameter /mm	40		
Punch profile radius /mm	4.5		
Boss diameter /mm	2.5		
Boss profile radius /mm	1		
Die inner diameter /mm	45.2		
Die profile radius /mm	2.5		—
Die profile radius /mm	—		7

Table 2. Lubrication.

Frictional condition	Lubricant area
Type A	Upper surface of blank
Type B	Upper and lower surface of blank

The relation between the punch load and the reduction in product bottom thickness is shown in Fig. 3. The variations of the boss height in the punch load are shown in Fig. 4. For the type B, with increasing the punch load, the reduction in product bottom thickness varies hyper sensitively and the boss height grows hardly. This trend was confirmed also by FEM simulation. This means that the lubrication condition of the type B is not suitable for the boss formation. For the type A, with increasing the punch load, the boss height grows regularly. This trend was confirmed also by FEM simulation. This means that the lubrication condition of the type A is suitable for the boss formation. It can be recognized that only 50 percent of the punch load for the backward extrusion is necessary for the bottom compression drawing just as predicted by FEM simulation.

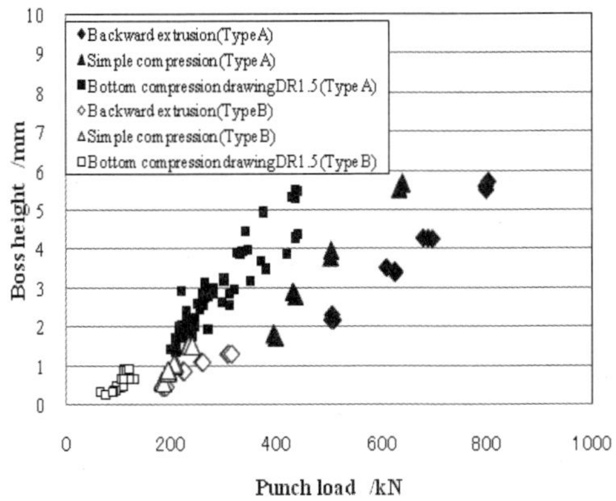

Fig. 3. Relationship between reduction in bottom thickness and load.

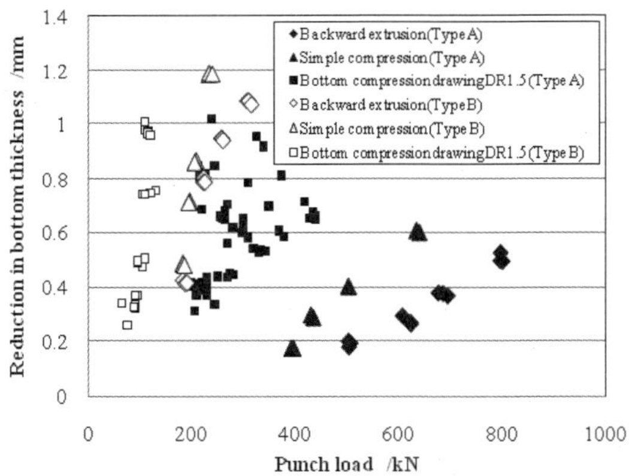

Fig. 4. Relationship between boss height and load.

5. Conclusion

A comparison was made between bottom compression drawing and backward extrusion in formation a solid boss by using FEM simulation and a series of experiment. The results indicated the following.

(1) Forming is possible with bottom compression drawing using about 50 percent of the load required in backward extrusion.

(2) The back surface of boss side should be kept a high frictional state to obtain the boss height.

References

1. Zhigang Wang: Plastic forming springtime defensecollection of papers, 2005, **423**

DEVELOPMENT OF BOTTOM COMPRESSION DRAWING PROCESS FOR SOLID BOSS FORMING

ZHIGANG WANG and MASANAO KARASAWA †

Department of Mechanical System Engineering, Gifu University, 1-1 Yanagido Gifu City, Gifu Prefecture 501-1193, Japan

Backward extrusion is the most common forming method of the solid boss integrated with product cases. However, backward extrusion requires a very large forming load and has the disadvantage of producing dimple defects when forming thin sheets. To overcome this disadvantage, the bottom compression drawing process has been developed to replace backward extrusion. The bottom compression drawing can form a solid boss with about 50% of the forming load required for backward extrusion and dimple defects are less likely to occur than with backward extrusion. In the present paper, the formability of boss influenced by lubrication and tool roughness will be described. It was recognized that the dimple defect occurs at very early stage when the counter punch is lubricated by lubricant. With increasing surface roughness of counter punch, the formable zone of solid boss expands. Adhesion on counter punch surface can be recognized with forming cycles repeating under dry condition but the irregularity in the product configuration converges to a befitting configuration after a few forming cycles.

Keywords: Plastic deformation; Sheet metal forging; Cold forging; Solid boss.

1. Introduction

Electric and electronic devises, mobile devises, and similar products are becoming lighter, smaller, and less expensive in recent years. In such devises the case or exterior accounts for a large proportion of the overall weight, and making cases lighter is a major issue. Studies are currently being conducted to form cases and internal bosses in a single piece. Because the use of bosses inside a case makes positioning and fastening easier and leads to a reduction in the number of parts, it should be possible with such boss to make the case more lightweight. Processes used in forming internal bosses include welding, casting, machining, powder metallurgy, and backward extrusion. Of these processes, backward extrusion is the most commonly used process based on considerations of accuracy and cost. However, backward extrusion requires a very large forming load. It also has the disadvantage of producing dimple defects when

† Corresponding author, email: o3122011@edu-u.ac.jp

forming thin sheets. To overcome this disadvantage, we investigated a method of bottom compression drawing to replace backward extrusion.

Bottom compression drawing is a technique in which, during normal drawing, back pressure from the counter punch that opposes the drawing punch applies a load on the blank, and a compressive force acts on the bottom face of the blank. This drawing can be used to form bosses at the same time the case is formed.

Tension produced by the drawing load acts on the bottom face of the blank in this processing method, and so it is anticipated that yield stress will be reduced, the influence of friction will be diminished, and the forming load will become smaller. Since molds are independent, it is easy to adjust the friction conditions for each section, and since compressed material flows to the punch R portion it should be possible to improve LDR depending on conditions. Another feature is that thinner material can be used because of improved rigidity of the bottom material. Further advantages of this method are that mold life can be extended because of the reduced forming load, and a press with lower power can be used. In this study, the feature of bottom compression drawing is investigated that formability of boss influenced by lubrication, roughness and adhesion on counter punch surface.

2. Principles of Bottom Compression Drawing

Figure 1 shows the principles of bottom compression drawing. Bottom pressure p is applied from the counter punch set in a position opposite the drawing punch. The die moves upward at velocity V and the work piece is drawn while keeping the bottom pressure. Tension T is generated by drawing and the punch load P is generated by drawing and bottom pressure. The yield condition for the bottom material is shown below.

$$p \pm \sigma_{\mathrm{T}} = 2k \qquad (1)$$

For bottom compression drawing, the tensile stress σ_{T1} from the tension produced by drawing acts on the bottom face of the product during processing. For backward extrusion, on the other hand, outward material flow is inhibited in the die R, and compression stress $-\sigma_{\mathrm{T2}}$ acts. Thus, with bottom compression drawing it is possible to reduce the forming load for $\sigma_{\mathrm{T1}} + \sigma_{\mathrm{T2}}$ in comparison with backward extrusion. In addition, because the required forming pressure is reduced with tension T, the friction resistance to the boss formation is reduces. Furthermore, compared with backward extrusion, the contact area between the test material and tool can be reduced and the forming load can be decreased.

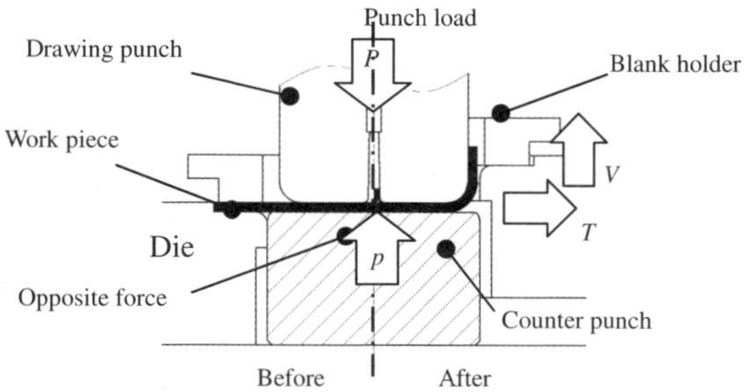

Fig. 1. Principle of bottom compression drawing.

3. Experimental Condition

A series of experiment was carried out with multi-action 1000kN hydraulic press. Product sets an example at Figure 2. The test material was A1050-O (by JIS) rolled sheet. The blank was 60mm in diameter and 2mm in thickness. A paraffinic mineral oil P480 was used to the surface of die, punch and counter punch.

In the present experiment, the punch load was adjusted by oil pressure of press. The punch load was measured by a load cell installed on press. Blank was drawn while keeping punch load.

All of tools were made of D2 die steel. The surface roughness of counter punch was prepared by using sandpaper and diamond paste and adjusted at the range 0.005 – 0.1μmRa measured by a noncontact 3-dimension shape measuring instrument. The surface roughness of punch and die was kept to 0.005μmRa constant.

Fig. 2. Product.

4. Experimental Results

4.1. *Forming defect by lubricant*

In the sheet forging process, a dimple defect occurs at back of boss. It's often experienced when bottom thickness is below limit thick.

When the counter punch is lubricated by lubricant, the dimple defect occurs at very early stage. An example is shown in Figure 3. In forming process, lubricant at the counter punch side concentrates to the center of counter punch. With the development of forming progress, the trapped lubricant generates high pressure and cause a local deformation at the back of boss. Similarly, in FEM analysis, the dimple defect occurs when pressure of 100MPa is operated in area of 1.25mm from center. In order to prevent of dimple defect, the experiments hereafter were carried out at the conditions that no lubricant is applied for the interface between the counterpunch and blank.

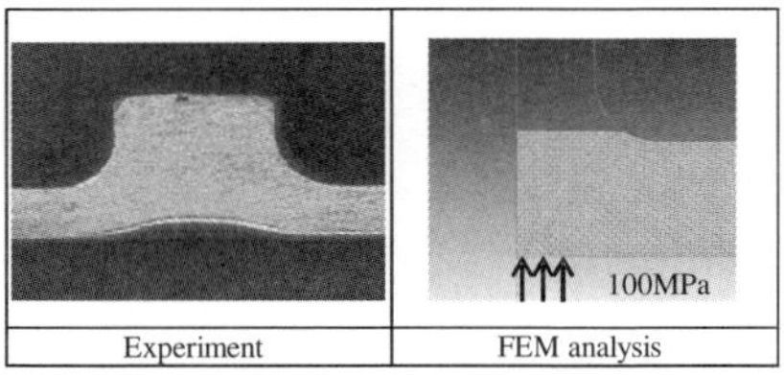

Fig. 3. Dimple defect.

4.2. *Boss forming limit*

As shown in Figure 4, crack occurs at the punch shoulder and fracture of blank can be recognized for unsuitable conditions.

In order to investigate the effect of the surface roughness of counter punch on the forming limit, the surface roughness of the counter punch was changed. The Relationship between the surface roughness of counter punch and the limit of punch load is shown in Figure 5. Success range expands greatly with an increase in surface roughness of counter punch.

a) Success b) Failure

Fig. 4. Example of fracture.

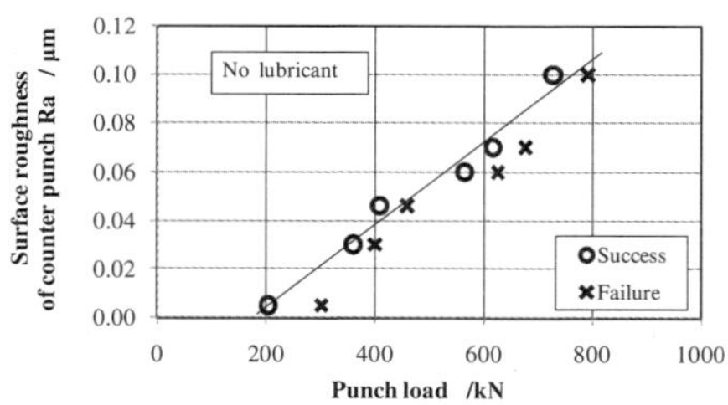

Fig. 5. Relationship between surface roughness of counter punch and limit of forming load.

4.3. *Adhesion on counter punch*

After repeating forming cycles, adhesion at macro scale on counter punch surface can be recognized.

Figure 6 shows adhesion zone on counter punch. It is observed that the center zone of 8mm in diameter has no adhesion (position A), the outside of center zone is black adhesion (position B) and further zone is white circular adhesion (position C). Figure 10 shows appearance of adhesion on counter punch shown zone A, B, C in Figure 9. The photo magnification was 100 times. At the zone A, any change in the surface appearance can be recognized with increase of forming cycle. At the zone B, C, few forming cycles cause a light adhesion in whole surface.

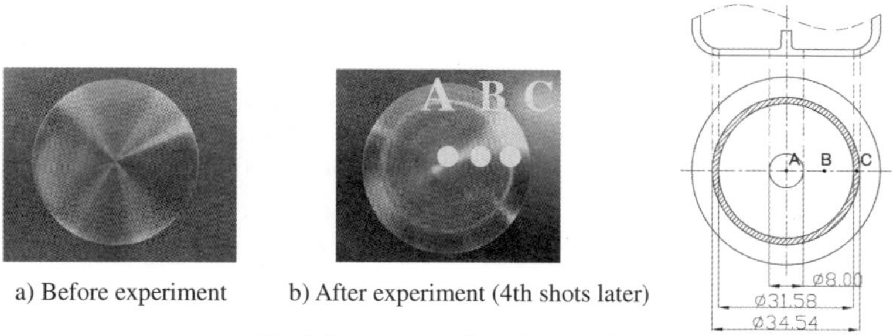

a) Before experiment   b) After experiment (4th shots later)

Fig. 6. Appearance of counter punch.

5. Conclusion

The results indicated the following.

1. With increasing surface roughness of counter punch, the formable zone of solid boss expands. When keeping surface roughness of counter punch constant, the boss height increases linearly with increasing punch load at the formable zone.
2. Adhesion on counter punch surface can be recognized with forming cycles repeating under dry condition, the irregularity in the product configuration converges to a befitting configuration after a few forming cycles.

References

1. K.Morishita and Z.Wang: Proceedings of the 3rd International Conference on Material and Processing, Illinois, USA, 183(2008).

A NEWLY DEVISED FORMING METHOD FOR TRIPLE CUP

ZHIGANG WANG, YASUHARU YOSHIKAWA AND ISAMU KATO[†]

Department of Mechanical Systems Engineering, Gifu University, 1-1 Yanagido Gifu, 501-1193, Japan

In the automobile manufacturing industry, it is demanded lighting the parts for protecting the environment from CO_2 and the decrease of the manufacturing cost and energy. This research aims at developing plastic deformation processes for the triple cup that has two bosses on upper and lower sides in the cup shape. A new sheet forging process is discussed in which a sheet metal is firstly worked to a drawn cup by deep drawing process and then deformed to the target shape by compressing the cup edge. The process feature has been investigated by using FEM simulation and verified by a series of press forming test.

Keywords: Plastic deformation; Sheet metal forging; Triple cup; Cold forging.

1. Introduction

Recently, complicated shape, low cost, improvement of the forming technology to keep environmental consideration becomes the main needs of the automobile industry, and similar directionality is demanded to the technology development of the metal press working industry which is the prime partner. [1] This research aims at developing plastic deformation processes for the triple cup shape such as Fig.1. A new sheet forging process is discussed in which a sheet metal is firstly worked to a drawn cup by deep drawing process and then deformed to the target shape by compressing the cup edge.

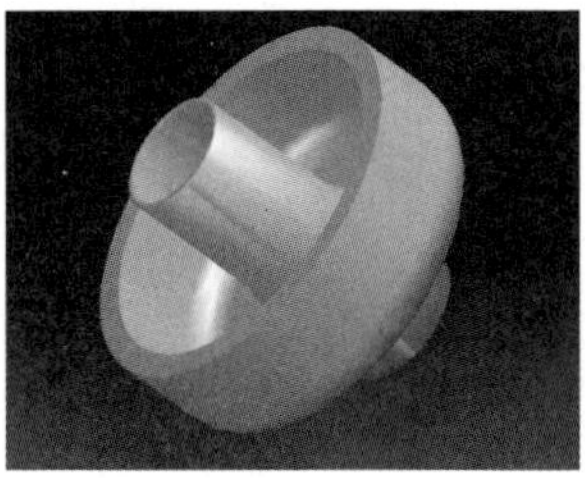

Fig. 1. The triple cup.

† Corresponding author, email: o3122008@edu.gifu-u.ac.jp

2. Principle of Forming Method

2.1. *Principle of Drawing – Edge Compression Forming process*

Fig.2 shows the principle of drawing-edge compression forming. Firstly a blank is drawn to a cup by the deep drawing process. After restriking and punching the central hole, the drawn cup edge is compressed by the ring punch while keeping pressing force by the sleeve punch and then the material is separated to the upper and lower bosses against the mandrel.

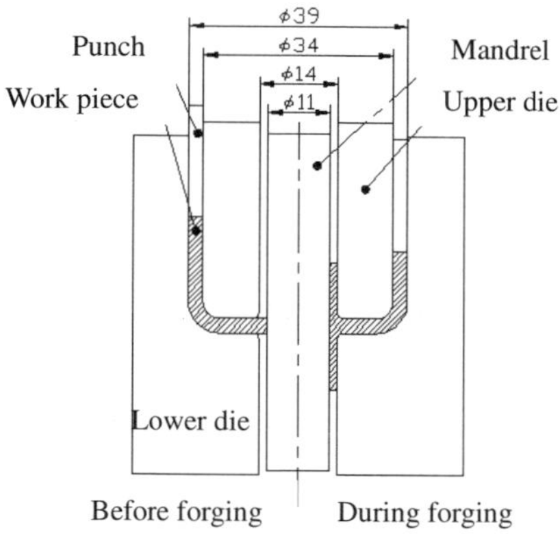

Fig. 2. Forming processes of the triple cup.

Fig.3 shows the triple cup dimensions of the present study. The shape A has added ironing to the above boss. The shape B was assumed that upper and lower bosses wall thickness was the same. The same tools was used for the two shape forming until the punching process, after punching process the forming tools corresponding to each shape was used, respectively.

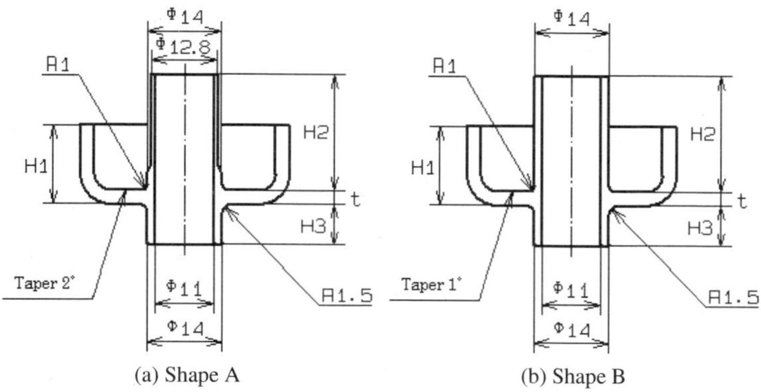

Fig. 3. Dimensions of the triple cup.

2.2. *FEM simulation condition*

The ring size is 6.4mm in thickness, 39mm in the outside diameter, and 11mm in the inside diameter. The analysis condition is shown in Table 1. The commercial FEM code DERORM2D was used. The forging mode is axial symmetry with constant temperature, the tools are assumed as rigid body. In this process, the mandrel and lower die are fixed. The sleeve is compressed by a constant pressure.

Table 1. Analysis conditions.

Work piece	SPCE270	Thickness	2.6 mm
		Tensile strength	270 MPa
Tool	Rigid body		
Friction condition	Frictional coefficient 0.05		
Working speed	10mm/s		
Temperature	20 °C		

3. FEM Simulation Result

Fig.4 shows the dimension variations during compression process. For both shapes, the FEM simulation gives the same trend as the experiment. This means the FEM simulation is valid.

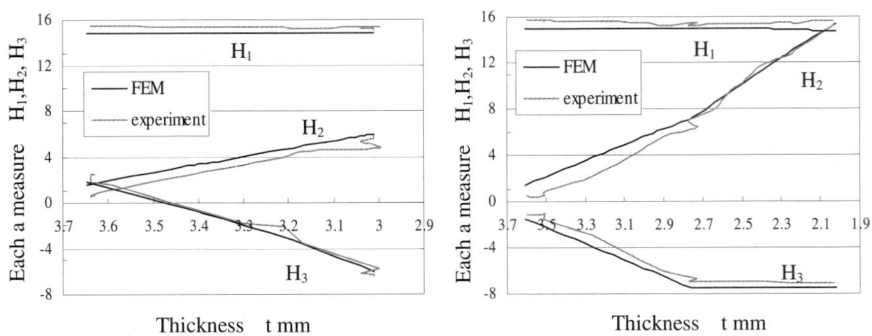

(a) Shape A (b) Shape B

Fig. 4. Variations in triple cup dimensions during compression process.

As shown in Fig.5, choose points on the interface between the billet and mandrel, chase the point position change during the compression process, and then measure the length between the points, so we can find the surface expansion ratio.

Fig.6 shows the variations of the surface expansion ratio in the contact position at the mandrel. The origin is the centre of billet material when the billet material begins to contact with the mandrel, plus means the upper boss direction, minus means the lower boss direction. In the compression process, the shear band causes a large surface expansion. For the shape A, the billet material flows into the upper and lower boss almost uniformly and thus the maximum surface expansion ratio appears at the origin. For the shape B, after the lower boss is filled, all of material flows into the upper boss and thus the maximum surface expansion ratio appears at the upper boss surface. The maximum surface expansion ratio for the shape A is 18.6 and 72.2 for the shape B. The difference due to the shape is very large.

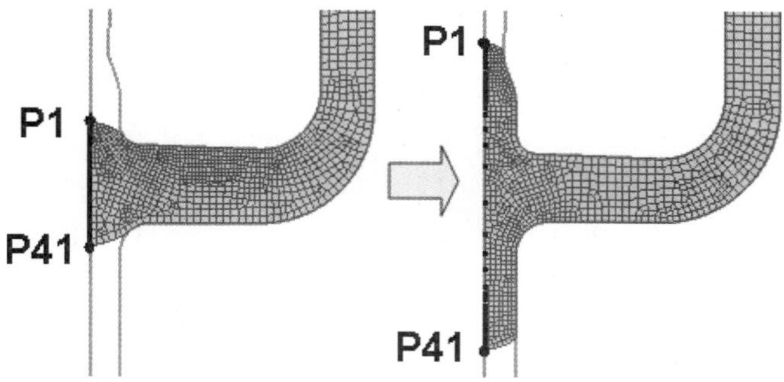

Fig. 5. Point tracking.

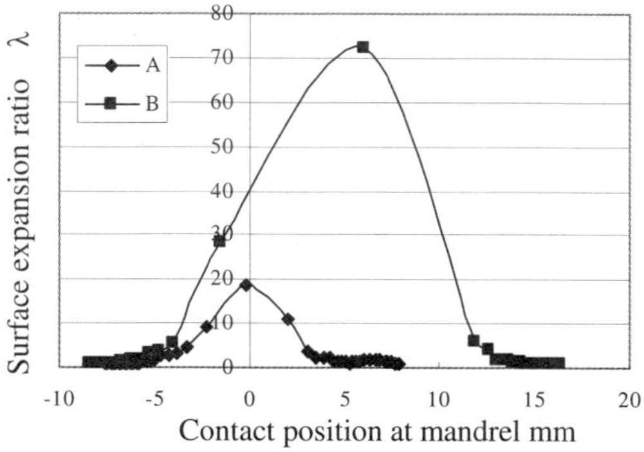

Fig. 6. Variations in surface expansion ratio with contact position at mandrel.

The sliding distance measured after the surface area is expanded to 10 times is defined as S, the maximum surface expansion ratio appeared at this sliding process is defined as λ_{max}, the variations of S and λ_{max} in the compression process are shown in Fig.7. It can be predicted that galling occurs due to the lubrication film breakdown over a distance 1.2mm for the shape A and 5.0mm for the shape B.

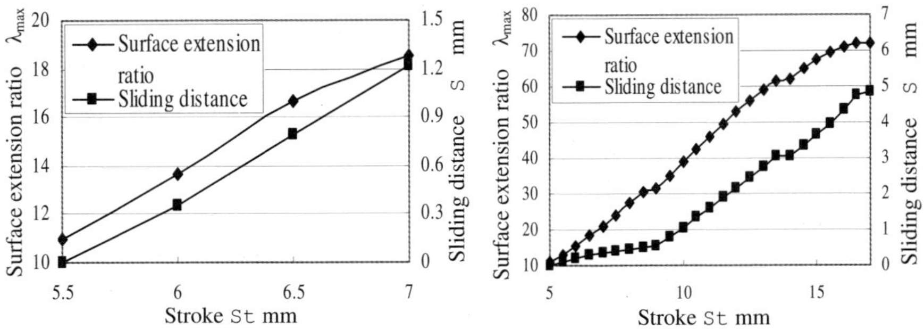

(a) Shape A (b) Shape B

Fig. 7. Variations in surface expansion ratio and sliding distance with stroke.

4. Conclusion

The feature of the triple cup forming process by sheet forging has been investigated by FEM simulation and verified by a series of press test. The results indicated the following.

(1) With increasing the punch stroke, the die pressure increases linearly at the first stage and rapidly after the lower boss filling finish or ironing addition.

(2) The surface expansion ratio for both shapes is large enough to cause galling due to lubrication film breakdown.

References

1. Timothy Schrader, Manas Shigaokar, Taylan Altan: *Materials Processing Technology*, 2007, **189**, 36–44.

THE DEFORMATION SIMULATION OF PLASTIC INJECTION MOLD BASED ON BOUNDARY ELEMENT METHOD

YUN ZHANG, LI DAI, HUAMIN ZHOU[†], WEIYI LI

State Key Laboratory of Materials Processing and Mold and Die Technology, Huazhong University of Science and Technology, 1037 Luoyu Road Wuhan, 430074, P.R. China

The deformation prediction is significant for plastic injection mold design. Whereas the existing deformation simulation base on finite element method (FEM) encounters many constraints on mesh discretizations and boundary condition definitions in actual injection mold production. A deformation simulation approach based on boundary element method (BEM) is proposed to overcome these constraints. In contrast to the FEM, only surface meshes of the mold are required. The cavity pressure and the contact constraints can be automatic defined on the mold boundary in the proposed approach. Therefore it saves a great deal of time for the mold engineers on laborious preprocessing task in deformation simulations.

Keywords: Plastic injection mold; deformation simulation; boundary element method.

1. Introduction

The plastic injection mold, which is usually made of steel, aluminum or other metals, will deform more or less under the filling or post-filling pressure. The deformation of the mold changes the dimensions of the cavity and decreases the precision of the final molded parts. It also increases the thermal resistance and influences the post-filling pressure[1,2]. Therefore, it is important for the mold engineers to improve the rigid of the mold and reduce the mold deformation by ways and means.

Numerical simulation has been proved to be an efficient engineering tool for mold designers to optimize mold design before the mold is built. However, the existing deformation simulations base on FEM softwares[3-5] encounter many constraints in actual injection mold design and production: (a) the FEM needs discretization over the whole interior body of the mold region, which is impossible sometimes because that the mold structures are extremely complicated and vary widely depending on the shape of the molded part and other constrains. (b) The pressure boundary condition on the cavity must be

[†] Corresponding author, email: hmzhou@hust.edu.cn

imported into commercial structure analysis software from filling simulation results, and contact constraints between mold components must be defined on boundaries manually. These imply that a laborious preprocessing task is inevitable in a mold deformation simulation and conflict with the increasingly shorter development period.

The work presented in this paper describes a deformation simulation approach of plastic injection mold based on boundary element method, which aims at overcoming the disadvantage in the FEM method mentioned above.

2. Force Analysis and Modeling

In an injection mold, the plastics melt is first injected into the gaps between the cavities and cores and then solidify under pressure for a specified time. In order to balance the pressure, the cavities and cores are clamped by the top and bottom clamping plate. The force analysis on the cavities and cores are modeled as shown in figure 1, the following conditions are assumed: (1) The deformation of the mold is elastic and small; (2) The thermal expansion of the mold is not taken into account; (3) The cavity pressure is not affected by the possible flash on the parting line; (4) The gravity and frictional force are neglected; (5) the top and bottom clamping plate are rigid bodies. Based on the above assumptions, an isotropic linear-elastic model can be applied to the deformation simulation. The boundary of the mold consists of Γ_{fix}, Γ_{free}, $\Gamma_{pressure}$, $\Gamma_{contact}$. The boundary conditions are defined as following: the displacements are zero on the surface Γ_{fix}, the forces are zero on the surface Γ_{free}, the cavity pressure are force loading on the surface $\Gamma_{pressure}$, and the contact constraints must be satisfied on the contact surface $\Gamma_{contact}$.

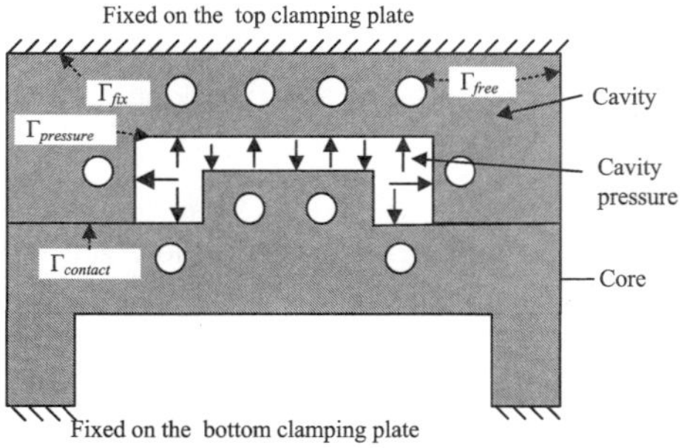

Fig. 1. The force analysis model on the cavities and cores.

3. The BEM-based Deformation Simulation of Injection Mold

The boundary integration equation for three-dimensional linear elastic problem is

$$C_{lk}u_k(\varphi)+\int_{\Gamma}u_k(\xi)\overset{*}{p}_{lk}(\varphi,\xi)d\Gamma \\ =\int_{\Gamma}p_k(\xi)\overset{*}{u}_{lk}(\varphi,\xi)d\Gamma+\int_{\Omega}f_k\overset{*}{u}_{lk}(\varphi,\xi)d\Omega,\ \ \forall\varphi\in\Omega, \tag{1}$$

where φ and ξ are source and field point, C_{lk} is generally a function of the geometry variation at the boundary point φ, respectively, $\overset{*}{p}_{lk}$ is the traction kernel function, and $\overset{*}{u}_{lk}$ is the displacement kernel function. It is worth to note that the last term of Eq.(1) can be discarded if the volume force is neglected. It means that the displacement of an arbitrary point in the whole mold body or on the mold surface can be calculated by integrations along the mold surface. Therefore only surface meshes of the mold are required and the spatial dimension of the computational domain can be reduced by one. Supposed that the boundary is discretized into NEL elements with each element has nod nodes, the integrals over the boundary can be replaced by the sum of the integrals over the surface of each element

$$C_{lk}u_k(\varphi)+\sum_{m=1}^{NEL}\int_{\Gamma_m}u_k(\xi)\overset{*}{p}_{lk}(\varphi,\xi)d\Gamma=\sum_{m=1}^{NEL}\int_{\Gamma_m}p_k(\xi)\overset{*}{u}_{lk}(\varphi,\xi)d\Gamma,\quad\forall\varphi\in\Omega. \tag{2}$$

The unknown functions u_k and p_k are approximated using shape functions, calculated by

$$u_k=\sum_{n=1}^{nod}\psi_n u_k^n,\ \ p_k=\sum_{n=1}^{nod}\psi_n p_k^n, \tag{3}$$

where u_k^n , p_k^n denote the displacement and traction on the n-th node of the element, ψ_n is the interpolation function. Substitute Eq.(3) into Eq.(2), and use the Gauss integration method, we can obtain a linear equation system

$$C_{lk}u_k^i+\sum_{m=1}^{NEL}\sum_{n=1}^{nod}h_{lk}u_k^n=\sum_{m=1}^{NEL}\sum_{n=1}^{nod}g_{lk}p_k^n \tag{4}$$

where h_{lk} and g_{lk} are both constants related to the positions of the source node point and the field element, calculated by

$$h_{lk}=\sum_{\alpha=1}^{m_1}\sum_{\beta=1}^{m_2}\left(\overset{*}{p}_{lk}\psi_n\left|J_{\xi\eta}\right|\right)_{\substack{\xi=\xi_\alpha\\ \eta=\eta_\beta}}\omega_\alpha\omega_\beta,\ \ g_{lk}=\sum_{\alpha=1}^{m_1}\sum_{\beta=1}^{m_2}\left(\overset{*}{u}_{lk}\psi_n\left|J_{\xi\eta}\right|\right)_{\substack{\xi=\xi_\alpha\\ \eta=\eta_\beta}}\omega_\alpha\omega_\beta, \tag{5}$$

where m_1, m_2 are Gauss integration points along direction ζ and η, respectively, $\omega_\alpha,\omega_\beta$ are corresponding weighting coefficients. Eq.(4) can be solved by either Gauss elimination method or block Jacobi preconditioned Krylov subspace iterative methods, such as GMRES, Bi-CGStab and so on.

4. Numerical Simulation Results and Discussion

A cantilever beam has been first analyzed by both FEM and BEM as a comparison study. The configuration of the cantilever beam is shown in figure 2. The FEM program used for analysis is ANSYS. The accuracy of a result is defined as the vertical displacement at the right end of the cantilever beam compared to the theoretical solution, which is obtained from ANSYS analysis by increasing the element number until the displacements of all nodes is convergence. The BEM program is written in C++ language several element types are used to model the cantilever beam.

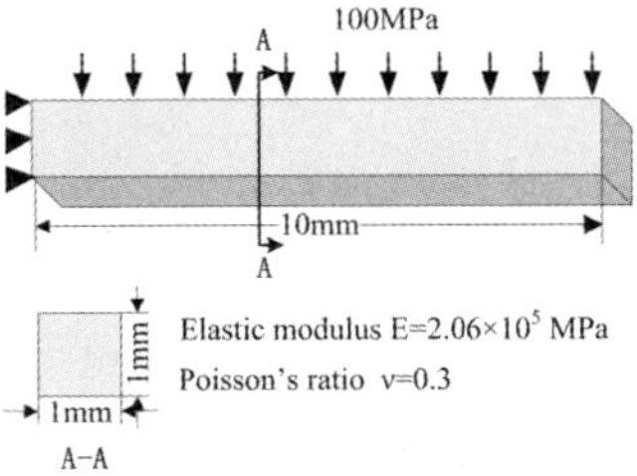

Fig. 2. Configuration of the cantilever beam.

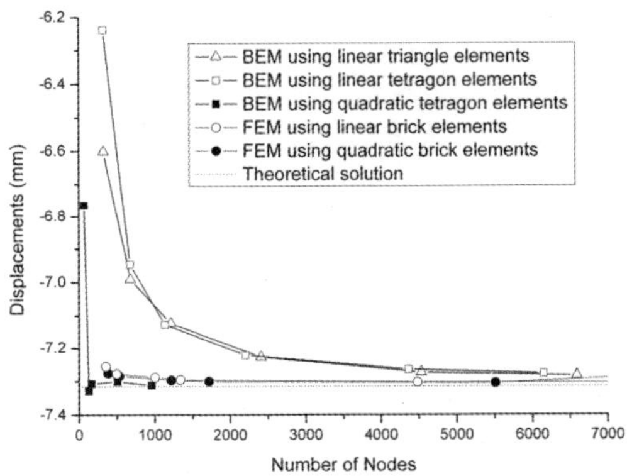

Fig. 3. Comparison of the simulation results of the cantilever beam.

The simulation results are compared in figure 3. It can be seen that the element type has an important impact on the result accuracy in simulations based on BEM. The accuracy is poor when either linear triangle or tetragon elements are used to model the mold, and is greatly improved while using quadratic tetragon elements. The accuracies of FEM simulations are not sensitivity as BEM to element type, but convergence slowly when approximating theoretical solution.

A test case is a mold of a box, as shown in figure 4. The cavity pressure distribution is simulated by the plastic injection simulation software HSCAE. The mold is modeled using quadratic tetragon elements and the pressure values are interpolated into mold meshes as the cavity boundary conditions. The displacements along Z direction of the mold are shown in figure 4(d). It can be seen that the bottom of the moving haft is the point with maximum displacement along Z direction, and it is about 0.00718 mm. Experiments to measure the displacements of the mold are in process and will be reported later.

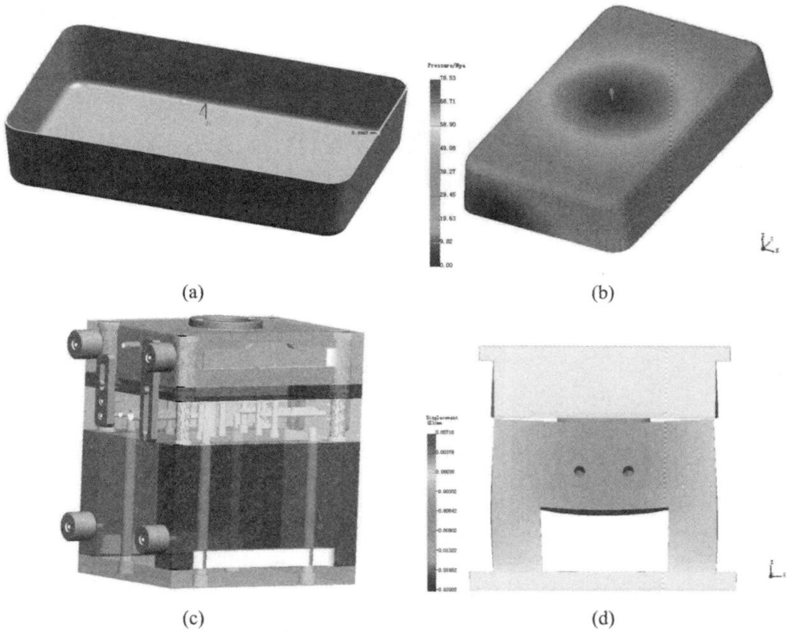

Fig. 4. The test mold of a box: (a) product, (b) the cavity pressure distribution, (c) mold, and (d) the cavity pressure distribution.

Acknowledgments

The authors would like to acknowledge financial support from the National Nature Science Foundation Council of China (No. 50875095) and the "863" Program of China (No. 2009AA03Z104).

References

1. R. Pantani, V. Speranza and G. Titomanlio: Polym. Eng. Sci., 2001. **41**, (11), 2022-2035.
2. D. Delaunay, P. Le Bot, R. Fulchiron, J. Luye and G. Regnier: Polym. Eng. Sci., 2000. **40**, (7), 1692-1700.
3. X. Li, G. Zhao, Y. Guan and H. Li: Int. J. Adv. Manuf. Tech., 2009. **45**, (3), 261-275.
4. J. Jia, Z. Ge and Y. Li: Advanced Materials Research, 2008. **44**, 85-89.
5. Z. Chen, F. Huang: Machinery Design & Manufacture, 2009. (3), 246-248.

ELASTO-PLASTIC DEFORMATION BEHAVIOR OF CEMENTED CARBIDE DIE IN COLD FORGING

ZHIGANG WANG, HAOJIE LI AND SHINYA NAKANO[+]

Department of Mechanical Systems Engineering, Gifu University, 1-1 Yanagido Gifu, 501-1193, Japan

A measurement system for predicting the stress state at the dangerous points of the die insert in the cold forging process has been proposed. In this measurement system, the hoop stress at the die approach can be conducted by measuring the strain at the die case. A FEM simulation showed that the relationship between the hoop stress at the dangerous point of the die insert and the strain measured at the die case is monotonous and sensitive. In this system, the hoop stress which is the major factor to control the fatigue behavior and die life is controlled by changing the press-fit interference. The die dimensions have been optimized by using FEM simulation. A series of forging experiments using the optimized die structure shows that the measured strain is the same as predicted by FEM simulation and the measurement system is useful.

Keywords: Plastic deformation; Cold forging; Tool life; Fatigue; Cemented carbide.

1. Introduction

Until now, many researches[(1)~(8)] have been conducted on the cemented carbide which is the most utilized die material for the cold forging. But it is still difficult to apply these study results to the actual forging die because the stress state of the die in the service is not clear and the failure process is unknown. In order to realize die life improvement, it is important to make breakthroughs in the fracture mechanism based on the failure process of die in the actual forging process [(9)]. In the previous paper [(10)], the variations of stress state of extrusion-upsetting forging die in the forging process had been investigated with FEM simulation by changing the extrusion die shape, extrusion ratio and forged product shape. It has been clarified that the hoop stress affects the fatigue behavior of the die greatly and the fatigue crack always concentrates around the inlet and outlet of the extrusion approach of die insert.

In this paper, we will conduct FEM simulation adapted to the actual forging process and discuss the optimum die structure to measure the stress state of die insert.

2. FEM Simulation

Fig.1 shows the die model used in this study. FEM simulation has been carried out under the conditions adapted to the actual forging, in order to check the relationship between the circumferential stress generated at the inlet (Point A) and outlet (Point B) and the circumferential strain at the measuring point Z in the actual forging experiment. As shown in Fig.1, the measuring point Z located at the die extrusion approach and 6mm away from the case outside.

In the present paper, a commercial code DEFORM PC-Pro was used for the FEM simulation. The deformation mode is axisymmetric and isothermal.

In the FEM simulation, the insert is firstly pressed into the sleeve and then pressed into the case and thus a compression hoop stress is generated in the insert. Two stages of forging process are simulated. First stage is a forward extrusion followed by a preliminary upsetting. The second stage is the final heading and the die insert failure at this stage is investigated. The shape and dimensions of the die and billet in each simulation are the same as that for the actual forging experiment.

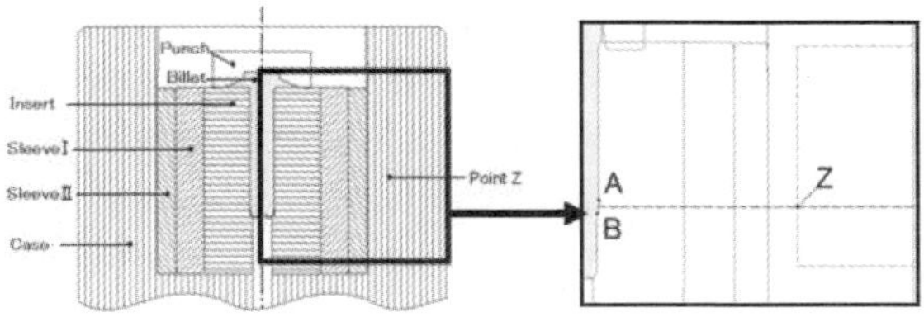

Fig. 1. The die model.

2.1. *Press-fit process simulation*

In the present paper, the interference of 0.43% was used to get a relatively early die failure. The friction coefficient was set to 0.05 in the press-fit. The simulation results show that a pre-compression stress of 600-650MPa in the hoop direction at the extrusion approach can be generated.

2.2. *The simulation of the first stage of forging*

The die assembled by press-fit was used in this simulation. The material of a billet was SWCH12A which is commonly used in cold forging.

Fig. 2 shows the change of hoop stress at point A, point B and the circumferential strain at the measured point Z. The hoop stresses at points both A and B increase gradually but the pre-compression effect was still remained until the end of the process. As for the strain at the point Z, no big change is observed and 15μs appeared at the end.

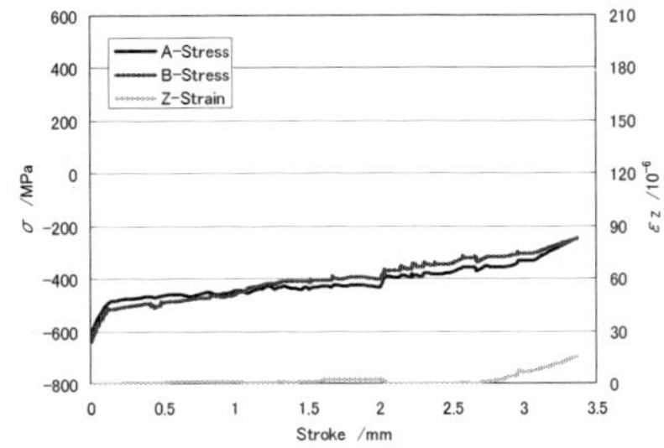

Fig. 2. Variations of hoop stress at points A and B and hoop strain at the measured point Z.

2.3. *The simulation of final stage forging*

In this stage, the billet is headed with keeping its extruded portion shape that was created at a previous stage. The die data at the end of press-fit process simulation was used and the data of the billet at the end of first stage was used.

Figure 3(A) shows the variations of hoop stress at the points A and B and hoop strain at the measured point Z. At the end of the process, about 500MPa tensile stress at the points A and B is generated and 150μs strain is developed at the point Z at the same time. When the material of billet gets contact to the punch shoulder, the hoop stress becomes to increase rapidly. It can be confirmed that the hoop stress in the die is closely related to the processing of the bolt head.

Figure 3(B) shows the relationship between the hoop stress at the point A and point B and hoop strain at the point Z. It can be recognized that the hoop strain at the point Z increases smoothly with an increase of the hoop stress at the point A and point B.

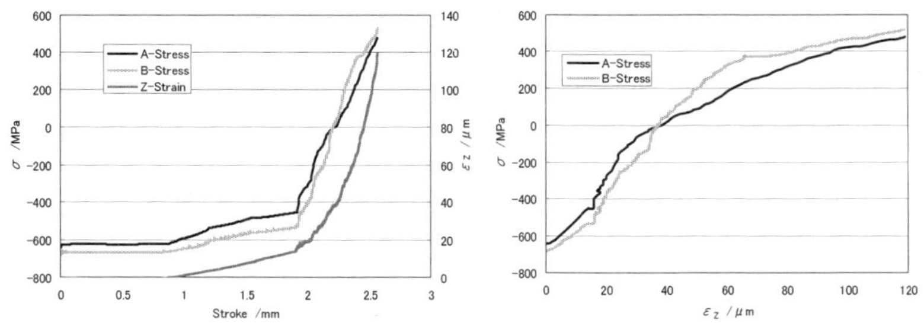

(A) (B)

Fig. 3. Variations and relationship of hoop stress at points A and B and hoop strain at the measured point Z.

3. Die Structure for Measuring Insert Stress during Forging

Figure 4 shows the simulation model which is used to investigate the die structure. For predicting the stress in the die exactly in the experiment, the strain at the measured point Z should change sensitively with the stress at the point A and B. In order to make the simulation simple, the insert and sleeve I, sleeve II and case are merged into one body respectively. According to the above mentioned results, the hoop stress at the points A and B increases rapidly and the measured strain at the point Z rises after the billet contacts to the punch shoulder. Here, the period that the billet becomes to contact to the punch shoulder until the forging finish is noticed because the die is dangerous in the period. The die structure is optimized to get a sensitive relation between the measured strain and the hoop stress at the dangerous point at the die insert surface.

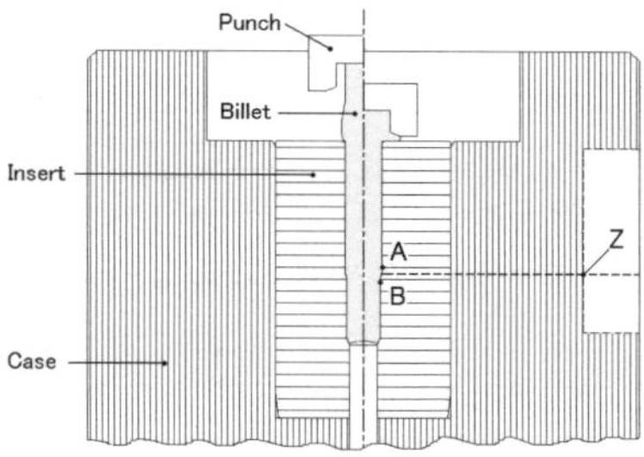

Fig. 4. The die model shape.

3.1. *Effect of press-fit interference*

The FEM analysis is performed by changing the press-fit interference to 0.01, 0.02, and 0.03 mm while keeping the insert diameter 10mm. Figures 5 show the relationship between the strain at the point Z and the stress at the point A and B, respectively. The results after the punch shoulder contacting the billet was plotted for every simulation step. It can be recognized that the hoop stress at the dangerous point at the insert surface could be controlled by changing the press-fit and could be measured by measuring the strain at the case surface.

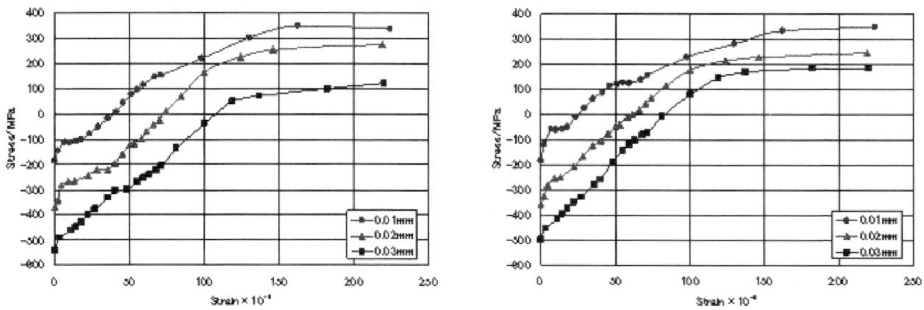

(A) (B)

Fig. 5. Relationship between the strain at point Z and the stress at point A and B.

4. Forging Experiment

A series of forging test was carried out using a micro former. The forging conditions and die structure were the same as the FEM simulation. An example of strain measurement results is shown in Figure 6. A little peak can be found at the position of the first forging stage and a large peak can be recognized at the position of the final forging stage. The amount of strain is also the same level just as predicted by FEM simulation. This means the measurement system is useful for investigate the stress state of die during the cold forging process.

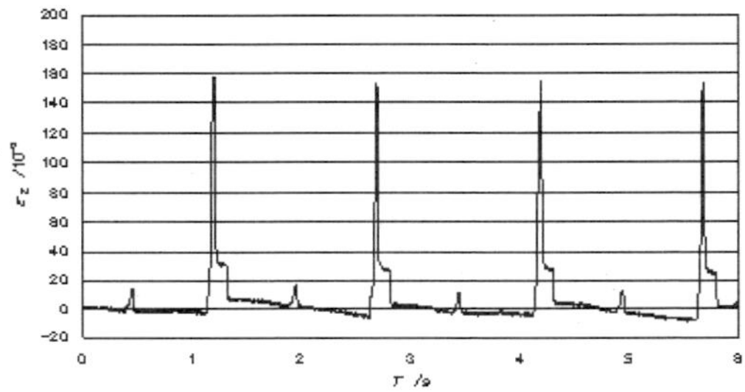

Fig. 6. Variations of the measured strain during forging process.

References

1. Otsuka, Tohgo, Sugawara, Ueda: *Material,*36-401,135(1987).
2. Otsuka, Tohgo, Sugawara, Ueda: *Material*, 36-404,462(1987).
3. Otsuka, Tohgo, Sugawara, Houjoh: *Material,* 36-411, 1361(1987).
4. Kin, Yoshikawa, Togura: *Journal of JSTP*, 343,1147(1989).
5. Satou, Honda: *Journal of JSME (Seris A),* 56-526, 1378(1990).
6. V. Wood: *Future developments of metals and ceramics (edited by. J. A. Charles)*, Vol.1, 235–239(1992).
7. M. H. Loretto: *Proc.of 2nd Int. Conf. on Research and development in net shape manufacturing*, Birmingham, UK (1999).
8. R.D.Niel: *Interfacial structures in intermetallic/steel joints after high temperature service*, Huntingdon, UK(2000).
9. Z.Wang, I.Nakashima, M. Ohhashi, T.Mastumoto: Proc.of 8th ICTP, Verona, Italy (2005).

Part III

Rock and Soil Mechanics

MODEL TEST AND NUMERICAL SIMULATION FOR DYNAMIC CHARACTERISTICS OF ROCK SLOPE UNDER BLASTING LOADS

ZHONG DONG-WANG †

College of Science, Wuhan University of Science and Technology, Wuhan, 430081, China

WU LIANG

Hubei Province Key Laboratory of Systems Science in Metallurgical Process, Wuhan, 430081, China

To study on dynamic characteristics of rock slope under blasting loads, both an indoor experiment method of concrete slope model and finite element model of that are adopted in this paper. The results show that not only propagation laws but also attenuation characteristics of the detonating earthquake waves on slope are different to those of level ground. That is to say, the attenuation of particle peak vibration on slope is much smaller along with blasting source distances than those of level ground, and there are vibration amplification effects at the top particles of the slope; On the other hand, not only can the damping ditch reduces the strength of peak particle vibration greatly, but also improves stress states of the slope under blasting loads. In addition, by comparing experiment results, the results of simulation are consistent with those of model test of the slope under blasting loads well.

Keywords: Slope engineering; Blasting vibration; Dynamic characteristics; Model test; Numerical simulation.

1. Introduction

It is well-known, during the demolition excavation, influences to the side slope mainly display in 2 aspects[1]: First, the vibration force of inertia increases side slope Sliding Possibilities; Second, the unceasing demolition vibration action increases shearing stresses of the rock side slope, which give rise to primary structural planes, structure structural planes, and original cracks to expand, even produce new demolition cracks and micro cracks. So its original physical mechanics performances are decreased, and affecting the side slope overall stabilities. At present, under the explosive loads, the scene monitor and

† Corresponding author, email: 498519502@qq.com

numerical simulation are the rock side slope's response characteristic research methods mainly[2]. By taking both the observation and the analysis to creep slope's demolition vibration, the weakening empirical formula of demolition earthquake waves has been obtained[3]. By time interval methods, the side slope dynamical stabilities have been analyzed under the explosive loads. In the analysis of both the sliding surface dynamic coupling processes and the detonation waves in the slippery bodies, the slippery bed's dissemination rules have been considered[4]. Under the vibration loads, rock high side slope dynamic responses have been carried on computations. According to the ultimate tensile stress criterion, both the reasonable spots and the safety thresholds of slope demolition vibration control have been reported[5].

By means of both the indoor concrete side slope model experiment method and the finite element method, under the impulsive loads, the analytical study has been conducted for the side slope dynamic performances.

2. Side Slope Model Tests

2.1. *Geometric Models and Material Parameters*

There are two non-damping ditch side slope models and two-damping ditch side slope models in the same sizes respectively. See figure 1

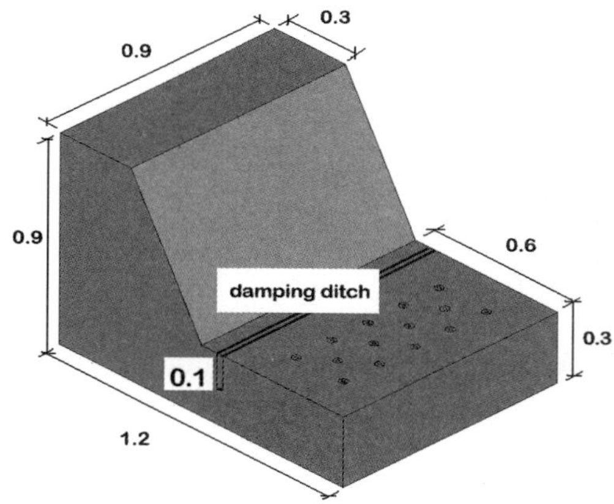

units： M

Fig. 1. Damping ditch side slope model.

2.2. *Experimental Plans*

By means of the single–hole demolition, the blasting explosive was of the 8# paper case instantaneously industrial fuse, the dose 0.6g.

2.3. *Test Results and Analysis*

The demolition tests have been carried on several dozen groups. Because of affecting factors (the model sizes, blasting cap's biases, jamming differences, partial watery artillery holes as well as concrete non-uniformity), the test data have showed discreteness to some degree, but the basic rules have existed. Sadaovsk empirical formula has been used to carry on the regression analysis separately to two groups of typical test data. According to the 6# artillery–hole test results, the following Eq. (1) has been attained,

$$\begin{aligned} V_{Horizontol} &= 3.21(\sqrt[3]{Q}/R)^{0.42} \\ V_{Vertical} &= 11.57(\sqrt[3]{Q}/R)^{0.76} \end{aligned} \tag{1}$$

The results show that both propagation laws and attenuation characteristics of the detonating earthquake waves on the slope are different to those of the level ground. That is to say, the attenuation of particle peak vibrations on the slope is smaller along with blasting source distances than that of the level ground relatively. At the same time the vertical and the horizontal particle peak attenuation index is 0.76 and 0.42 respectively. Compared measuring points on the top of the slope with those on the slope surface, the vertical and the horizontal particle peak values of the former are all greater than those of the latter. Therefore, there are vibration amplification effects at the top particles of the slope obviously.

In order to reduce the harm which the demolition vibration amplification effects bring, the damping ditch is used in this experiment plans.

3. Numerical Calculations

3.1. *Material Properties*

Select LS–DYNA (Mat–Plastic–Kinemetic) to simulate the rock medium. Because many actual materials are the combination of the follow–up strengthen and the isotropic strengthen, in the numerical simulation, hardening coefficient β has been taken as 0.5 to display the hardened characteristics of this real material.

3.2. *Explosive Equation of States*

In numerical simulation of detonation field, because the detonation wave's fluctuating scopes are very large, it is very difficult to find an equation of state to suit all pressure scopes, the equation JWL is used.

3.3. *Computation Models*

The solid164 has been used with non–damping groove model, 68024 elements and 74910 nodes; damping groove model, 62364 elements and 69619 nodes.

As analysing the rock side slope demolition stabilities, it is need to consider not only the demolition vibration frequencies and the side slope dynamic performances but also the demolition earthquake wave distributions and the propagation rules. After the trial, the Rayleigh constants are taken as $\alpha = 4$ and $\beta = 0.002$ respectively.

3.4. *Results and Analysis*

From the result magnitude, the simulation and test particle vibration velocities are in the identical magnitude, at the same time the simulation and the test data regression equation also do. Moreover, particle vibration velocities on top of slope also increase along with the blasting source distances. Therefore, obviously the numerical results can be consistent with demolition vibration rules well. By regression analysis the 6# artillery–hole simulated results have been obtained in Eq. (2)

$$\begin{aligned} V_{vertical} &= 13.97\,(\sqrt[3]{Q}/R)^{0.96} \\ V_{horizontal} &= 15.71\,(\sqrt[3]{Q}/R)^{0.99} \end{aligned} \tag{2}$$

And the typical particle velocity time interval curves are of Figure 2.

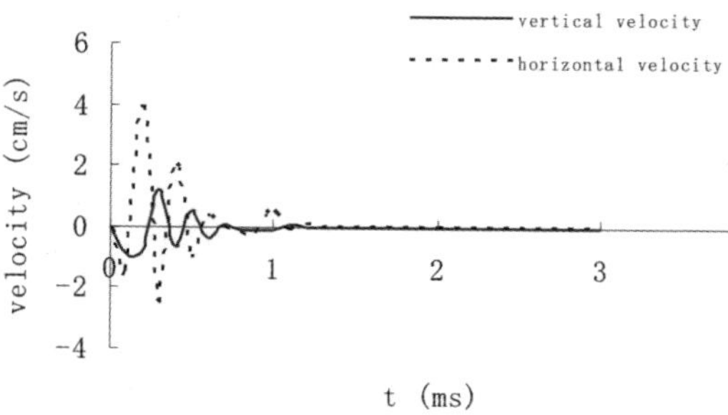

Fig. 2. Blasting vibration velocity of 3# measuring point for 14#hole demolition (damping ditch).

4. Conclusion

LS-DYNA (Mat-Plastic-Kinemetic) of FEM has been taken to simulate the experimental model in this paper, so that some conclusions are confirmed as follows:

- Experimental results state clearly that both propagation laws and attenuation characteristics of the detonating earthquake waves on slope are different to those of level ground. Such as: the attenuation of particle peak vibration on slope is much smaller along with blasting source distances than that of level ground, and there are vibration amplification effects at the top particles of the slope; On the other hand, not only can the damping ditch reduces the strength of peak particle vibration greatly, but also improves stress states of the slope under blasting loads; compared with particle peak vibration velocities of 3#measuring point in the windward surface and 4#measuring point in the back surface, the vertical direction's peak velocity has been decreased 72.7%,at the same time the horizontal direction's one has been decreased 67.6%.

- Numerical simulation results imply that regression equation index and vibration velocity on FEM are the identical magnitude. In addition, the particle vibration velocities on the top of slope appear increasing along with the explosive source distances. In brief, the numerical simulation results are consistent with the blasting vibration regulations perfectly.
- Both experimental and numerical analysis report that the damping ditch can effectively improve the stress states of rock high slope under the detonate loads.

References

1. XU Hongtao, LU Wenbo and ZHOU Chuangbing: Time History Analysis Method for Evaluating Dynamic Stability of High Rock Slope under Excavation Blasting[J].Chinese Journal of Rock Mechanics and Engineering,2006, 25(11): 2213~2219. (in Chinese)
2. Liu Yaqun, Li Haibo and Li Junru: UDEC Simulation on Dynamic Response of Rock Slope of Huangmailing Phosphorite Mine under Explosion [J]. Chinese Journal of Rock Mechanics and Engineering, 2004, 23(21): 3659~3663. (in Chinese)
3. Xie Quanmin, Xia Yuanyou and Li Xinping: Study on Blasting Vibration Control of Creep Mass Slope of Longtan Hydropower Station [J]. Chinese Journal of Rock Mechanics and Engineering, 2003, 22(11): 1929~1932. (in Chinese)
4. HU Yi-hua, LI Ning and HAN Xin: Dynamic Stability Analysis of Accumulation Body Sslope of Xiaowan Hydropower Station under Blasting Condition [J]. Chinese Journal of Rock Mechanics and Engineering, 2005, 24(s2): 5932~5937. (in Chinese)
5. CHEN Ming, LU Wenbo and WU Liang: Safety Threshold of Blasting Vibration Velocity to High Rock Slope of Xiaowan Hydropower Station [J]. Chinese Journal of Rock Mechanics and Engineering, 2007, 26(1): 51~56. (in Chinese)

DEFORMATION OF SOFT SOIL ROADBED ON-SITE MONITORING AND FLAC SIMULATION

LI QINYUAN*

Department of Civil Engineering, Engineering College of Armed Police Force Xi'an 710086, China

REN JIANXI, LIUHUI, WANGYANG, SUNYANG

School of Civil Engineering and Architecture, Xi'an University of Science and Technology, Xi'an 710054, China

Under the engineering background of experimental section K28+830 treatmented by cement injection mixing pile method of express highway circling city in Hang Zhou, based on the method combining in-suit monitoring with FLAC numerical simulation, creep-consolidation deformation laws of soft soil are researched. In-suit monitoring scheme of roadbed deformation is given. Moreover, FLAC2D simulating calculation of soft soil roadbed. The results show that: settlement and horizontal displacement of soft soil roadbed are reduced obviously by cement injection mixing pile. The stability of soft soil roadbed have been strengthened. Therefore, cement injection mixing pile is suitable for consolidating saturated soft clay roadbed, especially for the section in which relatively hard layer is served as bearing stratum in bridgehead. FLAC2D calculation results are consistent with in-suit monitoring results, which can reflect development trend of roadbed deformation. Constitutive model of viscoplastic creep can be used to predict soft soil settlement. The model has some reference value to control soft soil settlement.

Keywords: Soft soil roadbed; cement injection mixing pile; deformation laws; monitoring; FLAC numerical simulation.

Foundation Settlement Prediction and calculation method of great value. Foundation Settlement Prediction of properties and important results have been achieved [1-5]. However, the complexity of the problem, soft soil subgrade deformation mechanism needs further investigation. In this paper, Hangzhou Ring Expressway North Line of soft soil on-site monitoring and FLAC simulation method, on the soft subgrade deformation study. Purpose is to provide a reliable pavement design Freeway basis.

* Liqingyuan, email:liqingyuan1976@163.com

1. Soft Soil Creep - Consolidation Model

Biot consolidation theory to calculate the main saturated elastic soil consolidation deformation, but can not consider the impact of secondary consolidation. Viscoelastic - viscoplastic rheological model and the non-linear viscoelastic rheological model can calculate the secondary consolidation of soil. But the model does not take into account the permeability of soil and pore water pressure dissipation. Coupled flow model and the consolidation model. This would calculate soil consolidation and secondary consolidation, but also consider the process of soil deformation dissipation of pore water pressure.

Biot consolidation theory will be extended to flow in soil, can be derived based on Biot consolidation theory of creep - consolidation equation:

Balance equation $$d\left(d_{ijkl}^{v} d\varepsilon_{ij}\right)_{,j} + u_{,i} = f_i \qquad (1)$$

Geometric equation $$d\varepsilon_{ij} = -\frac{1}{2}\left(dw_{i,j} + dw_{j,i}\right) \qquad (2)$$

Continuity equation $$\dot{w}_{i,i} - \frac{k}{\gamma_w} u_{,ii} = 0 \qquad (3)$$

d_{ijkl}^{v} - Rheological model for the stress-strain-time of the coefficient matrix.

2. Deformation of Soft Soil Monitoring Programs

2.1. *Project overview and monitoring points layout*

Test section of K28 +830 section is the bridge foundation. The sections are 1.9~2.0m below the surface of the overlying hard layers. The first layer of soft soil thickness 5.6~6.3m. The second layer of soft clay roof depth of 12.8~13.7m.

Ground by dry jet cement mixing pile in, pile diameter 500mm, pile length 11m, distance 1.3m, lime and fly ash than 15%.

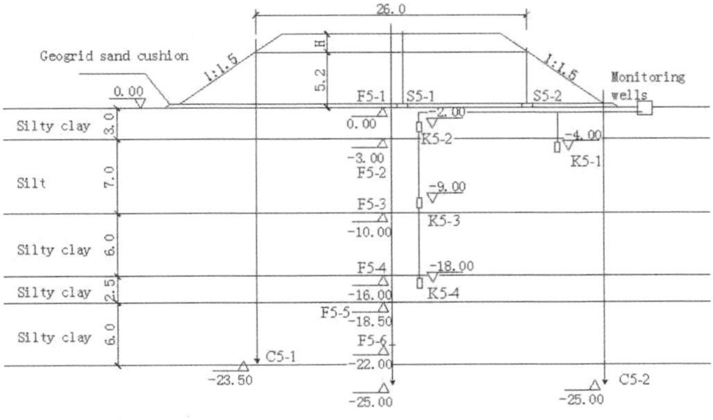

Fig. 1. Schematic diagram of K28+830 geologic section and monitoring instrument embedment.

Instrument conditions and K28 +830 section layout shown in Figure 1. C5-1 and C5-2, said inclinometer tube; S5-1 and S5-2, said the settlement plate; F5-1~F5-6 expressed sequence settlement ring; K5-1~K5-4 that the pore water stress gauges. As the embankment load is symmetrical, so telescopes basically laid in the embankment at the side.

2.2. *Monitoring results*

1. Surface subsidence: To unloading, the center of the cumulative maximum settlement of approximately 13.3cm, road shoulder about 12.4cm. Filling of which were 9.0cm and 10.0cm, preloading period were 4.2cm and 2.4cm. After unloading rebound 0.5cm, settlement stabilized.
2. Horizontal displacement: Road shoulder the maximum horizontal displacement of 3.7cm, the slope foot is 3.1cm. Maximum filling level of runway shoulder displacement of 2.79cm, the slope foot is 2.27cm. Preloading maximum horizontal displacement of Road Shoulder 0.91cm, the slope foot is 0.86cm. Maximum horizontal displacement of the initial filling up faster, then slower growth rate, and finally stabilized.

3. FLAC Simulation of Subgrade Deformation

3.1. *Computational model and computation parameters*

Soft soil with FLAC-2D of the WIPP viscoplastic creep model. Embankment by Mohr-Coulomb model. The following section of embankment subgrade using creep model. Powder spray cement mixing unit used to simulate the pile FLAC2D5.0. Powder spray surrounding soil cement mixing pile strength is increased by factor to consider pile on the role of soft soil foundation.

Foundation depth to take 30m, taking the scope of cross-sectional width of 120m. Subgrade level of restraint on both sides of the boundary conditions, under the bottom boundary condition for the horizontal and vertical constraints. The foundation soil and embankment, the finite difference mesh generation. Model consists of 8856 units, 9078 nodes

Table 1. Viscoplastic model of every soil layer in roadbed and other parameters.

Parameters	K (MPa)	k_ϕ (Pa)	q_ψ (Pa)	q_ϕ	G (MPa)	T (℃)	σ_t (kPa)
Soil ①	152	26891	0	1.261	0.57	16	15.0
Soil ②	190	28677	0	0.847	0.49	16	13.0
Soil ③	100	22440	0	0.693	0.25	16	18.7
Soil ④	120	20775	0	0.519	0.30	16	17.0
Soil ⑤	111	24320	0	0.588	0.24	16	20.0

According to the site of the original soil for VST and CPT results. K28 +830 section by the dry jet cement mixing pile after the deal with depth of (11m)

soft soil strength is 1.94 times the original. Thus the foundation soil parameters increased by 1.9 times to consider powder spray cement mixing pile reinforcement effect. WIPP model calculated parameters in Table 1.

3.2. *FLAC numerical simulation results with experimental results of comparative analysis*

To facilitate data analysis. In the Road middle, Road shoulder and slope foot to set calculation points. Specific location see Figure 2.

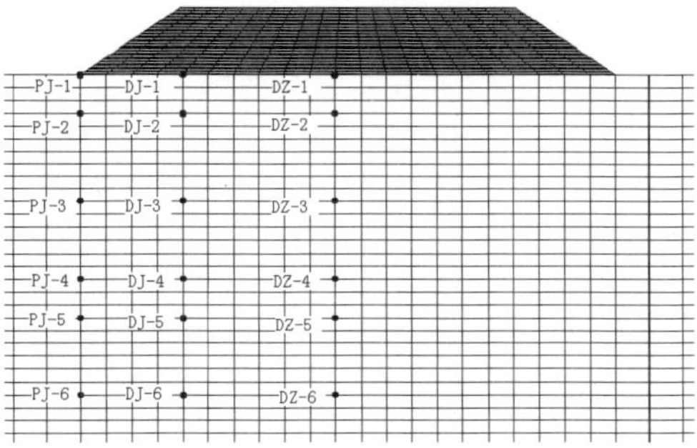

Fig. 2. Numbering graph of observation point position in K28+830 section.

Road middle foundation settlement and calculation of total points DZ-1 corresponds to the vertical displacement. Road shoulder foundation settlement and calculation of total point DJ-1 corresponds to the vertical displacement. Road shoulder point of maximum horizontal displacement and the calculation of the level of DJ-5 corresponds to the displacement. Slope foot horizontal displacement and PJ-5 corresponds to the horizontal displacement. FLAC results data and measured values of the contrast curve shown in Figure 3.

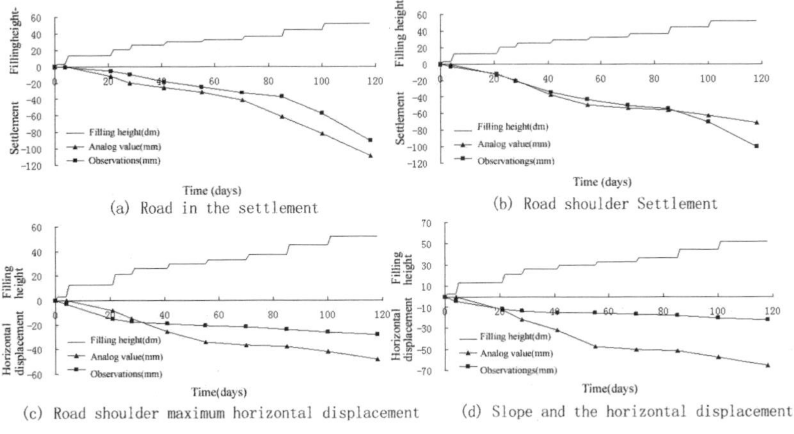

Fig. 3. Correlation curve of computation value and monitoring value.

Figure 3 known. Vertical settlement monitoring results and actual results, trend and value in very close. Horizontal displacement of the simulation results and the actual monitoring results relative differences are large. Simulated the horizontal displacement increased rapidly with the filling process. Starting from the third floor embankment, horizontal displacement increased rapidly.

4. Conclusion

In this paper, on-site monitoring and FLAC simulation method combining the powder spray cement mixing pile composite foundation of highway embankment caused by soft soil deformation. The main conclusions are:

1. Soft soil characteristics have obvious limitations. The deformation process of soil drainage is the process of consolidation and creep simultaneously. Analysis of deformation of soft soil should be considered when creep - Consolidation.
2. Powder spray after cement mixing pile significantly reduced soft soil settlement. Especially the horizontal displacement caused by loading was significantly reduced. Enhance the stability of the foundation.
3. FLAC results consistent with the results of field monitoring. That WIPP viscoplastic creep model is applicable to the deposition of soft soil. Viscoplastic constitutive model can be used to Foundation Settlement Prediction.

References

1. Yin zongze, Zhu Hong. Huning Expressway finite element analysis of foundation settlement.*Water Purification Technology*,1998,18 (2):22~26.
2. WU Da-zhi, LI Xi-bing. Expressway Settlement calculation. *Communication Science and Technology*, 2001, 27 (4): 4 ~ 6.
3. Duncan J.M., Limitations of Conventional Analysis of Consolidation Settlement, J. *Geotech.Engrg, ASCE*, 1993, 119(9):1333～1359.
4. Ng, C. W. W.; Lee, Cx TK. A three-dimensional parametric study of the use of soil nails for stabilizing tunnel faces. *Computers and Geotechnics*, 2002, 29(8):673~697.
5. Olson R.E., Settlement of Embankment on Soft Clays, J. *Geotech. Geoenviron. Engrg. ASCE*, 1998, 124(4):278～288.

NATURAL CRACK-FACE BOUNDARY CONDITIONS IN THE FRACTURE MECHANICS OF PIEZOELECTRIC SOLIDS

XIAN-CI ZHONG [†]
Key Laboratory of Disaster Prevention and Structural Safety, Guangxi University Nanning, Guangxi 530004, China
College of Mathematics and Information Science, Guangxi University Nanning, Guangxi 530004, China

KE-SHI ZHANG
Key Laboratory of Disaster Prevention and Structural Safety, Guangxi University Nanning, Guangxi 530004, China

Considering the piezoelectric effects, the natural crack-face boundary conditions are proposed and utilized to solve the problem of an opening Griffith crack in an infinite piezoelectric material. Numerical results for PZT-5H are carried out to present the effects of applied electric fields on the stress intensity factor for a vacuum or air crack. The obtained results reveal that the fracture toughness variations of piezoelectric ceramics are attributed to the signs of applied electric fields and the prescribed factor in the new crack-face boundary conditions. Then various experimental observations can be explained theoretically.

Keywords: Piezoelectric solid; Crack-face boundary conditions; Stress intensity factor.

1. Introduction

The fracture analysis for piezoelectric materials has attracted much attention[1-3], and one of the key issues is the effect of applied electric loadings on the crack propagation. According to the experimental results[4-6], the fracture toughness of piezoelectric materials is affected significantly by applied electric fields. But the contradictory observations have been observed. Based on the local domain switching, Zhu and Yang[7] have considered that the change of a relative permeability parameter associated with the crack-face boundary conditions may affect the fracture toughness. However, from the theory of linear piezoelectricity, the fracture toughness variations of piezoelectric materials are still lacked to be explained satisfactory, although several crack-face boundary conditions have

[†] Corresponding author, email: xczhong@gxu.edu.cn; zhongxianci@163.com

been proposed such as permeable, impermeable, semi-permeable and energetically consistent to consider the effects of the medium inside a crack.

In the present paper, we attempt to illustrate the different experimental results through a theoretical analysis. By considering the piezoelectric effects, the new crack-face boundary conditions are proposed. The variations of the stress intensity factor on applied electric fields are discussed and shown in a graphic for a vacuum or air crack in an infinite piezoelectric material.

2. Natural Crack-face Boundary Conditions

Consider that a Griffith crack is embedded in a piezoelectric body under electromechanical loadings E_0 and σ_0 perpendicular to the crack faces. The adopted coordinate system is xoz with the poling direction of a piezoelectric material as z -axis. The crack is located at the segment $-a < x < a$. Following the idea of Hao and Shen[8], it is assumed that the electric displacement on the crack faces can be calculated as

$$D^c = -\varepsilon^c \frac{\Delta\phi}{\Delta u_z} \tag{1}$$

where $\varepsilon^c = \varepsilon_r \varepsilon_0 (\varepsilon_0 = 8.85\times10^{-12}\,\mathrm{Fm^{-1}})$. $\Delta\phi$ and Δu_z are the jumps of the electric potential and electric potential across the crack respectively. In addition, from the physical knowledge, the electric displacement normal to the interface between the piezoelectric material and the dielectric of crack interior is continuous, which means that the normal component of the electric displacement in piezoelectric materials near the crack faces is D^c . It is further noted that the electric displacement D^c is proposed to be a sufficiently good approximation to the electric displacement perpendicular to the dielectric crack faces. Similar to Wang and Singh[5], the stress produced by the piezoelectric effects in a piezoelectric solid can be computed as

$$\sigma_{zz} = -\frac{e_{33} D^c}{\varepsilon_{33}} \tag{2}$$

As shown in McMeeking[9], for an impermeable crack, the stress will be zero. However, a real opening crack is full of a certain dielectric interior and not truly impermeable. A non-uniform strain in the vicinity of the crack will be generated. Furthermore, the stress concentration will be created by the electric field concentration near the dielectric crack tip. When the stress of Eq. (3) exists in a piezoelectric solid, it is similar to the case that a stress is applied to the piezoelectric solid. On the other hand, from the typical linear fracture mechanics, it is seen that when a mechanical loading is applied on the crack faces, the stress will concentrate at the crack tip. In analogy to this, to model the stress concentration induced by the electric field concentration at the crack tip, it is convenient to assume that a stress induced by the piezoelectric effects is acting on the crack faces and associated with that shown in (3). Hence we suppose that

$$\sigma^{p} = \kappa_{p}\frac{e_{33}D^{c}}{\varepsilon_{33}} \tag{3}$$

where κ_p is a prescribed and non-dimensional factor. The introduction of the parameter is mainly based on the following several reasons. First, the nonlinear electromechanical behavior may affect the stress concentration near the crack tip such as domain switching[7] and electric field yielding[10] when an electric field is applied. Moreover, the experimental observations have revealed that a negative (positive) electric field may increase or decrease the crack initiation loading [11]. It is seen that the effect of an applied electric field on the crack growth exhibits some uncertainties. The parameter may reveal the different levels of the stress concentration induced by the electric concentration near the crack tip to a certain extent. Second, an applied electric field will have influences on the crack-face free energy which includes the elastic energy and the electrostatic energy. In a ferroelectric solid, different elastic and electrostatic fields occur indeed due to the domains of different polar directions[12]. The introduced parameter may reflect the changes of the free energy on the crack faces in some extent. Generally speaking, for the complexity of a crack in a piezoelectric solid, it is difficult to determine the exact values of the introduced parameter only through numerical computations.

Furthermore, for a linear dielectric crack, the new crack-face boundary conditions can be stated as

$$\text{Case A:}\quad \sigma^{c} = \kappa_{p}\frac{e_{33}D^{c}}{\varepsilon_{33}},\qquad D^{c} = -\varepsilon^{c}\frac{\Delta\phi}{\Delta u_{z}} \tag{4}$$

Clearly, the electrostatic traction induced by the medium inside the crack is not considered in Case A. If introducing the electrostatic traction acting on the crack faces[13, 14], one has

$$\text{Case B:}\quad \sigma^{c} = \frac{1}{2}\frac{(D^{c})^{2}}{\varepsilon^{c}}\left(1-\frac{\varepsilon^{c}}{\varepsilon_{33}}\right)+\kappa_{p}\frac{e_{33}D^{c}}{\varepsilon_{33}},\qquad D^{c} = -\varepsilon^{c}\frac{\Delta\phi}{\Delta u_{z}} \tag{5}$$

It should be pointed out that the electrostatic traction in Case B may be not completely correct since the piezoelectric effects have been considered here. The new crack-face boundary conditions (4) and (5) should be also verified by other experimental data. As we know, the permeable and impermeable crack models cannot correctly describe a real dielectric crack in piezoelectric solids. Although the semi-permeable crack model has considered the effects of the dielectric inside the crack, the crack-face stress boundary condition is still assumed to be zero. The energetically consistent boundary conditions have shown the stress effects of the dielectric of crack interior on the crack faces. However, the basic reason of the crack growth in ferroelectric solids induced by applied electric fields should be attributed to the electromechanically coupling effects[2]. The new crack-face boundary conditions have reflected the influences of the piezoelectric effects on the crack growth. Then one can see that the new crack-face boundary conditions (4) and (5) are more natural and physical than the permeable, impermeable, semi-permeable and energetically consistent boundary conditions.

3. Numerical Results

Within the theory of linear piezoelectricity, the stress intensity factor for a dielectric crack in an infinite piezoelectric solid by using (4) or (5) can be computed as

$$K_\sigma = (\sigma_0 - \sigma^c)\sqrt{\pi a} \tag{6}$$

In the following computations, a vacuum or air crack embedded in a PZT-5H material[10, 15] is considered and the corresponding material properties are neglected here.

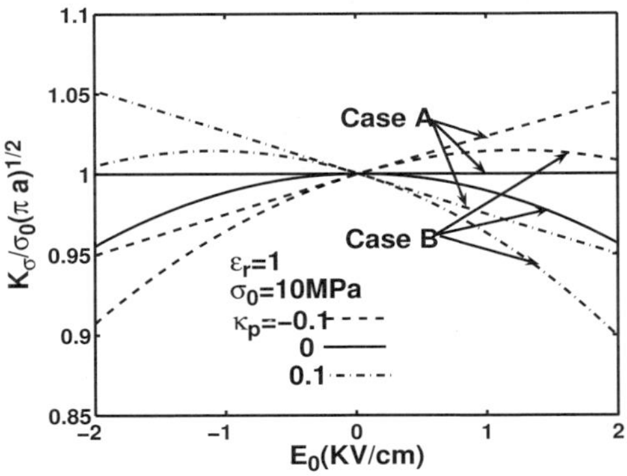

Fig. 1. The normalized stress intensity factor versus applied electric fields for a vacuum crack.

The variations of the normalized stress intensity factor versus applied electric fields are shown in Figure 1 for Cases A and B respectively. When $\kappa_p = 0$, it is seen that Case A reduces to the semi-permeable crack model and one obtains that the normalized stress intensity factor equals unity [8, 15]. Case B corresponds to the energetically consistent boundary conditions[13] and one obtains that applied electric fields always decrease the stress intensity factors[14,16]. However, for $\kappa_p = -0.1$, one can see that if the applied electric field is negative, the normalized stress intensity factor is greater than unity. And if the applied electric field is positive, the normalized stress intensity factor is less than unity. The obtained results show that an applied negative electric field enhances the crack propagation and an applied positive one retards the crack propagation, which is in agreement with the experimental results given by Wang and Singh[5]. While $\kappa_p = 0.1$, the effects of applied electric fields on the stress intensity factor are reversed to those for $\kappa_p = -0.1$. This agrees well with the experimental observations given in Park and Sun[4]. On the other hand, the experimental results of Fu and Zhang[6] have revealed that the applied positive and negative electric fields always reduce the fracture toughness of piezoelectric materials, which also can be explained by the combination of $\kappa_p > 0$ for $E_0 < 0$ and $\kappa_p < 0$ for $E_0 > 0$. The obtained results reveal that an applied electric field increases or decreases the fracture toughness of piezoelectric solids, depending on the signs of applied electric fields and the prescribed parameter respectively.

Acknowledgments

The work was supported by China Postdoctoral Science Foundation funded project (20100470897) and the Scientific Research Fund of Guangxi Provincial Education Department (200911MS15).

References

1. M. Kuna: *Eng. Fract. Mech.*, 2010, **77**, 309–326.
2. W. Yang: 'Mechatronic Reliability', 2001, Tsinghua University Press, Springer, Beijing.
3. D.N. Fang, J.X. Liu, 'Fracture Mechanics of Piezoelectric and Ferroelectric solids', 2008, Tsinghua University Press, Beijing.
4. S. Park and C.T. Sun: *J. Am. Ceram. Soc.*, 1995, **78**, 1475-1480.
5. H. Wang, R.N. Singh, *J. Appl. Phys.*, 1997, **81**, 7471-7479.
6. R. Fu, T.Y. Zhang, *J. Am. Ceram. Soc.,* 2000, **83**, 1215-1218.
7. T. Zhu, W. Yang, *Acta Mater.*, 1997, **45**, 4695-4702.
8. T.H. Hao, Z.Y. Shen, *Eng. Fract. Mech.*, 1994, **47**, 793-802.
9. R.M. McMeeking, *J. Appl. Math. Phys.*, 1989, **40**, 615-627.
10. H. Gao, T.Y. Zhang, P. Tong, *J. Mech. Phys. Solids*, 1997, **45**, 491-510.
11. Y. Shindo, F. Narita, K. Horiguchi, Y. Magara, M. Yoshida, *Acta Mater.*, 2003, **51**, 4773-4782.
12. Z. Suo, *Adv. Appl. Mech.*, 1997, **33**, 193-294.
13. C.M. Landis, *Int. J. Solids Struct.*, 2004, **41**, 6291-6315.
14. A. Ricoeur, M. Kuna, *Int. J. Fract.*, 2009, **157**, 3-12.
15. X.F. Li, K.Y. Lee, *Phil. Mag.*, 2004, **84**, 1789-1820.
16. X.C. Zhong, K.S. Zhang, *Int. J. Eng. Sci.*, 2010, **48**, 612-623.

ANALYSIS OF CUMULATIVE PLASTIC DEFORMATION OF ROAD ON SOFT SUBSOIL INDUCED BY VEHICLE TRAFFIC LOADINGS

LU ZHENG†, YAO HAILIN, LIU JIE, HU MENGLING

State Key Laboratory of Geomechanics and Geotechnical Engineering, Institute of Rock and Soil Mechanics, Chinese Academy of Sciences, Wuhan 430071, China

The characteristics of cumulative plastic deformation of road on soft subsoil induced by vehicle traffic loadings are studied. The total deformation is considered as consisting of two parts: one is the deformation of the road structure layers themselves, and the other is the deformation of the underlying soft ground. The two parts are derived respectively through laboratory full-size model experiment and a empirical equation of cumulative deformation in which the dynamic deviator stress, the number of repeated loading and the properties of soft subsoil are considered.

Keywords: Plastic deformation; Traffic loading; Soft subsoil.

1. Introduction

During the last decades, the analysis and design of road pavement and subgrade has traditionally relied on past experience and empirical methods which are on the basis of the static mechanical conditions. However, the permanent deformation of road structures is obviously due to the application of moving repeated traffic loading. Li and Selig[1] established a power equation with exponential form, in which the constants are determined by the physical state and strength of subsoil. Li and Selig[2] provided the values of the constants of several cohesive soils by employing cyclic loading experiment, and showed some successful applications of the equation to predict the long-term settlement of cohesive soils under train loading. Chai and Miura[3] proposed an empirical equation to calculate the permanent settlement of road on soft subsoil with a low embankment. Fujikawa et al.[4, 5] assumed a triangular distribution pattern of traffic-load-induced stress increments based on field measurement. Although this method is simple, the dynamic characteristics of soils are not taken into consideration, and the traffic-load-induced stress is not calculated explicitly.

† Corresponding author, email: lzwhrsm@163.com

In this paper, a new method is proposed to calculate the cumulative deformation of road on soft subsoil subjected to vehicle traffic loads. The total cumulative plastic deformation of road is considered as consisting of two parts: one is the deformation of the road structure layers themselves, which is calculated according to laboratory model experiment directly. The other is the deformation of the underlying soft ground, and it is determined by the empirical equation of cumulative deformation in which the dynamic deviator stress, the number of repeated loading and the properties of soft subsoil are considered.

2. Model Experiment on Layered Road Structure under Traffic Loads

2.1. *Introduction of the model experiment*

A laboratory experiment at full-scale which represents the typical road structures with multi-layers has been built in order to simulate the dynamic response of road subjected to vehicle traffic loads as shown in Fig. 1. The loading equipments are fixed on a reaction beam which is bolted on the steel frame. All components of the experiment are located on a reinforced concrete foundation able to support the important weights of these components.

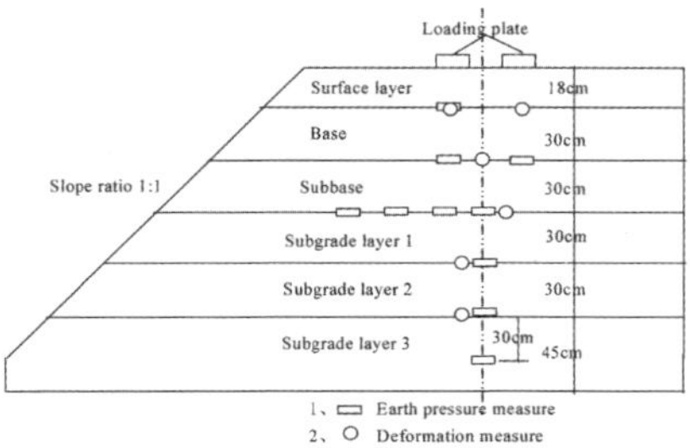

Fig. 1. Experimental model.

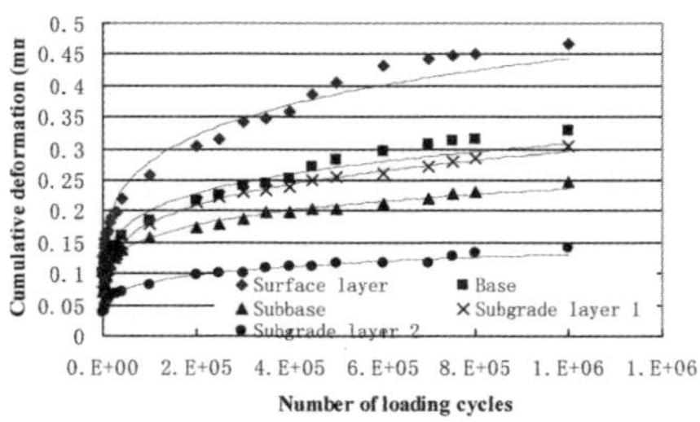

Fig. 2. Variation of cumulative deformation with N.

The detailed size of the model is as follows: the longitudinal length is 2.3m, the horizontal width of the top is 2.0m, the horizontal width of the bottom is 3.6m, and the slope ratio is 1:1. Different transducers are installed in the model of the road structures in order to measure stresses and displacements. All of the transducers are installed at the bottom of road structure layers as shown in Fig. 1. These transducers and their cables are protected by aluminum tubes to prevent damage from cyclic loadings. In order to obtain the cumulative deformation of road structures under long-term repeated vehicle loads, an experiment is performed by choosing the dynamic load amplitude as 75KN and frequency as 7Hz, and it consists of 10^6 cycles.

2.2. *Model experiment results*

The dynamic load with signal of semi-sinusoidal wave is applied in this experiment. The effects of load amplitude and the number of loading cycles on cumulative deformation of road structure layers are also investigated, and are shown in Fig. 2. As can be seen, the cumulative deformation of each layer has the same tendency to increase with the number of cycles till one million cycles. The cumulative deformation is very fast for a short number of cycles and becomes much lower at other moments. These fast deformations come from the reorganization of grains of the road structure layers.

3. The Total Cumulative Deformation

3.1. *The composition of cumulative deformation*

From the model experiment conducted in Section 2, we found that the total cumulative deformation of road structures is not great after one million repeated loadings. However, for a road constructed on soft subsoil, the cumulative deformation generated by the soft subsoil could be great, especially for the road with low embankment constructed on soft soil since the dynamic deviator stresses of soft soil caused by moving vehicles are lager for low road embankment than other ones, and the compressible soft soil layer could also be thicker. Therefore, it is significant to study the traffic-load-induced cumulative deformation of road and its soft subsoil. The cumulative plastic deformation of road on soft subsoil S_p is calculated according to the following formula:

$$S_p = S_{p1} + S_{p2} = S_{p1} + \sum_{i=1}^{n} \varepsilon_{pi} H_i \tag{1}$$

in which, S_{p1} is the deformation of road structure layers themselves, it can be calculated according to the results of model experiment of Section 2, S_{p2} is the deformation of the soft subsoil, here, $\varepsilon_{p\,i}$ is the cumulative plastic strain of each compressible layer of subsoil, and H_i is the depth of each compressible layer of subsoil. An empirical equation for calculating the cumulative plastic strain of cohesive soils under repeated loading has been proposed by Li and Selig [3], and be employed to determine the cumulative plastic strain of each compressible layer $\varepsilon_{p\,i}$ in present study

$$\varepsilon_p = a(q_d / q_f)^m N^b \tag{2}$$

Where a, m and b are experimental constants which are related to the plasticity index of subsoil and the suggested values are the same as Li[3], q_d is the dynamic deviator stress of subsoil determined by the components of dynamic stresses which can be calculated according to multi-layered elastodynamic theory, q_f is the static failure deviator stress and can be measured by a field vane shear test or determined by the following empirical equation

$$q_f = 2c_{cu}\cos\varphi_{cu}/(1-\sin\varphi_{cu}) + (1+K_0)\sigma_{cz}\sin\varphi_{cu}/(1-\sin\varphi_{cu}) \quad (3)$$

where c_{cu} and ϕ cu are the cohesion force and internal friction angle of the soft subsoil respectively, K_0 is the horizontal earth pressure coefficient. In this paper, Shanghai clay deposit, China, is adopted for calculating the cumulative plastic deformation of road, and the subsoil parameters are the same as Ref. [6].

3.2. *Analysis of traffic-load-induced cumulative deformation*

In this section, some numerical results are derived through the method mentioned above. The road structures are divided into 5 layers. The subgrade is simplified as a single layer whose elastic modulus is 35MPa and thickness is 0.9m. The soft subsoil is considered as elastic half-space medium and its elastic modulus and density are 10MPa and 1500kg/m^3, respectively. The multi-layered elastodynamic model is firstly used to calculate the dynamic deviator stress of the soft subsoil, then, combing the results of experiment of Section 2 with the method in present Section, the traffic-load-induced cumulative deformation of road structure layers themselves and the underlying soft soil are obtained.

In order to explain the different deformation of road structure layers themselves (S_{p1}) and the soft subsoil (S_{p2}), the cumulative deformation S_{p1}, S_{p2} and S_p are plotted respectively versus the number of repeated loading applications until 5 million cycles, as shown in Fig. 3. It is seen that the traffic-load-induced cumulative deformation of road structure layers themselves is much smaller than that of soft subsoil, and the total cumulative deformation is mainly generated by the soft soils. We can also see form Fig. 3 that the cumulative deformation of road structure is very fast for a short number of cycles and becomes much lower with the increasing of repeated loading number.

Providing that the loads moving along the centerline of road, Fig. 4 presents the distribution of cumulative deformation of crossroad profile after 1 million cycles, where different depth including 0.0m, 1.0m, 1.5m and 2.5m below the subgrade surface are considered. As can be seen, the values of cumulative deformation of different depth decrease sharply along the horizontal direction, then gradually decrease to zero when far away from road centerline more than 5 meters. It should be noted that the maximum cumulative deformation of 2.5m below the subgrade surface is significantly smaller than that of subgrade surface, which indicates that the traffic-load-induced cumulative deformation is mainly generated by the superficial layer of soft subsoil.

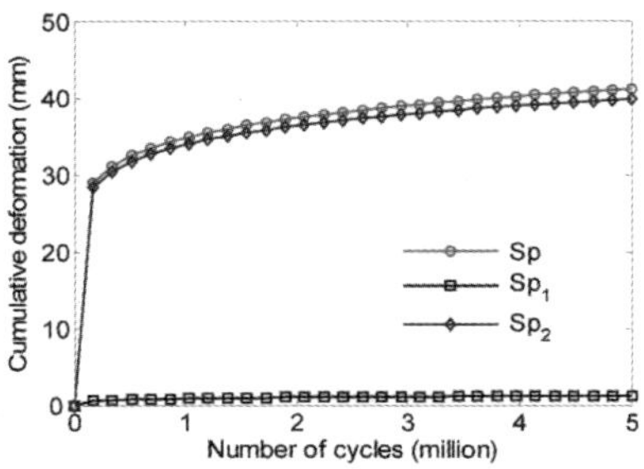

Fig. 3. Cumulative deformation variation with N.

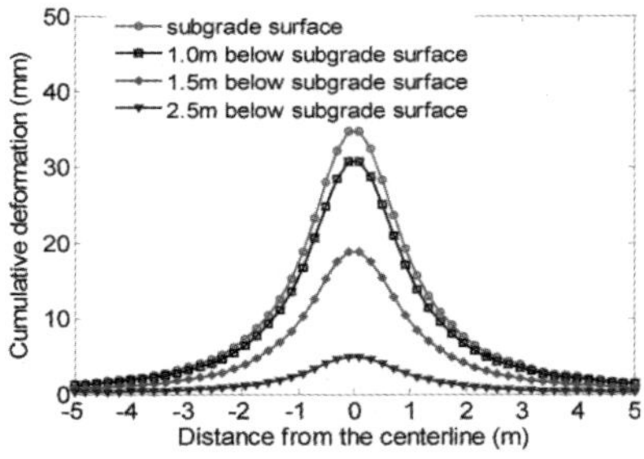

Fig. 4. Distribution of cumulative deformation.

4. Conclusion

The proposed method in present paper is reasonable in predicting the traffic-load-induced cumulative deformation and can be used to calibrate the design of road on soft subsoil with a low embankment. The analysis also revealed that, for the road constructed on soft soil, the traffic-load-induced cumulative deformation of road structure layers is mainly generated from the soft subsoil.

Acknowledgments

Financial support from the Foundation of State Key Laboratory of Geomechanics and Geotechnical Engineering under approved grant No.SKLQ016 is gratefully acknowledged.

References

1. D.Li and E.T.Selig: *J. Geo. Eng.,* 1996, **122**, (12), 1006-1013.
2. D.Li and E.T.Selig: *J. Geo. Eng.,* 1998, **124**, (4), 323-329.
3. J.C.Chai and N. Miura: *J. Geotech. Geoenvir. Eng.*, 2002, **128**(11): 907-916.
4. K.Fujikawa: Proc. 2nd Int. 'On optimistic design of low embankment road on soft subsoil by considering the traffic-load-induced settlement'. Japan, March 1996, University of Saga, Paper 121.
5. K.Fujikawa, N.Miura and I.Beppu: *Soils. Foundations.,* 1996, **36**, (4), 147–153.
6. J. J. Li, M. S. Huang and Y. D. Wang: *China. J. High. Trans.,* 2006, **19**, (1), 1-5.

THE THEORETIC ANALYSIS AND EXPERIMENTAL RESEARCH ON THE TIME-DEPENDENCY OF TC PILES

Y.H.CHEN

Key Laboratory of Ministry of Education for Geomechanics and Embankment Engineering, Hohai University, Nanjing, Jiangsu 210098, China

L.CHEN

Research Institute of Geotechnical Engineering, Hohai University, Nanjing, Jiangsu 210098, China

X.D.ZHOU

Department of Engineering Mechanics, Hohai University, Nanjing, Jiangsu 210098, xingdezhou@hhu.edu.cn

X.Q.WANG†

Department of Civil Engineering, Zhejiang University City College, zhejiang, Hangzhou 310015, China

Research on the time-dependency of the pile is a key subject on the geotechnical engineering, and chiefly concentrate on the precast piles. The bearing capacity of the plastic tube cast-in-place concrete(TC) piles have also not been concerned to this day. As having not enough stiffness, the TC piles are driven into ground by aid of a driving-tube, here, there exists a soil backfill process after the driving-tube is drawn. In the paper, an algorithm is suggested to simulate the process, the concept of influence area is proposed to modify the original plasticity extent determined by Hole Expansion Theory(HET). Meanwhile, in order to acquire a better calculation results, a fractional order model is introduced to calculate the excess pore water pressure(EPWP). In the last, an example is utilized to demonstrate the effectiveness of proposed method.

Keywords: AEPA 2010; TC pile; soil backfill; time dependency; fractional order; EPWP.

1. Introduction

Most soil deposits in Zhejiang Province of China are soft in nature. Therefore, pile foundations are used to support highway bridges and other structures to safely transfer the superstructure loads deeply into the ground. The plugging

---

† Corresponding author, email: *wangxq@zucc.edu.cn*

effect of precast pile certainly will produces a horizontal soil displacement and excess pore water pressure (EPWP) which can be determined by HET[1]. The bearing capacity of pile takes on a gradually incremental trend in saturated soil[2]. At present, the chief advance in bearing capacity calculation is that the influence of soil self-weight is considered into EPWP formula[3].

A driving-tube is needed to help the TC piles into ground, there exists a soil backfill process after the driving-tube is drawn. The presented time-dependency analysis method aim chiefly at precast pile, and unfit for the TC pile of existing a soil backfill process. In the paper, both the soil backfill model and the influence area concept are presented, and based on these hypotheses, a linear expression of EPWP within the influence area is suggested. The linear expression is improved further as a fractional order one for getting a better calculation results. In the last, a practical engineering is used as the demonstration to display the correction of the proposed method.

2. The Simulation of Soil Backfill Process

In the paper, the formula presented by Wang[3-4] are adopted to calculate the initial EPWP and dispersal process. The main process of the TC pile construction includes: 1) Driving the TC pile into ground by aid of tube and pile-head; 2) drawing the tube, here, a backfill process is happened and the pore water pressure near the pile certainly will be change.

Three hypotheses are make as follows: 1) soil is in saturated state; 2) the pore water pressure of the soil ektexine is zero at the instant of that the tube is drawn; 3) the radial dimension of the influence area because of the soil backfill is ‘b’.

2.1. *Initial contact height of pile-soil*

Initial contact height of pile-soil l_0 (i.e. the length of BC) can be determined by F_0 (initial bearing capacity of pile) and σ_b (the normal stress of soil located at the bottom of the pile) acquired by experiment. Meanwhile, supporting the soil pressure of pile-side alter linearly in depth, then

$$\int_0^{l_0} k_0 y \gamma 2\pi r_0 dy = F_0 - \pi r_0^2 \sigma_b \tag{1}$$

Where γ is the specific weight of soil: $k_0 = 1 - \sin\varphi$; r_0 is the radius of pile.

By Eq.(1), we get: $l_0 = \sqrt{\left(F_0 - \pi r_0^2 \sigma_b\right) / \left(k_0 \gamma \pi r_0\right)}$ (2)

2.2. *The Determination of 'b'*

The value 'b' can be determined by assuming the volume of backfill soil equal to the one of influence area, i.e. $V_b = V_I$, where l is the length of pile,

$V_b = \pi[(r_0 + a)^2 - r_0^2]l - \frac{1}{3}\pi[r_0^2 + 4ar_0 + a^2](l - l_0)$,

$V_I = \pi[(r_0 + a + b)^2 - (r_0 + a)^2]l$, one has

$b = \left(-\alpha_1 + \sqrt{\alpha_1^2 - 4\alpha_2}\right)/2$, Where, $\alpha_1 = 2(r_0 + a)$, $\alpha_2 = -V_b / \pi l$.

2.3. *Initial EPWP of influence area*

Taking the line EF as boundary of influence area (see Fig. 1), the initial EPWP of the soil lied the outside of the boundary is firstly determined. In area AEFCB, supposing that the initial soil EPWP lied line ABC is zero and alters linearly along the radial direction. The expression of the initial EPWP lied in radial line (such as line K_1M_1 and K_2M_2) can be written as

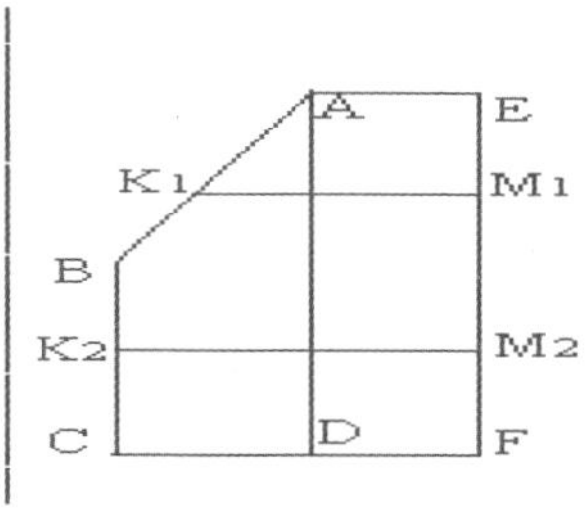

Fig. 1. Calculating model.

$$u = u_b r / (r_b - r_a) - u_b r_a / (r_b - r_a) \qquad (r_a \le r \le r_b) \tag{3}$$

where r_a is the distance from the point lied in line ABC to the pile axis; $r_b = r_0 + a + b$.

3. The Fractional Order Expression of Initial EPWP

The errors existed between simulation results and practical measurement ones are produced by the linear hypothesis of initial EPWP. The fractional order model may be a better consistency in model-built [5]. In order to construct the fractional order model, Eq.(3) firstly is transformed into ARX(1,1) model, i.e.

$$u'(t) + cu(t) = dr(t) \tag{4}$$

where c, d are unidentified parameters, t is a fictional time variable.

Let $t = 1, 2, \cdots, n$, and corresponding radial variables are $r = r_1, r_2, \cdots, r_n$, respectively, substituting r in Eq.(10), we get $u = u_1, u_2, \cdots, u_n$. The parameters c and d can be identified by (r_i, u_i) (see literatures [6]). In Laplace transform notation, Eq.(4) can be written as

$$sU(s)+cU(s)=dR(s) \tag{5}$$

The transfer function $G(s)$ is

$$G(s)=U(s)/R(s)=d/(s+c) \tag{6}$$

Introducing the fractional order to $G(s)$, i.e.

$$G(s)=U(s)/R(s)=d/(s^{\alpha}+c) \tag{7}$$

where α is a fractional degree, here, α=0.9.

s^{α} can be solved approximately by filter approximation. Let the fit frequency area is (ω_l,ω_h), ω_l, ω_h are the lowest and the highest fit frequency, respectively, a filter is constructed as follows

$$G_f(s)=K_f\prod_{j=-N}^{N}\frac{s+\omega_j^a}{s+\omega_j} \tag{8}$$

where N is a positive integer; the order of the filter is $2N+1$;the expression of K_f,ω_j^a,ω_j can be acquired in [7]. s^{α} is replaced by $G_f(s)$, so Eq.(7) is written as

$$G(s)=(s^{n_i}+a_1s^{n_i-1}+\cdots+a_{n_i-1}s+a_{n_i})/(s^{n_j}+b_1s^{n_j-1}+\cdots+b_{n_j-1}s+b_{n_j}) \tag{9}$$

where both n_i and n_j are positive integer.

4. Numerical Analysis

In the paper, the TC pile lied in a highway in Zhejiang Province of China is demonstrated, the soil parameters see table1, pile: the diameter is 0.16m, the length is 15m, E = 20GPa, $\mu=0.25$, $\gamma=2500$ (kg/m^3) ; the outer diameter of the tube is 0.203m. Calculation model: the elastic model is used for pile; the Mohr-Coulomb is adopted for soil. Meanwhile, α_f =1.096, let $C_a=C_u$, the friction coefficient of pile-soil is 0.2. the weight radius of plastic area is $R_p=0.7414m$. The results of the initial EPWP are in table 2, here, R1-R4 express the radial distance and are 0.2m, 0.4m, 0.6m, 0.8m, respectively.

Table 1. Soil parameters.

Depth (m)	name	γ (kN/m^3)	$\varphi(^\circ)$	C_u (KPa)	E (MPa)	μ	**PV**	**PH**
0.5	Clay	18.90	30.5	21.20	4.27	0.25	1.877	3.760
1.5	Silt	17.80	15.0	19.50	2.76	0.25	1.576	1.916
8	Silt	16.60	12.6	7.70	1.84	0.25	1.596	3.256
10	clay	17.50	23.5	19.80	2.41	0.30	2.200	3.748

Note: PV, PH are vertical and horizontal permeability coefficient of soil ($10^{-4}m/d$), respectively.

Table 2. The initial EPWP (kpa)

Depth (m)	R1	R2	R3	R4	Depth (m)	R1	R2	R3	R4
3.5	90.87	50.03	33.47	23.76	10.0	199.21	104.19	69.58	50.85
4.5	107.54	58.36	39.03	27.93	12.5	240.87	125.03	83.47	61.26
6.5	140.87	75.03	50.14	36.26	14.0	265.87	137.53	91.80	67.51
8.5	174.21	91.69	61.25	44.60	15.0	282.54	145.86	97.36	71.68

the initial bearing capacity F_0 =8.13t, the bottom stress of pile σ_b =0.4642MPa, substituting them in Eq.(2), we get l_0 =4.231m, subsequently, we have b=0.019m. Here, the linearity expression of the initial EPWP in Eq.(3) is demonstrated as follows: Locating at the depth of 10m, $r_a = 0.08m$, $r_b = 0.121m$, $u_b = 278.3KPa$, so we have $u = 6787.8r - 543.0$. Choosing 1000 points in line (i.e. Eq.(4)), the identified ARX(1,1) is $G(s) = 348.8/(s + 0.0514)$, and its fractional model is $G(s) = 348.8/(s^{0.9} + 0.0514)$. In the paper, N=2, ω_l =0.001, ω_h =10000, then, $s^{0.9}$ can be expressed approximately as

$$G_f(s) = G_{f1}(s)/G_{f2}(s)$$

$$G_{f1}(s) = 3981s^5 + 1.94\times10^6 s^4 + 3.62\times10^7 s^3 + 2.68\times10^7 s^2 + 7.90\times10^5 s + 891.3$$

$$G_{f2}(s) = s^5 + 8.86\times10^3 s^4 + 3.01\times10^6 s^3 + 4.06\times10^7 s^2 + 2.18\times10^7 s + 4.47\times10^5$$

Substituting $G_f(s)$, we get the $G(s)$, the initial EPWP of influence area can be determined. The simulation results (see Fig.2) shown that the fractional order model is better than the integer order one.

fractional order model

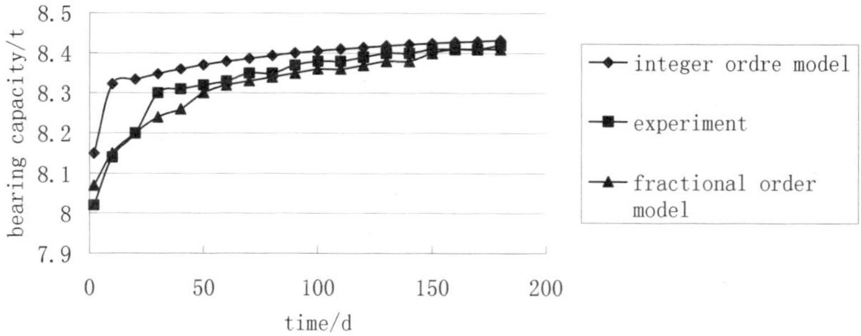

Fig. 2. The time dependency of bearing capacity of pile.

5. Conclusion

In the paper, the time dependency of bearing capacity of TC pile with soil backfill process is analyzed, a fractional model is presented, and the concept of influence area is suggested. The finite element simulation results have shown that the proposed method is effective.

References

1. Liu Y.H. Chen Z.Z. Analysis of Pile Driving Effect of Precast Tubular Pile Using Cylindrical Cavity Expansion Theory, Rock and Soil Mechanics, 2007,28(10):2167-2172
2. Zhang X.S., et al. Time-dependency Analysis of Soil Compaction Pile in Soft Clay Ground,. Chinese Journal of Rock Mechanics and Engineering. 2004; 23(19): 3365-3369
3. Wang W., Lu T.H. et al. Study on Ultimate Bearing Capacity of Static Pressed Pile Based on Excess Pore Water Pressure Dissipation,. Rock and Soil Mechanics.2005;26(11):1845-1848
4. Wang W., Zai J.M., Wang X.D. 3D Calculation of Excess Pore Water Pressure Due to Driving Pile and Its Application,. Rock and Soil Mechanics.2004;25(5):774-777
5. Nazarian P. Haeri M. Generalization of order distribution concept use in the fractional order system identification,. Signal Processing, 2010; 90(7):2243-2252
6. Duan L.W. Wu Z.H. Xu Z.W. Identification Research of the Piezoelectric Smart Structure System of Aircraft Wing Based on ARX Model, Piezoelectectrics & Acoustooptics 2008; 30(6):760-762
7. Oustaloup A, Levron F. Frequency-band Complex Noninteger differentiator: Characterization and Synthesis. IEEE Transaction on Circuit and Systems-I：Fundamental Theory and Applications，2000; 47(1)：25-39

THE IMPACT OF POLYPROPYLENE FIBRE CONTENT ON MECHANICAL PROPERTIES OF REINFORCEMENT SOIL

PAN HONGKE†

School of Civil Engineering and Architecture, Zhongyuan University of Technology, No.41 Zhong Yuan Middle Road, Zhengzhou, China
Zhengzhou, 450007, China

ZHU YANZHI

School of Civil Engineering and Architecture, Zhongyuan University of Technology, No.41 Zhong Yuan Middle Road, Zhengzhou, China
Zhengzhou, 450007, China

Soil has obvious plastic characteristic. In civil engineering, some methods, such as reinforcing with natural fibres or synthetic fibres, are often adopted to strengthen soil's bearing capacity or the ability of deformation resistance. Experiences and studies indicate: length, mixture ratio, toughness and fineness of synthetic fibre have rather great influence on the strength of reinforcement soil. In order to further realize the extent of the influence and achieve relatively explicit quantization relation, taking polypropylene fibre as an example, this paper studied the impact rule of the filling quantity of polypropylene fibre (mixture ratio) on mechanical property of reinforcement soil by doing intensity experiments to plain soil and several classes of reinforcement soils with different fibre contents. As the same time, the empirical equation about soil strength varying with fibre content was established by the method of regression analysis. In the end, by simulating reinforcing status with real fibre reinforcement soil and constituting mechanical model according to various fibre contents, the variation situation of stress and strain of soil under loading was numerically simulated based on the plastic assumption about constitutive relation of soil. The study results can be generalized to engineering practice of various polypropylene fibre reinforcement soils.

Keywords: Reinforcement soil; Fibre content; Mechanical property; Plastic.

1. Guidelines

In civil engineering, the problem is frequently encountered that foundation soils can not meet the required strength. The methods reinforcing it can be generally divided into two types as physical and chemical reinforcement. Fibre reinforced method belongs to the physical, of which the reinforcement method using natural plant fibers is relatively long history and the study to it is more, whereas the

† PAN Hongke, email: panhk_sh@126.com

research and application of synthetic fibers reinforcement to soils is relatively late[1-5]. At the same time, because of the different soil properties, fiber reinforcement process is different. This paper studied the reinforcing effect of mixing the sandy silts (clays) in one area with the current common using polypropylene fiber and the appropriate ratio and reinforcing methods. Prospect of such method should be promoted since the materials are relatively easy to obtain and the construction costs generally much lower than other reinforcement methods.

2. Experimental Study

In order to more clearly reflect the mechanical properties and reinforcing law of fiber reinforced soil, the typical foundation soil in one area was chosen (the silty clays from one deep foundation pit at the area) to be dealt with at laboratory and the basic physical and mechanical parameters were Measured. Then, by adding different ratios of polypropylene fibers evenly, various contents of fiber soil specimens were prepared (as cylinder shape) and strength tests done after interior curing for the purpose of obtaining the improvement discipline of soils' carrying capacity and deformation performance according to the contents of polypropylene fibers.

2.1. *Parameters of experimental materials*

Physical and mechanical parameters of soil samples were shown in table 1, while the ones of polypropylene fiber were shown in table 2.

Table 1. The basic physical and mechanical parameters of silty clays.

water content / %	Density / $g.cm^{-3}$	Liquid limit / %	plasticity index	optimum water content / %	maximum dry density / $g.cm^{-3}$
48	1.72	36	11.2	15.8	1.7

Table 2. The basic physical and mechanical parameters of polypropylene fiber.

Density / $g.cm^{-3}$	Diameter / mm	Average length / mm	Tensile strength / MPa	Elastic Modulus / MPa	Melting point / °C	Burning point / °C	Dispersion
0.9	0.02~0.05	12	⩾350	⩾3500	160	580	well

2.2. *Experimental methods and procedure*

There were four groups of specimens: plain soil; soil samples Containing 0.1%, 0.2%, 0.3% fiber volume, respectively. The samples in each group were produced 5 specimens (a total of 20). In the preparation process, the soil samples were firstly dispersed and air-dried, then matched well and mixed evenly as optimum moisture content, lastly maintained with 28 days (Note: only 5 days for plain soil). Specimens were made as cylinder shape with the 60mm height and 35mm diameter. All Specimens were done unconfined compressive strength tests and direct shear strength tests. Testing process was referred to the standard "GB/T 50123-1999 standard for soil test methods"[6].

2.3. *Experimental results and analysis*

The results of unconfined compressive strength test and direct shear strength test were shown in Table 3 and Table 4.

Table 3. Results of unconfined compressive strength experiments.

The type of specimens	unconfined compressive strength values of all specimens / MPa					Mean values of UCS test
	1	2	3	4	5	
Plain soil	0.21	0.23	0.22	0.24	0.21	0.22
0.1% fiber soil	0.30	0.31	0.33	0.35	0.33	0.32
0.2% fiber soil	0.37	0.36	0.36	0.38	0.39	0.37
0.3% fiber soil	0.43	0.41	0.42	0.40	0.43	0.42

Table 4. Results of direct shear strength experiments.

The type of specimens	Cohesive strength C / KPa						Internal friction angle ϕ / °					
	1	2	3	4	5	Mean values	1	2	3	4	5	Mean values
Plain soil	47.0	47.2	47.3	47.1	47.4	47.2	25.3	25.0	25.1	24.9	25.6	25.2
0.1% fiber soil	59.2	59.8	59.6	60.2	59.8	59.7	28.6	28.3	28.7	28.5	28.8	28.6
0.2% fiber soil	61.8	61.7	61.6	61.7	61.4	61.6	30.5	30.7	31.2	31.0	30.6	30.8
0.3% fiber soil	62.9	63.5	63.1	63.2	63.2	63.2	32.7	32.8	32.2	32.0	32.5	32.5

From the above results it can be seen that polypropylene fibers can significantly improve soil's unconfined compressive strength and shear strength, that is, which can also significantly improve the bearing capacity of foundation soil and the resistance ability to deformation. The figures show explicitly that the stress-strain curves are more flat, well plastic, higher maximum capacity points, or even towards the case of strain-hardening.

3. Numerical Analysis Studies on Actual Engineering

3.1. *Establishment of models*

Basing on the experiments mentioned above, we can obtain a variety of physical and mechanical parameters of fiber soils and the stress-strain curves. Because the fibers are mixed evenly with the soils, the fiber soils can be viewed as isotropic body from the angle of overall mechanical behaviour but viewed as nonlinear anisotropic when from stress-strain relationship. In order to further analyse and verify the impact of the fiber content on the mechanical properties of soils, the mechanical model was established to analyse the four types of soil samples. By applying a group of boundary force to do finite element numerical analysis to a sample, it can be observed the developing procedure of stress, strain (or cracks) so that the soil samples can be compared at their deferent bearing capacity, deformation resistance and destruction stages. Considering the scientific and convenience of the FLAC software to simulate the soils and underground rock engineering problems, the paper adopts this large-scale commercial software to do numerical analysis and calculation on various types of fiber soils.

Simulating the actual situation of reinforced fiber soils. To simplify the problem, analyzing a big rectangular fiber clod as the manner of plane stress problem, the dimension of the clod model was $8 \times 8\ m^2$, of which the upper 1.5 m consists of fiber soil. The model was applied with the load of 100 Kpa (equivalent to the load of five-story building) at the upper surface. Yield model was considered as the mode of tensile stress. The model parameters of four types of soil samples were separately selected as follows (combined with the experimental results):

Plain soils: E=15MPa, ν =0.3, f_y = 15 Kpa, C=0.0472 MPa, ϕ =25.2°;
0.1% fiber soils: E=36MPa, ν=0.26, f_y = 22 KPa, C=0.0597 MPa, φ=28.6°;
0.2% fiber soils: E=48MPa, ν=0.23, f_y = 25 KPa, C=0.0616 MPa, φ=30.8°;
0.3% fiber soils: E=65MPa, ν=0.20, f_y = 28 KPa, C=0.0632 MPa, φ=32.5°.

3.2. *Computer simulation results*

Numerical calculation of the models was done as the method of applying loads step by step. The change nephograms of inner force and deformation were shown in Fig.1 and Fig.2. (Due to limited space, only the nephograms of two types of soils were drawn here).

The simulation results show that, with the different fiber content, the fiber soils are different in stress, strain, displacement and the steps that load imposed at the time of yield point, which reflects the different bearing capacity and deformation capacity of various types of fiber soils. Apparently, the content addition of polypropylene fiber can improve the mechanical properties of soils.

syy graph vertical displacement vectograph

Fig. 1. Stress and displacement analog nephogram of plain soil model.

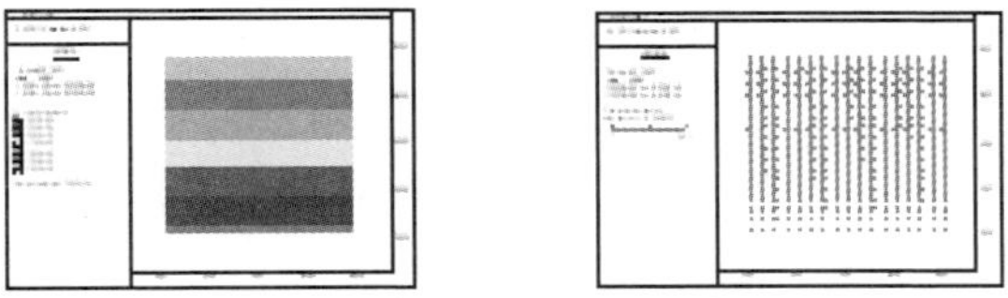

syy graph vertical displacement vectograph

Fig. 2. Stress and displacement analog nephogram of 0.2% fiber soil.

References

1. Li Guangxin, Chert Lun and Zheng Jiqin: Experimental study on fiber-reinforced cohesive soil. SHUILI XUEBAO, 1995, (6), 31–36.
2. Lei Sheng You: Theory and Technology of Modern Reinforced Soil. Beijing: China Communications Press, 2006.8.
3. Xiong Youyan: Texsol. CHINA JOURNAL OF HIGHWAY AND TRANSPORT, 1990,3(4), 84-88.
4. Jie Yuxin, Li Guangxin and Chen Lun: Study of centrifugal model tests on texsol and cohesive soil slopes. Chinese Journal of Geotechnical Engineering, 1998, 20(4), 12-15.
5. ZHANG Yan-mei, ZHANG Xu-dong and ZHANG Hong-ru: Test research of geotechnique textile soil reinforcement mechanism and engineering application. Rock and Soil Mechanics, 2005, 26(8): 1323-1326.
6. The People's Republic of China Ministry of Water Resources: The People's Republic of China national standard, GB/T 50123-1999 standard for soil test methods. Beijing: China Planning Press, 1999.9.

ON EFFECT OF NON-PERSISTENT JOINTS ON HYDRO-MECHANICAL BEHAVIOR OF JOINTED ROCK MASS IN UDEC

WU YUEXIU

Key Laboratory of Rock Mechanics in Hydraulic Structural Engineering, School of Civil and Architectural Engineering, Wuhan University, Wuhan, 430072, P.R. China

LIU QUANSHENG[1,2]

1. Key Laboratory of Rock Mechanics in Hydraulic Structural Engineering, School of Civil and Architectural Engineering, Wuhan University, Wuhan, 430072, P.R. China
2. State key laboratory of geomechanics and geotechnical engineering, Institute of Rock and Soil Mechanics, CAS Wuhan, Hubei, China

LIU XIAOYAN

Key Laboratory of Rock Mechanics in Hydraulic Structural Engineering, School of Civil and Architectural Engineering, Wuhan University, Wuhan, 430072, P.R. China

The existence of non-persistent joints will weaken the strength of rock mass and affect the mechanical behavior of rock mass which leads to change in the permeability of rock mass. A damage variable is introduced to account for weakened effect on strength of rock mass. It is defined as the ratio of the sum of actual length of all non-persistent joints in a block to the sum of length of all fictitious joints which are extensions of all non-persistent joints to the boundary of the block. According to the damage variable, the elastic modulus is deducted. Based on UDEC, a subroutine named damage_var is generated to calculate the damage variable and assign new properties to each block. Finally, numerical uniaxial compressive tests and the Hydro-Mechanical (HM) coupling calculations are carried out to study the effect of non-persistent joints.

Keywords: Non-persistent joint; Uniaxial compressive test; Hydro-Mechanical characteristic; Damage variable; Representative element volume; Permeability.

1. Introduction

There are non-persistent joints and connective joints in nature rock mass. The connective joints are the main flow channel of rock mass. The non-persistent joints are the main water storage zone and have little contribution on permeability of rock mass. Thus, non-persistent joints are deleted in most studies on seepage of rock mass. But, many laboratory tests showed that the existence of non-persistent joints will weaken the strength of rock mass.

Yunsheng Chen and Lixin Ma et. al[1,2] scanned the total process of uniaxial compressive test of samples with non-persistent joints with CT scanner and rock mass with non-persistent joints under different pressure is studied. Hao Li and Zhenyu Tao[3] gave out the damage constitutive relation with effective field method, considering the interaction among non-persistent joints. Shi Yan and Yunying Hang et.al.[4] proposed a model to analyze the equivalent elastic parameters of rock mass with non-persistent joints. Yuanming Liu and Caichu Xia et.al.[5,6] summarized the failure mechanism of rock mass with non-persistent joints and proposed the relative strength criterion. Aihua Liu et.al[7] proved that the existence of non-persistent joins will weaken the shear strength of rock mass through the laboratory tests on concrete samples. Kulatilak P.H.S.W et. al.[8] did research on determination of fiction joint stiffness, when there is some artificial joints in model in UDEC. According to the result, it is shown that adding the artificial joints into calculation models causes convergence and storage problems.

The existence of non-persistent joints will weaken the strength of rock mass and affect the mechanical behavior of rock mass which leads to change in the permeability of rock mass. Thus, the effect of non-persistent joints on HM behavior of rock mass is studied in this paper with UDEC.

2. Damage Variable D

The non-persistent joints are deleted in UDEC. To consider the effect of non-persistent joints on mechanical strength of rock mass, a damage variable D is defined as the ratio of the sum of actual length of all non-persistent joints in a block to the sum of length of all fictitious joints which are extensions of all non-persistent joints to the boundary of the block (the ratio of total length of solid lines to total length of dotted lines). According to the characteristic of joint network generated in UDEC, the connect joints are boundaries of blocks. So, there are only non-persistent joints in a block in theory.

$$D = \sum_{j=1}^{m} l_j \bigg/ \sum_{j=1}^{m} L_j \tag{1}$$

Where, m is the number of non-persistent joints in a block, l_j is actual length of the i_{th} non-persistent joint, and L_j is the length of the extension of the i_{th} non-persistent joint.

The elastic modulus of intact block is deducted by the damage variable:

$$E' = E_0(1-D) \tag{2}$$

A subroutine named damage_var is generated to calculate the damage variable and assign new properties to each block. Its flow chart is shown in Fig.1.

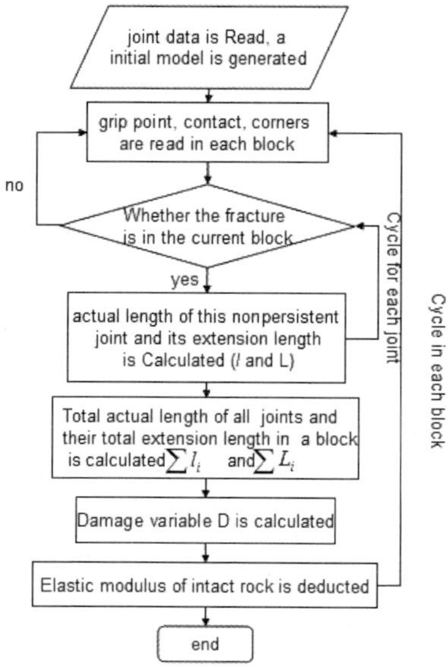

Fig. 1. Flow chart of calculation of *D* in a block (damage_var).

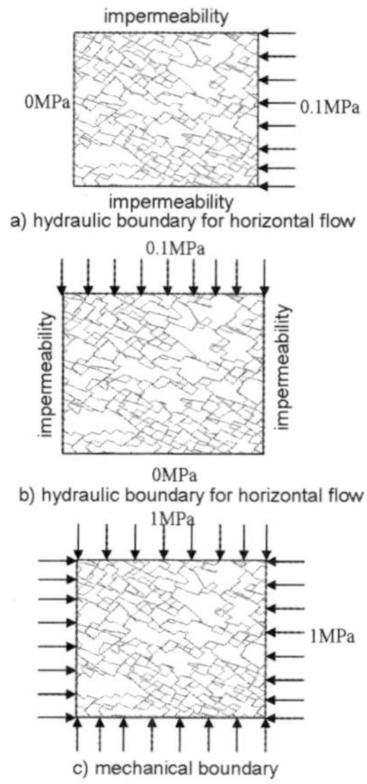

Fig. 2. Boundary conditions.

3. Calculation Model

To study the effect of non-persistent joints on HM coupling behavior of rock mass, discrete fracture network is generated with Monte-Carlo method. The joint middle point, trace length and dip are assumed to follow a normal distribution and the related parameters are shown in table 1. The damage_var subroutine is used to calculate the damage variable *D* for each block. Material parameters of intact rock and joint are list in table 2[9]. Two cases are considered (with considering non-persistent joints and without considering non-persistent joints). For study on the effect of non-persistent joints on HM coupling behavior of fractured rock mass, HM coupling calculation is done and its HM boundaries conditions are shown in Fig.2.

Table 1. Parameters of discrete fracture network.

Joint set		Trace length [m]	Dip [°]	Space [m]
1	average	4.0	150.0	2.0
	deviation	1.0	10	0.1
2	average	3.6	50.0	2.0
	deviation	0.7	7	0.1

Table 2. Material parameters of intact rock and joint.

Intact rock (M-C model)	SA model	
Density: 2500 kg/m3	Residual shear displacement: 6.25e-3 m	Critical stress: 8.0e7 Pa
Bulk modulus: 5.0×1012 Pa	maximum compression: 5.0e-3 m	shear stiffness:1.0e8 Pa
Shear modulus:4.5×1012 Pa	Cohesion of intact rock: 1.0e7 Pa	dilation angle: 8°
	Friction of intact rock: 31.5°	friction: 30°
	Normal stiffness: -5.0e10 Pa	Experiment parameters: k1=0.25, k2=0.81, b0=0.9

4. Results Analysis

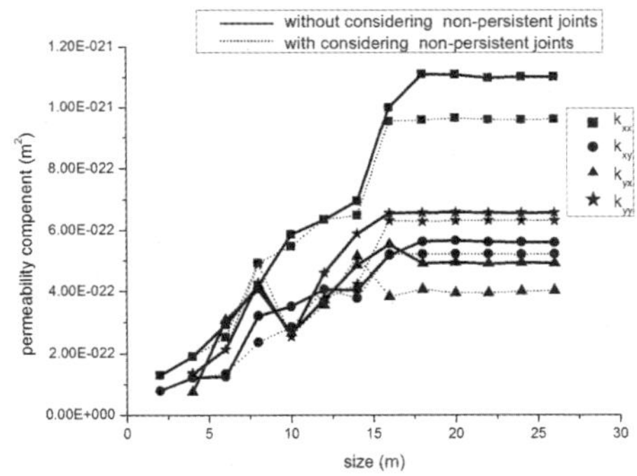

Fig. 3. The relationship between each component of permeability tensor and the model size.

Based on section 3, the HM coupling calculation is done for different models with different size (6~28m^2) and the results are shown in Fig.3. From Fig.3 (a), it is concluded that: 1) for two cases, each permeability component increases with the increasing of model size and is stable at a model size of 18m^2, which means that the REV of each case is about 18×18 m^2; 2) in general speaking, the permeability of fractured rock mass decreases when the non-persistent joints are considered. The elastic modulus decreases and the deformation of rock mass increases when the non-persistent joints are considered. Under the same pressure, joints are compressed and the aperture of joints will decrease. And then it causes the decreasing of permeability of fractured rock mass. From Fig. 3 (a) ~ (d), the permeability tensors of fractured rock mass without considering non-persistent joints ($[k]$) and with considering non-persistent joints ($[k']$) are:

$$[k]=\begin{bmatrix} 11.0 & 5.265 \\ 5.265 & 6.58 \end{bmatrix}\times 10^{-22},\quad [k']=\begin{bmatrix} 9.59 & 4.62 \\ 4.62 & 6.32 \end{bmatrix}\times 10^{-22}$$

5. Conclusion

From the results of HM coupling calculations of two cases (without considering the non-persistent joints and with considering non-persistent joints), it is concluded that: From the results of HM coupling calculations, REV of each case has little difference and is about 18×18 m^2. Considering the effect of non-persistent joints, the equivalent permeability tensor decreases. The elastic modulus decreases and the deformation of rock mass increases when the non-persistent joints are considered. Under the same pressure, joints are compressed and the aperture of joints will decrease. And then it causes the decreasing of permeability of fractured rock mass.

Based on the above analysis, it can be concluded that the existence of non-persistent joints has an important effect on HM coupling characteristic of rock mass. In nature engineering, the non-persistent joints should be considered. A new method to consider the non-persistent joints in UDEC is proposed at the end.

Acknowledgments

This study was supported by China National Natural Science Foundation (Grant No.: 50574087), and knowledge innovation project of Chinese Academy of Sciences (Grant No.: kzcx2-yw-116). Foundation of Second Secretary Science and Industry, State Administration of Science Technology and Industry for National Defense is also acknowledged (Grant No.127).

References

1. Yunsheng Cheng, Lixin Ma, Ning Li and Xin Han: Journal of Xi'an University of technology. 2006. 22(1): 1-4
2. Yunsheng Cheng, Ning Li, Aiguo Li, Yibing Pu and Quanrong Liao: Chinese journal of rock mechanics and engineering. 2000, 19(6): 702-706
3. Hao Li and Zhenning Tao: Journal of Wuhan University of hydraulic and electrical engineering. 2000. 33(3): 5-8
4. Shilin yan, yuying huang and chuanyao chen: Huazhong University of Science and Technology. 2001. 29(6): 64-67.
5. Yuanming Liu and Caichu Xia: Rock and soil mechanics. 2007. 28(8): 1719-1724
6. Yuanming Liu, Caichu Xia and Hongzhe Li: Chinese journal of underground space and engineering. 2007, 3(4): 682-693
7. Aihua Liu, Rongwu Luo, Hong Li and Yijin Zhang: Journal of Xi'an University of science and technology. 2009. 29(6): 726-730
8. Kulatilake P.H.S.W, Ucpirti H, Wang S, Radberg G, and Stephansson O: rock mechanics and rock engineering, 1992,25(4):253-274.
9. Wu Yuexiu, Modeling Rough Joint Network and Hydro-Mechanical Behavior of Fractured Rock Mass [Ph D thesis]. China: Institute of Rock and Soil Mechanics, Chinese Academy of Sciences, Wuhan, 2010.

THE CREEP BEHAVIOUR OF FRACTURE GRANITE UNDER HYDRO-MECHANICAL CONDITION

LIN-FANG SHEN [†], XIA-TING FENG, PENG-ZHI PAN, HUI ZHOU

State Key Laboratory of Geomechanics and Geotechnical Engineering, Institute of Rock and Soil Mechanics, Chinese Academy of Sciences, Wuhan, 430071, China

Experiments are conducted to investigate the creep and hydraulic behaviours of single fractured granite under hydro-mechanical condition. The results show that, both the axial strain and lateral strain develop with time. Axial strain becomes stable after the early hundreds hours, while lateral strain of seepage sample always increases during the whole experiment time. It is concluded that hydraulic pressure greatly influences the lateral strain variation, which shows some special properties that are different from the common traxial creep curve. Based on the test, anisotropy elastic-visco-plastic creep model is established, which is integrated into a 3D cellular automaton modeling system ($EPCA^{3D}$-EVP). With this system, anisotropy creep characters of fractured rock under test condition can be well modeled.

Keywords: Sample creep behaviour; fractured granite; HM coupling; EPCA3D-EVP.

1. Introduction

Complex behaviour of engineer rock mass always depends on intricate surrounding conditions, which usually connect with time variation. When suffered to mechanical field, seepage field and even the chemical field, rock mass shows more complicated creep property, especially anisotropy character. Even if the isotropic rock, it will exhibit anisotropic mechanical character in the case of engineer disturb. Hence, investigating the anisotropic creep property of fractured rock is of significant scientific meaning. As to the creep property of fractured rock, researchers developed a lot of deep and broad work at home and abroad. Important progresses of rock creep studies have been made both in laboratory and in basic theory, most of which are suitable to rock matrix without fracture [1-8]. Few considerations are made to the joint rock creep under coupled condition. Because of the lack of effective test method, theoretical study and practical application far lag behind. This paper focuses on the seepage-mechanical coupled creep test of fractured granite, and analyses the water disturbing effect on anisotropy deformation of fractured rock. Based on the test,

[†] Corresponding author, email:shenlinfang@gmail.com

anisotropy elasto-visco-plastic creep model is established, which is implanted into a 3D cellular automaton modeling system [9,10]. With this system, anisotropy creep characters of fractured rock under test condition can be well modeled.

2. Study on Triaxial Creep Test of Fractured Granite

Figure 1 is axial creep strain and lateral creep strain of fractured granite under two creep test conditions which are dry without seepage and with seepage condition (the detailed description of these tests can be referred to reference11). Both the axial strain and lateral strain develop with time (compression as positive). Axial strain becomes stable after the early hundreds hours, while lateral strain of seepage sample always increases during the whole experiment time.

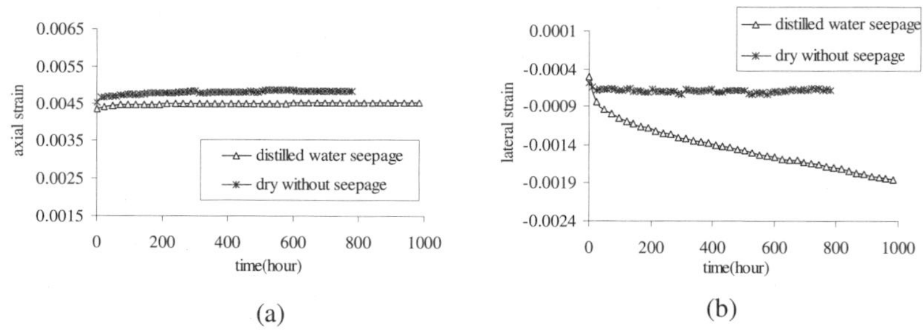

Fig. 1. Creep strain of fractured granite. (a) axial creep strain; (b) lateral creep strain.

Figure 1.a shows that the axial creep deformation under distilled water seepage is close to dry condition. The initial creep rate is small and then enters into steady stage rapidly, deformation doesn't develop, and creep rates approach to null. Initial rate is 10 times as larger as steady rate.

Lateral creep strain variations are shown in Figure 1.b. Distinct differences can be observed between samples with seepage and without seepage conditions. Lateral strain is almost at constant stage for the sample without seepage. While as to the seepage sample, the lateral strain develops at a steady rate with time. This phenomenon indicates the water has a great impact on lateral strain of fractured rock. The initial lateral strain rate of dry sample without seepage is $3.64\times10^{-6}h^{-1}$, the minimum lateral strain rate during steady stage is $0.35\times10^{-6}h^{-1}$. Hence, the lateral deformation doesn't change too much during the whole creep process of dry sample. As to the seepage sample, the initial lateral strain rate is $0.38\times10^{-5}h^{-1}$. The minimal lateral strain rate during steady stage is $6.93\times10^{-6}h^{-1}$. Both of them are much larger than that of dry sample

Thus it can be seen, because of the fracture existence and different hydraulic condition, anisotropy deformation becomes significant.

3. Visco-Plastic Flow Coefficient Tensor

Sun has studied the visco-plastic coefficient η_2 of geomaterials[1]. Under the condition of nonlinear creep, visco-plastic coefficient is related to viscosity, stress lever and time *etc.* In order to reflect the rock mass anisotropic creep character, visco-plastic coefficient tensor η_{2ij} is proposed based on visco-plastic coefficient. This tensor is the function of stress and time:

$$\eta_{2ij} = \eta_{2ij}(\sigma, t) \tag{1}$$

Hence, visco-plastic flow coefficient can be obtained in the form of tensor:

$$\gamma_{ij} = \eta_{2ij}^{-1} \tag{2}$$

Meanwhile, it is also the function of stress and time:

$$\gamma_{ij} = \gamma_{ij}(\sigma, t) \tag{3}$$

Let the diagonal element satisfies the below equation:

$$\gamma_{2ii} = at^b + c \tag{4}$$

$\gamma_{2ij} = \dfrac{(\gamma_{2ii} + \gamma_{2jj})}{2}$, i,j=x, y, z, i、j do not sum here.

In the above equation, t is time, a, b, c are constant parameters determined by test. These parameters include the effect of stress and some other factors (such as chemical erosion and water soften) on visco-plastic flow coefficient. Hence, they can well describe the anisotropic creep character of fractured rock.

Substitute the visco-plastic flow coefficient into viso-plastic strain rate equation, we have

$$(\dot{\varepsilon}_{vp})_{ij} = \gamma_{ij} < \varphi(\boldsymbol{F}) > \frac{\partial \boldsymbol{F}}{\partial \sigma_{ij}} \tag{5}$$

where i, j does not sum. Eq.(5) can be written as the following matrix form,

$$\begin{bmatrix} \dot{\varepsilon}_{xx} & \dot{\varepsilon}_{xy} & \dot{\varepsilon}_{xz} \\ \dot{\varepsilon}_{yx} & \dot{\varepsilon}_{yy} & \dot{\varepsilon}_{yz} \\ \dot{\varepsilon}_{zx} & \dot{\varepsilon}_{zy} & \dot{\varepsilon}_{zz} \end{bmatrix} = \langle \varphi(F) \rangle \begin{bmatrix} \gamma_{2xx}\dfrac{\partial F}{\partial \sigma_{xx}} & \gamma_{2xy}\dfrac{\partial F}{\partial \sigma_{xy}} & \gamma_{2xz}\dfrac{\partial F}{\partial \sigma_{xz}} \\ \gamma_{2yx}\dfrac{\partial F}{\partial \sigma_{yx}} & \gamma_{2yy}\dfrac{\partial F}{\partial \sigma_{yy}} & \gamma_{2yz}\dfrac{\partial F}{\partial \sigma_{yz}} \\ \gamma_{2zx}\dfrac{\partial F}{\partial \sigma_{zx}} & \gamma_{2zy}\dfrac{\partial F}{\partial \sigma_{zy}} & \gamma_{2zz}\dfrac{\partial F}{\partial \sigma_{zz}} \end{bmatrix} \tag{6}$$

This expression can be easily implanted into program. A numerical system ($EPCA^{3D}$-EVP) to simulate the elasto-visco-plastic behaviour of rocks is developed based on $EPCA^{3D}$ in Visual C^{++} environment[10].

4. Modeling of Fracture Granite Creep Deformation with EPCA3D-EVP

Through the analysis of different effect of visco-plastic parameters on axial and lateral deformation, and combine with test results, parameters in visco-plastic flow coefficient can be chosen in the case of dry sample without seepage and seepage-creep coupled test respectively based on back analysis. These modeling results are shown in figure 2. and figure 3. It can be seen that these two results fit well both in magnitude and in tendency.

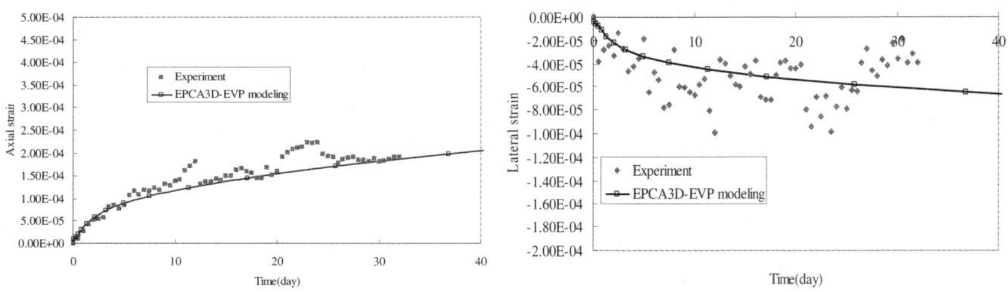

(a) (b)

Fig. 2. Comparisons between experiment and $EPCA^{3D}$-EVP modeling for fractured granite without seepage. (a) axial creep deformation; (b) lateral creep deformation.

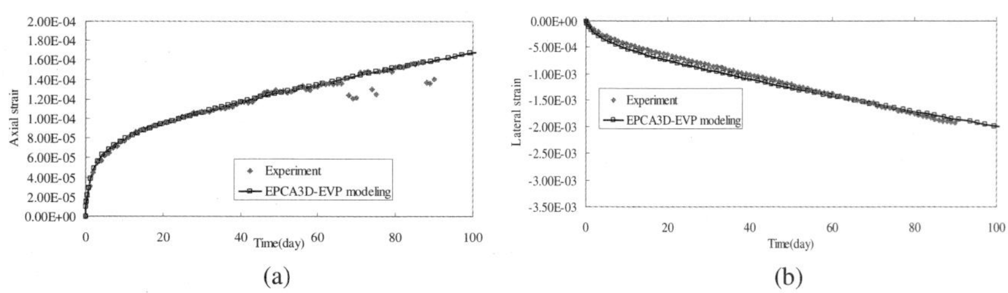

(a) (b)

Fig. 3. Comparisons between experiment and $EPCA^{3D}$-EVP modeling for fractured granite with hydro-mechanical coupling. (a) axial creep deformation; (b) lateral creep deformation.

5. Conclusion

Experiments are conducted to study the creep and hydraulic behaviours of single fractured granite under hydro-mechanical condition. The results show that, for the fracture existence and different hydraulic condition, anisotropy deformation becomes significant. Based on the test, anisotropy elasto-visco-plastic creep model is established, which is implanted into a 3D cellular automaton modelling system ($EPCA^{3D}$-EVP). With this system, anisotropy creep characters of fractured rock under test condition can be well modeled.

Acknowledgments

This work was supported by a grant from the National Basic Research Program of China (No. 2010CB732006) and the National Natural Science Foundation of China (Nos. 10872209, 10972231)

References

1. J. Sun: *Chinese J. Rock Mech. Engng.*, 2002, **21,** (6), 822–827.
2. H.H. Zhu and B. Ye: *Chinese J. Rock Mech. Engng.*, 2002, **21,** (12), 1791–1796.
3. F.S. Zhao, B.Y. Zhang and J.B. Peng: *Chinese J. Rock Mech. Engng.*, 2002, **21,** (10), 1527–1532.
4. Z.Y. Wang, S.T. Guo and Y.P. Li: *Rock & Soil Mech.*, 2006, **27,** (12), 2122–2126.
5. B. Liang and P. Li: *Mech In Engng.*, 2006, **28,** (5), 69–72.
6. Y.L. Zhao, P. Cao, H. Lin and Y.K. Liu: *Chinese J. Rock Mech. Engng.*, 2008, **30,** (4), 511–517.
7. X.H. Cui, Z.L. Fu: *Chinese J. Rock Mech. Engng.*, 2006, **25,** (5), 1021–1024.
8. Z.L. Fu, Y.F. Gao, W.Ning and J.P. Xu: *J. Mining & Safety Engng*, 2007, **24,** (03), 353–356.
9. X.T. Feng, P. Z. Pan and H. Zhou: *Int J. Rock Mech. and Min. Sci.*, 2006, **43,** (7), 1091–1108.
10. P.Z. Pan, X.T. Feng and J.A. Hudson: *Int J. Rock Mech. and Min. Sci.*, 2009, **46,** (4), 674–685.
11. L.F. Shen, X.T. Feng, P.Z. Pan and H.Zhou: *Chinese J. Rock Mech. Engng.*, 2010, **29,** (7), 1379–1388.

NUMERICAL ANALYSIS OF A TRANSMISSION TOWER WITH COMPOUND FOUNDATION UNDER HORIZONTAL GROUND MOTION

JU-KE BAI†

IESDP, School of Naval Architecture and Ocean and Civil Engineering, Shanghai Jiao tong University
Shanghai 200240, China

School of Architectural Engineering, Liaoning Technical University
Fuxin, Liaoning 123000, China

LONG-ZHU CHEN

IESDP, School of Naval Architecture and Ocean and Civil Engineering, Shanghai Jiao tong University
Shanghai 200240, China

Composite foundation can be considered as an isolation base of the upper structure. Coulomb friction model was introduced to simulate the isolation of the composite foundation. Frequency analysis shows that the mode shapes were greatly affected by the composite foundation. It is difficult to take too many modes into account. So, numerical transient dynamic method was employed to analyze the effects of the existence of composite foundation on a steel transmission tower. Two seismic waves, with the same peak acceleration 0.4g, were chosen for the analysis of the tower with composite foundation and with fixed end. The results show that composite foundation can apparently change the frequency characteristic of the tower and decrease the responses of the steel tower.

Keywords: Numerical analysis; Composite foundation; Finite element; Seismic wave; Mode shape.

1. Introduction

Seismic isolation systems are effective in protecting buildings from the earthquake damage.[1-2] This statement was confirmed by the actual behavior of seismic isolated buildings and bridges as reported by Mayes, Fujita.[3] In recent years, numerical analysis and experiments on isolation of transmission towers are promoted.[4-6]

---

† Corresponding author, email: baijuke@gmail.com

For structures in soft soil areas, composite pile foundations are commonly used to develop the potential capability of the pile foundation to support the upper structures. Generally, a cushion is inducted to isolate the upper structure from the piles so that the vertical loads of the upper structures can be evenly transmitted to the piles and the surrounding soil. [7] The cushion acts as the isolation device of the upper structure when suffering from the probable earthquake. [8] In this paper, numerical study is executed to verify the effectiveness of the cushion on the dynamic responses of a transmission tower.

2. Finite Element Model of the Tower

The finite element model of a 18-meter-high transmission tower is shown in Fig. 1(*a*). The tower is composed of angle irons and weights 1300kg. It is supported on the composite pile foundation and we call it a research tower. In order to study the function of the cushion, a tower supported by pile-shaft foundation is imported for comparison, which we called reference tower when mentioned in the later paragraphs.

Fig. 1. 1st mode shapes of the towers.

Finite element software abaqus is employed to study the dynamic properties and responses of the upper structure. Considered that the rigidity of the cushion is much smaller than the structure and the foundation, the foundation is simplified to an elastic lab and the cushion is regard as the friction contact between the composite foundation and the baseboard of the tower. The reference tower is considered as fixed end structure.

3. Results

3.1. *Frequency analysis*

In order to understand the functionality of the isolation cushion, frequency analysis was applied to obtain the dynamic properties of the tower. [9] The first mode shapes of the two towers with different foundation are shown in Fig. 1(*b*) and (*c*) and the detailed parameters of the first 5 modes are listed in Tab. 1 and Tab. 2.

Table 1. Frequencies of the reference tower.

mode	Frequency (rad/s)	Participation factor	Effective mass(kg)
1	43.6	1.68	614.5
2	43.8	-.328	22.18
3	62.9	2.28e-2	3.495e-2
4	68.9	0.823	27.20
5	71.5	0.927	35.44

Table 2. Frequencies of the research tower.

mode	Frequency (rad/s)	Participation factor	Effective mass(kg)
1	1.18	5.20e-3	3.427e-2
2	1.18	1.00	1300
3	1.89	3.68e-4	3.864e-5
4	29.3	-1.29e-5	3.969e-8
5	41.3	-6.95e-6	5.091e-9

From the above two tables, we can find that the fundamental frequency changed greatly by the isolation of the tower. The first mode shape of the reference tower has no distinction with the original shape and this means it has very little deformation except in the cushion.

3.2. *Dynamic time-history analysis*

Dynamic time history analysis is applied to quantitatively know the effects of the cushion on the responses of the tower under horizontal seismic waves. Two horizontal seismic waves, El Centro and Loma wave are chosen for the analysis. The peak values of accelerations of the two waves are normalized to 0.4g to compare the reactions of the towers under different seismic waves, where g is the acceleration of gravity. For the sake of calculating expenditure, the time duration is set as 10 seconds. The friction coefficient of the cushion between the tower baseboard and the foundation is assumed to be 0.1. Stiffness damping is considered in the calculation and the material damping ratio is 0.02.

Responses of three points are studied of the research tower. Point 141 is on the top of the tower, point 281 is on the center of the raft and point 1 is on the ground, respectively. For the reference tower, two points are selected to study and their positions are the same as point 141 and 281.

Through dynamic analysis, three accelerations of the reference points of the research tower are obtained and shown in Fig.2 and Fig.3. From the figures we know that the peak accelerations of the reference points on the rafts are much smaller than that of the ground motion. The peak responses of the reference points on the top of the tower are larger than that of the corresponding points on the rafts and approximately the same magnitude as the ground motion.

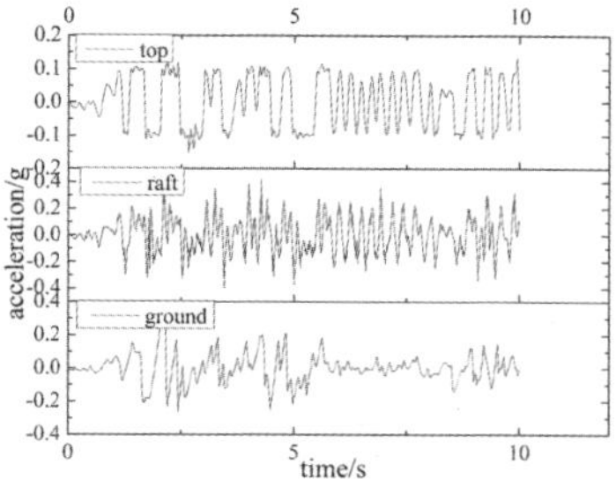

Fig. 2. Accelerations of the research tower under El Centro wave.

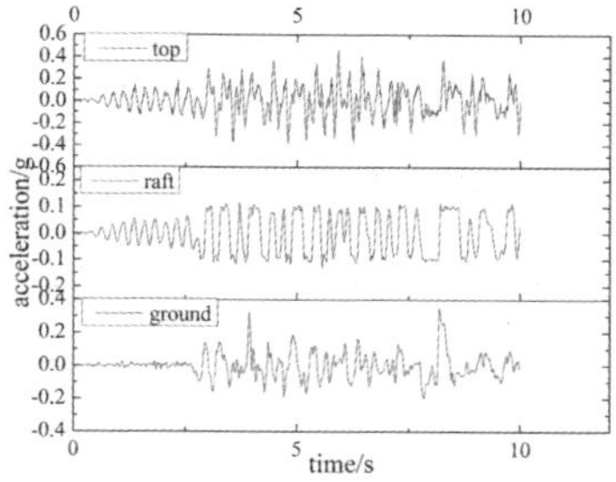

Fig. 3. Accelerations of the research tower under Loma wave.

Accelerations of the corresponding reference points of the reference tower are also studied and can be seen in Fig 4 and Fig.5. For the reference tower, the raft is fixed with the foundation and the responses are the same as the ground motion. Accelerations on the reference points are much larger than that of the raft and it is approximately double the peak value of the ground motion while it is subjected to the two seismic waves.

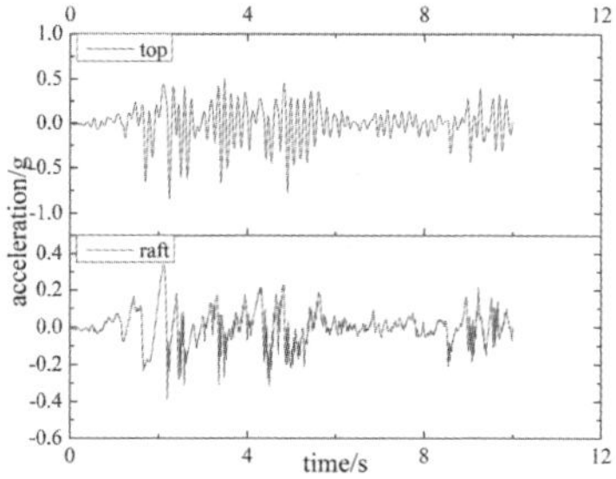

Fig. 4. Accelerations of the reference tower under El Centro wave.

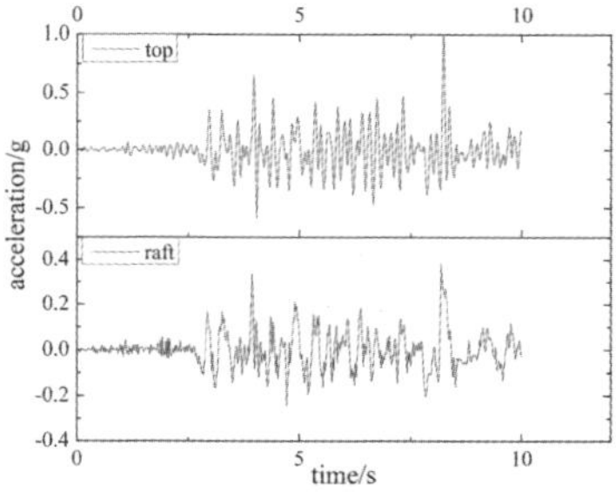

Fig. 5. Accelerations of the reference tower under Loma wave.

4. Conclusion and Discussion

Finite element analysis shows that the cushion between the baseboard of the upper tower and the composite foundation can effectively isolate the ground motion from influencing the responses of the upper tower. The peak value of the acceleration reduced about 50% by the cushion. The time-history curves of the top floors are greatly affected by the cushion and their shapes are not familiar with the seismic waves. Under horizontal seismic ground motion, the isolated tower moves like transverse motion of a rigid body.

There are still many problems in using Coulomb friction model to simulate the isolation of the interaction between the composite foundation and the upper structure. It can reflect the decay of the seismic energy in the soft soils. More complicated models, such as pile-soil-structure interaction systems, are expected to be supposed to simulate the model in the following research.

References

1. R. R. Martel. The effects of earthquakes on buildings with a flexible first story. Bulletin of the seismological Society of America **19**(3), 1929, 53-60
2. N. B. Green. Flexible first story construction for earthquake resistance. Transactions, American Society of Civil Engineers 100, 1935, 182-190
3. M. O. Moroni, M. Sarrazin and R. Boroschek. Experiments on a base-isolated building in Santiago, Chile, Engineering Structures **20**(8),1998, 720-725
4. L. Hongnan, B. Haifeng, State-of-the-art review on studies of disaster resistance of high-voltage transmission tower-line systems, China Civil Engineering **40**(2), 2007, 39-46
5. X. Zhonggen, W. Cuikun, L. Chen, Shaking table test of a highrise steel structure model, Spatial Structures **8**(4), 2002, 36-45.(in Chinese)
6. L. Shuguo, Z. Jihua, W. Lizheng, Analysis of dynamic characters of electrical transmission tower-line system with a big span, Earthquake Engineering and Engineering Vibration **23**(6), 2003, 63-69.(in Chinese)
7. G. Xiaonan. Generalized composite foundation theory and engineering application. Chinese Journal of Geotechnical Engineering **29**(1) , 2007, 1-13.(in Chinese)
8. Z. Shao-wei, D. Ming-yuan, G. Rong, et al. An experimental study of isolating properties of sand cushion under the foundation by shaking table. Journal of Hebei University of Technology **34**(3), 2005, 92-97.(in Chinese)
9. L. Hongnan, W. Qianxin, Dynamic characteristics of long-span transmission lines and their supporting towers , China Civil Engineering Journal **30**(5) , 1997 , 39 - 41.(in Chinese)

FINDINGS IN STRESS EVOLUTION OF SURROUNDING ROCKS IN GOB-SIDE ENTRY RETAINING

ZHANG NONG[†]

State Key Laboratory of Coal Resource and Mine Safety, School of Mines, China University of Mining and technology, Xuzhou, Jiangsu 221116, China

KAN JIA-GUANG, LI GUI-CHEN, ZHENG XI-GUI

State Key Laboratory of Coal Resource and Mine Safety, School of Mines, China University of Mining and technology, Xuzhou, Jiangsu 221116, China

Development and wider application of gob-side entry retaining have been stagnated by inadequate understanding of the dynamic stress evolution of its surrounding rocks. Numerical simulation and field measurement are combined to reveal its stress evolution in roadway excavation, gob-side entry retaining and entry stabilization under three typical roof conditions respectively. Results show that in roadway excavation, with increase in thickness of the immediate roof, supporting pressure increases. In the mining-induced stress adjusting period, with decrease in thickness of the immediate roof, the abutment stress concentration coefficient increases and maximum abutment stress occurs farther from the work face. These findings are of great value to roadway bolting.

Keywords: Gob-side entry retaining; Stress evolution; Roof condition; Numerical simulation; Fill-in pack.

1. Introduction

From the1950s, many experimental researches have been done worldwide to study stress regime, support resistance and control technology of gob-side entry retaining. Significant achievements have been made[1,2]. But most of them are inflexible for the varying geological and mining conditions in China. Lack of thorough study on similarities and differences of stress distribution under various roof conditions has constrained further development of gob-side retaining[3,4]. Further development of gob-side retaining has been constrained.

2. Classification of Roof Condition

Roof condition greatly influences stress evolution of surrounding rocks[5]. According to ratio between thickness of the immediate roof and mining height,

[†] Corresponding author, email: zhangnong@126.com

roof is generally classified as thick immediate roof, thin immediate roof and zero- immediate roof for practical convenience.

Thick immediate roof refers to roof with large thickness mainly composed of collapse prone strata, timely fall of which can fully fill the gob when support chocks are removed. The collapsed rocks would be gradually compacted as work face advances and make upper remnant rock to be re-supported to certain extent.

Thin immediate roof is when thickness of the immediate roof is 0.3 times bigger but 3–5 times smaller than mining height. Partial compaction of the gob by collapsed rocks makes most of the overlying strata suspending, so the weight of them will be passed into coal, pillar or fill-in pack via the beam or plate structure.

Due to too thin or vertical no immediate roof, main roof of the so–called zero- immediate roof won't fall with moving of support chocks. Then a cantilever beam is formed and tilts rapidly with strong stress regime when it is broken.

3. Numerical Simulations

Numerical simulation FLAC^{3D} is employed. Three groups of parameters are inputted for calculation to study stress characteristics in both roadway excavating and entry retaining under the above three roof conditions respectively.

3.1. *Stress distribution in roadway excavation*

Vertical stress distribution of the immediate roof along face direction under different roof conditions can be seen in Figure 1. Supporting pressure increases with increase in thickness of the immediate roof in roadway excavation. As roadway advances, lateral supporting pressure increases, maximum value of which occurs farther from the work face. Lateral supporting pressure occurs on both sides of the roadway. Its distribution can be identified as three zones: stress-decreasing zone (3 m in coal), stress-increasing zone (3–30 m in coal, peak value at about 5 m) and original stress zone (30 m deeper in coal). Location of the roadway is Y-coordinate of between 45 m and 50 m.

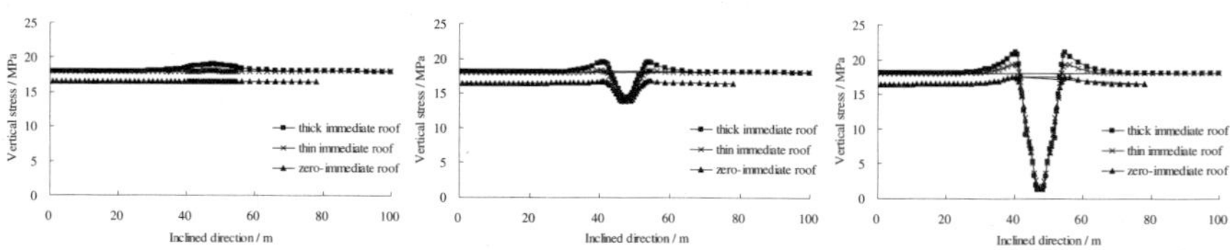

a. In front of work face 5 m

b. Driving work face

c. 10 m behind work face

Fig. 1. Vertical stress distribution in roadway excavating.

3.2. *Stress evolution in entry retaining*

Figure 2 shows vertical stress distribution along face direction under different roof conditions. In front of the work face as shown in Figure 2.a, supporting pressure under zero-immediate roof is lower than that under thick and thin immediate roof. Supporting pressures under the later two roof conditions are almost the same. Behind the work face as shown in Figure 2.b, with decrease in thickness of the immediate roof, abutment stress coefficient increases. Maximum pressure also occurs farther from the work face and extends deeper into the coal. Distribution of supporting pressure on solid coal sides shows that rear supporting pressure is stronger than advance supporting pressure. Location of the fill-in pack is Y-coordinate of between 49.5 m and 52 m.

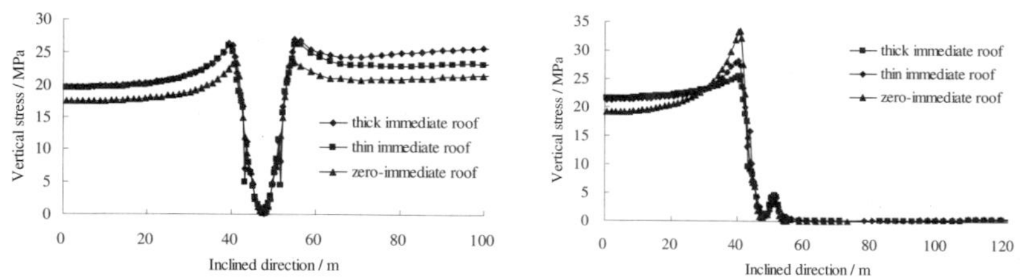

a. 40 m in front of work face
b. 40 m behind of work face

Fig. 2. Vertical stress distribution under different roof conditions in entry retaining.

4. Field Measurements

4.1. *Stress measurement methods*

Take Panel 1115(1) in Guqiao mime as an example, observation stations are set up under thin immediate roof and zero-immediate roof conditions. A concrete pressure gauge is buried into the to-be-filled wall (0.5 m away from the roof). Relationship between the stress on fill-in pack and its distance from the work face can reveal its stress evolution. Borehole stress-meter is applied to measure the stress in solid coal. Stress variation inside the hole can be transformed into liquid pressure within the stress pillow. Then pressure is delivered to a pressure gauge through pipeline, and the internal stress variations are indicated by the pressure needle.

4.2. *Stress evolution in fill-in pack*

Stress evolution in fill-in pack under different roof conditions is in shown in Figure 3. Under thin immediate roof, pressure varies between 3.2–5.4 MPa one day after the concrete pressure gauge being buried. It reaches maximum up to 5.7 MPa at 20 m behind the work face and then begins to decline quickly. Under zero-immediate roof, pressure varies between 4–11.2 MPa one day later. But as face advances, the pressure sharply drops to zero at 40 m behind the face. Then the concrete pressure gauge is manually pressurized to 3.4 MPa, making the declining rate slows down somehow.

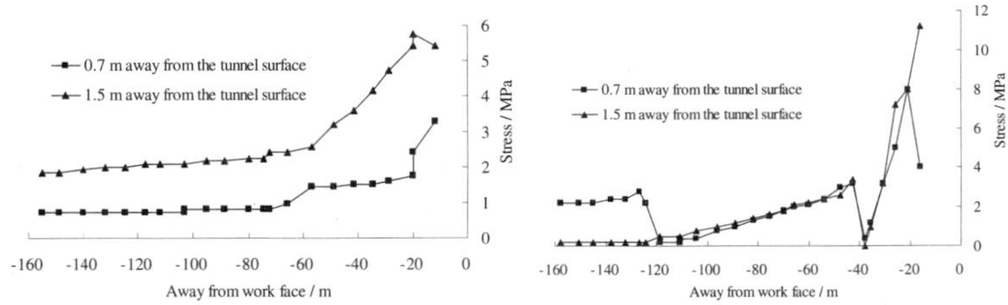

a. Thin immediate roof

b. Zero- immediate roof

Fig. 3. Stress curves of fill-in pack.

Observation results show that deformation and stress on fill-in pack are closely related to roof activities. Stress maximum occurs behind work face in a periodic weighing, and then begins to decrease as face advances. This can be attributed to the fact that when the panel is mined out, the lateral roof breaks, tilts and possesses rather high bearing stress behind work face in a periodic weighing. Under zero-immediate roof, due to violent tilting of roof structure and higher stress abutment; fill-in pack has to bear larger pressure and undergo more rib slough and serious deformation than that under thin immediate roof. However under either roof conditions, fill-in pack still maintains certain integrity and timely provides supporting resistance for the roof.

4.3. *Stress distribution in side coal*

As shown in Figure 4, peak stress (up to 21.2 MPa) in solid coal under thin immediate roof occurs close to the work face, which has been proved by both field measurement and numerical simulation. Peak stress is found at the 10 m testing points.

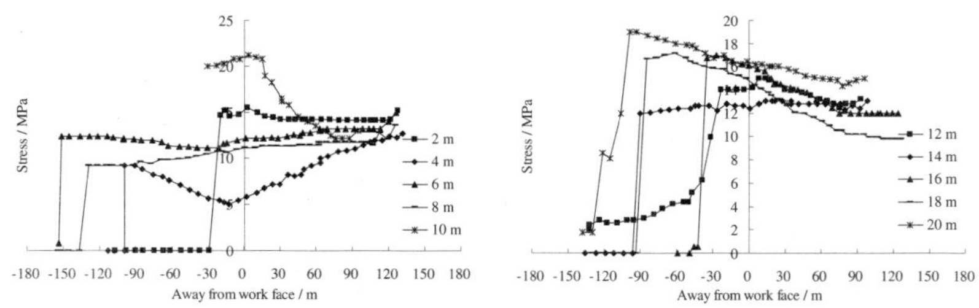

a. Stress of measure points between 2–10 m

b. Stress of measure points between 12–20 m

Fig. 4. Stress curves of coal under thin immediate roof.

Before mining-induced influence, stress in solid coal increases as approaching the work face; after mining-induced influence, the limit equilibrium zone extends deeper into coal seam with decreasing abutment stress coefficient.

5. Conclusion

Movement of overburden strata is a main reason to induce intensive stress and displacement of a retained entry. The combination of three-dimensional numerical simulations and field measurements has revealed stress development and evolution of surrounding rock under three typical roof conditions in gob-side entry retaining. Results show that in roadway excavating, supporting pressure increases with increase in thickness of the immediate roof while during the mining-induced stress adjusting period, with decrease in thickness of the immediate roof, the abutment stress concentration coefficient increases and peak abutment stress occurs farther from the work face. These findings are of great value to roadway bolting. However, more work is needed to better our understanding of stress distribution in gob-side entry retaining with ever-increasing mining depth.

Acknowledgments

We express our appreciation to State Key Laboratory of Coal Resources and Safe Mining (CUMT) (SKLCRSM09X03), the National Basic Research Program of China (2007CB209408) and the New Century Excellent Talents in University (Program 06-0478) for their financial support.

References

1. S. L. Lu: 'Mine Pressure Behavior without Chain Pillars', 1st edn, 124; 1982, Beijing, China Coal Industry Press.
2. H. H. Sun and B. L. Zhao: 'Theory and Practice of Gob-side Roadways', 1st edn, 256; 1993, Beijing, China Coal Industry Press.
3. W. B. Xie: *Chinese Journal of Rock Mechanics and Engineering*, 2004, 23, (18), 3059–3065.
4. L. Q. Ma, D. S. Zhang, T. Chen and G. W. Fan: *Chinese Journal of Rock Mechanics and Engineering*, 2007, 26, (3), 544–550.
5. J. G. Kan. 'Mechanism Analysis and Control Technology of Gob-side Entry Retaining Roadway of Typical Roof Structure', Ph. D. dissertation; Xuzhou, China University of Mining and Technology, 2009.

EFFECTS OF TEMPERATURE AND WHEEL LOADING ON RUTTING DEFORMATION OF ASPHALT MIXTURE

XINHUA YANG, JIN SHANG, ANYI YIN [†], SHENGFENG YANG, YONG YE

School of Civil Engineering and Mechanics, Huazhong University of Science and Technology, Wuhan, Hubei 430074, China

Hooke's law and the Bailey-Norton formulation are respectively used to characterise elastic and viscous deformations of asphalt mixture. A nonlinear and time-dependent numerical frame is constructed. The rutting deformations in standard cuboid specimens of asphalt mixture under different temperatures and cycled wheel loads are respectively calculated, and effects of temperature and wheel loading on rutting deformation are primarily evaluated.

Keywords: Asphalt mixture; elastic and viscous deformation; creep; rut depth.

1. Introduction

As an important pavement material, asphalt mixture is widely used in high-grade highway, because of its distinct advantages over other road surface materials. Due to its flexibility, however, rutting deformation is one of the main failure modes of asphalt pavement. In order to improve safety and comfortability of highway, it is necessary to further study and understand rutting deformation mechanism of asphalt mixture. In recent years, experimental and numerical investigations on rutting deformation in asphalt pavement attracted more and more attention[1-3]. These researches are of considerable significance for rutting resistance analysis, life prediction and material modification of asphalt pavement.

Generally, the numerical modelling technique can offer a more cost-effective solution than expensive and time-consuming experiments, but it need a constitutive model. Early in 1985, Sides *et al.*[4] proposed a comprehensive constitutive model for asphalt mixtures in which elastic, plastic, viscoelastic and viscoplastic strains are evaluated individually. For simplicity, the viscoelastic and viscoplastic responses are not partitioned in this paper, so that asphalt mixtures are described as elastic and viscous materials. Hooke's law and the Bailey-Norton formulation are respectively used to characterise elastic and

[†] Corresponding author, email: d4yin@smail.hust.edu.cn

viscous deformations. Similar to Chen and He's work[5], a nonlinear and time-dependent finite element algorithm frame is constructed. A series of numerical computations are performed to evaluate the effects of temperature and wheel loading on rutting deformation of asphalt mixture.

2. Nonlinear and Time-dependent Algorithm Frame

The equilibrium equation in increment form for finite element analysis is expressed as

$$[K_{\mathrm{T}}]\Delta\{d\}^n + \Delta\{\tilde{f}\}^n = 0 \tag{1}$$

with

$$[K_{\mathrm{T}}] = \int_{\Omega} [B]^{\mathrm{T}} [D]^{-1} [B] \mathrm{d}\Omega \tag{2}$$

$$\Delta\{\tilde{f}\}^n = \Delta\{f\}^n - Apt^{p-1}[R]\Delta t \int_{\Omega} [B]^{\mathrm{T}} [D]^{-1} [R] \{\sigma^q\}^n \mathrm{d}\Omega \tag{3}$$

where $\Delta\{f\}^n$ is the increment of nodal external force matrix $\{f\}^n$ during the time interval $\Delta t = t_{n+1} - t_n$, Ω is the integration domain, $[B]$ is the geometry matrix, A, p, and q are the material coefficients depending on temperature, $[D]$ is the compliance matrix, and $[R] = E[D]$. The strain increment is related to the nodal displacement increment by

$$\Delta\{\varepsilon\}^n = [B]\Delta\{d\}^n \tag{4}$$

According to the simplified constitutive model characterised by Hooke's law and the Bailey-Norton formulation, the stress increment during Δt can be written as

$$\Delta\{\sigma\}^n = [C]^{-1} (\Delta\{\varepsilon\}^n - Apt^{p-1}[R]\{\sigma^q\}^n \Delta t) \tag{5}$$

After the nodal displacement increment is solved from Eq.(1), strain and stress increments can be evaluated by Eqs.(4) and (5). Accordingly, the stress and strain at the $(n+1)$th time step are approximately obtained from the following expressions.

$$\{\sigma\}^{n+1} = \{\sigma\}^n + \Delta\{\sigma\}^n, \quad \{\varepsilon\}^{n+1} = \{\varepsilon\}^n + \Delta\{\varepsilon\}^n \tag{6}$$

The calculated strain and stress are then treated as the initial conditions for the next iterative time step, and the computation is repeated in the same way until the whole loading time is covered.

For asphalt mixture used in the paper, Poisson's ratio is treated as a constant of 0.3, and the other material parameters are determined by a series of uniaxial creep tests carried out under stress levels of 0.1MPa, 0.3MPa, and 0.5MPa, and at temperatures of 20℃, 40℃, and 60℃, respectively.

3. Evaluation of Temperature and Loading Effects

In this section, the aforementioned numerical algorithm frame is used for rutting deformation analysis under conditions of different temperature and wheel loading in order to evaluate effects of temperature and wheel loading on longterm deformation behaviours of asphalt mixture. The 300mm×300mm×50mm rutting specimen is discretized with 8-node hexahedral elements. The bottom and four sides of specimen are fixed, and the cycled loading with frequency of 0.7Hz is applied in the middle 50mm×20mm rectangular region of the specimen top surface as the wheel load. The computations are respectively conducted under three temperatures of 20℃, 40℃, and 60℃, and three wheel loading levels of 0.7MPa, 0.85MPa and 1.0MPa.

3.1. *Temperature effect*

Because of temperature dependence of constitutive parameters, asphalt mixture is a typical temperature-dependent material, so that seasonal variations in temperature can cause drastic changes in material properties of asphalt mixture used in pavements. Accordingly, it is very important for engineering applications of asphalt mixture to investigate influence of temperature on rutting deformation behaviours of asphalt mixture.

Fig.1 plots the rut depth vs time curves for different temperatures at a selected wheel loading level, (a) for 0.7MPa, (b) for 0.85MPa and (c) for 1.0MPa. For different wheel loads, the following common law can be found. There is little difference between the curves for 20℃ and 40℃. However, a saltation takes place in the curve for 60℃ and this curve is steeper than the curves for the other two temperatures. The most possible reason is that AH-70 bitumen used in this study has the softening point of 47℃. After the softening point, the mixture becomes softer and both elastic and viscous deformations remarkably enlarge.

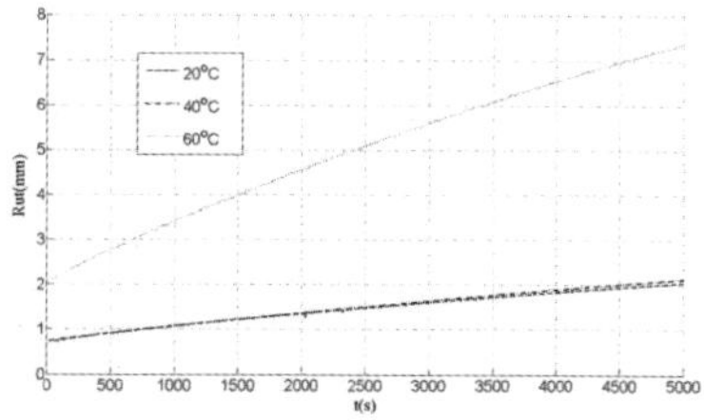

(a) 0.7MPa

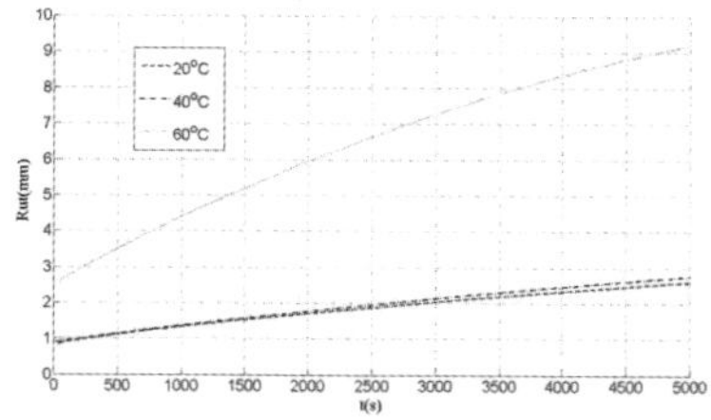

(b) 0.85MPa

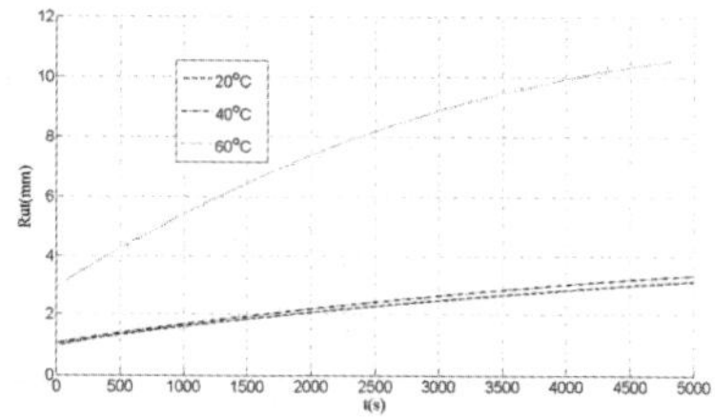

(c) 1.0MPa

Fig. 1. Comparison of rut depth vs time curves at different temperatures.

3.2. *Wheel loading effect*

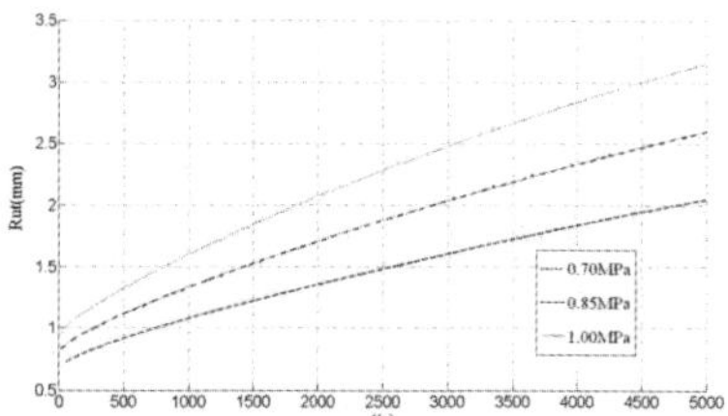

(a) 20℃

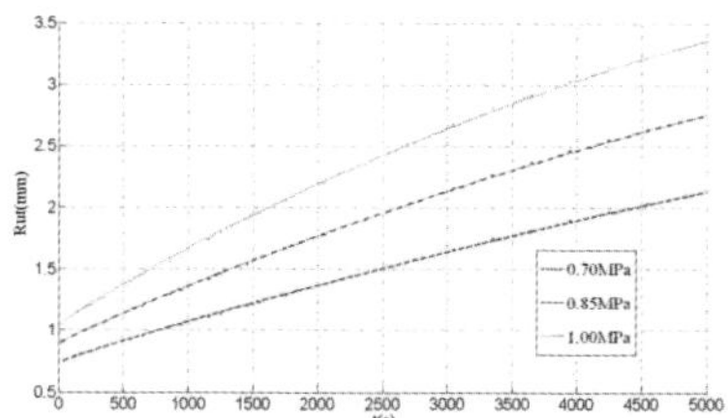

(b) 40℃

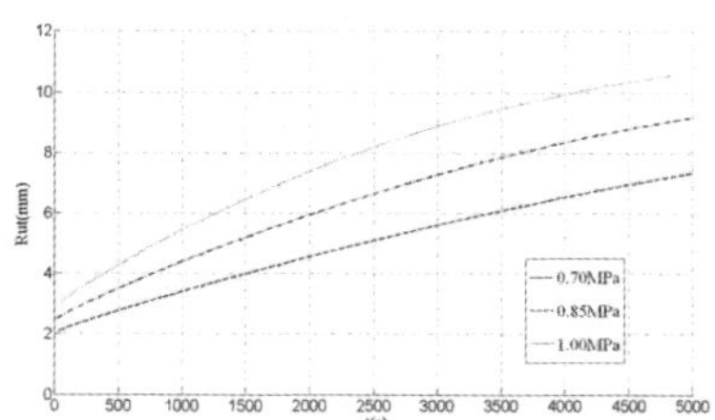

(c) 60℃

Fig. 2. Comparison of rut depth vs time curves under different stress levels.

In order to investigate the effect of wheel loading level on longterm deformations of asphalt mixture, the curves in Fig.1 are rearranged. The rut depth vs time curves under different wheel loading are drawn in Fig.2 with (a) for 20℃, (b) for 40℃ and (c) for 60℃, respectively. In the pictures, the gap between the curves for 0.7MPa and 0.85MPa almost equals that between the curves for 0.85MPa and 1.0MPa. For example, after 3500 loading cycles, the rut depths are respectively about 2.05mm for 0.7MPa, about 2.60mm for 0.85MPa, and about 3.15mm for 1.0MPa at 20℃. There is a common increment of about 0.55mm. It reveals that the rut depth is approximately directly proportionate to wheel loading level.

4. Conclusion

Based on the simplified constitutive model, a numerical algorithm is constructed to analyse rutting deformation in asphalt mixture under different tempratures and cycled wheel loads in order to evaluate effects of temperature and wheel loading on longterm deformation behaviours of asphalt mixture. The following conclusions are obtained.

1. The simplified model can be used to characterise elastic and viscous behaviours of asphalt mixture, and the numerical algorithm based on this model is feasible.
2. When the temperature is much less than the softening point of AH-70 bitumen, it has a little influence on rutting deformation. However, an obvious jump takes place in rutting deformations when temperature exceeds the softening point.
3. The wheel loading plays an important role in rutting deformation of asphalt mixture, and there is approximately a linear dependent relation between the rut depth and wheel loading within the ranges of stress and temperature discussed in this paper.

Acknowledgments

This work is supported by the National Natural Science Foundation of China (Nos. 10672063 and 10872073).

References

1. T. F. Fwa, S. A. Tan and L. Y. Zhu: *Journal of Transportation Engineering*, 2004, **130**, 675–683.
2. J. Yang, H. Z. Lu and H. R. Zhu: *Journal of Engineering Mechanics*, 2009, **135**, 510–516.
3. Y. Lu and P. J. Wright: *Computers and Structures*, 1998, **69**, 139–147.
4. A. Sides, J. Uzan and M. Perl: *Journal of testing and evaluation*, 1985, **13**, 49–59.
5. X. D. Chen and X.M. He: *Tribology International*, 2006, **39**, 1336-1341.